Dynamic Light Scattering

Applications of Photon Correlation Spectroscopy

Dynamic Light Scattering

Applications of Photon Correlation Spectroscopy

Edited by

Robert Pecora

Stanford University
Stanford, California

Plenum Press • New York and London

Library of Congress Cataloging in Publication Data

Main entry under title:

Dynamic light scattering.

Includes bibliographies and index.
1. Light beating spectroscopy. 2. Light—Scattering. I. Pecora, Robert, 1938–
QC454.L63D96 1985 535.8′4 84-24831
ISBN 0-306-41790-1

A Division of Plenum Publishing Corporation
233 Spring Street, New York, N.Y. 10013

Printed in the United States of America

Contributors

Victor A. Bloomfield
Department of Biochemistry, University of Minnesota, St. Paul, Minnesota

B. Chu
Department of Chemistry, State University of New York, Stony Brook, New York

N. C. Ford, Jr.
Langley-Ford Instruments, Amherst, Massachusetts

C. C. Han
National Bureau of Standards, Washington D.C.

Norman A. Mazer
Department of Physics, Massachusetts Institute of Technology, Cambridge, Massachusetts; and Department of Medicine, Brigham and Women's Hospital, Boston, Massachusetts

G. D. Patterson
Department of Chemistry, Carnegie–Mellon University, Pittsburgh, Pennsylvania

R. Pecora
Department of Chemistry, Stanford University, Stanford, California

P. N. Pusey
Royal Signals and Radar Establishment, Malvern, Worcestershire, England

D. W. Schaefer
Sandia National Laboratories, Albuquerque, New Mexico

Toyoichi Tanaka
Department of Physics and Center for Materials Science and Engineering, Massachusetts Institute of Technology, Cambridge, Massachusetts

R. J. A. Tough
Royal Signals and Radar Establishment, Malvern, Worcestershire, England

Karl Zero
Department of Chemistry, Stanford University, Stanford, California

Preface

In the twenty years since their inception, modern dynamic light-scattering techniques have become increasingly sophisticated, and their applications have grown exceedingly diverse. Applications of the techniques to problems in physics, chemistry, biology, medicine, and fluid mechanics have proliferated. It is probably no longer possible for one or two authors to write a monograph to cover in depth the advances in scattering techniques and the main areas in which they have made a major impact. This volume, which we expect to be the first of a series, presents reviews of selected specialized areas by renowned experts. It makes no attempt to be comprehensive; it emphasizes a body of related applications to polymeric, biological, and colloidal systems, and to critical phenomena.

The well-known monographs on dynamic light scattering by Berne and Pecora and by Chu were published almost ten years ago. They provided comprehensive treatments of the general principles of dynamic light scattering and gave introductions to a wide variety of applications, but naturally they could not treat the new applications and advances in older ones that have arisen in the last decade. The new applications include studies of interacting particles in solution (Chapter 4); scaling approaches to the dynamics of polymers, including polymers in semidilute solution (Chapter 5); the use of both Fabry–Perot interferometry and photon correlation spectroscopy to study bulk polymers (Chapter 6); studies of micelles and microemulsions (Chapter 8); studies of polymer gels (Chapter 9). In addition, the considerable advances made in the study of critical phenomena are reviewed (Chapter 7), as well as progress in the application of the depolarized dynamic scattering technique to a wide variety of systems. A survey of the uses of the light-scattering and laser Doppler velocimetry to study systems of biological interest is given in the final chapter (Chapter 10).

The current volume contains much introductory material for the beginner in light scattering. Chapter 2, for instance, should be especially useful for such readers and should be of aid to anyone contemplating setting up a dynamic light-scattering laboratory. The serious novice is, however, urged to consult the two monographs mentioned above for background information about time correlation functions, the electromagnetic theory of scattering, the elementary theory of Brownian motion, hydrodynamics, generalized hydrodynamics, molecular reorientation in liquids, dynamics of a single polymer chain, and nonequilibrium thermodynamics, as well as other topics in their relation to dynamic light scattering.

It is a pleasure to express my thanks to the authors, who have taken so much time from their busy schedules to contribute to this volume.

Stanford, California Robert Pecora

Contents

1

Introduction

R. Pecora

Department of Chemistry
Stanford University
Stanford, California 94305

Dynamic light scattering techniques are generally divided into two main classes—those that use photon correlation or related "time-domain" techniques to measure the frequency distribution of the scattered light and those that directly measure the frequency distribution by placing a monochromator ("filter") before the detection photomultiplier. This volume is concerned almost entirely with some of the currently important applications of photon correlation techniques to the measurement of dynamics of molecules in condensed systems.

Dynamic light scattering as a field of study may be said to date from 1914,[1,2] when Leon Brillouin published a short theoretical note describing the frequency distribution of the light scattered from thermally excited density fluctuations in an isotropic body. Brillouin predicted what is now known as the Brillouin doublet—the doublet symmetrically spaced around the incident frequency with frequency shift ω proportional to the sound velocity c in the medium and the propagation vector length q of the density fluctuation giving rise to the scattering:

$$\omega = \pm cq$$

The propagation vector length is simply related to the wavelength of light in the medium, λ and the scattering angle θ:

$$q = (4\pi/\lambda) \sin (\theta/2)$$

If q is varied by, for instance, varying the scattering angle, a doublet arising from a fluctuation of different propagation vector length is measured.

Gross[3] made the first experimental determination of the frequency distribution of light scattered from a liquid. In addition to the doublet predicted by Brillouin, he observed a central line, a portion of the intensity of scattered light with a maximum centered on the frequency of the exciting line. Landau and Placzek[4] in 1933 gave their now famous explanation of the central line as arising from density fluctuations produced by non-propagating fluctuations of entropy at constant pressure. According to their theory, the Brillouin doublet originates from density fluctuations produced by propagating fluctuations of pressure at constant entropy ("adiabatic sound waves"). They derived what is now known as the Landau–Placzek ratio for the ratio of the frequency integrated intensity of the central line to that of the Brillouin doublet:

$$\frac{I_c}{I_d} = \frac{C_p - C_v}{C_v}$$

where C_p and C_v are, respectively, the heat capacities at constant pressure and constant volume. The spectral widths of the doublets depend upon the sound attenuation coefficients, and the width of the central line depends on the thermal diffusivity.

Several research groups, in the years immediately following these important discoveries, extended the theory and experiments to include such subjects as scattering from viscoelastic bodies and surfaces and depolarized scattering. Much of this work was performed in India and the Soviet Union, and much of it, especially the Soviet work, is summed up in the monograph of Fabelinskii.[5] The relative lack of growth of this field in this period, despite its enormous potential as a source of information about dynamic properties of condensed matter, was due to the difficulty of obtaining reliable and precise experimental results using classical optical sources combined with photographic detection techniques. In effect, the same difficulties inhibited the closely related field of Raman scattering. Although the frequency changes involved in Raman scattering experiments are much larger than those involved in dynamic light scattering and hence should, in principle, be easier to resolve, the Raman scattered intensities are usually much lower.

It was the development of the laser in the early 1960s combined with the widespread adoption of photomultiplier detectors that has resulted in the rapid growth and development of this field (and of Raman scattering as well). The frequency stability and high collimated intensity of laser sources allowed relatively easy measurement of frequency changes and/or broadening of 10^8 Hz or more. The development of even more stable laser sources in the past few years and of sophisticated high-contrast, high-finesse Fabry–

Perot interferometers with data acquisition systems has made it practical (although still not easy) to measure frequency changes of less than several megahertz, to detect very weak scattering lines, and even to detect scattering lines "buried" in the tails of stronger scattering lines.

The interferometer (filter) techniques are now being applied in many research laboratories to the study of such diverse subjects as soft modes in solids, relaxation in polymer melts, sound propagation in polymer fibers, reorientation of small molecules in liquids and polymer systems, orientational relaxation and sound propagation in liquid crystals, far from equilibrium phenomena, and structural relaxation in a wide variety of materials.

The main subject of this volume, however, may be said to date from 1964, with the publication of theoretical and experimental work performed at Columbia University. In that year, the present author[(6)] published the major results of his 1962 doctoral thesis in chemistry. This thesis, prepared under the direction of Richard Bersohn, reviewed the long-neglected (at least by western researchers) work of Brillouin and the Soviet and Indian groups and, in addition, cast the theory into a modern correlation function format. More important, the possibility of measuring very tiny frequency shifts using the then newly developed laser was indicated. It was shown that the small frequency broadening in the light scattered from dilute solutions of macromolecules contained information about diffusion of the macromolecules. In addition, the application of dynamic light scattering to the study of critical phenomena was noted.

The experimental realization of these ideas depended not on directly measuring the small frequency changes involved by use of monochromators—they are usually much too small to be measured with even the best Fabry–Perot interferometers—but on the use of "optical-beating" techniques.

Optical-beating techniques were introduced by Forrester, Gudmunsen, and Johnson in 1955 as an alternative to conventional high-resolution spectroscopy.[(7)] Forrester *et al.* placed a classical mercury light source in a magnetic field. The emitted light contained Zeeman components separated by about 10^{10} Hz. Instead of using a monochromator to separate these components, they allowed the emitted light to impinge upon a photomultiplier where the Zeeman components "beat" with each other producing an oscillation in the photomultiplier output at the difference frequency ($\sim 10^{10}$ Hz). The signal-to-noise ratio in the experiment was very poor and the technique was not widely adopted at that time for conventional spectroscopy.

Cummins, Knable, and Yeh,[(8)] then working in the Columbia Physics Department, realized that variations of the technique discussed by Forrester[(9)] could be used to measure the tiny frequency changes predicted by Pecora in the light scattered by solutions of macromolecules. With a He–Ne laser as the light source and using what is now commonly called a "heterodyne" optical beating technique, they measured the diffusion coefficient of polystyrene latex spheres dispersed in water. Shortly thereafter, Ford and Benedek at MIT measured the thermal diffusivity of SF_6 near its liquid–vapor critical point using a closely related beating technique now usually called the "homodyne" technique.[10] Alpert, Yeh, and Lipworth[11] at Columbia University then studied the decay of concentration fluctuations of a binary liquid mixture (aniline–cyclohexane) near its consolute point. In the period from 1965 to 1970 notable contributions to this field were made by a relatively few research groups. Among these were the groups of Benedek (MIT), Berge (Saclay), Chu (SUNY at Stonybrook), Cummins (then at Johns Hopkins), Fujime (then at Nagoya, Japan), Mountain (National Bureau of Standards), Pecora (Stanford), Pike (Royal Radar Establishment, Malvern, U.K.), A. Wada (Tokyo), and Yeh (then at Lawrence Livermore Laboratory).

This pioneering theoretical and experimental work has led in the past 20 years to the current rich and diverse field to which this book is devoted. The earliest experiments utilized spectrum analyzers as postdetection filters to determine the frequency distribution of the scattered light. With the gradual replacement of spectrum analyzers by autocorrelators in most (nonvelocimetry) experiments, new terms came to describe the field—the most important of these are "intensity fluctuation spectroscopy" and "photon correlation spectroscopy." Photon correlation spectroscopy often denotes a "digital" technique for measuring intensity fluctuations in which the number of photons arriving at a detector at a set time interval is repeatedly counted and its time autocorrelation function computed. The Pike group at Malvern designed the first commercial digital autocorrelator. Digital correlation techniques greatly improved the signal-to-noise ratios obtainable in this type of dynamic light scattering experiment and most current work utilizes them. In fact, the field as a whole has come to be called photon-correlation spectroscopy, although the older more general synonym for dynamic light scattering—"quasielastic light scattering"—is still often used.

In the early 1970s the field expanded rapidly enough to warrant in the mid-1970s the publication of two monographs devoted to dynamic light scattering. Chu's monograph[12] emphasized the experimental aspects while that of Berne and Pecora[13] emphasized the theoretical. The lecture note volumes of two NATO Advanced Study Institutes on photon-correlation spectroscopy have also been influential and provide a cross section of the work on this subject in the mid-1970s.[14, 15]

The application of photon correlation techniques has been extended to such a wide variety of systems that no one volume could treat them all. In fact, it is anticipated that future volumes in this series will be devoted to applications to fluid mechanics and to applications to velocimetry in general (including electrophoretic light scattering, microorganism motility, protoplasmic flow, etc.). Other areas that are given scant attention in this volume are applications to liquid crystals and supercooled organic liquids.

Photon correlation techniques are usually applicable to the measurement of frequency changes in the approximate range from 1 to 10^6 Hz. At the upper end of this scale, photon correlation techniques overlap with Fabry–Perot techniques. The overlap region is a difficult experimental regime for both methods. The time scales involved in photon correlation studies correspond to what on a molecular scale are long-distance, long-time phenomena. Thus, these techniques are especially well suited for measuring dynamic constants associated with macromolecular and particulate systems. These include diffusion (translational, rotational, and intramolecular) in dilute solutions or suspensions, effects of interactions between the large species in more concentrated but still relatively dilute solutions, and interaction effects in semidilute and concentrated solutions. The polymer melt itself is also an important system which may be fruitfully studied by photon correlation spectroscopy. There are a wide variety of systems that fit into these categories—spherical latex particles, flexible polymers (such as linear polystyrenes), and rigid and semirigid rodlike macromolecules (such as myosin, collagen, and DNA). The solvents may be organic or aqueous and the macromolecules charged or uncharged. The polymers may even form gel networks. These topics are thoroughly treated in this volume (Chapters 3–5, 9, 10).

Often the measurement of dynamic constants such as translational and rotational diffusion coefficients may be used to obtain size and shape changes as conditions are altered. Important studies of this type for surfactant micelles and vesicles are being performed using photon correlation spectroscopy (Chapter 8).

For systems near thermodynamic critical points, the light scattering intensity increases strongly (critical opalescence) and the rate of decay of fluctuations decreases dramatically (critical slowing down), making photon correlation spectroscopy one of the best techniques for studying critical phenomena and phase transitions (Chapter 7).

One of the major problems in photon correlation spectroscopy occurs when the scattered intensity time autocorrelation function is known not to be a single exponential but the exact theoretical form is unknown. How, in this case, does one treat the data? It is, in fact, often important to determine whether a given measured autocorrelation function is a single exponential or, say, is the sum of two or more exponentials. Sophisticated data analysis techniques have been developed for treating these problems. Ford in Chapter 2 discusses the apparatus used in light scattering and some of these methods of data treatment. These recent advances have given impetus to the growing use of dynamic light scattering for the analysis of the distributions of macromolecular and particle sizes in polydisperse dispersions. A recently published volume is, in fact, devoted almost entirely to this aspect of photon correlation spectroscopy.[16]

REFERENCES

1. L. Brillouin, *C.R. Acad. Sci.* **158**, 1331 (1914).
2. L. Brillouin, *Ann. Phys. (Paris)* **17**, 88 (1922).
3. E. Gross, *Nature* **126**, 201 (1930); **129**, 722 (1932).
4. L. D. Landau and G. Placzek, *Phys. Z. Sowjetunion* **5**, 197 (1934).
5. I. L. Fabelinskii, *Molekulyarnoe Rasseyanie Sveta*, Nauka Press, Moscow (1965) [English Edition], *Molecular Scattering of Light*, Plenum, New York (1967).
6. R. Pecora, *J. Chem. Phys.* **40**, 1604 (1964).
7. A. T. Forrester, R. A. Gudmundsen, and P. O. Johnson, *Phys. Rev.* **99**, 1691 (1955).
8. H. Z. Cummins, N. Knable, and Y. Yeh, *Phys. Rev. Lett.* **12**, 150 (1964).
9. A. T. Forrester, *J. Opt. Soc. Am.* **51**, 253 (1961).
10. N. C. Ford and G. B. Benedek, *Phys. Rev. Lett.* **15**, 649 (1965).
11. S. S. Alpert, Y. Yeh, and E. Lipworth, *Phys. Rev. Lett.* **14**, 486 (1965).
12. B. Chu, *Laser Light Scattering*, Academic, New York (1974).
13. B. J. Berne and R. Pecora, *Dynamic Light Scattering*, Wiley-Interscience, New York (1976).
14. H. Z. Cummins and E. R. Pike, eds., *Photon Correlation and Light Beating Spectroscopy*, Plenum, New York (1974).
15. H. Z. Cummins and E. R. Pike, eds., *Photon Correlation Spectrometry and Velocimetry*, Plenum, New York (1977).
16. B. E. Dahneke, ed., *Measurement of Suspended Particles by Quasi-Elastic Light Scattering*, Wiley-Interscience, New York (1983).

2

Light Scattering Apparatus

N. C. Ford, Jr.

Langley-Ford Instruments
Amherst, Massachusetts 01002

2.1. INTRODUCTION

This chapter will introduce the reader to the various physical components of an experiment designed to obtain information from solutions of macromolecules by measurements of the time dependence of light scattered from the molecules. In some cases the components are readily available commercially and therefore only broad operation specifications will be discussed. Examples of items in this category include the laser and the light detector (photomultiplier). It will be assumed that devices of this type can be used satisfactorily as "black boxes." At the other extreme stands the light scattering spectrometer itself. While spectrometers are available commercially, many investigators prefer to assemble their own, and, in any case, their optimal use depends on a detailed knowledge of the principles of operation. Consequently, considerable effort will be expended to acquaint the reader with the physical principles involved in laser fluctuation spectroscopy.

We will begin with a brief introduction to the concepts involved in describing light as an electromagnetic wave. This will be followed by a description of the scattering process with the emphasis on the time dependence of the scattered intensity. In particular, we will show how to obtain such information as the translational diffusion constant, rotational diffusion constant, flow velocity, and electrophoretic mobility by making measurements of the spectrum of the scattered light.

With this material as background we will then describe the typical light scattering apparatus in considerable detail. Each of the major components of the overall apparatus will be the subject of a separate section where the considerations important to the selection and use of each component will be discussed. We will conclude the chapter with a brief mention of several types of specialized apparatus.

2.2. ELECTROMAGNETIC WAVES

In this section we review in an elementary way the concepts used to describe an electromagnetic wave,[(1)] a subject that includes radio waves, light, and X rays. All three types of radiation are propagating waves in which both electric and magnetic fields oscillate in perpendicular directions and in directions perpendicular to the direction of propagation of the wave. Thus the electric field, the magnetic field, and the direction of propagation are all mutually perpendicular. The only feature distinguishing the three types of radiation is the frequency of oscillation and, therefore, the wavelength since the frequency ν, wavelength λ, and velocity c of an electromagnetic wave in vacuum are related by $c = \lambda\nu$. Figure 1 shows these major features of the wave.

A light beam carries energy. The energy deposited on a one-square-meter area during one second is known as the intensity of a light beam, I, and is measured in watts per square meter (W/m^2) in the SI system of units. The instantaneous light intensity is related to the instantaneous electric field strength measured in volts per meter (V/m) by

$$I = \frac{E^2}{377}\ \text{W/m}^2 \tag{1}$$

However, since E is oscillating at a frequency of roughly 6×10^{14} Hz, it is

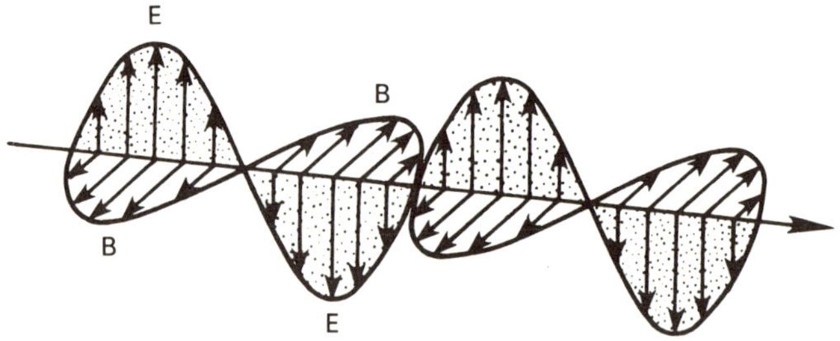

Figure 1. In a plane-polarized electromagnetic wave, electric and magnetic fields oscillate in perpendicular directions, both directions being perpendicular to the direction of propagation of the wave.

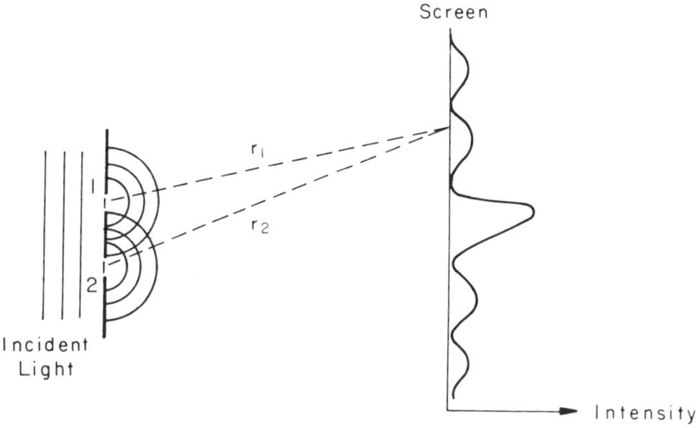

Figure 2. The interference pattern produced by two parallel slits is found to consist of alternating bright and dark regions. The bright regions occur when the electric fields from the two slits are in phase; the dark regions correspond to positions where the two electric fields are 180° out of phase.

possible to measure only the average intensity, which is just one half of the maximum instantaneous intensity

$$I_{av} = \frac{E_{max}^2}{2 \times 377} \text{ W/m}^2 \qquad (2)$$

For example, the sun's intensity at the earth's surface is about 10^3 W/m^2. The corresponding maximum electric field is consequently $E_{max} = 870$ V/m.

Many of the properties of scattered light we will discuss are easily explained using the rule that when two or more light beams fall upon the same surface, the resulting intensity is obtained by first adding the electric fields due to the individual beams and then squaring the sum to obtain the average intensity. The possible consequences of this rule may be discussed in terms of the two-slit interference experiment illustrated in Figure 2. We assume that a monochromatic beam of light (originating, perhaps, from a laser) falls upon two very narrow slits. A screen is placed far away from the slits and the intensity of the light is measured at points on the screen.

If the slits are sufficiently narrow the wave fronts emanating from them will be cylindrical in form as shown in Figure 2. In this case, at any point on the screen there will be two waves to be added, both having the same maximum electric field. However, they will, in general, have different relative phases because the distances to the two slits will be different. If the distances to slits 1 and 2 from point x are r_1 and r_2, respectively, the total electric field will be

$$E = E_{max}[\sin 2\pi\nu t + \sin(2\pi\nu t + \delta)] \qquad (3)$$

where $\delta = 2\pi[(r_2 - r_1)/\lambda]$. If δ is a multiple of 2π (or 0) the two sine waves will add and

$$I_{av} = 4\left(\frac{E_{max}^2}{2 \times 377}\right)$$

On the other hand, if δ is an odd multiple of π, the sine waves will subtract and

$$I_{av} = 0$$

Thus we can calculate the two-slit interference pattern.

It may seem strange that at some points the total intensity is twice as great as the combined intensities of the two original light beams. However, at other points, the combined intensity is less than that of the original beams, and if the intensity is averaged over the entire plane at the screen, it is found that the total power falling on the screen is exactly equal to the sum of the powers of the two original light beams.

Additional properties of light that may be understood using the wave theory are associated with polarization. The electric field can be in any direction perpendicular to the direction of propagation. Light from a conventional light source will at one time or another have the electric field pointing in all of these directions. If this light is then passed through a polarizer, the electric field will point in only one direction and we have polarized light. Note that two light beams polarized in different directions cannot interfere because there is no way of having the electric fields cancel to give zero.

While our discussion of the light scattering process will be based entirely on the wave theory of light as outlined above, we will have recourse to the photon theory of light when discussing light detectors. For our present purposes we can consider the wave theory to be modified by the ideas of the photon theory only to the extent that the energy transported by a light beam is carried in indivisible "chunks" called photons. Each photon has energy

$$E = h\nu \tag{4}$$

where $h = 6.63 \times 10^{-34}$ J sec is Planck's constant. Even given the large frequency of visible light (6×10^{14} Hz), the energy carried by a single photon is only about 10^{-19} J. Nevertheless, modern techniques allow us to detect single photons with a probability (called the quantum efficiency) lying between 5% and 50% depending largely on the frequency (and therefore, energy) of the photon.

2.3. LIGHT SCATTERING

2.3.1. Background

Light will be scattered by a molecule in solution if the molecule has a polarizability different from its surroundings. In this case, the oscillating dipole moment induced by the electric field of the incident light beam will radiate light in all directions. The intensity of the scattered light will be related to the direction of polarization of the incident light, scattering angle, and solution parameters. Assuming the incident light is linearly polarized and defining angles as in Figure 3, the scattered intensity due to a single molecule with dimensions much smaller than λ is given by[2]

$$I_s(1) = \frac{4\pi^2 M^2 (\sin^2 \phi)(dn/dc)^2 I_0}{N_A^2 \lambda^4 R^2} = \frac{S I_0}{R^2} \tag{5}$$

In this equation the angles are as defined in Figure 3; M is the molecular weight of the molecule in daltons; dn/dc is rate of change of index of

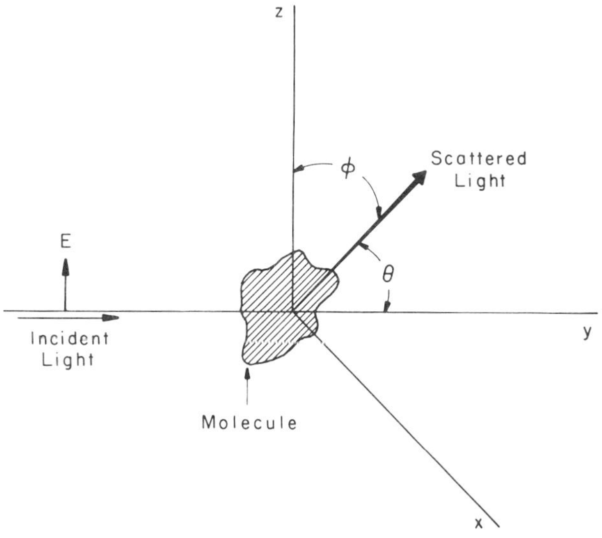

Figure 3. The angles required to give a mathematical description of light scattering are defined here. The electric field in the incident beam is along the z direction and the incident beam propagates in the y direction. ϕ and θ are the angles between the propagation direction of the scattered beam and the z and y axes, respectively.

refraction of the solution as the concentration of the solute is changed; I_0 is the incident light intensity; N_A is Avogadro's number; λ is the wavelength of light in the solution; and R is the distance from the scattering point to the observation point. The scattered light will be linearly polarized if the incident light is polarized. The direction of polarization will lie in the plane determined by the direction of the incident light beam and the scattering direction. Notice that, if these two directions are parallel, the angle ϕ is zero and the scattered intensity vanishes according to equation (5). Thus there is no ambiguity in determining the direction of polarization of the scattered light.

If a molecule is nonspherical or if it has unequal polarizabilities along two directions, there will be a component of the scattered light with polarization perpendicular to the direction defined above. While the intensity of this depolarized light (frequently given the symbol I_{VH}) is usually orders of magnitude lower than that of the polarized component (I_{VV}), useful information concerning the rotational diffusion constant of the scattering molecule can be obtained from measurements of I_{VH}.

2.3.2. Fluctuations

When more than one molecule is responsible for the light scattering (as is usually the case) we must first calculate the scattered electric field, E_{Si}, for each scattering molecule, then sum the electric fields and square the result to obtain the scattered light intensity.[3] If all of the molecules are identical, the electric field contributions due to the individual molecules will have the same magnitude but the phases will be different. The summation of the electric fields is most conveniently visualized using a two-dimensional plot in which each scattered electric field is represented by a line segment with length proportional to the magnitude of $\mathbf{E}_s$ and making an angle with the abscissa equal to the phase angle of $\mathbf{E}_s$ with respect to light scattered from the center of the scattering region. The vector sum of the line segments will then give the resultant total scattered electric field as shown in Figure 4.

The sum illustrated graphically in Figure 4 may be done explicitly. For the component of $\mathbf{E}_s$ in phase with the reference beam, we find

$$E_s^i = \mathbf{E}_s \sum_i \cos \delta_i \tag{6}$$

and for the component with a phase angle of 90°

$$E_s^0 = \mathbf{E}_s \sum_i \sin \delta_i \tag{7}$$

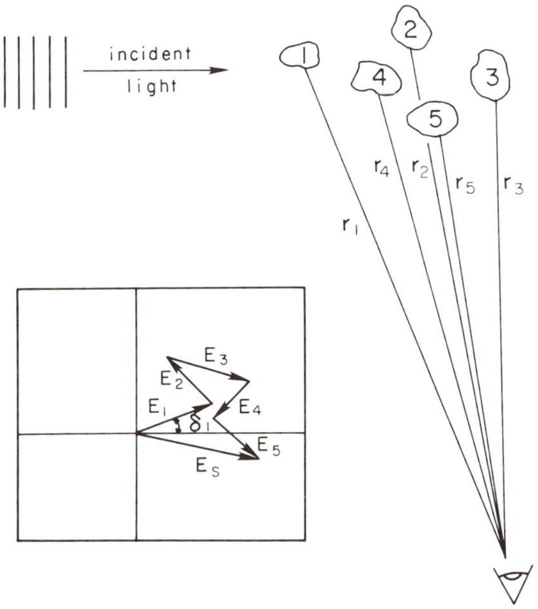

Figure 4. To calculate the intensity of light scattered from a collection of molecules, the contribution to the electric field of the scattered light (E_s) due to the individual molecules must be added taking into account the different phase angles. This process may be visualized by representing each contribution to E_s as a vector whose length gives the magnitude and whose angle, with respect to the x axis, gives the phase angle of that contribution to E_s. Calculating E_s is then equivalent to performing vector addition of the individual contributions as shown in the insert.

From equation (1) we obtain for the total scattered intensity

$$I_s = \frac{(E_s^0)^2 + (E_s^i)^2}{377} \text{ W/m}^2$$

$$= \frac{1}{377} E_s^2 \left[\left(\sum_{i=1}^{N} \cos \delta_i \right)^2 + \left(\sum_{i=1} \sin \delta_i \right)^2 \right] \tag{8}$$

where N is the number of molecules in the scattering volume. Equation (8) provides the starting point for a discussion of photon correlation spectroscopy. By proper study of this equation we can learn not only the principles behind photon correlation spectroscopy, but the factors that must be understood in order to properly apply the technique.

We can write equation (8) in a more meaningful form by first separating the sums into terms involving only one molecule and terms involving two molecules. The result is

$$\left(\sum_{i=1}^{N} \cos \delta_i\right)^2 + \left(\sum_{i=1}^{N} \sin \delta_i\right)^2$$

$$= \sum_{i=1}^{N} (\cos^2 \delta_i + \sin^2 \delta_i) + 2 \sum_{j>i=1}^{N} (\cos \delta_i \cos \delta_j + \sin \delta_i \sin \delta_j)$$

$$= N + 2 \sum_{j>i=1}^{N} \cos(\delta_i - \delta_j) \tag{9}$$

where the second step is the result of the use of the identities

$$\sin^2 \theta + \cos^2 \theta = 1$$

$$\cos(\theta_1 - \theta_2) = \cos \theta_1 \cos \theta_2 + \sin \theta_1 \sin \theta_2$$

We can then rewrite the total scattered intensity as

$$I_s = I_s(1)\left[N + 2 \sum_{j>i=1}^{N} \cos(\delta_i - \delta_j)\right] \tag{10}$$

Now recall that the phases of the various components of the electric field are changing with time because the molecules are moving about in the solution. Consequently the δ_i's and δ_j's depend on time with the result that the entire last term of equation (10) is time dependent. On the average this term is zero with excursions in both positive and negative directions. Therefore, the time average of the scattered intensity is

$$\langle I_s \rangle = N I_s(1) \tag{11}$$

which, taken together with equation (5), is the basis of the traditional molecular weight determination using light scattering intensities.

The instantaneous value of the last term in equation (10) is not zero. In fact, this term may be found more or less anywhere in the range from $-N$ to $+N$, leading to the result that the scattered intensity may be observed to fluctuate between zero and about twice its average value. The time required for a fluctuation between extremes is roughly equal to the time required by two molecules to move with respect to each other far enough to change the relative phase of the light scattered from each from 0 to π radians. This time will depend on the scattering angle as well as the size of the molecule;

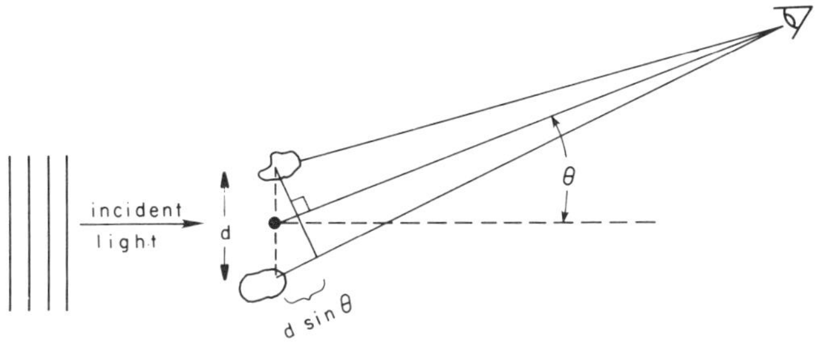

Figure 5. Geometry for calculating the intensity of light scattered from two molecules. The scattered intensity will depend on θ and d and therefore will depend upon the exact positions at which observations are made as well as on the relative positions of the molecules.

typical values range from a few microseconds for molecules with a molecular weight of 25,000 to many milliseconds for objects as large as a cell.

2.3.3. The Coherence Area

The intensity fluctuations described here are not usually observed for two reasons: first, they take place on a time scale faster than many photometers (and the human eye) will respond, and second, the fluctuations as described here take place only at a point. This fact is of great importance in properly designing an experiment and therefore will be examined in detail.

When we say that the fluctuations take place only at a point we mean that they may be observed at any point in the scattered field, but that the fluctuations at two different points will not be coincident unless the two points are very close together. To see why this is true, consider scattering from two molecules a distance d apart. As shown in Figure 5, the scattering pattern will have maxima at any angle for which $d \sin \theta$ is a multiple of λ. Let us define the angular separation between adjacent maxima as $\Delta\theta$. A simple calculation shows that if $\Delta\theta$ is small, then to good approximation $\Delta\theta = \lambda/d \cos \theta$. For example, if $\lambda = 500$ nm and $d = 0.1$ mm we find an angular separation of $\Delta\theta = 5 \times 10^{-3}/\cos \theta$ rad or $\Delta\theta = 0.3/\cos \theta$ deg between bright spots. Thus, a detector that subtends an angle significantly greater than this value would detect several maxima and several minima simultaneously. As the two molecules move about, the fluctuations in the detected signal will consequently be less than would be the case for a smaller detector.

The situation for the more realistic case of light scattered from many molecules distributed randomly throughout a small volume is more difficult to analyze, although the basic ideas are similar. If the scattering volume has radius a and the detector radius is b at a distance R from the scattering

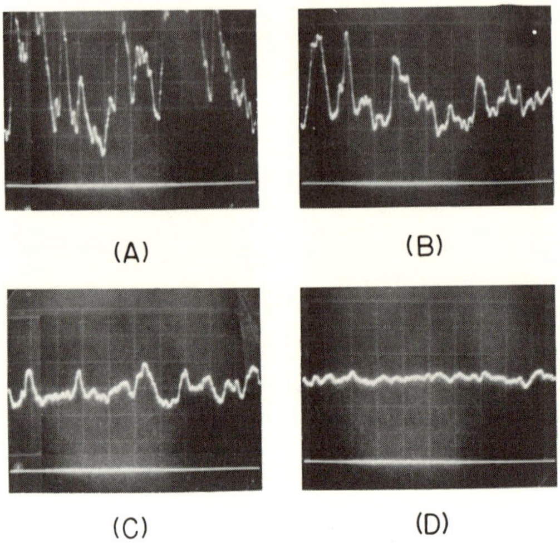

Figure 6. Effect of the number of coherence areas on the fluctuations in the scattered light intensity. All signals were recorded for the same sample of 0.085-μm polystyrene latex spheres. The laser intensity was adjusted so that the average scattered intensity was the same in all cases. The number of coherence areas was 0.064 in. (A), 2 in. (B), 8 in. (C), and 100 in. (D).

volume, the fluctuations will be substantially the same as they would be at a point if the detector area is one coherence area or less where the coherence area A_{coh} is defined as[4]

$$A_{\text{coh}} = \pi b^2 = \frac{\lambda^2 R^2}{\pi a^2} \tag{12}$$

The effect of the number of coherence areas on the character of the fluctuations is illustrated in Figure 6, where photographs of oscilloscope traces are reproduced for several values of N_{coh}. In each case the incident light intensity has been adjusted so that the average intensity detected is the same. Thus the fluctuations observed are influenced only by the number of coherence areas; the shot noise contribution caused by the detection of a finite number of photons is the same in each picture.

2.3.4. Time Dependence

The time required for the fluctuations to take place is the most important characteristic of the signal as that time contains information about the dynamic properties of the solute molecules. The simplest information obtained is the translational diffusion constant which, for a spherical molecule, is related to the radius, r, according to the Stokes–Einstein relation,[5]

$$D = \frac{k_B T}{6\pi\eta r} \tag{13}$$

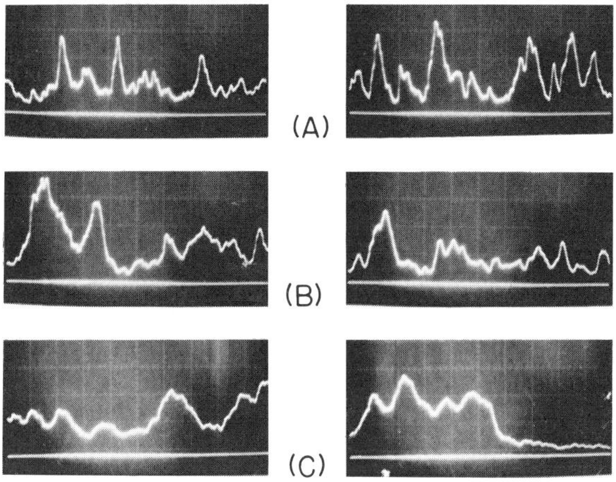

Figure 7. Effect of the size of PSL spheres on the fluctuations in the scattered light intensity. All signals were recorded at the same average light intensity and for 0.32 coherence areas. The diameter of the spheres was 0.085 μm (A), 0.220 μm (B), and 1.011 μm (C).

where k_B is Boltzmann's constant, T the temperature, and η the viscosity. Thus, a large molecule will have a smaller diffusion constant than a small molecule and therefore the fluctuations will take place more slowly as shown in Figure 7. As in Figure 6, the incident light intensity has been adjusted so that the average scattered intensity is the same for all pictures; the different character of the fluctuations is caused by the different sizes of the molecules.

The analysis of the fluctuations may be carried out using either a spectrum analyzer or a correlator. The more familiar of these instruments, the spectrum analyzer, obtains the power spectrum of a signal such as that of Figure 7. The spectrum for the case of translational diffusion of a monodisperse molecule is a Lorentzian

$$P(\omega) = \frac{2Dq^2/\pi}{\omega^2 + (2Dq^2)^2} \tag{14}$$

where $q = (4\pi/\lambda)\sin(\theta/2)$ is the scattering vector.

The use of correlation techniques is relatively new in signal analysis, although the mathematical form of the correlation function has been employed in theoretical treatments for many years. The correlation function is defined for a signal $I(t)$ as

$$G(\tau) = \lim_{T\to\infty} \frac{1}{2T} \int_{-T}^{T} I(t)I(t+\tau)\,dt \tag{15}$$

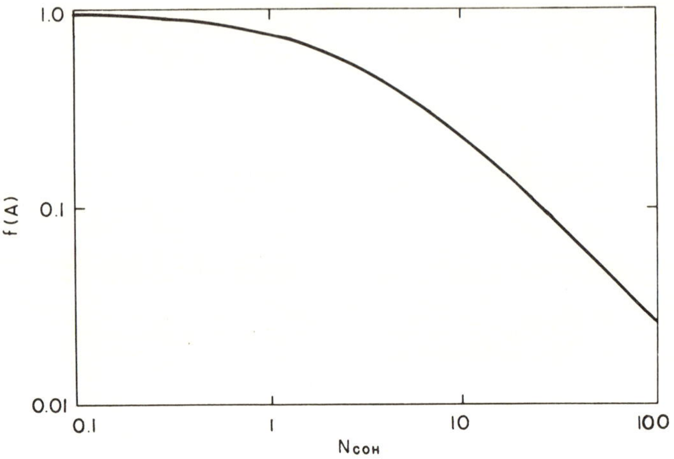

Figure 8. Dependence of $f(n_{coh})$ on the number of coherence areas. An approximate interpretation of this graph is that it gives the ratio between the amplitude of the fluctuations in scattered light intensity and the average scattered light intensity.

The correlation function has two features that are of particular interest to us here. First, it is easily measured, using modern digital techniques, for light signals of very low level; and second, it can be shown to be the Fourier transform of the power spectrum. Thus it is both experimentally accessible and theoretically interpretable in terms of the system parameters of interest. Indeed, the correlation function corresponding to equation (14) is

$$G(\tau) = 1 + e^{-2Dq^2\tau} \tag{16}$$

a waveform which is usually easier to analyze than is equation (14).

Under a more realistic condition in which the detector has an area exceeding a single coherence area, equation (16) must be replaced by

$$G(\tau) = 1 + f(n_{coh})e^{-2Dq^2\tau} \tag{17}$$

where $0 \leq f(n_{coh}) \leq 1$ is a factor which has been calculated by Jakeman *et al.*[4] and is plotted in Figure 8. At large values of n_{coh}, $f(n_{coh})$ can be approximated by

$$f(n_{coh}) = \frac{2.5}{n_{coh}} \tag{18}$$

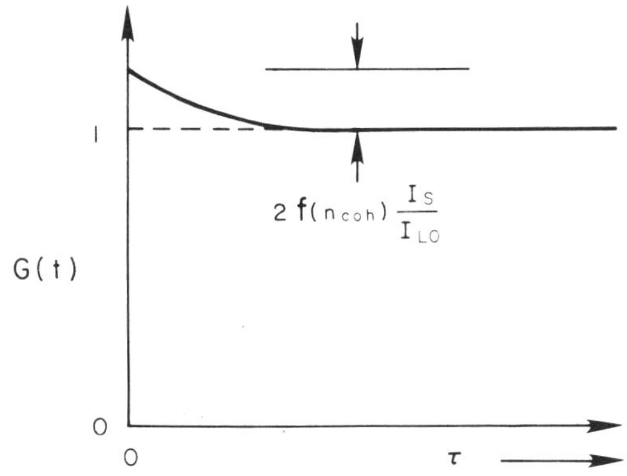

Figure 9. The correlation function in the limit of very large local oscillator.

2.3.5. Local Oscillator

It is sometimes desirable to include in the light incident on the detector light directly from the laser source that has not undergone the scattering process. This light serves as a local oscillator and permits several experiments to be done that would otherwise fail. For example, using a local oscillator it is possible to measure the velocity of a uniform flow throughout the sample volume, a quantity that would otherwise be undetectable. The presence of the local oscillator modifies the observed correlation function, even when simple diffusion is responsible for the light scattering. In the limit of a large local oscillator, equation (17) is replaced by

$$G(\tau) = 1 + 2f(n_{\text{coh}}) \frac{I_s}{I_{\text{LO}}} e^{-Dq^2\tau} \tag{19}$$

Here I_s is the intensity of the scattered light and I_{LO} that of the local oscillator. This result is illustrated in Figure 9. Notice that the decay rate observed for simple diffusion doubles in the absence of a local oscillator.

2.4. THE LIGHT SCATTERING EXPERIMENT

2.4.1. Introduction

In general terms all light scattering experiments contain the same components; it is only in specific details that experiments designed to measure different aspects of the dynamic properties of a solution are distinguished. The major components, as illustrated schematically in Figure 10, are the light source (which is almost always a laser, but does not need to be); the

Figure 10. A block diagram of the complete light scattering apparatus. Each of the blocks is discussed in detail in the text.

spectrometer, which contains an optical system to define the scattering angle and also to limit the number of coherence areas; the detector, usually a photomultiplier; and the signal analyzer, which may be either a spectrum analyzer or a correlator. It is often the case that a computer is used to abstract information from the spectrum or correlation function obtained by the signal analyzer.

In the remainder of this chapter we will take up each of the four major components of Figure 10. We will then discuss some aspects of the problem of data analysis so that the function of the last box in Figure 10 will be covered.

2.4.2. The Light Source

It is commonly believed that the light source employed in photon correlation spectroscopy must be a laser in order to meet spectral purity requirements. The argument is based on the fact that changes in frequency of the incident light of only a few kilohertz (or even a few tens of hertz) are detected, and therefore the frequency width of the incident beam must be only a few hertz to obtain the maximum resolution. This argument, however, is, fallacious. To show the proper line of reasoning, consider light scattered from two parts of the scattering volume located in such a way that the difference in path lengths for the two beams is the maximum that can occur for any pair of points in the sample volume. If the scattering volume is spherical in shape with radius r, the maximum possible path difference is

$$\Delta l = 2\sqrt{2}\, r \tag{20}$$

The criterion for a successful correlation experiment is that light traveling the longer path must maintain the same phase relative to light traveling the shorter path to within, say, $\pi/4$ rad. While detailed arguments are complicated, it is reasonable to say that in order to satisfy the above criterion using a conventional light source with a spectrum of half-width $\Delta\nu$ centered at frequency ν we require

$$-\frac{\Delta\lambda}{\lambda} = \frac{\Delta\nu}{\nu} \approx \frac{1}{8}\frac{\lambda}{\Delta L} \tag{21}$$

For example, if $\Delta L = 0.1$ mm, a value typical of many experiments, we would obtained $\Delta\lambda \approx 0.3$ nm. Spectral lines of pure elements are generally much narrower than this, so that in theory conventional light sources meet the spectral purity criterion for photon correlation experiments. That this is indeed the case has been demonstrated experimentally by Jakeman, Pusey, and Vaughn,[6] who measured the fluctuations in light originated by a mercury arc lamp and scattered from a liquid crystal. The results obtained compared well with results using a laser light source.

It is not likely, however, that conventional light sources will replace lasers in photon correlation applications because of the relatively low intensities obtained when the light is focused to a point. In the experiment mentioned above the scattered light detected using the mercury arc source was a factor of 10 lower than was detected when a 10^{-4}-W laser was substituted for the arc source. Since the cost of a mercury arc lamp source exceeds that of a 10^{-3}-W He–Ne laser, it is clear that the laser is by far the most cost-effective source. We therefore restrict the remainder of this discussion to lasers.

The spectrum of light emitted by a laser is as shown in Figure 11. The laser itself consists of a cavity formed by two spherical mirrors separated by distance L. With the space between filled with an active medium providing gain at optical wavelengths, the laser will oscillate (lase) at wavelengths satisfying the relation

$$n\frac{\lambda}{2} = L \tag{22}$$

and for which the gain of the medium–mirror combination exceeds unity. The laser will produce light in from one to 20 or more peaks depending on its length and construction.

The intensity patterns of the spectral lines represented by equation (22)

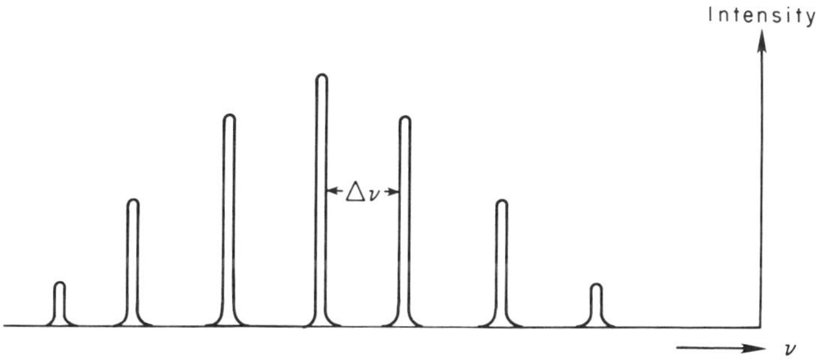

Figure 11. Representation of the spectrum of light emitted from a laser.

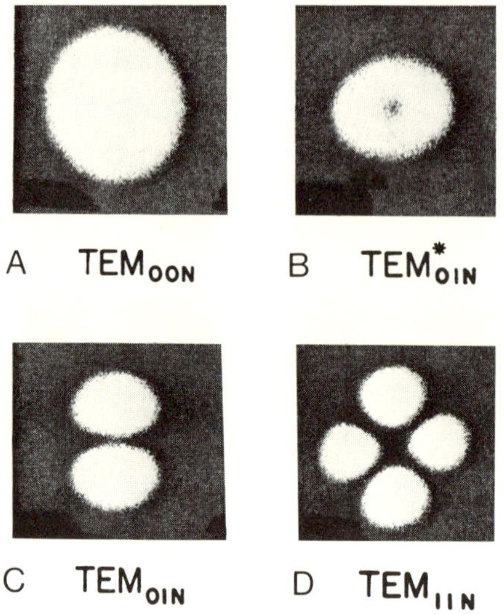

Figure 12. Photograph of the pattern of a laser beam when the laser operates in various modes. Only the TEM_{00n} mode is suitable for light scattering experiments. The TEM^*_{01n} is a mixture of TEM_{01n} and TEM_{10n} modes.

are circularly symmetric and are called TEM_{00n} modes. Modern laser design attempts to eliminate other modes of oscillation of the laser called off-axis modes, but occasionally they will be found. Unfortunately, off-axis modes can cause undesired contributions to the analyzed signal and therefore should be eliminated. The easiest test for off-axis modes is to expand the laser beam with a simple lens and observe the expanded beam on a card. A TEM_{00n} mode will appear as in Figure 12A while off-axis modes are shown in Figure 12B–12D. The off-axis modes, designated TEM_{lmn}, will have wavelengths clustered about but slightly different from the wavelength of the axial model TEM_{00n}.

The spectral lines in Figure 11 can beat against each other and will, in principle, give a contribution to the correlation function having a minimum frequency equal to the separation of adjacent lines

$$\Delta\nu = \frac{C}{2L} = \frac{1.5 \times 10^8}{L} \sec^{-1} \qquad (L \text{ in meters}) \tag{23}$$

The maximum frequency of any importance in real experiments is 5×10^6 Hz so that beating between TEM_{00n} lines will be important only if $L > 30$ m. Laboratory lasers are in fact more than an order of magnitude smaller than this, so no effects may be expected due to beating between axial modes.

The situation with regard to off-axial modes is entirely different. Here the spacing may be only a few kilohertz and beating between the various

TEM_{lmn} modes, all having the same n, can produce spurious components in the correlation function. Consequently it is always desirable to eliminate the off-axial modes. Some lasers, particularly medium- and high-power Ar^+ lasers, have an intracavity iris that may be adjusted to eliminate such modes. Most low-power modern lasers will not support off-axis modes. Nevertheless, the laser beam should be examined periodically to check its mode purity. This is especially important when the correlation function has an unexpected shape.

In selecting a laser there are four factors that must be taken into consideration. These are

- Wavelength;
- Power;
- Polarization;
- Fluctuations in power intrinsic to the laser.

The first two factors are coupled because the power required to achieve a given signal-to-noise ratio depends on the wavelength. The fourth factor enters in a negative sense: any fluctuations in laser power should be avoided if at all possible.

The wavelength to be selected is governed to a great extent by the nature of the experiments to be performed. The first requirement is that any absorption bands in the sample must be avoided. Otherwise substantial local heating of the sample will take place leading to convective flow of the solution, a thermal lens effect, and difficulty in interpreting the data. Similarly, it is undesirable to excite fluorescence lines in the sample, although unwanted light from this source could be removed with filters. Beyond these requirements there is relatively little scientific reason to choose one wavelength over another. As we shall see in Section 2.5, the signal-to-noise ratio is, except for photomultiplier efficiency, independent of wavelength despite the fact that the scattering power is proportional to $1/\lambda^4$. It is therefore preferable to work at the wavelength giving the best photomultiplier quantum efficiency if low light levels are anticipated. If there is an adequate amount of scattered light, cost both in purchase price and operating expenses is the dominant consideration. At the present time the best choices are:

- For low light scattering levels: up to 200-mW AR^+ laser. Any greater power will not greatly increase the range of possible experiments. The principal lines are at 488.0 and 514.5 nm.
- For high light scattering levels: 1- to 50-mW He–Ne laser. The principal line is at 632.8 nm.

Because of the relationship between the polarization of the incident light and intensity of the scattered light [see equation (5)], it is essential that the laser be capable of producing plane polarized light. The light with "random" polarization produced by some inexpensive lasers has in fact two components which are linearly polarized in perpendicular directions. Alternate lines in Figure 9 will have one polarization, the remainder the perpendicular polarization. The relative intensities of the two components as well as the directions of polarization will change with time, particularly during the first half hour of operation. Consequently, there will be fluctuations in the intensity of the scattered light that are related to the laser rather than the sample under study.

Another source of intensity fluctuation intrinsic to the laser is the plasma oscillation found in some large (50-mW) He–Ne lasers. The laser intensity will be modulated at the plasma frequency, about 100 kHz in a typical laser. Lasers subject to this instability usually have an optional rf excitation that will reduce the plasma oscillations to an insignificant level. This option is an absolute necessity if offered. Any large He–Ne laser without the option is suspect and should be carefully tested before using.

There is a more mundane source of intensity fluctuations not associated with the sample under study. Under some circumstances vibrations in various parts of the apparatus or of one part with respect to another can cause severe intensity fluctuations at the detector. Experiments that are prone to interference from vibration are those in which light scattering from a surface (either liquid or membrane) is studied or any experiment employing a local oscillator. In these cases relative motion of two parts of the experiment by a quarter wavelength of light or less can cause large, entirely erroneous, signals. A successful experiment will frequently require use of a vibration isolation table, elimination of mechanical pumps, enclosing the experiment in a sound-deadening environment, and construction of experimental apparatus using massive materials.

Simple light scattering experiments from solutions of macromolecules in which no local oscillator is employed are almost immune to this type of problem. The reason is that all the sources of light which will ultimately interfere at the detector are located within the relatively small sample volume (these sources are the molecules themselves, of course), and because of the relatively incompressible nature of the solvent, typical laboratory sources of vibration are incapable of causing relative motion of the various macromolecules by sufficient amounts to be important. Note, however, that if the laser beam and total sample move with respect to each other by an amount comparable to the total sample size, and if this happens in a time on the order of or less than the duration of the correlation function, the true sample will consist of different sets of molecules at different times and one can then expect to observe effects in the correlation function. To avoid this type of effect it is a good idea to attach the laser and light scattering spectrometer to a single plate.

A laser with an unstable cavity will be affected by vibration. The problem is most severe in lasers having water-cooled plasma tubes. The flow of water, particularly if there are trapped bubbles, will cause vibrations that lead to objectionable laser intensity fluctuations. Reduction of the flow rate to the minimum consistent with the cooling needs, elimination of trapped bubbles, and careful alignment of the cavity mirrors for maximum power are the only recourses.

We have discussed the elimination of laser intensity fluctuations, but one might ask, "How much is too much?" To answer this question we suppose that both the laser intensity $I_0(t)$ and scattering factor $S(t)$ are functions of time but that fluctuations in $I_0(t)$ and $S(t)$ are uncorrelated. The scattered intensity autocorrelation function is given by

$$\begin{aligned} G(\tau) &= \langle I_s(t)I_s(t+\tau)\rangle = \langle I_0(t)S(t)I_0(t+\tau)S(t+\tau)\rangle \\ &= \langle I_0(t)I_0(t+\tau)\rangle\langle S(t)S(t+\tau)\rangle \\ &= G_I(\tau)G_s(\tau) \end{aligned} \tag{24}$$

so that the observed correlation function is the product of the correlation function representing the laser intensity fluctuations and the correlation function representing the sample under study. $G_S(\tau)$ is given for a simple example in the absence of a local oscillator in equation (17) and in the presence of one in equation (19). The form of $G_I(\tau)$ may be deduced under the assumption that the laser intensity $I_0(t)$ may be divided into its average value $\langle I_0\rangle$ and a fluctuating part of amplitude i_0. Denoting the normalized correlation function by $g_I(\tau)$ we find

$$G_I(\tau) = I_0^2[1 + (i_0/I_0)^2 g_I(\tau)] \tag{25}$$

Combining equations (25) and (17) or (19) with equation (24) and keeping only the most important terms giving a time dependence to $G(\tau)$ we find that if the interesting portion of $G(\tau)$ is to dominate over the laser fluctuations

$$f(n_{\text{coh}}) \gg (i_0/I_0)^2 \tag{26a}$$

if there is no local oscillator or

$$2f(n_{\text{coh}})(I_s/I_{\text{LO}}) \gg (i_0/I_0)^2 \tag{26b}$$

if a local oscillator is present. Equations (26) show clearly the importance of

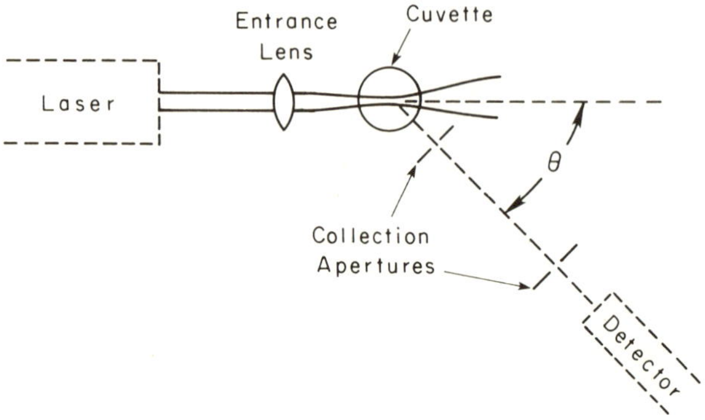

Figure 13. The elements of the spectrometer.

reducing laser fluctuations to a minimum, particularly when a local oscillator is employed.

It is possible to reduce the effect of laser fluctuations by modulating the duration of each correlator sample time during which the correlator is able to accept input pulses. This may be done by gating the input pulses so that they are turned "on" at the beginning of each sample time and turned off after a detector monitoring the laser intensity has counted to a preset level. In this way the detected light is normalized during each sample time. An instrument capable of accomplishing this task is available commercially.

2.4.3. The Spectrometer

The design and construction of a spectrometer suitable for use in correlation spectroscopy applications is not difficult, as is evidenced by the large number of "home built" instruments to be found in research laboratories. Indeed, the wide range of experiments currently being performed make it almost impossible to design a single commercial spectrometer capable of performing all of the experiments. Those currently on the market are suitable only for studies of solutions and have no provisions for the introduction of local oscillators, electric fields, or other variations that are of interest.

While the design of a spectrometer is not complex, there are several principles that must be employed if optimum results are to be obtained. We begin our discussion with a diagram (Figure 13) of a spectrometer showing the components necessary for correlation spectroscopy. The important decisions which must be made concern the focal length of the entrance lens, the design of the collection apertures, and the sample cell. An additional difficulty is introduced if a local oscillator is to be used.

The selection of the entrance lens is governed by the need, when weakly scattering samples are to be investigated, to scatter as much light as possible into a single coherence area. Combining equation (5) for the scattered intensity of a single molecule, equation (11) to account for N molecules in the scattering volume, and equation (12) defining the coherence area we find that the detected power is

$$P_s = NS \frac{\lambda^2}{\pi a^2} I_0 \tag{27}$$

If we assume that the entrance lens focuses the laser beam to a diameter $2a$, then

$$I_0 = \frac{P}{\pi a^2} \tag{28}$$

where P is the laser power. The number of molecules in the scattering volume is

$$N = C_n V \tag{29}$$

where C_n is the concentration of molecules in molecules/unit volume and V is the volume of the scattering region. Assuming the scattering volume is spherical with radius a (a slight underestimate), $V = \frac{4}{3}\pi a^3$ and we find

$$P_s = \frac{4}{3} C_n \frac{S\lambda^2 P}{\pi a} \tag{30}$$

The conclusion is immediate: in order to maximize the detected signal we must make a as small as possible.

For a simple lens of focal length f the radius of the beam at the focal point will be determined by the intrinsic divergence of the original laser beam, α, which is usually about 10^{-3} rad. The radius of the laser beam at the focal point is $f\alpha$ and we see that the power scattered into a single coherence area may be increased by decreasing the focal length of the entrance lens.

The increase in scattered power achieved by decreasing the scattering volume does have one undesirable side effect. As the size of the scattering volume is decreased, the knowledge of the scattering wave vector becomes less and less precise. The error in the scattering vector q is given approximately by

$$\Delta q \approx \frac{2\pi \cos(\theta/2)}{2a} \tag{31}$$

This is not a serious matter if the correlation function to be expected is exponential because the contributions in positive and negative Δq will tend to average to the central value. However, when sinusoidal signals are expected, as in surface wave scattering, the broadening can contribute erroneously to the damping of the correlation function.

The sample cell may take on a wide variety of shapes and sizes depending on the specific application. Square cells with path length as little as 1 mm have been successfully used as have round cells 1 cm or more in diameter. It is common to use ordinary spectrophotometer cuvettes with all four sides polished. Even the inexpensive plastic cuvettes may be used in a number of cases.

The chief problem arising in the use of square cells is that stray light may be generated by scratches or dirt on the outside of the cell and eventually find its way into the detector where it forms a weak local oscillator. The resulting change in the measured correlation function [see equations (17) and (19) for an example] leads to a loss in accuracy of the measurements. Otherwise, as long as one makes measurements only at a scattering angle of 90° or corrects carefully for refraction at the cell surface, square cells may be easily used.

When it is desired to make measurements at a number of different angles, a cylindrical cell would seem to offer many advantages. It is, however, difficult to achieve the alignment necessary to realize the potential for accuracy inherent in the light scattering method of measuring diffusion constants. The problem arises because of refraction at the cell surface and is particularly severe at small scattering angles. It is possible to show that the fractional error in the measured time constant ($\Delta\Gamma/\Gamma$) due to an error in alignment ΔX of either the laser beam or the detector is

$$\frac{\Delta\Gamma}{\Gamma} = \frac{\Delta X}{r}\left(1 - \frac{1}{n}\right)\cot\frac{\theta}{2} \tag{32}$$

where r is the cell radius, n the index of refraction of the solution, and θ the scattering angle. It is assumed that the cell wall is very thin. For example, to achieve a 1% error at a scattering angle of 20° and using a 1-cm-radius cell would require $\Delta X \approx 0.007$ cm or less. Alignment accuracy of this quality is difficult to achieve.

Convection currents in a cell can lead to a substantial contribution to the signal for the same reason as mentioned earlier in connection with the effects of vibration. They are particularly prevalent when a thermal control bath is poorly designed so that the bottom of the cell is warmer than the top. However, proper design of the temperature control system together with the use of cells with the smallest practical interior dimensions virtually eliminates the problem. Notice, however, that this makes scattering angles other than 90° (using a square cell) almost impossible.

The collection apertures should be designed to limit the detector to one or two coherence areas. One can think of the first aperture as the pinhole of a pinhole camera. The second aperture should then have a diameter equal to the diameter of the image of the laser beam at that point. If this is done then the laser beam size, the first aperture size, and the distance from the laser beam to the first aperture will determine the number of coherence areas according to equation (12).

2.4.4. The Detector

2.4.4.1. The Photomultiplier. The detector† or, more properly, the detection system usually consists of a photomultiplier tube and a pulse amplifier–discriminator. The photomultiplier contains a cathode usually made of one or more alkali metals which will absorb a photon and immediately emit an electron. The electron is accelerated by an electric field and then collides with a sheet of metal (a dynode) knocking out several electrons. The group of electrons is again accelerated, collides with a second dynode, and so forth until after 9 to 14 dynodes the single electron has been "multiplied" into 10^5 to 10^7 or more electrons. This group of electrons originating from the capture of a single photon forms the output of photomultiplier.

There is a type of electron multiplier called the Channeltron, which, instead of discrete dynodes, employs a continuous multiplication process. These devices are not suitable for correlation spectroscopy because the gain is markedly decreased immediately after a pulse and recovers to the original value only after about 100 μsec. Thus a photon arriving during this time will be detected with a smaller probability and a strong correlation function will develop reflecting only the properties of the detector.

Some light scattering experiments provide sufficient light that the photomultiplier output may be used directly as an analog signal either with a spectrum analyzer or an analog correlator. (Indeed, one experiment has so much scattered light that a photodiode serves as the light detector.) However, in most cases single photon events are processed by a digital correlator. The first step is to convert the relatively small single-photon signal (rarely more than 50 mV high and 50 nsec in duration) into a pulse of the proper amplitude and duration for employment by the correlator. At the same time, very small photomultiplier pulses are rejected as most likely

† This section assumes a rudimentary knowledge of the operation of a correlator. A brief reading of pp. 36–40 will provide the necessary information.

arising from sources other than the detection of a photon. These two functions are normally combined in a single unit called a pulse amplifier–discriminator (PAD).

The properties of a photomultiplier that are of interest to us here are the quantum efficiency (probability of detecting a photon) at the wavelength to be employed, the gain of the tube, width of the anode pulse, degree of afterpulsing, and dark current. In addition, there are two major types of construction, the side window and end window tubes. Since the construction has a major influence on the properties of a tube, we shall take it up first.

The side window tube has a totally enclosed photocathode. The photoelectrons leave the cathode from the front and are usually (but, unfortunately, not always) attracted to the first dynode. The construction of this tube leads to a nonuniform quantum efficiency over the active area of the photocathode. This is not a serious drawback in photon correlation applications because the light to be detected falls on a small area (a few coherence areas at most), and as long as this is selected to be the most sensitive area of the photomultiplier the nonuniform quantum efficiency will not be a serious disadvantage. The advantages of the side window tube are a low cost, high gain at modest accelerating voltage, and fast pulse rise time.

The end window photomultiplier was developed for applications in which a uniform quantum efficiency over a large area is required. The photocathode is deposited on the inside of the flat end of the tube and is less than 100 nm thick. Photons are absorbed on the front side and photoelectrons ejected from the back side of the cathode. These tubes have been widely employed in correlation spectroscopy at least partly because some manufacturers offer tubes specially designed or selected for low afterpulsing effects. However, there seems to be no reason for rejecting the side window tube, particularly in less critical applications.

Returning to the list of properties mentioned above, the tube should be chosen to have a quantum efficiency that is as large as possible at the wavelength, or range of wavelengths, to be employed. The photomultiplier companies publish curves giving the necessary information for all of the tubes they make.

The gain of the tube and width of the anode pulse are important in that they must be matched to the PAD. Most tubes have maximum gains somewherc in the range of 10^5 to 10^7, and the gain of an individual tube can be varied by a factor of 5 or more simply by changing the cathode potential. The width of the output pulse should be less than 20 nsec for applications involving correlator sample times in excess of 100 nsec. A tube with a correspondingly reduced output pulse width should be chosen if shorter correlator sample times are anticipated.

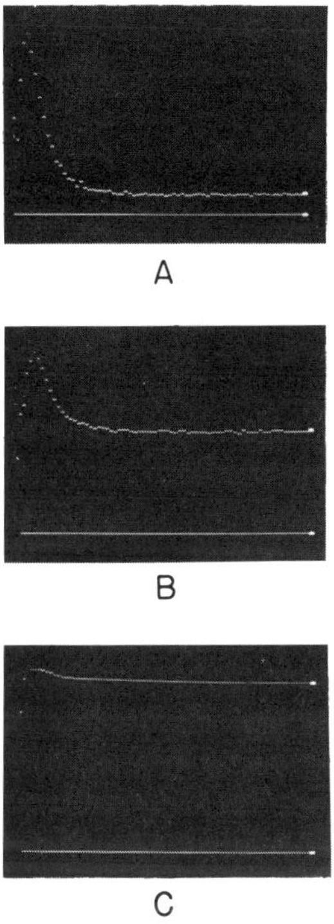

Figure 14. The contribution of afterpulsing to the correlation function at average photon rates of (A) 800 counts/sec, (B) 6000 counts/sec, and (C) 80,000 counts/sec. The correlator sample time in all three cases is 1.0×10^{-7} sec. The total probability of afterpulsing is calculated from equation (35) to be 0.2%. Notice that the first channel appears high at the low count rate and low at the higher rates. This is caused by competition between instrumental dead-time effects depressing the first channel and pulse amplifier–discriminator double pulsing increasing the first channel. The photomultiplier is a Hamamatsu R928P, the correlator a Langley-Ford Instruments model 1096, and the PAD a Langley-Ford Instruments model PAD-1.

Afterpulsing is the name given to the phenomenon in which a positive ion generated at some time during an electron cascade returns to the cathode or an early dynode and initiates a second electron pulse. The second pulse follows the first by several tenths of a microsecond and results in a correlation function with a peak such as is shown in Figure 14. The probability of having an afterpulse, P_A, is independent of the pulse rate. If a total of N photons are detected during time T, the absolute area ΔA of the afterpulse peak in the correlation function will be

$$\Delta A = NP_A \text{ counts} \tag{33}$$

These counts will be distributed over about 10 channels of the correlation function if the sample time, ΔT, is 0.1 μsec. The base-line count rate is, assuming the photons are uncorrelated,

$$G(\tau) = \left(\frac{N}{T}\,\Delta T\right)^2 \frac{T}{\Delta T} \tag{34}$$

where the term in parentheses is the average number of photons detected in a single sample time and $T/\Delta T$ is the number of samples taken. Dividing equation (33) by equation (34) we find for the ratio of the area of the afterpulse peak to the DC portion of the baseline

$$\frac{\Delta A}{G(\tau)} = \frac{P_A T}{N \Delta T} \tag{35}$$

which shows that the afterpulse peak is reduced in significance as the photon detection rate (N/T) or sample time increases. At any rate, if ΔT exceeds 1 μsec, the effect can be eliminated by ignoring the first channel. At faster times it is possible to use careful measurements of the shape of the afterpulse peak and equation (31) to correct measured correlation functions.

It would be, of course, best to eliminate the afterpulse problem at its source. This may be done to an extent by buying photomultipliers selected for and/or designed for a low afterpulse rate. The extra cost is warranted, however, only if correlator sample times less than 1 μsec are to be used. Finally, it should be pointed out that not all white light correlations can be traced to the photomultiplier. As we shall discuss below, the PAD can also generate a correlation function that has nothing to do with the detection of photons.

For a variety of reasons photomultipliers produce pulses indistinguishable from photon pulses even in the dark. This contribution to the anode current, known as dark current, is often reduced by one or two orders of magnitude by cooling the tube to from 0° to −30°C. This is done only if the dark current forms a significant part of the overall signal. The dark current for tubes sensitive to light from a He–Ne laser (632.8 nm) corresponds to about 300 photons/sec/cm^2 of cathode area. Therefore, a tube with a 2 cm^2 or less photocathode can be expected to produce 600 dark pulses/sec or less. This pulse rate is small compared to those obtained in virtually all photon correlation experiments. Consequently, there is no need to cool a photomultiplier for these applications as long as the tube is properly selected.

Having selected a tube it is necessary to obtain a housing and power supply. Excellent housings containing the tube socket and a resistor divider to obtain the proper dynode voltages from the cathode voltage are readily available. It is necessary to use the type containing rf shielding and, if a front window tube is selected, magnetic shielding as well.

2.4.4.2. Photomultiplier Power Supply. A power supply capable of delivering the maximum voltage and current requirements must also be acquired. It is extremely important that the ripple and short-term voltage fluctuations be reduced to an absolute minimum because of the very strong

Table 1. Typical Specifications for a Photomultiplier High-Voltage Power Supply

	Side window photomultipliers	Front window photomultipliers
Voltage	Adjustable from 500 to 1500 V	Adjustable from 500 to 2500 V
Current	10 mA	10 mA
Regulation—line	0.002%	0.002%
Regulation—load	0.002%	0.002%
Ripple	< 10 mV	< 10 mV

dependence of the photomultiplier gain on voltage (the gain is proportional to V^6 to V^{10} depending on the number of dynodes). Since the probability of detecting a photon depends on the gain, any fluctuation in cathode voltage on a time scale comparable with the total correlation time will show up in the measured correlation function. Minimum specifications for the power supply are listed in Table 1.

2.4.4.3. Pulse Amplifier–Discriminator. The interaction between the photomultiplier and the PAD presents the most difficult electronics problem in the light scattering system. The photomultiplier output is small, short, negative-going pulses, while the PAD must produce pulses conforming to the specifications of Table 2. The overall gain of the PAD is frequently of the order of 1000 or more, resulting in the possibility that a coupling from output to input can generate a spurious pulse, and therefore spurious correlations.

The signal produced by the photomultiplier is a pulse of charge equal to the gain of the tube, γ, multiplied by the electronic charge, e. This charge arrives at the anode spread out over a time equal to the pulse width, t_ω (about twice the pulse rise time). Thus, the current flowing during the pulse is

$$i = \frac{\gamma e}{t_\omega} \tag{36}$$

The equivalent circuit of the photomultiplier anode input to the PAD is shown in Figure 15. The resistor is the input resistor to the amplifier. The

Table 2. Specifications for the Three Major Standard Logic Levels

	TTL	NIM	ECL
Width	> 20 nsec	> 3 nsec	> 5 nsec
Allowed "1" level	2–5 V	> −0.1 V	> −0.98 V
Allowed "0" level	0–0.8 V	< −0.6 V	< −1.63 V
Impedance	50 Ω	50 Ω	50 Ω

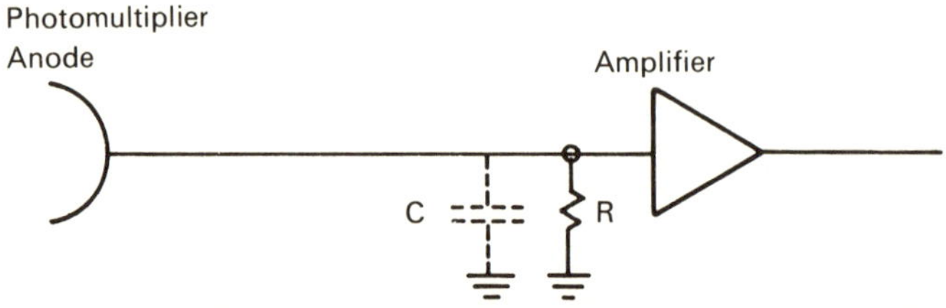

Figure 15. The input circuit of a pulse amplifier–discriminator. The capacitor represents stray capacitance due to input leads and the input circuit of the amplifier. This capacitance should be reduced to as small a value as possible by physically locating the PAD near the photomultiplier.

capacitor includes the input capacitance of the amplifier, about 10 pF, and the capacitance of the cable connecting the tube to the PAD if R is not equal to the characteristic impedance of the cable. If the cable is terminated in its characteristic impedance (usually 50 Ω) its capacitance need not be included in C. Consequently, there are two philosophies in designing PADs. The input impedance may be set at 50 Ω, which yields the fastest response time and allows the use of a longer cable between the photomultiplier and the PAD but requires a higher gain amplifier; or the imput resistance may be set higher, making a less expensive unit but stretching the pulses and requiring a very short distance between tube and PAD.

We can gain an appreciation for the requirements on the PAD by computing the input voltage to the amplifier under each of the design types. If $R = 50\ \Omega$ we have for the voltage

$$V = iR = \frac{\gamma e}{t_\omega} R \tag{37}$$

If $\gamma = 10^6$ and $t_\omega = 10$ nsec we find the peak voltage is 8×10^{-4} V. On the other hand, if R is much larger, the input voltage is limited by the capacitance to

$$V = \frac{\gamma e}{c} \tag{38}$$

which, for $\gamma = 10^6$ and $c = 30$ pF gives a peak voltage of 5×10^{-3} V, almost an order of magnitude larger. The added signal amplitude is an enormous advantage and PADs based on the "capacitance-limited" signal sell at less than half the price of the impedance-matched PADs. If carefully designed, they are, in fact, superior for correlation spectroscopy applications in which sample times of 100 nsec or greater are used.

The most difficult aspect in designing a PAD adequate for correlation spectroscopy is to reduce the PAD correlations to an acceptable level. Correlations have two sources. If the amplifier output does not return to zero immediately following a pulse, the probability of detecting the next pulse will be either enhanced or diminished depending on whether the amplifier output is above the baseline or below. Some amplifiers exhibit "ringing" following a pulse, a condition that leads to very interesting but meaningless correlation functions. In general, ac coupled amplifiers are prone to problems of this type. In particular, the baseline output level of the amplifier (and therefore, the discrimination level) depends on the pulse rate. Consequently, PADs containing ac coupled amplifiers should be avoided in all but the least critical applications.

The second source of unwanted correlations is the tendency for the trailing edge of one discriminator output pulse to initiate a second pulse because of a weak inductive coupling between the output and input of the PAD. This problem may be detected by the tendency of the first correlator channel to receive counts at a faster rate than the second channel at short sample times (it should count slower due to the dead time of the PAD or correlator input section). The problem is particularly severe for low-cost–high-gain PADs, but no unit is completely immune. The best one can do to minimize the problem is to use a high-gain photomultiplier (at least 10^6 and preferably 10^7) and a PAD designed specifically for use in correlation work.

2.4.5. Signal Analyzers

The signals at the output of the photomultiplier are similar to those shown in Figures 6 and 7. At first sight the signals appear to be random noise, but careful analysis shows that the spectrum of the signal contains information about the system responsible for the light scattering.

Early experiments were performed using a single-channel wave analyzer which was capable of obtaining a spectrum in about an hour provided the photomultiplier signal was large. In this system the photomultiplier output was used as an analog signal—no attempt was made to use the ability to detect single-photon events.

Technical advances soon provided two new instruments, both capable of obtaining the spectrum of the scattered light in a time as short as a few seconds. Both instruments have continued to be developed and are widely used in correlation spectroscopy today. They are the real time spectrum analyzer and the correlator.

2.4.5.1. The Spectrum Analyzer. The real time spectrum analyzer obtains the spectrum of an analog input signal using digital techniques. In a typical instrument the input signal will be digitized at 512 sequential equally spaced points in time. During the time the second 512 samples are being taken, 256 Fourier components of the first set are obtained and a 200-point spectrum is presented. The results from a number of samplings are summed to provide a smooth spectrum. If the instrument can perform the Fourier transform in a time less than that required to obtain the second set of 512 samples, the instrument will operate in real time and no information will be lost. The maximum frequency at which real time operation is obtained is usually in the range of 1 to 5 kHz.

The spectrum analyzer is useful in experiments in which there is a high detected light level as, for example, when an optical local oscillator is used. A typical application is in electrophoretic light scattering experiments. Here the fact that the spectrum is obtained directly is an aid in interpreting data as each species having a different electrophoretic mobility results in a distinct spectral peak.

2.4.5.2. The Correlator. We turn now to correlators, instruments capable of obtaining the correlation function of an electrical signal. Mathematically a correlation function is defined as

$$G(\tau) = \lim_{T \to \infty} \frac{1}{2T} \int_{-T}^{T} I(t)J(t + \tau)\, dt \tag{39}$$

where $I(t)$ and $J(t)$ are signals which depend upon time. $G(\tau)$ is called the autocorrelation function or cross-correlation function depending on whether $I(t)$ and $J(t)$ are the same or different signals. Instruments designed to obtain an approximation to equation (35) for a number of values of τ are available from several manufacturers. Some are designed primarily to accept analog input signals; others accept pulse trains similar to those obtained from the output of a single-photon detector. We will describe the operation of the latter correlator and mention some of the properties of the former.

A block diagram of a typical digital correlator is shown in Figure 16. The timing and operation of the correlator is controlled by the Sample Time Generator, which divides time into intervals of equal duration, $\Delta\tau$. The number of pulses at input A occurring during each sample time is counted by the Shift Register Counter. This situation is illustrated in Figure 17, where $n_0, n_1, \ldots,$ are the number of pulses appearing at input A and counted by the Shift Register Counter.

At the end of each sample time the number in the Shift Register Counter is entered into the first stage of the Shift Register, the number that was in the first stage is shifted to the second, the number that was in the second stage is shifted to the third, etc. As a consequence, after the correla-

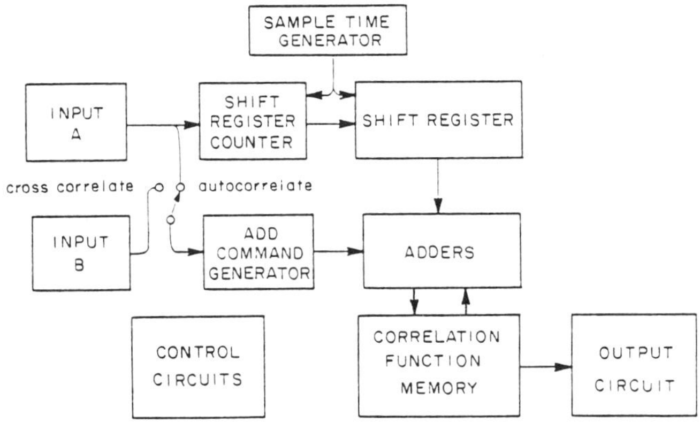

Figure 16. Block diagram of a correlator. The principles of operation are discussed in the text.

tor has been in operation for a brief period of time, the first stage contains $I(t - \Delta\tau)$, the second $I(t - 2\Delta\tau)$, the third $I(t - 3\Delta\tau)$, and the kth stage contains $I(t - k\Delta\tau)$.

During the "present" sample time each pulse appearing at input A (when in the autocorrelate mode) or input B (when in the cross-correlate mode) is processed by the Add Command Generator and instructs all Adders in the correlator to add each of the numbers stored in the Shift Register to the number stored in the associated channel of the Correlation Function Memory. As an example, consider the pulse sequence in Figure 17. During the sample time interval 2, the product $n_2 n_3 = 0$ is added to Correlation Function Memory channel 1, the product $n_2 n_4 = 6$ is added to channel 2, $n_2 n_5 = 0$ is added to channel 3, etc.

Thus, the correlator will accumulate in the first channel

$$G(\Delta\tau) = n_0 \tilde{n}_1 + n_1 \tilde{n}_2 + n_2 \tilde{n}_3 + \cdots$$
$$= \sum_{i=0}^{N-1} n_i \tilde{n}_{i+1} \tag{40}$$

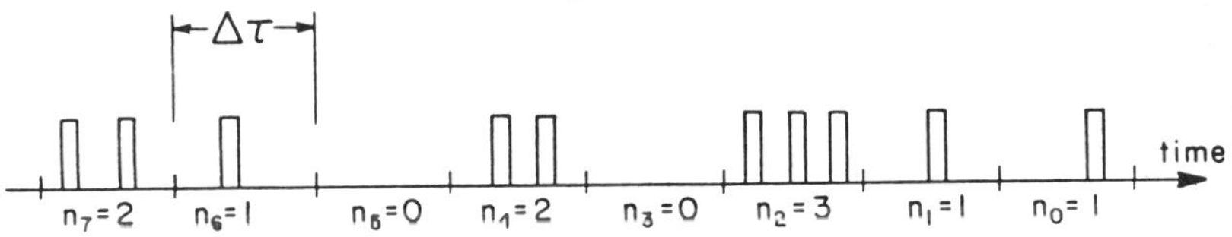

Figure 17. The train of pulses processed by a correlator. The small marks represent the sample time clock which divides time into increments of ΔT.

and in the second channel

$$G(2\Delta\tau) = n_0 \tilde{n}_2 + n_1 \tilde{n}_3 + n_2 \tilde{n}_4 + \cdots$$
$$= \sum_{i=0}^{N-1} n_i \tilde{n}_{i+2} \tag{41}$$

In general, the kth channel will contain

$$G(k\Delta\tau) = n_0 \tilde{n}_k + n_1 \tilde{n}_{k+1} + n_2 \tilde{n}_{k+2}$$
$$= \sum_{i=0}^{N-1} n_k \tilde{n}_{i+k} \tag{42}$$

which is a good approximation to the true correlation function whenever the change in the value of the correlation function during the time $\Delta\tau$ is small. In each of these expressions the numbers n_i represent the number of times the content of each stage of the shift register is added to its respective correlation function memory channel and the numbers $\tilde{n}_{i+k}$ are the numbers stored in the shift register.

The characteristics of a correlator that are important in light scattering experiments are the efficiency of operation; the capacity of the shift register counter and, therefore, of the shift register; the range of sample times available; and the number of channels. We will discuss each of these characteristics bearing in mind the requirements imposed by the majority of experimental conditions.

The efficiency with which a correlator performs the sums of equations (40)–(42) depends critically on the architecture of the correlator. The instrument illustrated in Figure 16 employs a separate adder and multibit counter for each channel. This is not as extravagant as it sounds as each adder must handle only a 4-bit by 4-bit sum (assuming the shift register has a 4-bit capacity) and this may be accomplished with a single integrated circuit. Multibit counters are also relatively inexpensive. If the adder is capable of doing a sum every 10^{-7} sec, the rate of obtaining sums in a 64-channel correlator is 64×10^7 sums/sec. An instrument with these abilities will operate at nearly 100% efficiency at sample times of 100nsec or longer.

An alternate approach frequently used in instruments intended for analog input signals is to use a single arithmetic unit and a large-scale memory. For each cycle of data updating this requires (again for a 64-channel instrument) 64 memory reads, 64 multiplications and additions, and 64 memory writes. This can be done with full efficiency only if the sample time is long (10^{-3} sec or longer in typical cases). For shorter sample times the instruments operate in a batch mode in which only a fraction of the terms of equation (42) are included. As a result there is a substantial loss of efficiency. At sample times of 10^{-7} sec the efficiency is frequently less than 1%. The greatly reduced efficiency is particularly serious when light scattering experiments at low light levels are contemplated. Under the best of conditions the duration of a measurement is increased from a minute to well over an hour. In many cases the longer measurement time makes the experiment impossible because of changes in the sample occurring over a period of an hour or more.

The capacity of the shift register is measured by the number of bits of information used to represent the intensity during each sample time. Correlators intended for use with fluctuation spectrometers generally use either one or four bits. The one bit or clipped correlator introduces a zero into the shift register if the number of photons detected during a sample time is less than or equal to a preset number (called the clipping level) and a one if the photon count exceeds the clipping level. While it may appear that the correlation function obtained in this way would bear little relationship to the true correlation function, it is possible to show that the clipped correlation function is proportional to the true correlation function if the signal is random in the proper way. In particular, the clipped correlation function is proportional to the true function if the signal obeys Gaussian random statistics. This is expected to be the case whenever the signal originates from a large number of independent scatterers, and is independent of the form of the correlation function. However, if the sample contains only a few scatterers or the scatterers are coupled together, the clipped correlation function may be distorted.

Correlators having 4-bit shift registers obtain an accurate correlation function for a wider range of input signals, although it is still conceivable that certain classes of input signals will exceed the operating range of the correlator. For example, a signal having two (or more) greatly different intensities each with small but important fluctuations, would yield erroneous results. In addition to permitting a wider range of input signals, the 4-bit correlator is easier to use because the adjustment of the clipping level is either easier or unnecessary.

The sample times available to the user are for all practical purposes unlimited at the long end. The shortest sample time is consequently the factor of importance. For most fluctuation spectroscopy applications a minimum sample time of 100 nsec is satisfactory, and a number of instruments are available at this speed at moderate prices. Several correlators in the 20 nsec or faster category are available at much higher prices. They are largely single-bit instruments intended for use in high-speed velocity measurements and do not offer any significant advantage in the present application.

The number of channels required depends on the nature of the correla-

Table 3. Specifications for Correlator Suitable for Light Scattering Applications

Minimum sample time	100 nsec
Shift register capacity (each step)	4 bits
Conditions for batch mode operation	None
Number of channels	64

tion function to be studied. If the correlation function consists of a single exponential or a narrow distribution of exponentials, a 64-channel instrument will provide all the detail in the correlation function that can sensibly be used. However, if a broad range of exponential time constants (greater than a factor of 5 between maximum and minimum time constants) or if an oscillatory correlation function is involved, a larger instrument is warranted. In any case, a correlator should have several channels that may be delayed a substantial amount in order to establish a base line.

The characteristics of a correlator suitable for most applications in fluctuation spectroscopy are summarized in Table 3.

2.5. SIGNAL-TO-NOISE RATIO

2.5.1. Introduction

Consideration of the signal-to-noise ratio is complicated by the large number of factors that enter into the final answer. Some factors (dust, for example) can be dealt with only in the most general terms while others (such as photon shot noise) are susceptible to a precise theoretical description. Because of the complexity of the subject, and the presumed interest of the reader in getting on with the experiments, the present discussion will be limited to "factor of 2" calculations, thus avoiding some of the complex details required by an exact analysis. Given the nature of some of the imponderables, the results will probably be as close to reality as would be possible even with a more exact theory.

We recognize at the outset three sources of noise that limit our ability to measure the properties of the scattered light with arbitrarily high precision. They are effects due to the finite intensity of the scattered light; effects due to a finite duration of the experiment; and effects due to light scattered by unwanted effects (dust, for example). We take up each of these contributions.

2.5.2. Effects due to Finite Intensity

A contribution to fluctuations in the scattered light intensity is caused by the fact that the number of photons detected during each sample time is

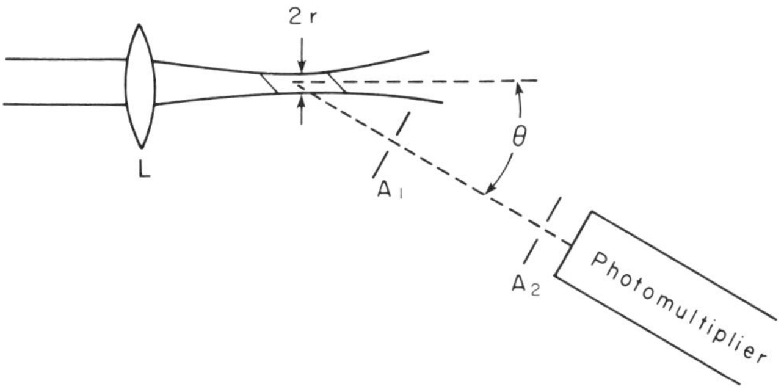

Figure 18. Detailed figure showing the scattering geometry.

finite. If the instantaneous intensity of the scattered light corresponds to N photons/sample time, we expect most of the time to detect from $N - \sqrt{N}$ to $N + \sqrt{N}$ photons during a sample time. The number detected will obey a Poisson distribution law. The expected uncertainty in the correlation function due to this effect is given by $\sqrt{G}$, where G is the number given by equation (40). This contribution to the noise may be reduced by increasing the laser intensity or solute concentration, by scattering at a smaller angle, or by focusing the laser beam to a smaller diameter.

To understand the reasons for these actions we calculate the number of photons scattered into a single coherence area each time constant of the exponentially decaying correlation function. If this number is much less than 1, the measures just mentioned will improve the signal-to-noise ratio. If, however, it is much greater than 1, the signal-to-noise ratio is limited by an entirely different process which will be described below.

The geometry of the scattering region is illustrated in Figure 18. The laser beam is focused to a beam of radius r by lens L. In most cases r is determined by the intrinsic divergence of the laser beam and the focal length f of lens L. For our purposes we can take $r = 10^{-3} f$. A system of apertures, A_1 and A_2, defines a scattering region which we will assume is a cylinder of radius r and length $2r/\sin\theta$. In this way, the scattering region as viewed from the photomultiplier will be a square with sides $2r$. The volume of the scattering region will be $(\pi r)^2(2r/\sin\theta)$ and the number of solute molecules in that volume is

$$N = \frac{2\pi r^3}{\sin\theta} C_n \tag{43}$$

where C_n is the number density of solute molecules.

In order to find the intensity of the scattered light, I_S, at the detector we use equation (5), $I_s = I_0 S/R^2$. For a total laser power P_0 we have $I_0 = P_0/2\pi r^2$. The total power incident on a single coherence area at the detector is $P_d = I_s A_{\text{coh}}$. Combining all these factors and using equation (12) for A_{coh} we finally find for the power incident on the detector

$$P_d = P_0 \frac{C_n S\lambda^2}{\pi r \sin\theta} \tag{44}$$

We must now multiply P_d by the exponential time constant obtained from equation (16), $(2Dq^2)^{-1}$, where $q = (4\pi/\lambda)\sin(\theta/2)$, in order to obtain the total energy incident on the detector each time constant,

$$E = P_0 \frac{C_n S\lambda^4}{32\pi^3 Dr \sin\theta \sin^2(\theta/2)} \tag{45}$$

For small scattering centers $S \propto \lambda^{-4}$. Thus, equation (45) shows directly that an increase in P_0 or C_n or decrease in r or $\sin\theta \sin^2(\theta/2)$ will all increase the light energy detected by each exponential decay constant.

It might appear from equation (45) that arbitrarily small values of the solute concentration C_n could be studied by increasing the detected light energy through the other three factors. There is, however, a practical limit achieved when the light scattered by the solvent is roughly equal in intensity to that scattered by the solute molecules. For a solute molecule with molecular weight 25,000 this will occur at a concentration of about 0.5 mg/ml.

The importance of focusing the laser to a small diameter must also be emphasized. A 50-mW laser focused into a cell with a 10-cm focal length lens is no more effective than a 5-mW laser focused with a 1-cm focal length lens. The cost of the second option is a small fraction of the cost of the first.

2.5.3. Effects due to Finite Experiment Duration

A second limitation in accuracy is due to the fact that data are collected for a finite number of decay times of the correlation function. If the correlation function decays as

$$G(\tau) = 1 + e^{-\Gamma\tau} \tag{46}$$

and the total duration of an experiment is T, then the number of decay times during the experiment is ΓT. The corresponding signal-to-noise ratio, even if the detected light level is infinite, is

$$S/N = (\Gamma T)^{1/2} \tag{47}$$

The only way to improve this contribution to the signal/noise ratio (other than increasing the duration of the experiment) is to *increase* Γ which requires that light be scattered at a larger angle.

2.5.4. Effects due to Unwanted Scattered Light

The presence of unwanted signals in the scattered light provides the third major limitation to the quality of the light scattering results. This topic is frequently neglected in discussions of signal-to-noise ratios, perhaps because of its very complexity, and yet it provides the ultimate limit in signal-to-noise ratio for the vast majority of real experiments. Included in this category are such effects as

- Fluctuations in laser intensity;
- Unwanted laser light due to reflections or flare that has not been scattered but acts as a local oscillator;
- Convection currents in the scattering cell;
- Dust, air bubbles, glass particles, bacteria, and other foreign matter in the solution,
- Light scattered at the wrong angle present because of reflections in the cell;
- Molecules or other artifacts resulting from improper or inadequate sample preparation;
- Light scattered by the solvent.

The principal difficulty with many of these contributions to the noise signal is that systematic effects take place so that the measurements are consistent from experiment to experiment but unfortunately give the wrong answer. The clearest example of this phenomenon is provided by the second effect. When a small amount of light (small compared to the real scattered light) is unscattered and is able to act as a local oscillator, the correlation function will contain two exponentials, one with a decay rate of $2Dq^2$ and another at Dq^2 proportional in amplitude to the intensity of the local oscillator. If the resulting correlation function is fitted to a single exponential, the calculated decay rate will differ from the correct result according to[7]

$$\frac{\Delta\Gamma}{\Gamma} = -\frac{16}{9}\frac{I_{LO}}{I_s} \tag{48}$$

Equation (48) shows that extreme caution must be taken to avoid small amounts of local oscillator if the greatest possible accuracy is to be obtained. Of the remaining effects, much the same philosophy must be adopted: do whatever is possible to eliminate the cause.

The light scattered by the solvent cannot, of course, be eliminated. Fortunately it usually does not contribute to systematic effects because it has a very broad spectrum. The one example in which this is not true is when relatively large molecules are included in the solvent as, for example, when glucose is added to a solution to alter the viscosity. In this case the correlation function deriving directly from the solvent will cause a significant systematic error. If in doubt it is best to study correlation functions of the solvent alone. They should be flat when using sample times of relevance to the experiment. If they are not flat, it may be possible to subtract their effect from the final correlation function using careful curve-fitting procedures.

A few attempts have been made to eliminate the effects of large aggregates and dust particles by gating off the correlator when the scattered intensity is too large. This technique may be used to advantage particularly when studying samples that spontaneously form a few very large aggregates which may not be removed by any of the standard techniques. There are difficulties, however, in that any procedure for selecting the scattered intensity at which the correlator is gated off will introduce a new component in the correlation function. Consequently, this technique should be used only with extreme caution and after a thorough understanding of the possible consequences is obtained.

The elimination of dust from a solution is considered by many to be a black art. This may be because each sample requires slightly different procedures. For this reason, it is not possible to give a single prescription that is guaranteed to lead to clean solutions. There is, however, a method that should lead to clean solutions if combined with sufficient effort on the behalf of the experimenter.

There are, in fact, only three methods for cleaning a sample once it has been prepared. These are distillation (usually used only with the solvent), filtering, and centrifugation.

Without question, the first step is to learn to clean the solvent to the point that a laser beam focused into a cuvette full of the solvent and examined with a 10 power (or higher) magnifying system in a dark room will reveal no bright scatterers (or perhaps, one occasionally). Since water is one of the most difficult solvents to clean, the system successfully used in the author's laboratory for a number of years will be described. It is possible that a simpler system would be as effective; however, since this one works, we have chosen to devote our efforts to more interesting problems rather than trying to improve on it.

Tap water is first run through a deionizing column into an all-glass

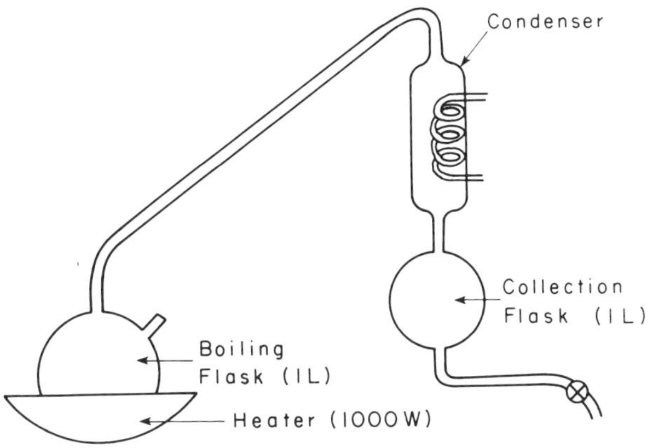

Figure 19. Still suitable for preparing very clean water.

commercial still. After the first distillation water is fed into a stainless steel tank and pressurized with dry nitrogen. Following filtration through a 0.025-μm mixed cellulose acetate and nitrate membrane, the water is distilled a second time using the all-glass still shown in Figure 19. In this still three special precautions are taken: to reduce the possibility of glass dust contaminating the water, care is taken to avoid glass-to-glass joints by using neoprene gaskets; water from the condenser is led down a glass rod to the inner surface of the collection flask in order to avoid air bubbles introduced when a drop of water falls from the condenser to the water surface; and water is withdrawn from the bottom of the collection flask through a plastic tube fitted with a plastic stopcock. The resulting water is always sufficiently clean for light scattering purposes.

The second step in preparing a sample for light scattering is to clean the cuvette so that, when filled with clean solvent, no bright scatterers are seen when tested with the focused laser beam. A dirty cuvette (that is, one with the residue of earlier experiments on the inner walls) should be cleaned in acid according to the cuvette maker's instructions. It should next be soaked for a day or more in a solution of water and detergent. Finally, the cuvette should be rinsed in many volumes of clean water using one of the vacuum jet devices available for this purpose. The final rinses should use the very clean water prepared as described above, while earlier rinses may use water from the first still. Alternatively, the cuvettes may be stored in ethyl alcohol immediately after using. They are then rinsed with many volumes of very clean water and dried in a warm oven before reuse.

The third step is to prepare buffers, if necessary. Use only the very clean water and the best grade of chemicals available. (It is best to maintain bottles of chemicals that are used only for preparation of light scattering buffers. It is surprising how much foreign matter is introduced into chemical bottles in a busy laboratory.) After preparation the buffer must be filterd using the smallest pore size available until it passes the usual cleanliness test.

The fourth and final step is the solubilization of the sample and filtering or centrifugation of the resulting solution to remove any large light scatterers. This step is by far the most varied and, therefore, most difficult of the four. Both the filter composition and pore-size must be chosen to match the sample. Considerable experimentation is frequently necessary to find the proper combination so that the necessary task is accomplished without removing or denaturing an excessive fraction of the sample. It is a good idea to check the sample concentration both before and after filtering.

A properly prepared sample will scatter substantially more light than the buffer, but the scattering will have a very smooth appearance. If there are bright spots in the laser beam, the sample is still dirty, unless the sample consists of particles that are about 0.5 μm or more in diameter. These particles are large enough to be seen individually.

The above four steps, while far from providing a cookbook approach to sample preparation, should give enough of an outline to the diligent worker to enable successful light scattering experiments.

2.6. DATA ANALYSIS

2.6.1. Introduction

The correlation functions we have discussed in this chapter have been simple exponentials resulting from the translational diffusion of a monodisperse molecule or particle. In the other chapters of this book many other processes will be discussed that also lead to exponential contributions to the correlation function. There are, in addition, contributions that may be damped cosine functions or Gaussian functions. Given the diversity of wave forms that are observed, it is clearly impossible to give a single data analysis scheme that will encompass all experiments. We can, however, discuss the steps that must be taken.

2.6.2. Selecting the Theoretical Form

The first step in data analysis is thus the selection of the form (or combination of forms) expected in the correlation function. This selection depends on the purpose of the experiment under consideration and, to a certain extent, prejudices the outcome of the experiment. Fortunately, if judicious use of the well known chi-squared test is made, it is possible to determine if the choice of the theoretical expression used to fit the correlation function is adequate for the job. On the other hand, it is not always possible to prove that the suspected physical process is responsible for the details of the correlation function. This is especially true of the many physical processes known to lead to exponential terms in the correlation function.

Having determined the form of the expected correlation function, the usual procedure is to write a general expression containing one or more parameters that may be adjusted to make the theoretical expression have as nearly as possible the same values as the experimental correlation function. For example, if the correlation function is expected to have the form of a single exponential, we would write as the theoretical expression

$$G_{th}(\tau) = A + Be^{-C\tau} \tag{49}$$

and vary A, B, and C to make $G_{th}(\tau)$ resemble the measured correlation function, $G_{ex}(\tau)$ as closely as possible at all values of τ.

The theory of determining the "best" values of the parameters has been extensively developed.[8] Without discussing the details of the arguments, we simply state that the "best" values are those which minimize χ^2 (chi squared) defined as

$$\chi^2 = \sum_{i=1}^{N} \frac{[G_{th}(\tau_i) - G_{ex}(\tau_i)]^2}{\sigma_i^2} \tag{50}$$

where the sum is to be taken over all values of τ for which $G_{ex}(\tau)$ has been measured and σ_i is the expected error in $G_{ex}(\tau_i)$. The required minimization is almost always performed by a computer using one of several standard techniques.

The quantity σ_i is the standard deviation for the quantity $G_{ex}(\tau_i)$. If $G_{ex}(\tau_i)$ were measured many times, and a plot made showing the number of times $G_{ex}(\tau_i)$ takes on each value, a bell-shaped curve should be obtained with a peak at the true value of $G_{ex}(\tau_i)$ and a half-width at half-height of σ. Roughly 68% of all points would lie in the range between $G_{i\,\mathrm{peak}} - \sigma_i$ and $G_{i\,\mathrm{peak}} + \sigma_i$. Using an ideal correlator, the value of σ_i is $(G_{i\,\mathrm{peak}})^{1/2}$. This value will hold for a 4-bit correlator operated in a mode in which the shift register is rarely saturated.

2.6.3. Use of the χ^2 Test

Having obtained the desired values of the parameters it is appropriate to return to the question, "Is the initial form chosen for $G_{th}(\tau)$ consistent with the experimental data?" (Notice that we have not asked if the chosen form is correct. Even in the best of circumstances we can determine only that the form is consistent because there are always an infinite number of expressions that will fit the data to an arbitrary accuracy. Presumably only one of them is "correct".) The answer is obtained with the aid of χ^2.

The quantity χ^2 defined in equation (50) may be normalized by dividing by $N - p$, where p is the number of parameters in the expression to be fitted to the data. The resulting quantity is expected to be close to 1, particularly when $N - p$ is large (i.e., there are many more experimental points than parameters). The probability $P(x)$ that the quantity $x = \chi^2/N - p$ will be in the range $X \pm \Delta X$ diminishes rapidly as x increases beyond 1. The theoretical prediction for $N - p = 60$ is shown in Figure 20. If an experiment is performed a large number of times and x calculated for each run, a plot of the number of times x has values between 0 and 0.1, 0.1,

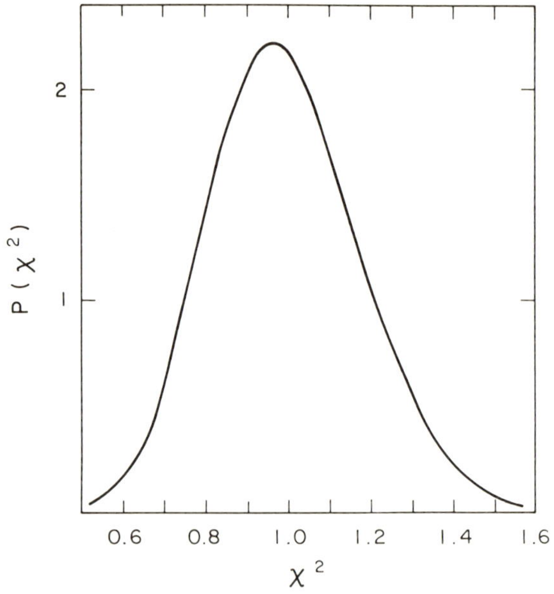

Figure 20. $P(\chi^2)d\chi^2$ gives the probability that χ^2 lies between χ^2 and $\chi^2 + d\chi^2$ if the form of the equation chosen is correct. This curve is calculated assuming there are 64 channels in the correlation function and three free parameters in the equation to be fit. A histogram showing the number of times each value of χ^2 is obtained in a real experiment should reproduce this curve. An experimental curve peaking at a higher value of χ^2 indicates the chosen function is not correct.

Table 4. Theoretical Form of Correlation Functions Corresponding to Various Physical Processes

Physical process	Theoretical form of $G(\tau)$
1. Translational diffusion of single species	$A + Be^{-2Dq^2\tau}$
2. Translational diffusion of polydisperse sample	$A + [\int B(M)e^{-D(M)q^2\tau}\, dM]^2$
3. Rotational diffusion of single species	$A + \{\sum_{l \text{ even}} B_l \exp[-(Dq^2 + l(l+1)D_R)\tau]\}^2$
4. Flexing molecule	$A + \left\{\sum_{m,n} P_{m,n} \exp\left[-\left(Dq^2 + \frac{m}{\tau_n}\right)\tau\right]\right\}^2$
5. Motile objects	$A + Be^{-\tau^2/2\tau_{1/2}}$
6. Directed motion; propagating waves on a surface	$A + Be^{-\Gamma\tau} \cos \omega\tau$

and 0.2, etc. vs. x should resemble Figure 20. If the resulting graph is significantly different from this figure, it may be presumed that the chosen form for the correlation function is inadequate to represent the data. The presumption becomes stronger as the value of $p(x)$ at large values of x increases and as the number of runs increases.

2.6.4. Summary of Possible Forms

Thus far we have discussed, admittedly in rather general terms, the concepts involved in obtaining information of interest from the measured correlation function. The process is in fact rather mechanical in nature and researchers unfamiliar with the appropriate computer techniques can almost always find help from their local computer experts. True scientific insight is required, however, in the first step in the process, the selection of the form the correlation function is expected to take. Examples of some of the more common forms expected for specific cases are listed in Table 4.

The last two forms for a correlation function are relatively easy to recognize and analyze if present in a pure form. More often there will be an exponential contribution due, in the case of motile objects for example, to the presence of a certain number of " dead " objects. (Indeed, the fraction of dead object can be determined by studying the ratio of the amplitudes of the Gaussian term to the exponential term. With this analysis it is possible to determine the fraction of beef sperm in a given sample that is still active.[9]) In any case, the presence of a small exponential term in addition to the expected forms in lines 5 and 6 does not constitute a serious barrier to the analysis of data.

In contrast, the second, third, and fourth forms are easily confused and it sometimes is difficult to give convincing evidence that the process under investigation is really responsible for the observed shape of the correlation function. Fortunately, both polarization effects and the effect of scattering angle can sometimes be used to discriminate between the various contributions. In these experiments the chances of success are greatest if the sample is monodisperse.

2.6.5. Polydispersity

If the sample is polydisperse, the analysis of data becomes very complex. The perfect analysis would give the complete molecular weight distribution but the chance of attaining this ideal in any but the simplest of cases is virtually nonexistent. A great deal of effort has been expended in obtaining a limited amount of information about the molecular weight distribution resulting in three general classes of analysis.

The simplest method conceptually is to assume a form having a small number of adjustable parameters to describe the molecular weight distribution and then calculate the expected correlation function.[10] This requires, of course, a knowledge of the relationship between the molecular weight and correlation function derived, perhaps, from experiments made on well-fractionated samples. The adjustable parameters are then varied until the calculated and experimental correlation functions agree as closely as possible. The parameters then give, typically, the central molecular weight and the width of the distribution. The results obtained are, unfortunately, dependent upon the initial molecular weight distribution chosen.

A second approach is the method of cumulants,[11] in which the correlation function is expanded as

$$\frac{1}{2}\ln[G(\tau) - G(\infty)] = C_0 - K_1\tau + \frac{1}{2!}K_2\tau^2 - \frac{1}{3!}K_3\tau^3 + \cdots \tag{51}$$

and the numbers K_1, K_2, K_3, etc. known as the cumulants are interpreted in terms of the average diffusion constant, the width of the distribution of diffusion constants, skewness of the distribution, etc. The relationship between the cumulants and the moments of the distribution of the decay rates (not, unfortunately, the moments of the distribution of molecular weights, or some other useful quantity) is

$$\begin{aligned}
K_1 &= \langle\Gamma\rangle \\
K_2 &= \langle(\Gamma - \langle\Gamma\rangle)^2\rangle \\
K_3 &= \langle(\Gamma - \langle\Gamma\rangle)^3\rangle \\
K_4 &= \langle\Gamma - \langle\Gamma\rangle)^4\rangle - 3\langle(\Gamma - \langle\Gamma\rangle)^2\rangle
\end{aligned}$$

Here we assume that the correlation function contains many decay rates and $\langle \Gamma \rangle$ is the average decay rate. The symbol $\langle \ \rangle$ signifies the average value of the enclosed quantity. While the determination of cumulants is absolute, the interpretation remains dependent on the model chosen.

Finally, a histogram[12] method has recently been developed in which the molecular weight distribution is represented by a small number of discrete molecular weights each with an adjustable fraction of the total weight. Again, the correlation function is calculated with an assumed distribution and the adjustable parameters varied to give the best agreement between calculated and experimental correlation functions. This method and a method devised by Provencher[13] offer the best hope of determining the molecular weight distribution without independent knowledge of the form of the distribution function.

The above discussion has been very general in nature and is intended to point out the various approaches that have been taken in data analysis rather than attempt to give the details of any single method. The literature cited will help the interested reader explore the subject in greater detail.

2.7. SPECIAL APPARATUS

The apparatus we have discussed in earlier sections will be found in one variation or another in the majority of laboratories engaged in experiments using the principles of correlation spectroscopy. There are, in addition, a variety of instruments built in research laboratories having various unique applications. Because these instruments have somewhat more limited appeal they will be described only briefly.

2.7.1. Electrophoretic Light Scattering

Perhaps the most commonly employed special technique is electrophoretic light scattering[14] used to obtain information about the electrophoretic mobility of a molecule or cell. The method obtains results in a small fraction of the time required using conventional gel-electrophoresis. Further, the data are obtained on a sample undergoing free electrophoresis and consequently the interpretation in terms of an electrophoretic mobility is simplified.

If an electric field E is imposed on a solution, the macro ions will attain a terminal velocity v given by

$$v = \mu E \tag{52}$$

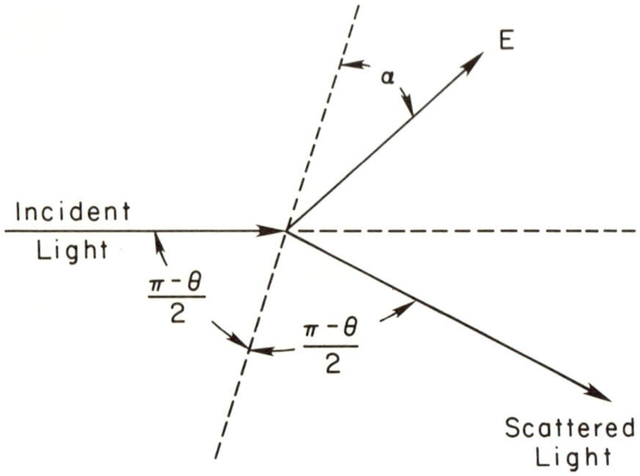

Figure 21. Geometry for an electrophoretic light scattering experiment.

where μ is a characteristic of the macro ion under study, the electrophoretic mobility. Light scattered from the moving ions will experience a doppler shift. The scattered light will have a frequency differing from that of the incident light by an amount ν_D which depends on the geometry of the scattering experiment. Referring to Figure 21 to define the angles of importance, the doppler shift will be

$$\nu_D = \frac{2\mu E}{\lambda} \sin \frac{\theta}{2} \cos \alpha \tag{53}$$

where λ is the wavelength of light in the solution. By beating the scattered light against light that is specularly reflected into the detector (a local oscillator or reference beam), the doppler frequency itself will appear in the detector signal. The signal may be analyzed either with a spectrum analyzer or a correlator. The result is, when account is also taken of diffusion

$$g(\tau) = A + Be^{-Dq^2\tau} \cos(2\pi\nu_D \tau) \tag{54}$$

where $q = (4\pi/\lambda)\sin(\theta/2)$. It is clear that both the electrophoretic mobility and diffusion constant can in principle be obtained.

In more complicated cases a separate doppler frequency will be observed for each species of molecule having a different mobility. A spectrum analyzer will then show a peak for each different mobility.

If the experimental procedure were as easy to follow as the idea seems to suggest, the technique would enjoy much greater popularity than it does at present. The major difficulties are concerned with the construction of the light scattering cuvette. Great care is required to prevent the formation of bubbles at the electrodes, and the electric field may be applied only in relatively short pulses (about 0.1 sec long); otherwise joule heating will cause convection processes that would of themselves produce a doppler shift. Finally, the mounts for the optical elements must be very stable. To illustrate this point we note that periodic change in the path lengths of the reference beam measured relative to the path of the scattered light of less than a half wavelength of light will produce a large periodic signal independent of the characteristics of the sample. This type of effect can easily be produced by vibration of the apparatus. Elimination of the effect requires construction similar to that used in holography experiments. Even air currents in the light beam can cause an undesirable level of noise in the detected signal.

An alternative use of the light scattering method in an electrophoresis experiment is to do free electrophoresis in a cuvette beginning with a sample layered in a thin band. After the electrophoresis, the cuvette is translated through the sample region of a light scattering apparatus and the signal observed as each electrophoretic band passes through the laser beam.[15] A similar technique has been used as a detector in ultracentrifugation experiments.[16]

2.7.2. Fabry–Perot Interferometers

In a few experiments (for example, the measurement of rotational diffusion constants of small macromolecules) the spectrum becomes difficult or impossible to observe using a beating technique and correlator for one or both of two reasons. First, the time scale of interest becomes too short for processing in even the fastest of correlators, and second, the signal-to-noise ratio even for an arbitrarily fast correlator improves so slowly as to require an impractically long signal averaging time. Both of these problems can be overcome using a direct spectral measurement with a Fabry–Perot interferometer.

A Fabry–Perot interferometer consists of two mirrors that may be either spherical or flat. If two flat mirrors are separated by a distance, L, it can be shown that light parallel to the axis of the mirrors and having wavelength λ satisfying the relation

$$n\lambda = 2L \tag{55}$$

will be transmitted through the mirrors. Light with other wavelengths will be reflected. It is clear from equation (55) that light having many different wavelengths will be transmitted with a given mirror separation. Two neigh-

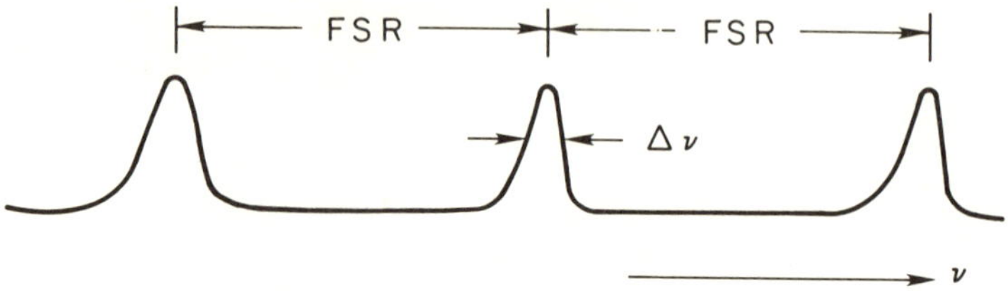

Figure 22. Response of a Fabry–Perot interferometer to light of different frequencies. In the range shown light will be transmitted by the interferometer at three different light frequencies. Thus, the fact that the interferometer transmits light shows only that the light has one of a number of different frequencies. Additional information is required to determine real spectral information.

boring spectral components are said to be separated by the free spectral range of the interferometer as given by

$$\mathrm{FSR} = \frac{C}{2L} \tag{56}$$

in Hz. It is common to use $L = 10$ cm so that the free spectral range for a flat plate interferometer may be 3×10^9 Hz.

If the flat plates are held fixed and the frequency of light slowly varied while the transmitted intensity is measured, the results of Figure 22 are obtained. The width of the transmission peak is broadened by a number of factors including surface roughness, finite diameter, and finite reflectivity of the mirrors. A measure of the quality of an interferometer is the finesse defined as

$$F = \frac{\mathrm{FSR}}{\Delta\nu} \tag{57}$$

Values of F greater than 40 are difficult to obtain in practice for a flat plate interferometer. The instrumental width of $\Delta\nu = \mathrm{FSR}/F$ presents a practical limit to the spectral broadening due to a scattering process that can be resolved using a Fabry–Perot interferometer. Thus a broadening of the order of 10^8 Hz may be resolved using a flat plate interferometer.

The final feature of importance in describing a Fabry–Perot is the light gathering power or etendue. Defining the admission solid angle of the interferometer as Ω, the flat plate instrument obeys the approximate relation

$$\frac{\nu}{\Delta\nu}\Omega \approx 2\pi \tag{58}$$

Thus, as better resolution is demanded the acceptance solid angle diminishes as does the amount of light collected.

The spherical Fabry–Perot requires that two mirrors ground with spherical surfaces of radius R be mounted with a separation R. The free spectral range is

$$\text{FSR} = \frac{C}{4L} \tag{59}$$

[compare with equation (56)] and, whereas the etendue for the flat plate instrument is proportional to $\Delta\nu$, it is inversely proportional to $\Delta\nu$ for the spherical instrument. Consequently, the spherical Fabry–Perot interferometer has substantial advantage over one with flat plates for resolving very narrow linewidths. For example, the spherical instrument able to resolve a 20-MHz linewidth will gather about 60 times as much light as an instrument of the same resolving power using 1-in. diameter flat mirrors.

In order to observe the width of a spectral line the transmitted wavelength of the Fabry–Perot must be scanned across the wavelengths present in the spectral line. This is done either by physically changing the separation of the mirrors using piezoelectric elements or by changing the pressure (and therefore the index of refraction) of the gas between the plates. Maintaining a known relationship between the Fabry–Perot and the wavelength of the laser light source during a scan constitutes a major problem in using this type of spectrometer. The usual technique involves frequent (every few seconds) recalibration of the interferometer using direct laser light.[17] When this is done carefully it is possible to measure a 20-MHz linewidth with an accuracy of about 10%.

With this brief background it is now possible to compare the characteristics of the Fabry–Perot spectrometer to those of the correlation spectrometer. The two most important features are the limits of frequency accessible to an instrument and the rate at which the signal-to-noise ratio is improved.

Taking first the frequency domain accessible to each instrument, the comparison must be made between the instrumental width of a Fabry–Perot interferometer Δf and the minimum sample time of a correlator Δt. We may say approximately that if $\Delta f = 1/2\pi\Delta t$ the two instruments are of equal utility; that if the frequency broadening is large compared to Δf the Fabry–Perot is useful; and that if $1/2\pi\Delta f$ is small compared to ΔT the correlator is useful. If we have available a Fabry–Perot with $\Delta f = 10^7$ Hz and a correlator with $\Delta T = 10^{-7}$ sec, we would use the Fabry–Perot for a frequency $f \gg 10^7$ Hz and the correlator if $f \ll 1.6 \times 10^6$ Hz. We see that if the frequency lies between roughly 10^6 and 10^7 Hz neither instrument is really suitable. This is, in fact, a very difficult spectral range to investigate. While both the sample time of a correlator and the instrumental linewidth of a Fabry–Perot can be reduced, the technology is costly and the experiment must be very important to justify the expense.

In a sense the frequency arguments just given are adequate to uniquely define the domain of usefulness of the two instruments. However, even supposing that ΔT and Δf could be reduced arbitrarily there is an additional factor relating to the rate at which the signal-to-noise ratio improves that will favor one or the other instrument. We notice that a single photon is sufficient to give a signal in a Fabry–Perot. Consequently, the signal will grow in proportion to the number of photons available for detection and the signal-to-noise ratio grows as the square root of that number. In correlation spectroscopy, however, we require the occurrence of two photons at different times to contribute to the signal. The signal, therefore, grows as the square of the number of photons detected and the signal-to-noise ratio is porportional to the number of photons detected. As a result, at very low light levels the Fabry–Perot will develop a useful signal more quickly, while at higher light levels the correlation technique will be superior.† Of course, as pointed out earlier, at very high light levels the statistics of the scattering process become the limiting factor and both instruments will in theory give the same signal-to-noise ratio.

2.7.3. Software Correlators

A number of correlators have been assembled using either a standard multichannel scaler or a small dedicated computer in conjunction with a small amount of home-built circuitry.[18] These instruments are economical alternatives to full correlators for researchers who already have access to the necessary multichannel scaler or computer and possess the electronic skills required in such a project. The result will perform satisfactorily in experiments with a high light level and slow decay rate, but loses efficiency rather badly if the scattered light is weak or the decay rate is fast.

2.7.4. Cross-Correlation Experiment

Recently a new type of experiment has been reported[19] in which two detectors were used to observe light scattered in opposite directions (each 90° from the incident light direction) from a solution of rod-shaped particles, tobacco mosaic virus. Because of the large size of the virus, the intensity of light scattered in a given direction depends on the orientation of

† This statement is true only on the assumption that the Fabry–Perot transmission frequency is swept from well below to well above the central frequency of the scattered light. Consequently, only a small fraction of the available scattered light is detected.

the virus. Consequently, light scattered by a single particle into one detector would be of different intensity from light scattered into the other detector. By studying the cross-correlation function of the two signals the authors were able to show that a signal related to the rotational diffusion constant of the virus was obtained. The technique is certainly of interest in that it demonstrates our ability to do complex experiments requiring a detailed knowledge of the scattering process. Whether it will be useful in more routine studies of a solution remains to be seen. Certainly a technique capable of giving the rotational diffusion constant with the ease and convenience we now have in measuring the translational diffusion constant would be of great value.

2.8. CONCLUSIONS

In this chapter we have attempted to give the novice to light scattering a firm understanding of the tools of the trade. At the end we have described very briefly several of the many more advanced or specialized techniques that have been developed in recent years. In contrast to what we hope was a thorough treatment of the basic instrumentation, the description of advanced techniques can serve only to whet the appetite for new methods. There are many to choose from; only a few were mentioned here. The field offers many opportunities to the researcher interested in developing new techniques. It is the belief of the author that many useful and exciting advances remain to be made.

REFERENCES AND NOTES

1. For an introduction to this subject see any elementary general physics text such as: D. Halliday and R. Resnick, *Fundamentals of Physics*, John Wiley and Sons, New York (1970), Chaps. 34–40.
2. C. Tanford, *Physical Chemistry of Macromolecules*, John Wiley and Sons, New York (1961), Chap. 5.
3. For a more complete treatment of this theory see: B. Chu, *Laser Light Scattering*, Academic Press, New York (1974); B. Berne and R. Pecora, *Dynamic Light Scattering*, John Wiley and Sons, New York (1976).
4. E. Jakeman, C. J. Oliver, and E. R. Pike, *J. Phys. A* **3**, L45 (1970).
5. F. Reif, *Fundamentals of Statistical and Thermal Physics*, McGraw-Hill, New York (1965), Section 15.6.
6. E. Jakeman, P. N. Pusey, and J. M. Vaughan, *Opt. Commun.* **17**, 305 (1976).
7. N. C. Ford, Jr., *Chem. Scr.* **2**, 193 (1972).
8. P. R. Bevington, *Data Reduction and Error Analysis for the Physical Sciences*, McGraw-Hill, New York (1969).
9. T. Craig, R. R. Hallett, and B. Nickel, *Biophys. J.* **28**, 457 (1979).

10. N. C. Ford, Jr., R. Gabler, and F. E. Karasz, *Adv. Chem.* **125**, 25 (1973); Y. Tagami and R. Pecora, *J. Chem. Phys.* **51**, 3293 (1969).
11. This technique has been extensively discussed in the literature. A theoretical treatment was given by: D. E. Koppel, *J. Chem. Phys.* **57**, 4814 (1972); and experimental work has been done by: C. B. Bargeron, *J. Chem. Phys.* **60**, 2516 (1974); **61**, 2134 (1974).
12. B. Chu, E. Gulari, and E. Gulari, *Phys. Sci.* **19**, 476 (1979); E. Gulari, E. Gulari, Y. Tsunashima, and B. Chu, *J. Chem. Phys.* **70**, 3965 (1979); *Polymer* **20**, 347 (1979).
13. S. W. Provencher, *Makromol. Chem.* **180**, 201 (1979); *Comput. Phys. Commun.* **27**, 213, 229 (1982).
14. Reviews are given by: B. R. Ware, *Adv. Colloid Interface Sci.* **4**, 1 (1974); B. A. Smith and B. R. Ware, *Contemp. Top. Anal. Clin. Chem.* **2**, 29 (1978). See also: D. D. Haas and B. R. Ware, *Anal. Biochem.* **74**, 175 (1976), for a design of an electrophoretic light scattering chamber.
15. T. K. Lim, G. J. Baran, and V. A. Bloomfield, *Biopolymers* **16**, 1473 (1977).
16. M. A. Loewenstein and M. H. Birnboim, *Biopolymers* **14**, 419 (1975).
17. A. Asenbaum, *Appl. Opt.* **18**, 540 (1979). For further references see: W. Hayes and R. London, *Scattering of Light by Crystals*, John Wiley and Sons, New York (1978).
18. A. Lindgård, R. Moss, and J. Oxenboll, *Comput. Chem.* **1**, 7 (1976); R. W. Wijnaendts von Resandt, *Rev. Sci. Instrum.* **45**, 1507 (1974); W. Lempert and C. H. Wang, *ibid.* **51**, 380 (1980).
19. W. G. Griffin and P. N. Pusey, *Phys. Rev. Lett.* **43**, 1100 (1979).

3

Dynamic Depolarized Light Scattering

Karl Zero† and R. Pecora

Department of Chemistry
Stanford University
Stanford, California 94305

3.1. INTRODUCTION

Most photon correlation experiments, as may be determined from a perusal of the articles in this book, measure time correlation functions of the polarized scattered light. The depolarized scattered light is, however, a rich source of dynamic and structural information which is often not readily obtainable by other techniques. The major reason for this relative neglect of the depolarized scattering is that depolarized signals are often much less intense than polarized signals and are thus often obscured by interfering signals due to optical imperfections in lenses and cells, multiple scattering, polarizer leakage, and solvent scattering. In addition, as will be described in somewhat more detail below, the depolarized time correlation functions often decay on a fast time scale (a few microseconds), rendering photon correlation difficult because of phototube afterpulsing and the signal-to-noise problems encountered when the scattered signals display a small number of photons per correlation time. Of course, correlation functions which decay on the nanosecond or faster time scales may usually be obtained by measuring the spectral distribution of the scattered light intensity using a Fabry–Perot interferometer as a monochromator. A troublesome regime occurs in the region approximately between 0.05 and 1 μsec where both Fabry–Perot interferometry and photon correlation are difficult. The difficulties associated with measureing the depolarized signals are,

† Present address: Allied Corporation, Box 1021 R, Morristown, New Jersey 07960.

in view of the importance of the information that may often be obtained, well worth the effort that is demanded to overcome them.

In this article we first review some of the basic principles of depolarized dynamic light scattering (DDLS) and then describe some of the currently important areas of application with emphasis on applications to the study of macromolecules. Work on dynamic depolarized scattering from small molecules in liquids of ordinary viscosity is mentioned only briefly since this subject has been adequately reviewed elsewhere[1–3] and the experiments usually fall well outside the range of the photon correlation technique.

3.2. PRINCIPLES OF DEPOLARIZED SCATTERING

3.2.1 Scattering Configurations

It is useful to specify a particular scattering geometry in order to precisely define the terminology used in light scattering experiments in which intensities and time correlation functions associated with various combinations of the initial and final polarization directions are measured. A convenient scattering geometry that is often used is shown in Figure 1. The incident beam with wave vector $\mathbf{k}_0$ and polarization $\hat{\varepsilon}_0$ propagates in the x–y plane and the scattered radiation with wave vector $\mathbf{k}$ propagates in the x direction. The angle that the polarization of the incoming beam makes with the z axis is called μ, i.e., $\hat{\varepsilon}_0 \cdot \hat{z} = \cos \mu$. Thus, for vertically polarized incident radiation $\mu = 0$. The angle that the polarization, $\hat{\varepsilon}$, of the scattered beam makes with the z axis is called ψ. Thus, $\hat{\varepsilon} \cdot \hat{z} = \cos \psi$. The scattering angle is, as usual, θ and the scattering vector $\mathbf{q} = \mathbf{k}_0 - \mathbf{k}$.

Sometimes, as in studies of depolarized intensities from polymeric films,[4] observations are made with incident polarization angle $\mu = 0$ but

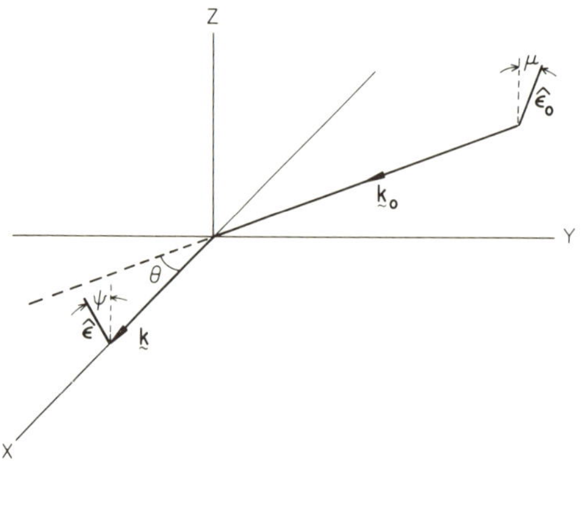

Figure 1. Scattering geometry. The incident light beam is propagating in the x–y plane with propagation vector $\mathbf{k}_0$ and polarization vector $\hat{\varepsilon}_0$. The angle between $\hat{\varepsilon}_0$ and the z axis is μ. The scattered beam propagates along the x direction with propagation vector $\mathbf{k}$ and polarization $\hat{\varepsilon}$. The angle between $\hat{\varepsilon}$ and the z axis is ψ. The scattering angle is θ.

with the scattered light propagating above or below the x–y plane. In the scattering geometry defined here in which both incoming and scattered beams propagate in the x–y plane (the scattering plane), observations of this type are equivalent to changing the incident polarization; that is, setting $\mu \neq 0$ and performing all observations in the x–y plane.

The most common combinations of μ and ψ are as follows.

Case 1: Polarized scattering, $\mu = 0$ and $\psi = 0$. The intensity of this component of the scattering is often given the symbol I_{VV} (for vertical–vertical polarizations of the incoming and scattered beams).

Case 2: Depolarized scattering, $\mu = 0$ and $\psi = 90°$. The intensity of this component is usually given the symbol I_{VH} and the correlation functions of this component are those most commonly measured in dynamic depolarized scattering experiments. The reason for this, as we describe below, is that this component usually depends solely on the optical anisotropy of the scattering centers (particles or molecules or parts of molecules) in the scattering medium and not on the average molecular polarizability.

Case 3: Horizontal–horizontal scattering, $\mu = 90°$ and $\psi = 90°$. The intensity associated with this component is, naturally, denoted by I_{HH}. It is mainly useful when dealing with large structures and especially so in the measurement of depolarization ratios.[5]

Most prelaser and many postlaser light scattering experiments used no polarizers at all and interpreted the results in terms of the "isotropic scattering," that part of I_{VV} that depends only upon the average polarizability of the scattering centers in the scattering medium. In dealing with structures that are not large compared to the reciprocal of the scattering vector length (q^{-1}), the contributions of terms containing the optical anisotropies of the scattering centers to I_{VV} are relatively small and this approximation is often valid. Note, however, that I_{VH} effectively isolates these terms so that although they may be small, they may still often be readily measured.

3.2.2. Physical Principles

A formal treatment of DDLS is given in detail in Chapter 7 of Berne and Pecora.[1] In this section, some of the underlying physical principles and main results are briefly reviewed.

Consider a light scattering experiment in the I_{VH} geometry described above. The incident light is polarized along the $\hat{z}$ axis ($\mu = 0$) and the scattered electric field component in the x–y plane ($\psi = 90°$) is selected.

The scattering medium is considered to be composed of "scattering elements" (they could be particles, macromolecules, small molecules, or parts of macromolecules) each with a constant element fixed polarizability tensor, $\boldsymbol{\alpha}$. In this approximation, fluctuations in $\boldsymbol{\alpha}$ due to interscatterer interactions are ignored since these fluctuations usually relax on time scales much faster than those normally measured in photon-correlation experiments. They are often referred to as "collision-induced" scattering and have been extensively reviewed elsewhere.[6]

In our approximation the scattered electric field in the VH configuration, E_y, is proportional to

$$E_y \propto \sum_j \alpha^j_{yz}[\Omega(t)]e^{i\mathbf{q}\cdot\mathbf{r}_j(t)} \tag{1}$$

where $\alpha^j_{yz}[\Omega(t)]$ is the yz component of the polarizability tensor of element j in the *laboratory fixed* frame defined in Figure 1. Since the scattering elements are assumed to be optically anisotropic, the α^j_{yz} depend on components of the element fixed $\boldsymbol{\alpha}$ and functions of the angles defining the element orientation, $\Omega(t)$, at time t. The exponential term in equation (1) contains the scattering vector $\mathbf{q}$ and the position of the center of mass of element j at time t. Note that molecular rotations modulate the amplitude of the scattered electric field through the dependence of the α^j on the element orientation.

The time correlation function of $E_y(t)$ is proportional to

$$S_{\mathrm{VH}}(q, t) = \left\langle \sum_{i,j} \alpha^j_{yz}[\Omega(t)]\alpha^i_{yz}[\Omega(0)]\exp\{i\mathbf{q}\cdot[\mathbf{r}_j(t) - \mathbf{r}_i(0)]\}\right\rangle \tag{2}$$

where the $\langle \;\rangle$ denote a statistical average.

This correlation function may be written in terms of "self" ($i = j$) and "distinct" ($i \neq j$) correlation functions

$$S_{\mathrm{VH}}(q, t) = S^s_{\mathrm{VH}}(q, t) + S^d_{\mathrm{VH}}(q, t) \tag{3}$$

where the self part

$$S^s_{\mathrm{VH}}(q, t) = \left\langle \sum_{i} \alpha^i_{yz}[\Omega(t)]\alpha^i_{yz}[\Omega(0)]\exp\{i\mathbf{q}\cdot[\mathbf{r}_i(t) - \mathbf{r}_i(0)]\}\right\rangle \tag{4}$$

and the distinct part

$$S^d_{\mathrm{VH}}(q, t) = \left\langle \sum_{i\neq j} \alpha^j_{yz}[\Omega(t)]\alpha^i_{yz}[\Omega(0)]\exp\{i\mathbf{q}\cdot[\mathbf{r}_j(t) - \mathbf{r}_i(0)]\}\right\rangle \tag{5}$$

Note that in general: (1) S^s_{VH} and S^d_{VH} contain dynamical information about translation and rotation; (2) S^d_{VH} includes dynamical correlations in both orientation and position between all pairs of molecules; (3) $S^{s}_{VH}(q, t)$ includes information on the dynamic coupling between translation and rotation of a single scattering element.

A common application of DDLS is to a dilute solution of rigid molecules that are small compared to q^{-1}.[7] This may in fact include some molecules that are quite large since

$$q = \frac{4\pi n}{\lambda} \sin\left(\frac{\theta}{2}\right)$$

can be made small by making scattering observations at low (or even zero) scattering angle. The solution is assumed to be so dilute that $S^{d}_{VH}(q, t)$ does not contribute appreciably to the DDLS. In addition, DDLS from the solvent molecules must be assumed to be negligible compared to that from the solute.

The most common model for these systems is that the molecules undergo *independent* rotational and translational diffusion. For molecules of arbitrary symmetry the rotational diffusion is characterized by a second-rank tensor, the rotational diffusion tensor $\boldsymbol{\Theta}$. This tensor reflects the fact the rotational motions about different molecular axes could experience different frictional resistances. Pecora[8] has shown that within this "anisotropic" rotational diffusion model, $S^{s}_{VH}(q, t)$ consists of a sum of *five* exponentials in the most general case. Five exponentials appear since there are five laboratory fixed "anisotropic" components of a symmetric second-rank tensor such as the polarizability tensor. Each of these components decays with its distinct decay time, giving five exponentials in the correlation function. There are two molecular second-rank tensors in this problem, the rotational diffusion tensor and the polarizability tensor. If these tensors are simultaneously diagonalizable, then the number of exponentials in the correlation function is reduced from five to two.

It has been further shown[8] that if the molecule contains a rotation axis of fourfold symmetry or higher (e.g., symmetric top molecule, rod-shaped molecule, ellipsoid of revolution) then the $S^{s}_{VH}(q, t)$ correlation function in the rotational diffusion model consists of one decaying exponential

$$S^{s}_{VH}(q, t) = N\beta^{2} \exp[-(q^{2}D + 6\Theta)t] \tag{6}$$

where N is the number of molecules in the scattering volume and β is the optical anisotropy of a molecule; i.e., the difference between the polarizabilities parallel and perpendicular to the molecular symmetry axis. The decay time of the correlation function depends on the rotational diffusion coefficient for reorientation of the molecular symmetry axis Θ and on the translational diffusion coefficient D multiplied by the squared length of the scattering vector q^2. Equation (6) and its generalizations to molecules of lower symmetry[8] are appropriate for the analysis of photon-correlation experiments on macromolecules reorienting in dilute solution.

In the case of smaller molecules reorienting on picosecond time scales, the decay time of the correlation function is usually too fast to observe by correlation techniques. One thus measures the spectrum of scattered light, given by the frequency Fourier transform of equation (6)

$$\begin{aligned} S^s_{\mathrm{VH}}(q, \omega) &= \frac{1}{2\pi} \int_{-\infty}^{+\infty} dt\, e^{-i\omega t} S^s_{\mathrm{VH}}(q, t) \\ &= \frac{N\beta^2}{\pi} \frac{6\Theta}{\omega^2 + (6\Theta)^2} \end{aligned} \tag{7}$$

where ω is the difference between the laser and scattered light frequencies. In equation (7), the q^2D contribution to the line width has been ignored since it is usually negligible in cases where the spectrum of scattered light is directly measured.

Equation (7) and its generalizations to molecules of arbitrary symmetry have been extensively applied to the interpretation of Fabry–Perot experiments on small molecules in dilute solution.[1–3] DDLS experiments on macromolecules are fewer and are discussed in Sections 3.3–3.5.

The correlation function obtained from DDLS experiments, equation (3), contains a contribution from dynamic correlations between pairs of molecules. In the approximation of independent rotational and translational diffusion, Keyes and Kivelson[9] have proposed a theory to relate the relaxation time of S_{VH} (the "collective" relaxation time) to that of S^s_{VH} (the "single-molecule" relaxation time). Their theory as applied to symmetric top molecules predicts that for a solution of one anisotropic component (mole fraction X) in a solvent composed of optically isotropic molecules that

$$S_{\mathrm{VH}}(q, t) = X\beta^2 g_2 \exp\left[-6\Theta\left(\frac{J_2}{g_2}\right)t\right] \tag{8}$$

where g_2 and J_2 are measures of static and dynamic orientational correlations, respectively. The collective relaxation time in equation (8) is related to the single-molecule time in equation (6) by the factor g_2/J_2.

In systems in which pair orientational correlations are important DDLS measures S_{VH} while other techniques such as Raman band shape analysis and some NMR and EPR relaxation techniques measure the decay constant of S^{s}_{VH}. Many groups have used DDLS (Fabry–Perot interferometry) with dilution techniques or DDLS in combination with other spectroscopic techniques to study the relation between the collective and single-particle reorientation times for small molecules in neat liquids and solutions.[1–3] There have to date been no comparable studies for non-spherical macromolecules in solution.

Equations (6) and (7) may be directly applied to dilute solutions of rigid macromolecules exhibiting rodlike or spherical symmetry. Applications to these systems as well as some discussion of the experimental techniques are given in Section 3.3. In more concentrated solutions, however, long, thin rodlike macromolecules will interact, causing a coupling between the rotation and the translation of the molecule. The theory and experiments that apply to these systems are discussed in Section 3.4.

In the case of flexible macromolecules, the basic scattering element may be taken to be a "segment" of the molecule. Intramolecular motions may then contribute to the depolarized correlation function. Applications to flexible macromolecules are given in Section 3.5.

Rods that are rotating in amorphous or glassy polymer systems should also exhibit some variation from equations (6) and (7). Such systems are briefly discussed in Section 3.6.

Finally, the total depolarized intensity from dilute rodlike molecules depends on β^2. Often it is important to maximize β^2 in order to obtain sufficient signal to measure the depolarized scattering from a particular species. The use of resonance enhancement techniques for this purpose is discussed in Section 3.7.

3.3. RIGID MACROMOLECULES IN DILUTE SOLUTION

3.3.1. Hydrodynamics of Rigid Macromolecules

Polarized DLS may be routinely used to obtain translational diffusion coefficients of macromolecules in dilute solution. DDLS may often be used to obtain the macromolecular rotational diffusion coefficient in addition. Due to the small number of scattering particles, the intensity of depolarized scattering for dilute solutions of macromolecules is usually quite low. Consequently, a poor signal-to-noise ratio is obtained and precisions of only 10% to 20% are common. In addition, long data collection periods, ranging from hours to even days, are necessary. However, unlike most other methods of measuring the rotational diffusion constant, depolarized light scattering requires no modification of the macromolecular structure (such as the addition of a fluorophore) and observes the equilibrium state directly (unlike electric birefringence). DDLS also has the advantage over many other methods of studying rotational relaxation in that it may be used to study relaxation over times extending from seconds to picoseconds.

One can determine the approximate hydrodynamic dimensions of a particle by comparing the rotational and translational diffusion coefficients obtained from dynamic light scattering with expressions derived from hydrodynamics. For solvated spheres, the translational and rotational diffusion coefficients are given by the Stokes–Einstein relations:

$$D = kT/6\pi\eta_0 a \tag{9}$$

$$\Theta = kT/8\pi\eta_0 a^3 \tag{10}$$

where η_0 is the viscosity of the solvent and a is the radius. Note that Θ has a much greater dependence on the particle size ($1/a^3$ versus $1/a$) than D; thus, small size changes should be much easier to detect by measurement of Θ. Spheres, of course, are usually optically isotropic and, consequently, have very little depolarized scattering. Anisotropic particles, however, have a similar dependence on the largest dimension. For example, Perrin's relations[10] for ellipsoids involve multiplying equations (9) and (10) by a shape-dependent factor; i.e., for rotation of the symmetry axis,

$$D = \frac{kT}{6\pi\eta_0 a} G(\rho) \tag{11}$$

$$\Theta = \frac{kT}{8\pi\eta_0 a^3} \left\{\frac{3}{2}\left[\frac{(2-\rho^2)G(\rho) - 1}{(1-\rho^4)}\right]\right\} \tag{12}$$

where ρ is the axial ratio ($\rho \equiv b/a$), a is the length of the major semiaxis, and b is the length of the minor semiaxis. For prolate ellipsoids ($\rho < 1$), the "shape" factor $G(\rho)$ has the form

$$G(\rho) = (1-\rho^2)^{-1/2} \ln\left[\frac{1+(1-\rho^2)^{1/2}}{\rho}\right], \qquad \rho < 1 \tag{13}$$

and for oblate ellipsoids ($\rho > 1$) has the form

$$G(\rho) = (\rho^2 - 1)^{-1/2} \arctan[(\rho^2 - 1)^{1/2}], \qquad \rho > 1 \tag{14}$$

Since Θ and D have different size and shape dependences, comparison of their measured values with these relations should yield both the size (a) and the axial ratio (ρ), assuming the particle can be approximated by an ellipsoid of revolution.

Starting from Kirkwood and Riseman's pioneering approach,[11] several different expressions have also been obtained for the diffusion coefficients of rods.[12–14] One of the most commonly used set of expressions, although not necessarily the best,[14] are Broersma's relations for rods of length L and diameter d ($L/d > 4$)[12, 13]:

$$D = \frac{kT}{3\pi\eta_0 L}\left[6 - \frac{1}{2}(\gamma_\parallel + \gamma_\perp)\right] \tag{15}$$

$$\Theta = \frac{3kT}{\pi\eta_0 L^3}(\delta - \zeta) \tag{16}$$

where

$$\delta = \ln(2L/d)$$

$$\gamma_\parallel = 1.27 - 7.4(1/\delta - 0.34)^2$$

$$\gamma_\perp = 0.19 - 4.2(1/\delta - 0.39)^2$$

$$\zeta = 1.45 - 7.5(1/\delta - 0.27)^2$$

For wormlike chains of persistence length Q and contour length L, Hearst has obtained expressions for D and Θ at the limits of small and large Q.[15–17] For large Q ($\lambda L \ll 1$, where $\lambda = 1/2Q$)

$$D = \frac{kT}{3\pi\eta_0 L}\left[\ln\left(\frac{L}{a}\right) + 0.166\,\lambda L - 1 + \frac{a}{d}\right] \tag{17}$$

$$\Theta = \frac{kT}{\pi\eta_0 L^3}\left\{3\ln\left(\frac{L}{a}\right) - 7.00 + 4\left(\frac{a}{d}\right) + \lambda L\left[2.25\ln\left(\frac{L}{a}\right) - 6.66 + 2\left(\frac{a}{d}\right)\right]\right\} \tag{18}$$

where a is the monomer length and d is the monomer diameter. For small Q ($\lambda L \gg 1$)

$$D = \frac{kT}{3\pi\eta_0 L}\left[1.843(\lambda L)^{1/2} - \ln(\lambda a) - 2.431 - \frac{a}{d}\right] \tag{19}$$

$$\Theta = \frac{kT\lambda}{\eta_0 L^2}\left[0.716(\lambda L)^{1/2} - 0.636\ln(\lambda a) - 1.548 + 0.640\,\frac{a}{d}\right] \tag{20}$$

Expressions also exist for once-broken rods,[18, 19] rings, lollipops, and dumbbells.[20] Bloomfield and co-workers have presented a theoretical formulation for calculating the diffusion coefficients for molecules of arbitrary shape.[20, 21] For the case of small molecules, where slip boundary conditions are necessary, Youngren and Acrivos give a method to determine the diffusion coefficients of arbitrarily shaped molecules.[22] A general early review of the application of hydrodynamics to conformation problems is given by Benoit *et al.*[23]

3.3.2. Interferometric Studies

For frequency shifts greater than $\sim 10^6$ Hz, the spectrum of the scattered light can be measured directly with a Fabry–Perot interferometer. This is usually true for molecules with a molecular weight less than $\sim$ 50,000. Since even small macromolecules are near the upper end of this molecular weight range, the frequency shifts obtained are extremely small. Consequently, for these small macromolecules, a large interferometer plate spacing is needed to analyze the spectrum of the depolarized scattered light. To facilitate alignment, confocal spherical plates are normally used.[24] Although the free spectral range of these plates cannot be varied, as is the case for flat plates, they are much easier to maintain in alignment for larger plate spacings. The smallest frequency shift measurable is about a factor of 2 greater than the instrumental linewidth. The instrumental half-width is usually limited by the spectral width of the light source, which for a laser can be as small as a megahertz.

For macromolecules small enough to observe with interferometry, the translational component (q^2D) is negligible compared to the rotational component (6Θ) and the rotational diffusion coefficient can be extracted directly from the half-width of the Lorentzian, equation (7).

One of the earliest interferometric studies was performed by Dublin *et al.* on hen egg white lysozyme.[25] Due to the presence of a large molecular weight impurity, a sharp spike was observed in the depolarized light scattering spectra. Such a spike can also result from depolarized scattering from dust particles. Dublin *et al.* were able to correct their spectra for the presence of the spike and consequently obtained a decay time identical to that obtained in a subsequent study on ultrapure lysozyme.[26] The time obtained (10 ± 0.5 nsec at 20°C), when combined with the translational diffusion coefficients, is consistent with the rotational diffusion of a prolate ellipsoid with solvated dimensions of 55×33 Å. Thus, although the presence of dust should certainly be minimized, usable spectra can often be obtained due to the great difference in the time scale of the motions.

Both carbon-13 NMR and depolarized dynamic light scattering experiments were performed on muscle calcium binding protein (MCBP).[26] MCBP has a molecule weight of 12,000 and x-ray crystallographic dimensions of approximately 30 × 36 Å. The α-carbon region was found to give a single NMR reorientation time of 11–14 nsec, while the depolarized scattering gave a value of 12 ± 1 nsec. Because of this similarity in reorientation times, it is highly likely that the α-carbons move as an integral part of a rigid body.

As noted above, because rotational diffusion coefficients have a stronger dependence on size than translational diffusion coefficients, small size changes are more evident in the depolarized light scattering than in the polarized light scattering. Michielsen and Pecora[27] have performed experiments on gramicidin, which exists in solution as a rodlike dimer with dimensions of 32 × 15 Å. They clearly observed that the rodlike dimer decreased in length by about 5 Å in the presence of potassium ions. This resulted in only a 7% change in the translational diffusion coefficient, whereas a 35% change in the rotational diffusion coefficient was observed.

3.3.3. Photon Correlation Studies

For frequency shifts less than a megahertz, the spectrum of the scattered light can no longer easily be measured directly. As a result, photon correlation is the usual method of data analysis for larger macromolecules.[1] In this method, the scattered light is allowed to impinge directly on the photocathode, either by itself (homodyning) or mixed with some unscattered light (heterodyning). In the heterodyne method, beat frequencies are produced between the scattered and unscattered light intensity fluctuations. Consequently, the output from the photomultiplier tube can be electronically filtered with a spectrum analyzer, yielding the frequency spectrum of the scattered light. More commonly, the time correlation function of the heterodyne depolarized scattered intensity is measured. In the homodyne method, the time correlation function obtained is usually proportional to the square of the one for heterodyning; consequently, the time constants obtained are usually a factor of 2 smaller.

Thus for dilute solutions of small rods with no translational-rotational coupling, the polarized $[I_{VV}(q, t)]$ and depolarized $[I_{VH}(q, t)]$ homodyne time correlation functions are given by the expressions[1]

$$I_{VV}(q, t) = A[e^{-q^2Dt}]^2 + B \tag{21}$$

$$I_{VH}(q, t) = A'[e^{-(q^2D+6\Theta)t}]^2 + B' \tag{22}$$

where A, A', B, and B' are constants for a given q and incident light intensity. Equation (22) is simply the square of equation (6) with an arbitrary amplitude (A') and baseline (B').

The lower limit on the correlation times which can be measured is determined by the properties of the photomultiplier tube and the electronics. Hardwire correlators are capable of measuring decay times down to a few hundred nanoseconds. However, afterpulsing from the phototube obscures most of the information in the 100-nsec range at the low signal levels obtained in depolarized scattering. Consequently, the decay times which can be reliably measured are normally greater than a microsecond. Thus, there tends to be a grey area between interferometry and direct photon correlation where neither technique is effective. However, photoelectron time-of-arrival techniques can drastically narrow this gap.[28] In addition, Phillies has shown that cross correlation between two detectors can virtually eliminate afterpulsing.[29]

For molecules with rotational times slow enough to measure by photon correlation, the q^2D term is often not negligible. As a result, the depolarized scattered light correlation functions are sometimes measured at zero angle (i.e., forward depolarized light scattering); thus, $q^2 = 0$ and only the rotational term remains. This also prevents complications due to molecular structure factors, which can change the single exponential into a sum of exponentials.[1, 7] Naturally, forward depolarized light scattering requires particularly good sets of polarizers and nondepolarizing optical surfaces to prevent the incident light from leaking through. Even then, enough depolarized incident light passes through to cause complete heterodyning to occur.[30] Because of the larger scattering volume, the depolarization of light by multiple scattering is particularly common for zero-angle scattering. Even isotropic particles, such as polystyrene latex spheres, depolarize light by multiple scattering.[30] Sorensen *et al.*[31] have shown that the time constant due to multiple scattering is roughly angle independent and approximately equal to the time constant for polarized scattering at 180° (i.e., backscatter). To help minimize these problems, the depolarized light scattering may be measured at higher angles. In the absence of structure factors, the I_{VH} and I_{HH} scattering for an anisotropic molecule have the same time dependence for an angle of 90°.[1] Consequently, only a single analyzing polarizer is usually needed for depolarized light scattering at 90°. Of course, the time constant obtained at 90° must be corrected for q^2D, which can be found from polarized dynamic light scattering. These correlation techniques have been applied to numerous systems, such as viruses (TMV, T7, and T4B),[32–35] collagen,[36] poly-p-phenylene-benzbisthiazole (PBT),[34] tRNA,[28] poly-*n*-hexyl isocyanante (PHI),[37] poly-γ-benzyl-L-glutamate (PBLG),[38] light meromyosin (LMM),[39] and 70S ribosomes.[40]

The pioneering experiments on the depolarized dynamic light scattering from macromolecules were performed by Wada *et al.*[32] on tobacco mosaic virus (TMV) and confirmed by several other authors.[33, 34] TMV has a molecular weight of $\sim 40 \times 10^6$ and is a rod with a length of 3000 Å and a diameter of 150–170 Å. Wada *et al.*[32] obtained the heterodyne power spectra of the forward depolarized scattering from a ~ 1 mg/ml solution of TMV in 0.1 *M* phosphate buffer. At 20°C, the value for Θ was found to be 350 ± 20 sec^{-1}, which is consistent with the values obtained from electric birefringence experiments and the size of the virus. More recent studies on T4B and T7 ($\Theta_{20°,\omega} = 258 \pm 12$ sec^{-1} and 4528 ± 100 sec^{-1}, respectively) bacteriophages were performed by Hopman *et al.*,[35] in which they accounted for the effect of multiple scattering on the forward depolarized scattering.

Thomas and Fletcher[36] performed forward depolarized light scattering experiments on solutions of rat tail collagen. Two distinct decay times were observed. The slow time was attributed to large nonspecific aggregates of collagen, while extrapolation of the fast time to zero concentration yielded a rotational diffusion coefficient of 1082 ± 30 sec^{-1} at 20°C. This time corresponds to an apparent length of 253 ± 7 nm, which is significantly shorter than the contour length observed by electron microscopy (294 ± 4 nm). This difference was attributed to flexibility of the collagen rod, assuming a persistence length of 130 nm.

In the PBT studies, Crosby *et al.*[34] accounted for the effect of the molecular weight distribution on the depolarized spectra. In addition, Θ was related to the persistence length, assuming the wormlike chain approximation[16] [equation (18)]. The molecular weight distribution was derived from exclusion chromatography and used to calculate the time correlation function for a given persistence length. The persistence length was then adjusted to give the best fit to the experimental data. PBT was found to have a large, but finite, persistence length (640 ± 90 Å). In addition, Crosby *et al.* tried to reverse this procedure by performing a histogram analysis of the data; i.e., they attempted to extract the molecular weight distribution from the depolarized light scattering correlation functions. Unfortunately, the data were too noisy to determine anything more than the approximate width and skewness of the decay rate distribution.

Using photoelectron time-of-arrival techniques,[41] Patkowski *et al.*[27, 42] have obtained the depolarized time corelation functions for transfer RNA. Most commonly, this method uses a time-to-amplitude converter (TAC) coupled to a multichannel pulse height analyzer to yield a function which, for low count rates and fast decay times, is similar to the actual correlation function.[44] The TAC produces a pulse whose amplitude is proportional to the time interval between the arrival of two different photons at the detector. Similar results can be obtained by using a time-to-digital converter coupled to a multichannel scaler.[28] Due to the dead time of the electronics, the first 20 nsec of the correlation function are useless. Furthermore, there is a large amount of systematic noise below 50 nsec and an oscillatory, systematic signal. Thus, even after accumulating data for 12–24 hours, precisions of 30% or greater were typical.[28] By careful data analysis, including allowances for the systematic noise, Patkowski *et al.*[28] found the decay time of tRNA to be about 30 nsec. By comparing the rotational times with the translational times obtained from polarized light scattering, they followed the changes in apparent size and shape produced by varying NaCl concentration and temperature. The molecule was observed to become smaller and more spherical when NaCl concentration increased. On the other hand, high temperatures (70°C), as a result of partial melting, produced an extremely elongated molecule (180 × 22 Å).

3.4. ROD-SHAPED MACROMOLECULES IN SEMIDILUTE SOLUTIONS

As previously stated, there are many different relations used for calculating the diffusion coefficients of rods in dilute solution with a given length (L) and diameter (d), based mainly on corrections to the Stokes–Einstein equations for spheres [see equations (15) and (16)]. For long thin rods, the expressions for the rotational diffusion coefficient (Θ) and the translational diffusion coefficient (D) can be approximated by the Kirkwood–Riseman results[11]:

$$\Theta = 3kT \left(\ln \frac{L/d}{\pi \eta L^3} \right) \tag{23}$$

$$D = kT \ln \frac{L/d}{3\pi \eta L} \tag{24}$$

where k is Boltzmann's constant, T is the absolute temperature, and η is the viscosity of the solvent. These relations assume the rods are large enough so that stick boundary conditions hold; i.e., the particles behave like macroscopic rods in a viscous continuum.

As the concentration is increased, one would expect the motion of each rod to be hindered by its neighbors, resulting in a decrease in the rate of translational and rotational diffusion. Doi and Edwards[43] have presented a theory for the diffusion coefficients of rods with number concentrations (c) much greater than $1/L^3$ but less than $2\pi/dL^2$: i.e., solutions where the rods are still oriented randomly but where their motions are strongly hindered by steric repulsions between rods. In their treatment, Doi and Edwards assume that translational motion parallel to the rod's axis ($D_{\parallel}$) is unhindered (i.e., the rods are infinitely thin) and equal to its infinite dilution value while translational motion perpendicular to the rod's axis ($D_{\perp}$) is restricted to about the distance (a_c) to the nearest rod. Since a_c is much less than the length for these concentrations ($a_c \approx 1/cL^2$ and $c \gg 1/L^3$), $D_{\perp} \approx 0$ compared to $D_{\parallel}$; thus, $D = \frac{1}{3}(D_{\parallel} + 2D_{\perp}) = \frac{1}{3}D_{\parallel}$. Since $D_{\parallel} \approx 2D_{\perp}$ at infinite dilution, the translational diffusion coefficient in this region is roughly half the value at infinite dilution. In order for a particular rod (A) to rotate a significant amount, the nearest-neighboring rod (B) must translate a distance on the order of its length, thus temporarily releasing the constraint on rod (A) (see Figure 2). In order for the orientation of a rod to lose correlation with its initial orientation, it must go through this process many times. Thus, the rotational diffusion is strongly coupled to the translational diffusion and the rotational diffusion coefficient is predicted to be very small relative to its infinite dilution value. From this model and equations (23) and (24), Doi and Edwards[43] obtain an expression for the concentration dependence of the rotational diffusion coefficient in the semidilute region ($1/L^3 \ll c < 2\pi/dL^2$):

$$\Theta = \beta' kT \frac{\ln L/d}{\eta_s L^9 c^2} \tag{25}$$

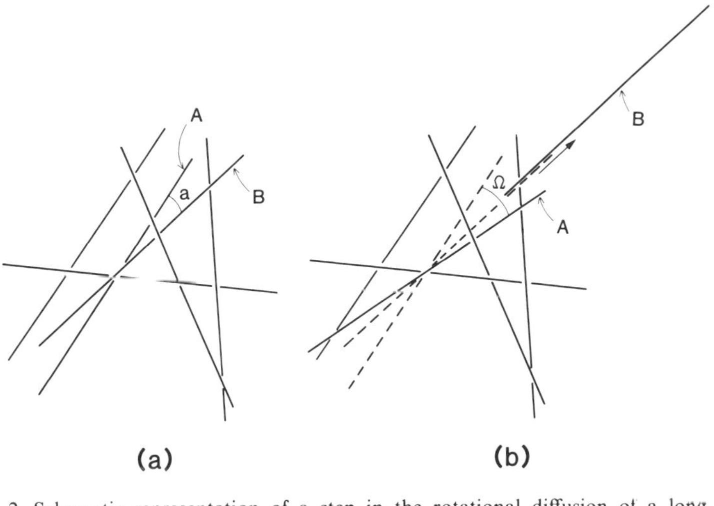

Figure 2. Schematic representation of a step in the rotational diffusion of a long rod in semidilute solution. (a) Rod A cannot rotate through a significant angle because of the presence of neighboring rods. (b) Rod A is able to rotate through the angle Ω after rod B translates a distance approximately equal to its length.

where c is the number concentration of rods and β' is a proportionality constant. Note that Θ now has an inverse c^2 dependence on concentration and an inverse L^9 dependence on length.

In this model the large translational diffusion anisotropy ($\Delta D = D_{\parallel} - D_{\perp}$) leads to a coupling between the rotation and translation which may be treated by writing a combined rotation–translation diffusion equation.[37] As a result of this coupling the polarized DLS correlation function contains contributions from the rotational motion. Thus, the simple results given in equations (21) and (22) for dilute solutions become more complex for long rods in semidilute solutions. To second order in γ, where $\gamma \equiv q^2\Delta D/\Theta$, the homodyne polarized and depolarized correlation functions may be shown to be[38]

$$I_{\mathrm{VV}}(q, t) = A\left(\sum_{i=1}^{2} a_i e^{-\Gamma_i^0 t}\right)^2 + B \tag{26}$$

$$I_{\mathrm{VH}}(q, t) = A'\left[\left(\sum_{i=1}^{2} b_i e^{-\Gamma_i^{21} t}\right)\cos^2\left(\frac{\theta_s}{2}\right) + \left(\sum_{i=1}^{2} c_i e^{-\Gamma_i^{22} t}\right)\sin^2\left(\frac{\theta_s}{2}\right)\right]^2 + B' \tag{27}$$

where

$$\Gamma_1^0 = q^2 D - \frac{2}{135}\Theta\gamma^2$$

$$\Gamma_2^0 = q^2 D + \frac{4}{21} q^2\Delta D + 6\Theta + \frac{2}{135}\Theta\gamma^2$$

$$\Gamma_1^{21} = q^2 D - \frac{2}{21} q^2\Delta D + 6\Theta - \frac{4}{1029}\Theta\gamma^2$$

$$\Gamma_2^{21} = q^2 D - \frac{34}{231} q^2\Delta D + 20\Theta + \frac{4}{1029}\Theta\gamma^2$$

$$\Gamma_1^{22} = q^2 D - \frac{4}{21} q^2\Delta D + 6\Theta - \frac{2}{1029}\Theta\gamma^2$$

$$\Gamma_2^{22} = q^2 D + \frac{16}{231} q^2\Delta D + 20\Theta + \frac{2}{1029}\Theta\gamma^2$$

$$a_2 = \left(\frac{2}{135}\gamma^2\right)\bigg/\left(1 + \frac{2}{63}\gamma + \frac{4}{135}\gamma^2\right), \qquad a_1 = 1 - a_2$$

$$b_2 = \left(\frac{4}{1029}\gamma^2\right)\bigg/\left(1 - \frac{2}{539}\gamma + \frac{8}{1029}\gamma^2\right), \qquad b_1 = 1 - b_2$$

$$c_2 = \left(\frac{2}{1029}\gamma^2\right)\bigg/\left(1 + \frac{10}{539}\gamma + \frac{4}{1029}\gamma^2\right), \qquad c_1 = 1 - c_2$$

Rods in semidilute solutions, where $\Delta D \approx D_{\parallel}$, $D \approx \frac{1}{3}D_{\parallel}$ and $\gamma = q^2 D_{\parallel}/\Theta$, should have a significant amount of translation–rotation coupling, particularly when Θ becomes small relative to $q^2 D_{\parallel}$. For real systems, the faster components in the depolarized correlation function (i.e., the $i = 2$ terms) would be difficult to observe experimentally, both because of their low amplitude and their rapid decay. Thus, equation (22) is still a good approximation to the depolarized correlation functions. However, the faster components should definitely make contributions to the polarized DLS. It should be emphasized that this theory still assumes that intermacromolecular correlations [equation (5)] are negligible.

The concentration dependence of Θ was determined by depolarized dynamic light scattering for both PBLG[38] and LMM.[39] PBLG is a synthetic polypeptide which forms α-helices in nonpolar solvents such as 1,2-dichloroethane. The lengths of these helices in the Zero and Pecora study[38] ranged from ~1000 to 1400 Å. For concentrations above ~5 mg/ml, a c^{-2} dependence was observed, as predicted. The length dependence, however, followed an inverse eighth power law (L^{-x}, where $x = 7.6 \pm 0.6$) rather than the predicted ninth power law. LMM ($L \approx 790$ Å), a rodlike constituent of myosin rod, also was found to have a c^{-2} dependence for Θ.[39] Maguire *et al.*[44] have obtained similar concentration and length dependences for Θ for semidilute solutions of long rodlike viruses using the electric birefringence decay technique.

3.5. FLEXIBLE MACROMOLECULES

Intramolecular motions which involve the tumbling or torsion of large, anisotropic groups should contribute to the depolarized spectra of nonrigid macromolecules. Numerous theoretical papers have dealt with the depolarized scattering from dilute solutions of random coil molecules with anisotropic segments; however, few experiments have been done. One of the earliest treatments is by Ono and Okano.[45,46] They obtain a spectrum that is a sum of equally weighted Lorentzians, each of which corresponds to a Rouse–Zimm mode.[47] Experimentally, Han and Yu[37] attempted to apply this theory to the forward depolarized scattering from isotactic polystyrene using a spectrum analyzer. Based on the instrumental bandwidth, they truncated the sum of Lorentzians to five equally weighted Lorentzians. However, due to the experimental precision, there was little difference in quality between a fit to the Lorentzians and one to a single Lorentzian. Depolarized interferometric studies by Bauer *et al.*[48] on solutions of atactic polystyrene yielded two Lorentzians of roughly equal area. The slow time was molecular weight dependent and was approximately equal to the relaxation time of the longest-wavelength Rouse–Zimm mode. The fast time (3–6 nsec) was relatively molecular weight independent and appeared to correspond to the local, correlated motion of phenyl groups about the backbone of the chain. Jones and Wang[39] performed similar experiments with polyethylene glycol; however, they detected only the slow, molecular weight dependent relaxation time.

Many authors have made corrections to the Ono–Okano theory in an attempt to explain the results of Bauer *et al.* Moro and Pecora[50, 51] showed that including some "local stiffness" into the polymer chain would tend to enhance the contribution from the longer-wavelength modes; however, the enhancement was not enough to explain the results for polystyrene. By explicitly assuming that the overall reorientation and local segmental motions can be separated, due to the large difference in time scales, Lin and Wang[52] have presented a theory that naturally yields two correlation functions. The first correlation function corresponds to the overall reorientation and would yield a single Lorentzian corresponding to the slow relaxation time in the polystyrene and polyethylene glycol experiments. The second correlation function corresponds to the local segmental motion modulated by overall rotation. Lin and Wang gave no explicit form for this second correlation function, although they indicated that a model of localized torsional oscillations may be used to calculate this correlation function. Although performed for a physically unrealistic system, calculations by Carpenter and Skolnick[53] indicate that including harmonic rotational interactions between segments would yield the observed two correlation times. Evans,[54] on the other hand, has shown that using a different model for the chain polarizability causes the Lorentzians to become unequally weighted. By including dipole–induced-dipole interactions between substituents and carbon backbone atoms, Evans obtained a spectrum consisting of a superposition of Lorentzians with one high-frequency term and a sum of weighted Lorentzians dominated by the low-frequency Rouse–Zimm modes. Cornelius's Monte Carlo simulations of the dynamics of chains on a tetrahedral lattice with optically anisotropic side groups also gave depolarized spectra that are qualitatively similar to those observed for polystyrene.[55]

Recent depolarized light scattering studies of myosin rod have also detected two relaxation times.[39] The slower time, 35 ± 6 μsec, corresponds to the overall reorientation of the molecule. The faster time, 3.1 ± 0.8 μsec, was attributed to a hingelike bending of the rod near its center.

If a joint is inserted into the center of a rod, one would expect a contribution to the DDLS due to motion of the segments relative to the joint. For dilute solutions of once-broken rods with no coupling between translation, overall rotation, and bending at the joint, the DDLS homodyne correlation functions can be approximated by[56, 57]

$$I_{\mathrm{VH}}(q, t) = A'\left(\sum_{i=1}^{2} C_i e^{-\Gamma_i t}\right)^2 + B' \tag{28}$$

where

$$\Gamma_1 = q^2 D + 6\Theta + \nu_1 D_\beta$$

$$\Gamma_2 = q^2 D + 6\Theta + \nu_2 D_\beta$$

The amplitudes (C_i) and the factors ν_i depend on what model is chosen for the bending motion. D_β is related to the rotational diffusion coefficient (D_R) of an unconnected segment. Due to hydrodynamic interaction between segments and differences in the center-of-mass positions, $D_\beta \approx \frac{2}{3} D_R$.[18, 19]

Most theories for the once-broken rod assume completely free motion at the joint.[18, 19] For this special case, $C_1 \approx \nu_1 \approx 0$ and $\nu_2 \approx 6$; thus, almost all the contribution to the DDLS is from the independent motion of the segments. Therefore, one would expect a single exponential with a time constant close to that for a solution of unconnected segments. One correction to this model is to impose a maximum bending angle (θ^0). For $(\theta^0/2) \leq 80°$, this model results in the following expressions[56]:

$$\nu_1 = 0$$

$$\mu_0 = \cos(\theta^0/2)$$

$$C_1 = (1/4)\mu_0^2(1 + \mu_0)^2$$

$$C_2 = (1/20)(4 - \mu_0 - 6\mu_0^2 - \mu_0^3 + 4\mu_0^4)$$

$$\nu_2 = \nu_2^0(\nu_2^0 + 1)$$

where ν_2^0 is a complicated function of $\theta^0/2$ tabulated in Table 1. From Table 1, it can be seen that the bending mode decays faster and decreases in amplitude as the maximum bending angle is decreased.

Another model assumes that any bending angle is possible, provided the segments do not pass through each other, but a cosine potential exists

Table 1. Parameters for the Once-Broken Rod with a Maximum Bending Angle (θ^0)

$\theta^0/2$	ν_2^0	C_1	C_2	C_2/C_1	ν_2
0	∞	1	0	0	∞
15	14.12	0.901	0.00084	0.00093	213.5
30	6.83	0.653	0.0117	0.0179	53.5
45	4.4	0.364	0.0470	0.129	24.4
60	3.2	0.141	0.106	0.752	13.4
75	2.4	0.0265	0.167	6.30	8.2
90	2	0	0.2	∞	6

which influences motion of the segments. In other words a restoring force F_β is exerted on the segments proportional to one-half the bending angle θ_β, i.e.,

$$F_\beta = f \sin(\theta_\beta/2)$$

where f is a force constant. The bending angle is defined so that at zero bending angle the segments are "parallel" and form a rod of length equal to twice the segment length. At $\theta_\beta = 180°$ the segments are "antiparallel." Since the segments are not allowed to pass through each other in this model, the maximum value of $\theta_\beta = 180°$, where the restoring force is a maximum.

A Smoluchowski equation describing the combined bending and overall molecular rotational diffusion may be written down and solved for the pertinent time correlation functions that occur in DDLS.[57] For values of f/kT less than six, these correlation functions are of the two-exponential form given in equation (28). The amplitudes (C_i) of the two exponentials and factors ν_i are functions of f/kT. Numerical values of these quantities are given in Table 2 for some values of f/kT less than six. In this model, the

Table 2. Parameters for the Once-Broken Rod with a Cosine Potential

f/kT	ν_1	ν_2	C_1	C_2	C_2/C_1
0	0	6	0	1	∞
0.5	0.751	7.212	0.009	0.935	104
1.0	1.505	8.468	0.031	0.868	28
2.0	3.038	11.102	0.095	0.748	7.9
3.0	4.610	13.891	0.163	0.660	4.05
4.0	6.226	16.860	0.220	0.603	2.74
5.0	7.884	20.076	0.264	0.573	2.17
6.0	9.583	23.729	0.294	0.572	1.95

bending motion contributes to both exponentials which decay faster with increasing f/kT.

From the myosin rod studies,[39, 56] $C_2/C_1 = 1.2 \pm 0.4$. Comparison with the maximum bending angle model yields a maximum angle of 128° (ranging from 121° to 132°) and a value of 24 ± 6 krad/sec for D_β. Takahashi's electron microscopy data[58] yield a similar maximum angle (145°), while D_β is ~69% of the average D_R for the two segments free in solution. The cosine potential model yields amplitudes that are too small and time constants that are too fast to explain the experimental results.

Little work has been done on the concentration dependence of the depolarized light spectra for nonrigid molecules. One study by Delsanti *et al.* on poly-α-methylstyrene solutions showed that the depolarized relaxation times of semidilute solutions have a third power dependence on concentration and an inverse dependence on molecular weight.[59] From the theory of deGennes,[60] one would predict the polymers in these solutions to be highly entangled and to have relaxation times dependent on the distance between entanglement points. Although the concentration dependence is consistent with this view, no molecular weight dependence is expected for such entangled solutions.

3.6. ROTATION OF SMALL MOLECULES IN VISCOUS MEDIA

As was noted in the Introduction, the reorientation times of small molecules in liquids are usually on the order of picoseconds so that measurements are performed using Fabry–Perot interferometry. However, there are many systems in which, although the system is still a "liquid," its viscosity may be quite large. For these systems both the polarized (I_{VV}) and depolarized (I_{VH}) time correlation functions are no longer single exponentials and exhibit decays ranging over very wide time ranges. Examples of such systems are organic liquids such as benzyl benzoate[61] and *o*-terphenyl[62] in the supercooled liquid state, dihydroxy and polyhydroxy alcohols[63, 64] and small molecules reorienting in amorphous and glassy polymers [such as chlorobenzene reorienting in poly(methyl methacrylate)].[65]

A major difficulty with the interpretation of experiments of this type is the fitting of the raw data. A popular method for fitting functions which decay over many time decades is the use of the Williams–Watts equation described by Patterson elsewhere in this volume.

For many polymer systems such as the chlorobenzene–poly(methyl methacrylate) system studied by Ouano and Pecora,[65] it is clear that the decay extends into the picosecond range where the faster relaxation times may be studied by Fabry–Perot interferometry. For this system, these authors have interpreted the rotational relaxation data of the chlorobenzene using the rotation-in-a-cone model (also called the "restricted rotational diffusion model").[66, 67] In this model the chlorobenzene undergoes rotational diffusion restricted to a conical region with a rotational diffusion coefficient that is not very different from that of chlorobenzene in a liquid of low viscosity. The "walls" of the cone are made up of polymer chains which restrict further fast motion of the chlorobenzene. Further rotational relaxation of the chlorobenzene must then await motion of the polymer chains which is relatively slow compared to the initial reorientation. Thus, the depolarized spectrum consists of a "wide" line (with a width corresponding to picosecond relaxation times) and a "spike" corresponding to the slow final motion of the chlorobenzene coupled to the polymer chains. From the ratio of areas under the wide line to that under the spike it is possible to estimate the "cone" angle[67] and to then study its variation with temperature and concentration.

3.7. RESONANCE-ENHANCED DEPOLARIZED DYNAMIC LIGHT SCATTERING

A major advantage of DLS over many other techniques is its ability to study relaxation times over a very wide range (seconds to picoseconds). Most other techniques are limited to smaller relaxation time ranges. A major drawback, however, of DLS is its lack of selectivity and relatively low signal strength, making it difficult to study substances present in low concentration or to study one component in a complex mixture of comparably sized molecules.

This disadvantage may, in some cases, be circumvented by selecting the frequency of exciting light to be near that of an absorption band of the molecule of interest. In this region the polarizability of the molecules and, hence, the scattered intensity can become quite large. This phenomenon is known as "resonance-enhanced Rayleigh scattering" and is predicted by the same theory that predicts resonance-enhanced Raman scattering.

The first study of resonance-enhanced Rayleigh scattering from molecules in solution was a study of diphenylpolyenes by Bauer *et al.*[68] These authors were able to measure the depolarized spectrum (via interferometry) of some polyenes at concentrations as low as 10^{-5} *M*. Subsequently, Miller[69] presented a macroscopic fluctuation theory of resonance-enhanced scattering and Anglister and Steinberg[70, 71] have presented experimental studies of resonance-enhanced Rayleigh intensities and depolarization ratios and detailed comparisons with theory. Stanton *et al.*[72] have formulated a semiclassical molecular theory of the effect and have obtained experimental Rayleigh enhancement data on diphenylpolyenes, coumarins, and dinitrophenols. In a subsequent work,[73] Stanton *et al.* have used the Rayleigh enhancement to study the reorientation of small anions in solution at millimolar concentrations.

It is probable that resonance enhancement techniques will prove valuable in the study of molecular dynamics via depolarized light scattering. Stanton *et al.*[72] point out that the depolarization ratio is larger near resonance than far from resonance, indicating a larger resonance enhancement of the depolarized scattering as a percentage of the total solute scattering. Furthermore, the background due to polarized scattering from the solvent is usually much greater than that due to depolarized scattering. As a result, resonance enhancement effects are easier to observe in the depolarized component than in the polarized component. Stanton *et al.*[72] give some general guidelines for selecting molecules for resonance enhancement studies. They conclude that the best dynamic spectrum is obtained when the scattering cross section is large and the absorption cross section is small. Such a condition occurs on the red "preresonance" side of a sharply rising absorption band. It is, of course, preferable for such molecules to also be photochemically stable and nonfluorescent.

REFERENCES AND NOTES

1. B. J. Berne and R. Pecora, *Dynamic Light Scattering*, Wiley-Interscience, New York (1976).
2. D. R. Bauer, J. I. Brauman, and R. Pecora, *Ann. Rev. Phys. Chem.* **27**, 443 (1976).
3. D. Kivelson and P. A. Madden, *Ann. Rev. Phys. Chem.* **31**, 523 (1980).
4. R. S. Stein and M. B. Rhodes, *J. Appl. Phys.* **31**, 1873 (1960).
5. S. R. Aragon and R. Pecora, *J. Colloid Interface Sci.*, **89**, 170 (1982).
6. G. C. Tabisz in *Molecular Spectroscopy*, Vol. 6, G. Barrow and D. Long, eds., Chemical Society, London (1979).
7. For a treatment of some cases where intramolecular interference affects the dynamic scattering see S. R. Aragon and R. Pecora, *J. Chem. Phys.* **66**, 2506 (1977).
8. R. Pecora, *J. Chem. Phys.* **49**, 1036 (1968).
9. T. Keyes and D. Kivelson, *J. Chem. Phys.* **56**, 1057 (1974).
10. S. H. Koenig, *Biopolymers*, **14**, 2421 (1975).
11. J. Riseman and J. G. Kirkwood, *J. Chem. Phys.* **18**, 512 (1950).
12. S. Broersma, *J. Chem. Phys.* **32**, 1626, 1632 (1960).
13. J. Newman and H. L. Swinney, *J. Mol. Biol.* **116**, 593 (1977).
14. M. M. Tirado and J. G. de la Torre, *J. Chem. Phys.* **71**, 2581 (1979); **73**, 1986 (1980).
15. J. E. Hearst and W. H. Stockmayer, *J. Chem. Phys.* **37**, 1425 (1962).
16. J. E. Hearst, *J. Chem. Phys.* **38**, 1062 (1963).
17. H. Yamakawa, *Modern Theory of Polymer Solutions*, Harper and Row, New York (1971).

18. M. Fujiwara, N. Numasawa, and N. Saito, *Rep. Prog. Polymer Phys. Jpn.* **23**, 531 (1980).
19. W. A. Wegener, R. M. Dowben, and V. J. Koester, *J. Chem. Phys.* **73**, 4086 (1980).
20. J. G. de la Torre and V. A. Bloomfield, *Biopolymers* **16**, 1747, 1765, 1779 (1977).
21. V. A. Bloomfield, W. O. Dalton, and K. E. Van Holde, *Biopolymers* **5**, 135, 149 (1967).
22. G. K. Youngren and A. Acrivos, *J. Chem. Phys.* **63**, 3846 (1975).
23. H. Benoit, L. Freund, and G. Spach in *Biological Macromolecules: Poly-α-Amino Acids*, G. D. Gasman, ed., Marcel Dekker, New York (1967).
24. M. Hercher, *Appl. Opt.* **7**, 951 (1968).
25. S. B. Dubin, N. A. Clark, and G. B. Benedek, *J. Chem. Phys.* **54**, 5158 (1971).
26. D. R. Bauer, S. J. Opella, D. J. Nelson, and R. Pecora, *J. Am. Chem. Soc.* **97**, 2580 (1975).
27. S. Michielsen and R. Pecora, *Biochemistry* **20**, 6994 (1984).
28. A. Patkowski, S. Jen, and B. Chu, *Biopolymers* **17**, 2643 (1978).
29. G. D. J. Phillies, in *Measurement of Suspended Particles by Quasi-Elastic Light Scattering*, B. Dahneke, ed., Wiley-Interscience, New York (1983).
30. A. Wada, K. Soda, T. Tanaka, and N. Suda, *Rev. Sci. Instrum.* **41**, 845 (1970).
31. C. M. Sorensen, R. C. Mockler, and W. J. O'Sullivan, *Phys. Rev. A* **14**, 1520 (1976).
32. A. Wada, N. Suda, T. Tsuda, and K. Soda, *J. Chem. Phys.* **50**, 31 (1969).
33. J. M. Schurr and K. S. Schmitz, *Biopolymers* **12**, 1021 (1973).
34. C. R. Crosby, N. C. Ford, F. E. Karasz, and K. H. Langley, *J. Chem. Phys.* **75**, 4298 (1981).
35. P. C. Hopman, G. Koopmans, and J. Greve, *Biopolymers* **19**, 1241 (1980).
36. J. C. Thomas and G. C. Fletcher, *Biopolymers* **18**, 1333 (1979).
37. C.-C. Han and H. Yu, *J. Chem. Phys.* **61**, 2650 (1974).
38. K. Zero and R. Pecora, *Macromolecules* **15**, 87 (1982).
39. S. Highsmith, C.-C. Wang, K. Zero, R. Pecora, and O. Jardetzky, *Biochemistry* **21**, 1192 (1982).
40. J. Bruining and H. M. Fijnaut, *Biophys. Chem.* **9**, 345 (1979).
41. S. Chopra and L. Mandel, *Rev. Sci. Instrum.* **43**, 1489 (1972).
42. A. Patkowski and B. Chu, *Biopolymers* **18**, 2051 (1979).
43. M. Doi and S. F. Edwards, *J. Chem. Soc. Faraday II* **74**, 560 (1978).
44. J. F. Maguire, J. P. McTague, and F. Rondelez, *Phys. Rev. Lett.* **45**, 1891 (1980); **47**, 148 (1981).
45. K. Ono and K. Okano, *Jpn. J. App. Phys.* **9**, 1356 (1970).
46. T. Norisuye and H. Yu, *J. Chem. Phys.* **68**, 4038 (1978).
47. B. H. Zimm, *J. Chem. Phys.* **24**, 269 (1956).
48. D. R. Bauer, J. I. Brauman, and R. Pecora, *Macromolecules* **8**, 443 (1975).
49. D. R. Jones and C. H. Wang, *J. Chem. Phys.* **66**, 1659 (1977).
50. K. Moro and R. Pecora, *J. Chem. Phys.* **69**, 3254 (1978).
51. K. Moro and R. Pecora, *J. Chem. Phys.* **72**, 4958 (1980).
52. Y.-H. Lin and C. H. Wang, *J. Chem. Phys.* **66**, 5578 (1977).
53. D. K. Carpenter and J. Skolnick, *Macromolecules* **14**, 1284 (1981).
54. G. T. Evans, *J. Chem. Phys.* **71**, 2263 (1979).
55. C. W. Cornelius, Depolarized Rayleigh scattering studies of the rotational motion of small molecules and polymers, and of lattice model polymer chains, Ph.D. Thesis, Stanford University (1979).
56. K. Zero and R. Pecora, *Macromolecules*, **15**, 1023 (1982).
57. K. Zero, The dynamics of rodlike macromolecules in solution, Ph.D. Thesis, Stanford University (1982).
58. K. Takahashi, *J. Biochem.* **83**, 905 (1978).
59. M. Delsanti, C.-C. Han, and H. Yu, *Polym. Prepr.* **22**, 76 (1981).

60. P.-G. de Gennes, *Scaling Concepts in Polymer Physics*, Cornell University Press, Ithaca, New York (1979).
61. P. Bezot, C. Hesse-Bezot, N. Ostrowsky, and B. Quentrec, *Mol. Phys.* **39**, 549 (1980).
62. G. Fytas, C. H. Wang, D. Lilge, and Th. Dorfmüller, *J. Chem. Phys.* **75**, 4247 (1981).
63. G. Fytas and Th. Dorfmüller, *J. Chem. Phys.* **75**, 5232 (1981).
64. Th. Dorfmüller, H. Dux, G. Fytas, and W. Mersch, *J. Chem. Phys.* **71**, 366 (1979).
65. A. C. Ouano and R. Pecora, *Macromolecules* **13**, 1167, 1173 (1980).
66. M. P. Warchol and W. E. Vaughan, *Adv. Mol. Relaxation Processes* **13**, 317 (1978). (1978).
67. C.-C. Wang and R. Pecora, *J. Chem. Phys.* **72**, 5333 (1980).
68. D. R. Bauer, B. Hudson, and R. Pecora, *J. Chem. Phys.* **63**, 588 (1975).
69. G. A. Miller, *J. Chem. Phys.* **82**, 616 (1978).
70. J. Anglister and I. Z. Steinberg, *Chem. Phys. Lett.* **65**, 50 (1979).
71. J. Anglister and I. Z. Steinberg, *J. Chem. Phys.* **74**, 786 (1981).
72. S. G. Stanton, R. Pecora, and B. S. Hudson, *J. Chem. Phys.* **75**, 5615 (1981).
73. S. G. Stanton, R. Pecora, and B. S. Hudson, *J. Chem. Phys.* **78**, 3365 (1983).

4

Particle Interactions

P. N. Pusey and R. J. A. Tough

Royal Signals and Radar Establishment
Malvern, Worcestershire
WR14 3PS, England

4.1. INTRODUCTION

It is well known that the diffusion coefficients of particles suspended in a liquid depend on the concentration of the suspension. The cause of such concentration dependence is, of course, interparticle interactions. In the early days of dynamic light scattering (DLS), interactions were generally regarded as a nuisance whose effects could be removed by extrapolation of data to zero concentration. Thus "free-particle" properties were obtained and such information as particle sizes and shapes could be deduced. More recently, however, there has been growing interest in interactions themselves, their nature and effects. This interest is partly academic since concentrated particle suspensions provide a challenging many-body problem to which modern statistical mechanical and hydrodynamical theories can be applied. The subject is also of considerable practical importance because many processes in industry (paints, foods, detergents, etc.) and biology involve suspensions in which the particles occupy a significant fraction of the total volume.

The main aim of this article is to review the contributions which experiments using the DLS technique have made to our current understanding of systems in which interparticle interactions play an important role. It should be admitted at the outset that this remains a confusing, and still somewhat confused, subject. While the quantities measured by DLS are usually quite well defined formally, it is not always obvious how they relate to more familiar properties of the suspension. Even when interpretation of the light scattering data can be made confidently in terms of, for example, a macroscopic collective diffusion coefficient, it is frequently difficult to obtain an intuitive understanding of how the interactions determine its behavior. Because of this, theories with the same ostensible aim often start from apparently different viewpoints, pursue different courses, and, sometimes, reach different conclusions. Here we will concentrate on areas where, in our view, a reasonable consensus exists (or is at least emerging) on the theoretical interpretation of the light scattering data. Where necessary, areas still in dispute will be identified as such.

Inevitably the content and layout of the paper are determined by our own interests (and prejudices); we will not attempt an exhaustive survey of the literature. Consideration will be limited to rigid, compact (roughly spherical) particles. In the case of charged particles (macroions) it will be assumed that the dominant role of the small ions (H^+, OH^-, etc.) in the suspension is to determine the form of the potential of interaction between the macroions. This "effective two-component" approach has proved remarkably successful. Nevertheless, in Section 4.6 we discuss briefly theories which treat the effects of the small ions more explicitly.

We start here with some general comments on the structure and dynamics of suspensions of interacting particles and their relationship to the quantities measured by dynamic light scattering.

In the absence of interactions (a hypothetical situation approached at very low concentrations) a particle can take up any position in, and diffuse through, the suspension without interference from other particles. Interactions lead to correlations in particle positions and motions. The consequent spatial ordering is best expressed in terms of the radial distribution function $g(r)$; if $\langle C \rangle$ is the average particle number density in the suspension, $\langle C \rangle g(r)\, dV$ is the average number of particles in a small volume element dV centered at a distance r from a given particle. Two important spatial scales in the suspension are $L \equiv \langle C \rangle^{-1/3}$, the typical spacing between neighboring particles, and R_I, the range of the spatial ordering, roughly speaking the range of r over which $g(r)$ is significantly different from one. In a weakly interacting system $R_I \ll L$, whereas for strong interactions R_I can be comparable to or greater than L. Experimentally, the average structure [described by $g(r)$] in a suspension can be determined by, for example, time-averaged light scattering or x-ray or neutron diffraction, while computer simulation and approximate analytical methods provide theoretical approaches based on model potentials. This is an interesting topic in its own right (see Reference 1 for various examples) but not one we will emphasize here where the main interest is dynamics.

In a two-component system composed of rigid particles and liquid, three types of force act on a particle. The first is the "Brownian" force arising from collisions with thermally agitated liquid molecules. This force is responsible for the Brownian motion of isolated particles. It gives rise to large rapid fluctuations in the particle velocity whose characteristic fluctuation time τ_B can be taken to be†

$$\tau_B = M/\zeta_0 \tag{1}$$

where M is the particle mass and ζ_0 the "free-particle" friction coefficient

$$\zeta_0 = 6\pi\eta a \tag{2}$$

η being the shear viscosity of the liquid and a the particle radius.[4] For particles of diameter less than one micrometer, $\tau_B \lesssim 5 \times 10^{-8}$ sec. This is much smaller than delay times $\tau(\gtrsim 1\ \mu\text{sec})$ typically used in DLS experiments which therefore operate in the regime $\tau \gg \tau_B$.

The second force is due to direct interactions between the particles, e.g., a shielded Coulombic force between charged particles or simply an "excluded-volume" or "hard-sphere" force between uncharged particles. Simple calculations show that the mean-square fluctuation in the particle velocity caused by direct interaction forces is generally many orders of magnitude smaller than that caused by the Brownian forces. However the velocity fluctuation time τ_I associated with the interaction forces is much greater than τ_B. Thus the strong rapidly fluctuating Brownian forces and the weaker but slowly fluctuating interaction forces can frequently produce effects in the particle dynamics of comparable magnitudes. The fluctuation time τ_I is hard to define exactly. In a system with long-ranged repulsive forces (e.g., Section 4.4) it can be viewed as some fraction of the lifetime of the shell of nearest neighbors surrounding a given particle and can be as large as 10^{-4} sec. Then it is possible to examine with dynamic light scattering both the short-time ($\tau_B \ll \tau < \tau_I$) behavior where a particle is temporarily trapped within its nearest-neighbor shell as well as the long-time behavior ($\tau > \tau_I$) where it diffuses through several nearest-neighbor shells. In a weakly interacting system τ_I can be taken as the duration of a "collision" between particles.

The third interaction is the coupling between particle motions transmitted indirectly by the flows they induce in the liquid. The time scale associated with these hydrodynamic interactions is (Reference 5, p. 91)

$$\tau_H \approx \rho L^2/\eta \tag{3}$$

† In this chapter we will not consider the "long-time tail" in the velocity autocorrelation function of an isolated particle since its effects are hard to observe experimentally (see, however, Refs. 2 and 3).

the time taken by a viscous shear wave to propagate across the typical distance L between particles in a liquid of density ρ; for $L = 0.5$ μm, $\tau_H \approx 2.5 \times 10^{-7}$ sec. This is somewhat smaller than typical photon correlation delay times and in this chapter we will assume that hydrodynamic interactions essentially act instantaneously (see, however, References 6 and 7 for further discussion).

In common with most scattering techniques, light scattering provides information in reciprocal rather than real space; as a consequence it is sometimes difficult to visualize the behavior of the light scattering correlation functions in simple terms. Roughly speaking, dynamic light scattering observes the growth and decay of fluctuations in a spatial Fourier component of refractive index of wavelength $2\pi q^{-1}$, where q is the scattering vector [equation (5)]. Associated with this spatial scale is a characteristic time $(D_0 q^2)^{-1}$, the time taken by a particle to diffuse freely a distance q^{-1}, equal also to the decay time of the light scattering correlation function for noninteracting particles; here D_0 is the "free-particle" translational diffusion constant [equation (37)]. As a rule of thumb, therefore, low-q (small scattering angle) measurements probe large-scale slow motions in the suspensions whereas high-q measurements probe small-scale rapid motions. In the so-called thermodynamic or hydrodynamic limit $q^{-1} \gg L$ (the interparticle spacing), $\tau \gg \tau_I$, DLS observes fluctuations of macroscopic spatial extent. In this limit the macroscopic collective diffusion coefficient obtained from the light scattering correlation function is expected to be the same as that which would be measured in conventional "boundary spreading" measurements. In the opposite limit $q^{-1} \ll L$, $\tau \ll \tau_I$, DLS observes the motion of individual particles over distances small compared to the interparticle spacing. In the regime intermediate between these extremes the technique probes spatial scales comparable to the interparticle spacing and time scales comparable to the collision time. Here, in strongly interacting systems, interesting and complicated phenomena are found. It should be noted that, while (in principle, though not necessarily in practice) the low-q regime can always be achieved by using a small enough scattering angle, an upper limit to q is imposed by the maximum scattering angle of 180°. Thus, in certain cases, concentrated suspensions of small particles, for example, the high-q ($q^{-1} \ll L$) regime is excluded and measurements at all angles are effectively "low-q" (see Section 4.5).

The organization of the paper is as follows. In the next section we define formally the quantities measured in light scattering experiments. Section 4.3 deals with the theory of the Brownian motion of interacting particles. Emphasis is on use of the generalized Smoluchowski equation, essentially a many-particle diffusion equation. We also discuss in some detail the nature and origin of hydrodynamic interactions. The treatment in this section is fairly formal. Since, in later sections, we will attempt, where possible, to describe experimental results in simple physical terms (as well as in terms of the formal theory), Section 4.3 could probably be omitted by the less theoretically inclined reader.

Discussion of specific systems starts in Section 4.4 with dilute aqueous suspensions of charged spherical particles. Here long-ranged Coulombic repulsive forces cause pronounced spatial ordering. Although these systems are of little practical importance, they constitute valuable experimental "models" which not only demonstrate clearly many important aspects of light scattering by interacting suspensions but have also stimulated the generation of new theoretical approaches. Two attractive features of these systems stem from their diluteness. Firstly, hydrodynamic interactions can, to a good degree of approximation, be neglected. Secondly, typical interparticle spacings L can be made comparable to the light wavelength (and therefore to typical reciprocal scattering vectors q^{-1}). Thus it is possible to probe in detail the interesting regime $q^{-1} \approx L$ as well as the high- and low-q limits.

In Section 4.5 we turn to more concentrated suspensions where hydrodynamic interactions become important. Here most experiments to date have been performed in the low-q limit. We show how the concentration dependence of the collective diffusion coefficient is determined by competition between the effects of hydrodynamic and direct interactions. The latter can be purely repulsive or can include an attractive part in the interparticle potential. These theoretical considerations are illustrated with experimental results on various systems including uncharged colloids, charged and uncharged proteins, and microemulsions.

Section 4.6 deals briefly but more explicitly with the effects of small ions in suspensions of charged particles. This is a topic where significant areas of doubt still exist. In Section 4.7, we summarize some of the important conclusions of the paper and highlight areas where, in our opinion, further work is needed. Finally, Section 4.8 comprises a brief addendum written after the main article.

It is a relatively recent realization that even a small degree of polydispersity, a distribution of particle size, can have profound effects on the scattering of light by strongly interacting systems.[8–10] In a suspension composed of different types of particle, the refractive index fluctuations which cause the scattering arise from two distinct types of particle fluctuation. Firstly there can be fluctuations in the number density of particles, regardless of type. Secondly, even at constant number density, there can be "polydispersity fluctuations," that is, more particles than average of one type in a given region of the suspension. In the presence of strong interactions, the time evolution of these two types of fluctuation can be very different. In general the theory is complicated because the two types of fluctuation are coupled. It appears, however, that, if the spread in particle size is small, the polydispersity fluctuations constitute an effective "tracer" mode which decays by self-diffusion. This allows the valuable possibility of measuring both collective and self-diffusion by DLS in a strongly interacting system. The relevant theory will be outlined in Section 4.2 and illustrated by experimental examples in Sections 4.4.4 and 4.5.4.

Much of the subject matter of this chapter has recently been covered more concisely in a valuable paper by Hess.[11] From the large theoretical literature we would also select as seminal references the Ph.D. thesis of Ackerson,[12] his two subsequent papers[13, 14] on application of the generalized Smoluchowski equation, and the work of Batchelor[15, 16] and Felderhof[17, 18] on hydrodynamic interactions.

4.2. QUANTITIES MEASURED BY LIGHT SCATTERING

4.2.1. Introduction

In the main the following assumptions will apply throughout this chapter:

- The scattering volume V contains a large number N of particles so that the amplitude E_s of the scattered electric field is a complex Gaussian random variable;
- The intensity of the light scattered by the liquid and the small ions is negligible compared to that scattered by the particles;
- The incident light is polarized perpendicular to the scattering plane and the scattered light has this same polarization;
- The suspension is sufficiently transparent that the first Born approximation can be applied and multiple scattering neglected (see, however, Section 4.4.1);
- The particles are either small and/or spherical so that their individual scattering amplitudes are independent of time.

With these assumptions the time-dependent part of the measured photon correlation function is $c[g^{(1)}(q, \tau)]^2$, where c is an apparatus constant (of order 1) determined by experimental conditions and $g^{(1)}(q, \tau)$ is the normalized temporal autocorrelation function of the scattered field amplitude:

$$g^{(1)}(q, \tau) \equiv \frac{\langle E_s(q, 0)E_s^*(q, \tau)\rangle}{\langle | E_s|^2\rangle} \tag{4}$$

Here τ is the correlation delay time and q the magnitude of the scattering vector

$$q = (4\pi/\lambda_0)n \sin(\theta/2) \tag{5}$$

where λ_0 is the wavelength of the light *in vacuo*, n the refractive index of the suspension, and θ the scattering angle. Aside from constants which disappear through normalization, the scattered field amplitude E_s can be written

$$E_s(q, t) = \sum_{i=1}^{N} b_i(q) \exp[i\mathbf{q} \cdot \mathbf{r}_i(t)] \tag{6}$$

where $b_i(q)$ is the field amplitude of the light scattered by particle i at angle θ and $\mathbf{r}_i(t)$ is the position of the center-of-mass of particle i at time t.

4.2.2. Monodisperse Systems

For identical particles, $b_i(q) = b_j(q) = b(q)$, the average scattered intensity, defined by

$$\langle I(q) \rangle \equiv \langle | E_s(q, t) |^2 \rangle \tag{7}$$

can be written

$$\langle I(q) \rangle = N b^2(q) S^I(q) \tag{8}$$

where the static structure factor $S^I(q)$ is given by

$$S^I(q) \equiv N^{-1} \sum_{i=1}^{N} \sum_{j=1}^{N} \langle \exp[i\mathbf{q} \cdot (\mathbf{r}_i - \mathbf{r}_j)] \rangle \tag{9}$$

The superscript I is used to indicate a monodisperse or "ideal" system. The structure factor can be written in terms of the radial distribution function $g(r)$ [19]

$$S^I(q) = 1 + 4\pi \langle C \rangle q^{-1} \int_0^\infty [g(r) - 1] r \sin qr \, dr \tag{10}$$

where $\langle C \rangle$ is, as before, the particle number density, $\langle C \rangle \equiv N/V$.

With use of (4)–(9), the scattered field autocorrelation function can be written

$$g^{(1)}(q, \tau) \equiv F^I(q, \tau)/S^I(q) \tag{11}$$

where

$$F^I(q, \tau) \equiv N^{-1} \sum_i \sum_j \langle \exp\{i\mathbf{q} \cdot [\mathbf{r}_i(0) - \mathbf{r}_j(\tau)]\} \rangle \tag{12}$$

and

$$F^I(q, 0) \equiv S^I(q) \tag{13}$$

In this chapter we will call $F^I(q, \tau)$ and $F^M(q, \tau)$ (see below) "dynamic structure factors" although, in the neutron scattering literature, this term is usually reserved for the temporal Fourier transform and $F(q, \tau)$ itself is called the "intermediate scattering function."

In the absence of interactions, cross-terms $i \neq j$ in (9) and (12) vanish to give

$$S^I(q) = 1 \tag{14}$$

and

$$F^I(q, \tau) \equiv F^I_S(q, \tau) = \langle \exp\{i\mathbf{q} \cdot [\mathbf{r}(0) - \mathbf{r}(\tau)]\} \rangle \tag{15}$$

where $F^I_S(q, \tau)$ is the ideal self-dynamic structure factor which describes single-particle motions.

4.2.3. Polydisperse Systems

In the case of nonidentical particles (8) and (9) can be generalized to

$$\langle I(q) \rangle = N\overline{b^2(q)}S^M(q) \tag{16}$$

where the "measured" static structure factor in a polydisperse system is defined by

$$S^M(q) = [N\overline{b^2(q)}]^{-1} \sum_i \sum_j \langle b_i(q)b_j(q)\exp[i\mathbf{q} \cdot (\mathbf{r}_i - \mathbf{r}_j)] \rangle \tag{17}$$

and the bar indicates an average over the particle type distribution. Defined in this way $S^M(q)$ again reduces to unity in the absence of interactions where the mean scattered intensity is simply $N\overline{b^2(q)}$.

Similarly the dynamic structure factor measured in such a system is given by

$$F^M(q, \tau) = [N\overline{b^2(q)}]^{-1} \sum_i \sum_j \langle b_i(q)b_j(q)\exp\{i\mathbf{q} \cdot [\mathbf{r}_i(0) - \mathbf{r}_j(\tau)]\} \rangle \tag{18}$$

In the absence of interactions (18) becomes

$$F^M(q, \tau) = [N\overline{b^2(q)}]^{-1} \sum_i \overline{b_i^2(q)}\langle \exp\{i\mathbf{q} \cdot [\mathbf{r}_i(0) - \mathbf{r}_i(\tau)]\}\rangle \tag{19}$$

the sum of self-terms which, for a small degree of polydispersity, will not be too different from an appropriately defined average self-term.[20]

In the case of arbitrary polydispersities and interactions it is not possible to simplify (18) and $F^M(q, \tau)$ remains a complicated sum of partial dynamic structure factors. However, valuable insight into the behavior of $F^M(q, \tau)$ can be obtained if it is assumed that the particles only differ in scattering power b_i and are otherwise identical in terms of size and interactions.[9, 10] Such a situation could be achieved, for example, with particles that differed only in refractive index. Then the average in (18) can be separated into independent averages, first of the scattering amplitudes over the particle type distribution and second the usual thermal average over particle positions:

$$F^M(q, \tau) = [N\overline{b^2(q)}]^{-1} \sum_i \sum_j \overline{b_i(q)b_j(q)}\langle \exp\{i\mathbf{q} \cdot [\mathbf{r}_i(0) - \mathbf{r}_j(\tau)]\}\rangle \tag{20}$$

Obviously the first average can be written

$$\overline{b_i b_j} = (\bar{b})^2 + [\overline{b^2} - (\bar{b})^2]\delta_{ij} \tag{21}$$

Substitution of (21) into (20) with use of (12) and (15) then gives

$$F^M(q, \tau) = (1 - x)F^I(q, \tau) + xF_S^I(q, \tau) \tag{22}$$

where

$$x \equiv 1 - \{\overline{[b(q)]}^2/\overline{b^2(q)}\} \tag{23}$$

Thus the effect of a distribution of scattering amplitudes is to make the measured dynamic structure factor the sum of two terms proportional, respectively, to the full- and self-dynamic structure factors appropriate for identical scattering amplitudes. Equation (22) behaves as expected in two trivial limits. In the absence of interactions $F^I(q, \tau) \rightarrow F_S^I(q, \tau)$ so that $F^M(q, \tau) = F_S^I(q, \tau)$ regardless of x. Second, for a monodisperse system $x = 0$ so that $F^M(q, \tau) = F^I(q, \tau)$.

The analysis of the previous paragraph is formally identical to that of neutron scattering where a distribution of scattering amplitudes can arise, for example, from a mixture of isotopes of a given element (see Reference

19, p. 218). In neutron scattering the first term in (22) is attributed to "coherent" scattering and the second term to "incoherent" scattering. As discussed in Section 4, coherent scattering is due to fluctuations in the number density of particles regardless of type, whereas incoherent scattering is caused by "polydispersity fluctuations," variations in the composition of the suspension at constant number density.†

The complete decoupling of the two types of fluctuation which allowed the derivation of (22) occurs only when particle sizes and interactions are identical. This can be seen by considering a suspension of hard spheres of differing sizes. If, for example, some region of the suspension contains instantaneously more small weakly scattering spheres than average it is also likely to contain more spheres altogether since small spheres can pack more closely. Thus $b_i b_j$ and $\mathbf{r}_i - \mathbf{r}_j$ in (18) are correlated. Nevertheless Weissman[9] has suggested that, because for particles much smaller than the wavelength of light b_i is proportional to a_i^3 (where a_i is the radius of particle i), a *small degree* of size polydispersity will have a much greater effect on the distribution of scattering amplitudes than on the mobilities and interactions of the particles. Thus, in such a system, it seems likely that equation (22) will hold as a first approximation with $x \ll 1$ and F^I and F_S^I as appropriately defined average dynamic structure factors. However, due to the coupling of the fluctuations, (23) is unlikely to hold exactly.‡

Finally we note from (22) and (13) that the ratio of the intensities of incoherent and coherent scattering is $x/(1-x)S^I(q)$. Thus, if $x \ll 1$, incoherent scattering will only be important if $S^I(q) \ll 1$, which occurs at low q for systems with strong repulsive interactions (Sections 4.4.4 and 4.5.4).

4.2.4. Discussion

With this formal background we can amplify statements made in Section 4.1 concerning the high- and low-q limits.

Consider first the high-q regime where it can be seen from (10) that rapid oscillations in $\sin qr$ cause the second term to be zero so that $S^I(q) = 1$. This is equivalent to neglecting cross-terms ($i \neq j$) in (9), and, on applying the same arguments to equation (12), we find

$$F^I(q, \tau)\big|_{\text{large } q} = F_S^I(q, \tau) \equiv \langle \exp\{i\mathbf{q} \cdot [\mathbf{r}(0) - \mathbf{r}(\tau)]\}\rangle \tag{24}$$

† These considerations show the error of previous statements by one of the authors[21, 22] that there is no light scattering equivalent of incoherent neutron scattering.

‡ Using the Percus–Yevick approximation, Vrij[23] has derived expressions for the static structure factor $S^M(q)$, equation (18), for polydisperse hard spheres. This approach could probably be extended to the dynamic structure factor.

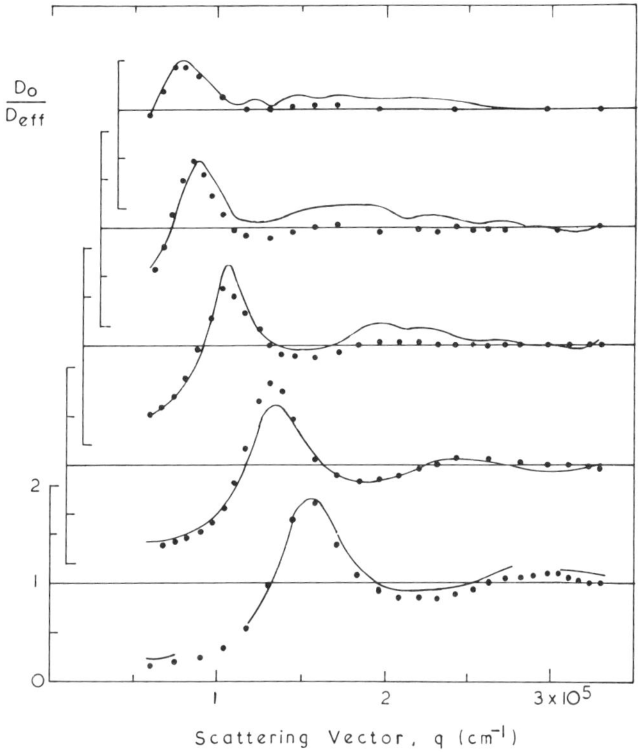

Figure 1. Structure factors $S(q)$ (solid lines) and relative reciprocal diffusion coefficients D_0/D_{eff} (data points) obtained from the first cumulant (Section 4.4.3) for polystyrene spheres of radius about 250 Å dispersed in water. Number concentration ranges from $\langle C \rangle \approx 8.5 \times 10^{12}$ cm^{-3} at bottom to $\langle C \rangle \approx 1.2 \times 10^{12}$ cm^{-3} at top. (Taken from Brown *et al.*[84])

where $F_S^I(q, \tau)$ is the self-dynamic structure factor. It should also be noted that there can be specific, smaller values of q where $S^I(q) = 1$ due to "accidental" cancellations in the sum of cross-terms (see, for example, Figure 1). At these values of q, since $F^I(q, 0) \equiv S^I(q)$, we clearly have $F^I(q, 0) = F_S^I(q, 0)$. Now if the sum of cross-terms in (12) is zero for $\tau = 0$ it seems unlikely that it will grow large at later times. Thus, as a general rule, we expect $F^I(q, \tau) \approx F_S^I(q, \tau)$ whenever $S^I(q) = 1$; however, exact equality only holds when q is sufficiently large. Under these conditions, therefore, DLS measurements provide information on self-diffusion (the motions of individual particles) which is relatively easy to visualize and model (for example, Section 4.4.2) even in strongly interacting systems.

In the low-q limit, where fluctuations of large spatial extent are studied, it is useful to write $F^I(q, \tau)$ in the form

$$F^I(q, \tau) = \langle a_{-\mathbf{q}}(0) a_{\mathbf{q}}(\tau) \rangle \tag{25}$$

where $a_{\mathbf{q}}$ is the spatial Fourier transform of the particle number density $C(\mathbf{r}, t)$

$$a_{\mathbf{q}}(t) = \frac{1}{\sqrt{N}} \int d^3r \exp(-i\mathbf{q} \cdot \mathbf{r}) C(\mathbf{r}, t) \tag{26}$$

$$C(\mathbf{r}, t) = \sum_{i=1}^{N} \delta[\mathbf{r} - \mathbf{r}_i(t)] \tag{27}$$

and δ is the Dirac delta function. Equation (25) shows explicitly the connection between light scattering and fluctuations in number density.

4.3. THEORY

4.3.1. Introduction

The dynamics of interacting Brownian particles, as they are manifest in light scattering experiments, can be analyzed in terms of a configuration space distribution function describing their motion over times long compared with τ_B. A second time scale defined by their motion is an interaction relaxation time τ_I, typical of the decay of correlations in the Brownian particle configuration. In the long-wavelength, long-time limit contact can be made between this microscopic description and the thermodynamic description of the system of particles in terms of a generalized Stokes–Einstein relation. Motions of the particles away from this limit but still over times much greater than τ_B are described by the generalized Smoluchowski equation (GSE) which explicitly incorporates the effects of direct interparticle interactions and the hydrodynamic coupling of their motions, treated in the low Reynolds number limit. The initial decay of the dynamic structure factor of the system is determined by motions taking place over times much less than τ_I and is described by the $t \to 0$ limit of the GSE (which corresponds to times still much greater than τ_B). To analyze the Brownian particle dynamics away from these limits a memory function equation for the dynamic structure factor can be constructed using the Mori projection operator technique which provides a useful starting point both for modeling $F(q, t)$ in a semiempirical fashion and for its calculation by more systematic approximations.

Although both full $F(q, \tau)$ and self $F_S(q, \tau)$ dynamic structure factors are discussed in this section, we will assume the particles to be monodisperse and will not use the superscript I introduced in Section 4.2.

4.3.2. Stokes–Einstein Relations

While a microscopic interpretation of light scattering by Brownian particles may give more insight into the interparticle interactions, their thermodynamic description in terms of generalized Stokes–Einstein relations has formed the basis of the interpretation of much experimental data. In this section we show how the microscopic description of the behavior of the dynamic structure factor is related to the macroscopic quantities of the thermodynamic approach.

For completeness we start by considering noninteracting particles for which the only important time scale is provided by the Brownian velocity fluctuation time τ_B, assumed to be much smaller than τ. Then, from equation (15), the dynamic structure factor can be written

$$F_S(q, \tau) = \langle \exp[-i\mathbf{q} \cdot \Delta\mathbf{r}(\tau)] \rangle \tag{28}$$

where

$$\Delta\mathbf{r}(\tau) \equiv \mathbf{r}(\tau) - \mathbf{r}(0) \equiv \int_0^\tau dt\ \mathbf{V}(t) \tag{29}$$

is the displacement of the particle in time τ and $\mathbf{V}$ the particle velocity. For $\tau \gg \tau_B$, $\Delta\mathbf{r}$ is a Gaussian random variable so that (28) becomes[24]

$$F_S(q, \tau) = \exp\left[-\frac{q^2}{6} \langle \Delta r^2(\tau) \rangle \right] \tag{30}$$

With use of (29) the mean-square displacement is given by

$$\langle \Delta r^2(\tau) \rangle = \int_0^\tau dt_1 \int_0^\tau dt_2\ \langle \mathbf{V}(t_1) \cdot \mathbf{V}(t_2) \rangle \tag{31}$$

$$\equiv 2 \int_0^\tau (\tau - t) \langle \mathbf{V}(0) \cdot \mathbf{V}(t) \rangle\ dt \tag{32}$$

For $\tau \gg \tau_B$ this becomes

$$\langle \Delta r^2(\tau) \rangle = 2\tau \int_0^\infty \langle \mathbf{V}(0) \cdot \mathbf{V}(t) \rangle \, dt \tag{33}$$

which can be compared with the well-known free diffusion result holding on the same time scale:

$$\langle \Delta r^2(\tau) \rangle = 6 D_0 \tau \tag{34}$$

Equation (30) can thus be written as

$$F_S(q, \tau) = \exp(-D_0 q^2 \tau) \tag{35}$$

where the free particle diffusion coefficient is given by

$$D_0 = \frac{1}{3} \int_0^\infty \langle \mathbf{V}(0) \cdot \mathbf{V}(t) \rangle \, dt \tag{36}$$

Comparison of (36) with the Stokes–Einstein equation

$$D_0 = k_B T / \zeta_0 \tag{37}$$

where k_B is the Boltzmann constant and T the absolute temperature, gives a microscopic expression for the "free-particle" friction coefficient

$$\zeta_0^{-1} = (3 k_B T)^{-1} \int_0^\infty \langle \mathbf{V}(0) \cdot \mathbf{V}(t) \rangle \, dt \tag{38}$$

Thus the dynamic structure factor of a set of independent Brownian particles decays exponentially at all times and scattering vectors attainable in a DLS experiment. Interparticle interactions bring about correlations in the positions and velocities of different particles which persist for times of the order τ_I. Consequently $F(q, \tau)$ no longer reduces to the simple form (35) and in general does not decay exponentially. However, exponential decay is still expected at sufficiently small scattering vector and at times much greater than τ_I.[(25)] In this limit, as discussed in Sections 4.1 and 4.2.4, DLS essentially studies fluctuations $a_{\mathbf{q}}(t)$ of particle concentration of large spatial extent. Such small fluctuations[(8)] are expected to decay by a linear diffusive process so that

$$a_{\mathbf{q}}(\tau) = a_{\mathbf{q}}(0) \exp(-D_C q^2 \tau) \tag{39}$$

and

$$F(q, \tau)|_{q \to 0, \tau \gg \tau_I} = S(0)\exp(-D_C q^2 \tau) \tag{40}$$

where $S(q) \equiv F(q, 0)$ is the static structure factor and, through the Onsager regression hypothesis, D_C can be identified as the same collective diffusion coefficient as would be measured in conventional gradient diffusion experiments.

D_C can be related to microscopic properties of the system by the following argument. Double differentation of (12) and use of the stationarity condition give, as a well-known exact expression for a classical system of interacting particles (see Reference 19, p. 229):

$$\frac{d^2}{d\tau^2} F(q, \tau) = -\frac{q^2}{3N} \sum_i \sum_j \langle \mathbf{V}_i(0) \cdot \mathbf{V}_j(\tau) \exp\{i\mathbf{q} \cdot [\mathbf{r}_i(0) - \mathbf{r}_j(\tau)]\}\rangle \tag{41}$$

A single integration now gives

$$\lim_{q \to 0} \frac{d}{d\tau} F(q, \tau)|_{\tau \gg \tau_I} = -\frac{q^2}{3N} \sum_i \sum_j \int_0^\infty \langle \mathbf{V}_i(0) \cdot \mathbf{V}_j(t)\rangle \, dt \tag{42}$$

Comparison of (40) and (42) allows us to identify D_C as

$$D_C = [3NS(0)]^{-1} \sum_i \sum_j \int_0^\infty \langle \mathbf{V}_i(0) \cdot \mathbf{V}_j(t)\rangle \, dt \tag{43}$$

This result was first obtained by Altenberger[26] by more formal arguments.

To make contact with the macroscopic properties of the system we first note that the static structure factor in the $q \to 0$ limit is related to the isothermal osmotic compressibility through[27]

$$S(0) = k_B T \left(\frac{\partial C}{\partial \Pi}\right)_T \tag{44}$$

where Π is the osmotic pressure. Particle motions can be characterized by a coarse-grained velocity $\mathbf{v}_i$, the effects of Brownian motion being averaged over the long times [$\tau \gg \tau_B$, $\tau \gg \tau_H$; see equations (1) and (3)] characteristic of the thermodynamic description. The velocity of particle i in this coarse-grained description can be related to the systematic forces $\mathbf{F}_i$ acting on it and the other particles in the suspension through the mobility tensor $\mathbf{b}_{ij}$ [16]

$$\mathbf{v}_i = \sum_j \mathbf{b}_{ij} \cdot \mathbf{F}_j \tag{45}$$

A linear response treatment of the many-particle Fokker–Planck equation relates the mobility tensor to the correlation function of the fluctuating particle velocity $\mathbf{V}_i$, $\mathbf{V}_j$ [11]

$$\mathbf{b}_{ij} = (k_B T)^{-1} \int_0^\infty \langle \mathbf{V}_i(0)\mathbf{V}_j(t) \rangle \, dt \tag{46}$$

Thus the average drift velocity resulting from the application of the same force $\mathbf{F}$ to each particle (as would occur in the sedimentation of identical particles under gravity) is given by

$$\bar{\mathbf{v}} = \mathbf{F}/\zeta \tag{47}$$

where we identify an inverse friction or drag coefficient

$$\zeta^{-1} = (3Nk_B T)^{-1} \sum_i \sum_j \int_0^\infty \langle \mathbf{V}_i(0) \cdot \mathbf{V}_j(t) \rangle \, dt \tag{48}$$

By combining (43), (44), and (48) the collective diffusion coefficient can now be shown to satisfy the generalized Stokes–Einstein relation:

$$D_c = \frac{(\partial \Pi / \partial C)_T}{\zeta} \tag{49}$$

Equations (43), (48), and (49) can be compared with (36), (38), and (37), to which they clearly reduce in the absence of interactions (where $\partial \Pi / \partial C = k_B T$).

While it is useful in establishing the connection between microscopic and macroscopic quantities, the above derivation of equation (49) provides little direct insight into the physical processes involved. The generalized Stokes–Einstein equation has been more conventionally derived by non-equilibrium thermodynamics[24] (see Section 4.6), where $\partial \Pi / \partial C$ appears as a "thermodynamic driving force" and ζ as a phenomenological friction coefficient. A semimicroscopic derivation of less general validity has also been given by Phillies.[29]

It is seen from equation (49) that the low-q, long time collective diffusion coefficient D_C is determined by competition between the thermodynamic factor and the hydrodynamic drag. In dilute suspensions of particles with long-ranged repulsive direct interactions (Sections 4.4.4 and 4.5.3), the former factor outweighs the latter and D_C can be many times greater than the free-particle value D_0. However, it appears that in more concentrated suspensions of particles with short-ranged (hard-sphere) interactions (Section 4.5.3) the effects of direct and hydrodynamic interactions more or less cancel and D_C is only weakly dependent on concentration.

The arguments taking us from the dynamic structure factor to the generalized Stokes–Einstein relation are at once too specialized, as they apply only in the $t \to \infty$, $q \to 0$ limit, and too detailed, in that the particle mobilities are given only by the microscopic expressions (46), to be of much practical use. The first of these constraints can be removed by developing an equation of motion for the particles' configuration distribution function valid at times greater only than τ_B, applicable away from the $q = 0$ limit and incorporating direct interparticle interactions. To overcome the second difficulty the particle mobilities can be found by assuming the solvent to be a continuum; the couplings between the particles' motions can now be found from hydrodynamic arguments. In the next two sections we will discuss this equation of motion of the distribution function, the so-called generalized Smoluchowski equation, and the hydrodynamic calculation of the particle mobilities.

4.3.3. The Generalized Smoluchowski Equation

To describe the motions of interacting Brownian particles over times measured by DLS it is sufficient to consider their configuration space distribution function $P(\mathbf{r}^N\ t)$, where $\mathbf{r}^N$ represents the set of position vectors of the N particles, whose relaxation to its equilibrium form can be discussed in terms of a simple thermodynamic picture, introduced by Einstein in his study of free particle Brownian motion[30] and recently generalized to interacting systems by Zwanzig,[31] Batchelor,[16] and Wills.[28] A nonequilibrium distribution $P(\mathbf{r}^N)$ is prevented from relaxing by the application to each particle of a force $\mathbf{F}_{dj}(\mathbf{r}^N)$ derived from a potential $\Psi(\mathbf{r}^N)$:

$$\mathbf{F}_{dj} = -\nabla_j \Psi(\mathbf{r}^N) \tag{50}$$

In this hypothetical equilibrium situation $P(\mathbf{r}^N)$ is given by the Boltzmann expression

$$P(\mathbf{r}^N) = \text{const} \times \exp\left\{ -\frac{1}{k_B T} [\Psi(\mathbf{r}^N) + U(\mathbf{r}^N)] \right\} \tag{51}$$

interparticle interactions and any external applied forces being represented by the potential $U(\mathbf{r}^N)$. Consequently $\mathbf{F}_{dj}$ is given by

$$\mathbf{F}_{dj} = \nabla_j U(\mathbf{r}^N) + k_B T \frac{\nabla_j P(\mathbf{r}^N)}{P(\mathbf{r}^N)} \tag{52}$$

If $P(\mathbf{r}^N)$ is now allowed to relax (the forces $\mathbf{F}_{dj}$ are "switched off") a force $-\mathbf{F}_{dj}$ will be driving the diffusive motion of the jth particle. We assume that the coarse-grained description of particle motion introduced in Section 4.3.2 is adequate on the timescale $\tau_H \ll \tau \ll \tau_I$ and write

$$\mathbf{v}_i = -\sum_j \mathbf{b}_{ij} \cdot \mathbf{F}_{dj} \tag{53}$$

[cf. (45)]. Following Zwanzig[31] we see that the conservation law for Brownian particles now leads to a generalized diffusion or Smoluchowski equation

$$\begin{aligned}\frac{\partial}{\partial t} P(\mathbf{r}^N, t) &= -\sum_i \nabla_i \cdot [\mathbf{v}_i P(\mathbf{r}^N, t)] \\ &= O[P(\mathbf{r}^N, t)]\end{aligned} \tag{54}$$

and

$$O[A(\mathbf{r}^N, t)] = \sum_{i,j} \nabla_i \cdot \mathbf{D}_{ij} \cdot \exp\left[-\frac{U(\mathbf{r}^N)}{k_B T}\right] \nabla_j \left\{\exp\left[\frac{U(\mathbf{r}^N)}{k_B T}\right] A(\mathbf{r}^N, t)\right\} \tag{55}$$

and the diffusion and mobility tensors are related through the Einstein relation

$$\mathbf{D}_{ij} = k_B T \mathbf{b}_{ij} \tag{56}$$

The generalized Smoluchowski equation (54) has the fundamental solution $P(\mathbf{r}^N, \mathbf{r}_0^N, t)$, the conditional probability that the system of particles wil! adopt the configuration $\mathbf{r}^N$ at time t, given that its configuration at time zero was $\mathbf{r}_0^N$. As

$$\begin{aligned}P(\mathbf{r}^N, \mathbf{r}_0^N, 0) &= \prod_i \delta(\mathbf{r}_i - \mathbf{r}_{0i}) \\ &= \delta(\mathbf{r}^N - \mathbf{r}_0^N)\end{aligned} \tag{57}$$

the time evolution of the system of particles can be described by the propagator[32]

$$P(\mathbf{r}^N, \mathbf{r}_0^N, t) = e^{Ot}\delta(\mathbf{r}^N - \mathbf{r}_0^N) \tag{58}$$

obtained by the formal solution of (54).

In the case of independent particles, $U(r^N) = 0$,

$$\mathbf{b}_{ij} = \mathbf{1}\,\frac{\delta_{ij}}{\zeta_0} \tag{59}$$

where $\mathbf{1}$ is the unit dyadic, and (54) reduces to the simple diffusion equation

$$\frac{\partial P(\mathbf{r}, t)}{\partial t} = D_0 \nabla^2 P(\mathbf{r}, t) \tag{60}$$

D_0 satisfying the Einstein relation (37). The effect of an external force on single-particle Brownian motion is described by the forced diffusion or Smoluchowski equation[33]

$$\frac{\partial P(\mathbf{r}, t)}{\partial t} = D_0 \nabla \cdot \left(\nabla P + P \nabla \frac{U}{k_B T} \right) \tag{61}$$

The generalized Smoluchowski equation provides a description of particle motion which, while adequate for a discussion of DLS, is nonetheless incomplete. The N-particle Langevin and Fokker–Planck equations can describe the relaxation of the Brownian particle velocity distribution over times less than τ_B while the evolution of the system of Brownian and solvent particles is governed by its Liouville equation. By using projection operator methods previously applied to single-particle Brownian motion by Mazur and Oppenheim,[34, 35] Deutch and Oppenheim[36] obtained the N-particle Langevin and Fokker–Planck equations from the more detailed Liouville equation, showing them to be valid provided that the Brownian particles are much heavier than the solvent particles and that fluctuations in Brownian particle density decay much more slowly than fluctuations of similar wavelength in the solvent particle density. The subsequent reduction of the N-particle Fokker–Planck equation to the generalized Smoluchowski equation has been discussed by Aguirre and Murphy,[37] Wilemski,[38] and Hess and Klein,[39] who identify the mobility tensor $\mathbf{b}_{ij}$ as the inverse of a friction tensor given in terms of the correlation functions of rapidly varying solvent forces acting on the particles. Altenberger[40] has also reduced the Liouville equation directly to a generalized Smoluchowski form, obtaining the mobility tensors in terms of velocity correlation functions [c.f. equation (46)].

4.3.4. Hydrodynamic Interactions

So far the mobility tensors $\mathbf{b}_{ij}$ have been identified only in terms of particle velocity correlation functions. We will now consider the motions of the solvent, which is treated as a continuum, induced by the coarse-grained motion of a single particle and obtain convenient expressions for $\mathbf{b}_{ij}$ in terms of particles' radii and separation and the solvent shear viscosity η, and valid for times $\tau \gg \tau_B, \tau_H$.

The motion of a Brownian particle induces a flow in the solvent, characterized by the pressure and velocity fields $p(\mathbf{R}, t)$, $\mathbf{u}(\mathbf{R}, t)$ where $\mathbf{R}$ is the position vector of some point in the fluid. In macroscopic hydrodynamics[5] the conservation laws for mass and momentum in an incompressible fluid are, expressed in differential form,

$$\nabla \cdot \mathbf{u} = 0 \tag{62}$$

$$\rho \frac{\partial \mathbf{u}}{\partial t} + \rho(\mathbf{u} \cdot \nabla)\mathbf{u} - \eta \nabla^2 \mathbf{u} + \nabla p = \mathbf{F} \tag{63}$$

Here ρ is the fluid density and $\mathbf{F}$ a body force. The ratio of the inertial force $\rho(\mathbf{u} \cdot \nabla)\mathbf{u}$ to the viscous force $-\eta\nabla^2\mathbf{u}$, the so-called Reynolds number of the flow, is given by

$$R_e = \frac{\rho|\mathbf{u}|^2}{l} \bigg/ \frac{\eta|\mathbf{u}|}{l^2} = \frac{\rho|\mathbf{u}|l}{\eta} \tag{64}$$

where l is a characteristic length (e.g., a typical dimension of the Brownian particle). In the case of steady slow motion of a macroscopically small particle through a viscous fluid the viscous forces are much greater than the inertial forces and the fluid motion is described by the steady-state, linearized Navier–Stokes equations

$$\nabla \cdot \mathbf{u} = 0$$

$$-\eta\nabla^2\mathbf{u} + \nabla p = \mathbf{F} \tag{65}$$

The solution of these equations has been the subject of much study[41]; for our immediate needs (the determination of the drag on the sphere and the pressure and velocity fields in the fluid) we refer to the elegant treatment of Landau and Lifshitz.[5] For a sphere of radius a at rest at the origin of a coordinate system in which the surrounding fluid has a velocity $\mathbf{u}_0$ and pressure p_0 at infinity

$$\mathbf{u}(\mathbf{R}) = \mathbf{u}_0 - \frac{3}{4}\frac{a}{R}\left[\mathbf{u}_0 + \frac{\mathbf{R}(\mathbf{u}_0 \cdot \mathbf{R})}{R^2}\right] - \frac{1}{4}\frac{a^3}{R^3}\left[\mathbf{u}_0 - 3\,\frac{\mathbf{R}(\mathbf{u}_0 \cdot \mathbf{R})}{R^2}\right] \tag{66}$$

and

$$p(\mathbf{R}) = p_0 - \frac{3}{2}\eta a \frac{\mathbf{u}_0 \cdot \mathbf{R}}{R^3} \tag{67}$$

where stick boundary conditions have been assumed. The viscous drag on the sphere is given by the Stokes result

$$\mathbf{F}_{\text{drag}} = 6\pi\eta a \mathbf{u}_0 \tag{68}$$

from which we identify the friction coefficient ζ_0 in equation (2). This simple treatment can be extended to give the force and torque $\mathbf{\Gamma}$ acting on a sphere moving with a velocity $\mathbf{v}$ (the coarse-grained velocity of Section 4.3.2) and angular velocity $\boldsymbol{\omega}$ at position $\mathbf{r}$ in a nonuniform flow, whose velocity field in the absence of the sphere is $\mathbf{u}_0(\mathbf{R})$. These important and useful results are known as Faxen's theorems and for stick boundary conditions take the form (see Reference 41, p. 67–78)

$$\mathbf{F}_{\text{drag}} = 6\pi\eta a[\mathbf{u}_0(\mathbf{r}) - \mathbf{v}] + a^3\pi\eta\nabla^2\boldsymbol{u}_0(\mathbf{r}) \tag{69}$$

and

$$\mathbf{\Gamma} = 8\pi\eta a^3[\tfrac{1}{2}\nabla_\wedge \mathbf{u}_0(\mathbf{r}) - \boldsymbol{\omega}] \tag{70}$$

the latter vanishing in the Stokes limit uniform velocity flow.

The Stokes law friction coefficient has also been obtained from the fluctuation–dissipation result[42]

$$\zeta_0 = \frac{1}{3k_B T}\int_0^\infty \langle \mathbf{f}(0) \cdot \mathbf{f}(t)\rangle\, dt \tag{71}$$

Equation (69) is used to evaluate the random force $\mathbf{f}$ due to the solvent in terms of the velocity fluctuations in the unperturbed fluid, their correlation function being evaluated using the Landau–Lifshitz theory of fluctuating hydrodynamics.[43]

If a force $\mathbf{F}$ is applied to a sphere at position $\mathbf{r}_i$ in a fluid previously at rest we see from (66) and (68) that the velocity field induced in the fluid is

$$\mathbf{u}(\mathbf{R}) = \left[\frac{3}{4}\frac{a}{R_i}\left(\mathbf{1} + \frac{\mathbf{R}_i\mathbf{R}_i}{R_i^2}\right) + \frac{1}{4}\frac{a^3}{R_i^3}\left(\mathbf{1} - 3\frac{\mathbf{R}_i\mathbf{R}_i}{R_i^2}\right)\right] \cdot \frac{\mathbf{F}}{6\pi\eta a} \tag{72}$$

where $\mathbf{R}_i = \mathbf{R} - \mathbf{r}_i$. Thus in the limit $(a \to 0)$ of a body force $\mathbf{F}$ acting at a point $\mathbf{R}'$ the velocity field induced at $\mathbf{R}$ is given by

$$\mathbf{u}(\mathbf{R}) = \mathsf{T}(\mathbf{R} - \mathbf{R}') \cdot \mathbf{F}(\mathbf{R}') \tag{73}$$

where $\mathsf{T}(\mathbf{R} - \mathbf{R}')$ is the Oseen tensor[31]

$$\mathsf{T}(\mathbf{R}) = \frac{1}{8\pi\eta R}\left(\mathbf{1} + \frac{\mathbf{RR}}{R^2}\right) \tag{74}$$

The corresponding pressure field is

$$p(\mathbf{R}) = p_0 + \mathbf{Q}(\mathbf{R} - \mathbf{R}') \cdot \mathbf{F}(\mathbf{R}') \tag{75}$$

where

$$\mathbf{Q}(\mathbf{R}) = \frac{1}{4\pi} \frac{\mathbf{R}}{R^3} \tag{76}$$

The Oseen tensor has been introduced here on the basis of a simple physical picture. In more formal terms it can be thought of as the Green's dyadic of equations (65) and is discussed as such by Oseen[44] and Happel and Brenner.[41] Felderhof and Deutch[45] also give a *post hoc* demonstration of the Green's dyadic property, showing that

$$\mathbf{u}(\mathbf{R}) = \int d^3R' \; \mathsf{T}(\mathbf{R} - \mathbf{R}') \cdot \mathbf{F}(\mathbf{R}')$$

and

$$p(\mathbf{R}) = p_0 + \int d^3R' \; \mathbf{Q}(\mathbf{R} - \mathbf{R}') \cdot \mathbf{F}(\mathbf{R}')$$

satisfy (65).

The fluid motions causing the hydrodynamic interaction between two steadily moving spheres have been studied in great detail. Exact solutions of equations (65) were obtained by Stimson and Jeffrey[46] for the special case of two spheres with stick boundary conditions moving with equal velocity along their line of centers in an unbounded fluid. Subsequent generalizations of this work (see References 15 and 17 for references and discussion) give an essentially complete description of the translation and rotation of two spheres with stick boundary conditions, subject to known forces and torques. These solutions are, however, extremely complex and much numerical work is needed in reaching explicit results, which makes them unsuitable for use in the generalized Smoluchowski equation. Approximate solutions, giving expansions of $\mathbf{b}_{ij}$ and its inverse, the friction tensor ζ_{ij}, in powers of ξ, the ratio of sphere radius to particle separation, can be obtained by the method of reflections (Reference 41, Chapter 6) and are better suited to use in (54). The work of Brenner[47] and Aguirre and Murphy[48] gave ζ_{ij} to order ξ^4; $\mathbf{b}_{ij}$ was then found by inversion. The Faxen theorems (69) and (70) allow $\mathbf{b}_{ij}$ to be calculated directly; this has been done for freely rotating spheres of different radii with mixed slip-stick boundary conditions by Felderhof.[49] Similar calculations for permeable spheres, through which the fluid motion is described by the Debye–Bueche equations, have also been carried out, the evaluation of the friction tensor and its subsequent inversion[50] again proving to be a more arduous route to the mobility tensor than its direct evaluation using the appropriate generalization of Faxen's theorems.[51, 52]

For our purposes it is sufficient to consider $\mathbf{b}_{ij}$ for two identical freely rotating spheres i, j with stick boundary conditions whose velocities $\mathbf{v}_i$, $\mathbf{v}_j$ are related to the forces $\mathbf{F}_i$, $\mathbf{F}_j$ acting on them through

$$\begin{pmatrix} \mathbf{v}_i \\ \mathbf{v}_j \end{pmatrix} = \begin{pmatrix} \mathbf{b}_{ii}(\mathbf{r}_i, \mathbf{r}_j) & \mathbf{b}_{ij}(\mathbf{r}_i, \mathbf{r}_j) \\ \mathbf{b}_{ji}(\mathbf{r}_i, \mathbf{r}_j) & \mathbf{b}_{jj}(\mathbf{r}_i, \mathbf{r}_j) \end{pmatrix} \begin{pmatrix} \mathbf{F}_i \\ \mathbf{F}_j \end{pmatrix} \tag{77}$$

$\mathbf{b}_{ii}$ represents the mobility of i modified by the presence of j and tends to the free particle value $\mathbf{1}/6\pi\eta a$ as $|\mathbf{r}_i - \mathbf{r}_j| \to \infty$, while $\mathbf{b}_{ij}$ determines the velocity imparted to i by a force acting on j. If now a force $\mathbf{F}_i$ acts on i, while no force acts on j, $\mathbf{u}_0(R)$, the velocity field due to i's resulting motion is given by (72). As no force or torque acts on j it must now move so as to exert no force or torque on the fluid. From (69) we see that, to first order, the velocity of j is

$$\mathbf{v}_{j1} = \mathbf{u}_{i0}(\mathbf{r}_j) + \tfrac{1}{6}a^2\nabla^2\mathbf{u}_{i0}(\mathbf{r}_j)$$

allowing us to identify the lowest-order, Oseen, contribution to $\mathbf{b}_{ij}$

$$\mathbf{b}_{ij}(\mathbf{r}_i, \mathbf{r}_j) = \frac{1}{8\pi\eta r_{ij}}\left(\mathbf{1} + \frac{\mathbf{r}_{ij}\mathbf{r}_{ij}}{r_{ij}^2}\right) + O\left(\frac{1}{r_{ij}^3}\right) \tag{78}$$

where $\mathbf{r}_{ij} = \mathbf{r}_i - \mathbf{r}_j$. Higher-order corrections to $\mathbf{b}_{ii}$, $\mathbf{b}_{ij}$ can be found by considering the velocity field $\mathbf{v}_{j1}(\mathbf{R})$ created by the force density induced on j by $\mathbf{v}_{i0}(\mathbf{R})$. This is turn acts on i, producing an increment in its velocity, again found from Faxen's theorem and the requirement that $\mathbf{v}_{j1}(\mathbf{R})$ exerts no force or torque on i. An interesting graphical representation of these complicated calculations has recently been given by Ball and Richmond.[53] By carrying through this procedure in detail Felderhof obtains the mobility tensors as the rapidly converging series[17]

$$\mathbf{b}_{ii}(\mathbf{r}_i, \mathbf{r}_j) = \frac{1}{6\pi\eta a}\left[\mathbf{1} - \frac{15}{4}\frac{a^4}{r_{ij}^4}\mathbf{P} - \frac{1}{16}\frac{a^6}{r_{ij}^6}(17\mathbf{I} - 105\mathbf{P}) + O\left(\frac{a^8}{r_{ij}^8}\right)\right] \tag{79}$$

and

$$\mathbf{b}_{ij}(\mathbf{r}_i, \mathbf{r}_j) = \frac{1}{6\pi\eta a}\left[\frac{3}{4}\frac{a}{r_{ij}}(\mathbf{1} + \mathbf{P}) + \frac{1}{2}\frac{a^3}{r_{ij}^3}(\mathbf{1} - 3\mathbf{P}) + \frac{75}{4}\frac{a^7}{r_{ij}^7}\mathbf{P} + O\left(\frac{a^8}{r_{ij}^8}\right)\right]$$

where $\mathbf{P} = \mathbf{r}_{ij}\mathbf{r}_{ij}/r_{ij}^2$. In many applications just the lowest-order (Oseen) terms in the expansion are retained:

$$b_{\alpha\beta}(\mathbf{r}_i, \mathbf{r}_j) = \frac{1}{6\pi\eta a}\left[\delta_{\alpha\beta}\mathbf{1} + (1 - \delta_{\alpha\beta})\frac{3}{4}\frac{a}{r_{ij}}(\mathbf{1} + \mathbf{P})\right] \tag{79a}$$

This mobility tensor is traceless in the sense that $\nabla_i \cdot \mathbf{b}_{ij} = 0$, a property not shared by the more complete results (79).

A microscopic calculation of the two-particle friction tensor similar to Zwanzig's[43] treatment of Stokes friction has been carried out by Deutch and Oppenheim,[36] who obtain a result which agrees with that of the hydrodynamic calculation to lowest order in (a/r_{ij}).

The two-particle mobility tensors (79) can now be used in (54), taking hydrodynamic couplings to be pairwise additive. This assumption should be valid in dilute suspensions; the intractability of the three-body hydrodynamic problem forces us to hope that it holds in more concentrated suspensions as well.

4.3.5. Short-Time Motions

The initial decay of the dynamic structure factor is determined by particle motions taking place over times much less than τ_I and is most conveniently analyzed in terms of the cumulant expansion

$$\ln\left[\frac{F(q, \tau)}{S(q)}\right] = \sum_n K_n \frac{(-\tau)^n}{n!} \tag{80a}$$

Thus

$$\begin{aligned} K_1 &= -\lim_{\tau\to 0}\frac{d}{d\tau}\ln\left[\frac{F(q, \tau)}{S(q)}\right] \\ &= -\lim_{\tau\to 0}\frac{1}{S(q)}\frac{dF(q, \tau)}{d\tau} \end{aligned} \tag{80b}$$

and

$$K_2 = \lim_{\tau \to 0} \frac{d^2}{d\tau^2} \ln\left[\frac{F(q, \tau)}{S(q)}\right]$$

$$= \lim_{\tau \to 0} \left\{\frac{1}{S(q)} \frac{d^2 F(q, \tau)}{d\tau^2} - \frac{1}{[S(q)]^2}\left[\frac{dF(q, \tau)}{d\tau}\right]^2\right\} \quad (80c)$$

It should be emphasized that, in these definitions, $\tau \to 0$ implies the short-time limit of the Smoluchowski equation, which is still, in fact, at $\tau \gg \tau_B$. The first cumulant K_1 defines an effective diffusion coefficient governing the initial decay of $F(\mathbf{q}, \tau)$

$$D_{\text{eff}} = K_1/q^2 \quad (81)$$

while the higher cumulants characterize its subsequent departure from exponential behavior. For a system of noninteracting particles the cumulants are simply [equations (35) and (80)]

$$K_1 = D_0 q^2, \qquad K_n = 0, \qquad n \geq 2$$

As a result of interparticle interactions the decay of $F(q, \tau)$ is, in general, nonexponential and the calculation of the lower-order cumulants is more difficult. By exploiting the separation of time scales in the particle motion it is possible, if hydrodynamic interactions are neglected, to calculate the first cumulant directly from (41).[54] A single integration gives, as an identity

$$\frac{dF(q, \tau)}{d\tau} = -\frac{q^2}{3N} \sum_i \sum_j \int_0^\tau dt \langle \mathbf{V}_i(0) \cdot \mathbf{V}_j(t) \exp\{i\mathbf{q} \cdot [\mathbf{r}_i(0) - \mathbf{r}_j(t)]\}\rangle \quad (82a)$$

In the limit $\tau_B \ll \tau \ll \tau_I$, where particle j has not moved significantly from its position at time $t = 0$, (82a) becomes

$$\left.\frac{dF(q, \tau)}{d\tau}\right|_{\tau_B \ll \tau \ll \tau_I} = -\frac{q^2}{3N} \sum_i \sum_j \int_0^\tau dt \langle \mathbf{V}_i(0) \cdot \mathbf{V}_j(t) \exp\{i\mathbf{q} \cdot [\mathbf{r}_i(0) - \mathbf{r}_j(0)]\}\rangle \quad (82b)$$

The velocity of each particle is now decomposed into a rapidly varying Brownian and a slowly varying interaction component (see Section 4.1)

$$\mathbf{V}_i = \mathbf{V}_{Bi} + \mathbf{V}_{Ii} \quad (83)$$

where

$$\mathbf{V}_{Ii} \equiv -\nabla_i U(\mathbf{r}^N)/\zeta_0 \tag{84}$$

and $\mathbf{V}_{Ii}$ represents the "drift velocity" of particle i in the force field created by the other particles.[55] As $\langle \mathbf{V}_{Bi}^2 \rangle \gg \langle V_{Ii}^2 \rangle$ (Sections 4.1 and 4.4.2), terms containing $\mathbf{V}_{Ii}$ may be dropped in the short-time limit of equation (82b). Finally, as $\mathbf{V}_{Bi}$ is determined by local particle–solvent interactions, we assume that (in the absence of hydrodynamic interactions)

$$\langle \mathbf{V}_{Bi}(0) \cdot \mathbf{V}_{Bj}(\tau) \rangle = \delta_{ij} \langle \mathbf{V}_{Bi}(0) \cdot \mathbf{V}_{Bi}(\tau) \rangle \tag{85}$$

whereupon (82b) reduces to

$$\left. \frac{dF}{d\tau}(q, \tau) \right|_{\tau_B \ll \tau \ll \tau_I} = -D_0 q^2 \tag{86}$$

the free-particle diffusion coefficient being identified through (36). Consequently the value of the first cumulant is

$$K_1 = D_0 q^2 / S(q) \tag{87}$$

We note, as an aside, that in the original derivation[54] of (87), it was assumed that $\mathbf{V}_{Bi}$ and $\mathbf{V}_{Ii}$ were not correlated. This is not correct since a random motion of particle i (through V_{Bi}) to a position of higher interparticle force will lead to a larger drift velocity V_{Ii} of the particle than a motion to a position of lower force. However, this incorrect assumption does not affect the validity of equation (87) since terms containing $\mathbf{V}_I$ were neglected in its derivation.

A similar calculation gives the first cumulant of the self-dynamic structure factor, defined by

$$\ln F_S(q, \tau) = \sum_n \frac{K_{Sn}(-\tau)^n}{n!} \tag{88}$$

as

$$K_{S1} = q^2 D_0 \tag{89}$$

A more systematic procedure for the calculation of cumulants in which the effects of hydrodynamic interactions are included has been developed by Ackerson.[12,13] The dynamic structure factor is written in the form [cf. (58)]

$$F(q, \tau) = \frac{1}{NZ} \sum_{i,j} \int d\mathbf{r}^N \, d\mathbf{r}_0^N \exp[i\mathbf{q} \cdot (\mathbf{r}_{j0} - \mathbf{r}_i)] \exp[-U(\mathbf{r}_0^N)/k_B T] e^{O\tau} \delta(\mathbf{r}^N - \mathbf{r}_0^N) \tag{90}$$

where

$$Z = \int d\mathbf{r}^N \exp[-U(\mathbf{r}^N)/k_B T]$$

Derivatives of $F(q, \tau)$ evaluated in the $\tau = 0$ limit corresponding to times $\tau_B \ll \tau \ll \tau_I$ can now be written as

$$\left.\frac{\partial^n}{\partial \tau^n} F(q, \tau)\right|_{\tau_B \ll \tau \ll \tau_I} = \frac{1}{NZ} \sum_{i,j} \int d\mathbf{r}^N \, d\mathbf{r}_0^N \exp[i\mathbf{q} \cdot (\mathbf{r}_{j0} - \mathbf{r}_i)] \times \exp[-U(\mathbf{r}_0^N)/k_B T] 0^n \delta(\mathbf{r}^N - \mathbf{r}_0^N) \quad (91)$$

which can be obtained in a more explicit form by repeated integrations by parts and a final integration over the delta function. This procedure, while it is straightforward in principle, is so complicated in practice (see, for example, the explicit calculations of Ackerson[12] and Allison *et al.*[56]) that as yet just the first two cumulants have been calculated, the second only in the limit in which hydrodynamic interactions are neglected. The first cumulant incorporating the effects of hydrodynamic interactions has been evaluated by Ackerson[12, 13] in the Oseen approximation (79a) and by Allison *et ai.*,[56] who also did not make use of the explicit expressions for the mobility tensors published subsequently by Felderhof.[17] By following Ackerson's procedure and incorporating Felderhof's result (79) we find that[57]

$$D_{\text{eff}} = \frac{K_1}{q^2} = \frac{D_0}{S(q)} [1 + \langle C \rangle (\delta_0 + \delta_D + \delta_A + \delta_S)] \quad (92a)$$

where $\langle C \rangle$ is the mean particle concentration

$$\delta_0 = 6\pi a \int_0^\infty dr \, r[g(r) - 1] \left[\frac{\sin qr}{qr} + \frac{\cos qr}{(qr)^2} - \frac{\sin qr}{(qr)^3} \right] \quad (92b)$$

$$\delta_D = 4\pi a^3 \int_0^\infty \frac{dr}{r} [g(r) - 1] j_2(qr) + \frac{4\pi a^3}{3} \quad (92c)$$

where j_2 is a spherical Bessel function and a the particle radius,

$$\delta_A = 4\pi \int_0^\infty dr \, r^2 g(r) \left(\frac{9}{8} \frac{a^6}{r^6} - \frac{5}{4} \frac{a^4}{r^4} \right) \quad (92d)$$

and

$$\delta_S = 75\pi a^7 \int_0^\infty \frac{dr}{r^5} g(r) \left[\frac{\sin qr}{qr} + \frac{2 \cos qr}{(qr)^2} - \frac{2 \sin qr}{(qr)^3} \right] \quad (92e)$$

Ackerson's result being the special case

$$\frac{K_1}{q^2} = \frac{D_0}{S(q)}\,(1 + \langle C \rangle \delta_0) \tag{93}$$

A similar calculation gives the first cumulant of the self-dynamic structure factor as[57]

$$\frac{K_{S1}}{q^2} = D_0(1 + \langle C \rangle \delta_A) \tag{94}$$

this being the one term in (92) which does not vanish in the high-q limit. The second derivatives of $F(q, \tau)$ and $F_S(q, \tau)$, neglecting the effects of hydrodynamic interactions, are given by[12, 13]

$$\left.\frac{d^2 F(q, \tau)}{d\tau^2}\right|_{\tau_B \ll \tau \ll \tau_I} = D_0^2\Bigg(q^4 + \frac{1}{Nk_B T} \times \sum_{i,j} \langle \exp[i\mathbf{q}\cdot(\mathbf{r}_i - \mathbf{r}_j)](\mathbf{q}\cdot\nabla_i)(\mathbf{q}\cdot\nabla_j)U(\mathbf{r}^N)\rangle\Bigg) \tag{95a}$$

and

$$\left.\frac{d^2 F_S(q, \tau)}{d\tau^2}\right|_{\tau_B \ll \tau \ll \tau_I} = D_0^2\left\{q^4 + \frac{1}{(k_B T)^2}\langle[\mathbf{q}\cdot\nabla_1 U(\mathbf{r}^N)]^2\rangle\right\} \tag{95b}$$

which, if the interparticle potential can be written as a sum of pair potentials

$$U(\mathbf{r}^N) = \sum_{i<j} u(|\mathbf{r}_i - \mathbf{r}_j|) \tag{96}$$

reduces to

$$\left.\frac{d^2 F(q, \tau)}{d\tau^2}\right|_{\tau_B \ll \tau \ll \tau_I} = D_0^2 q^4\left\{1 + \frac{\langle C \rangle}{(k_B T)q^4}\int d\mathbf{r}\; g(r)[1 - \cos(\mathbf{q}\cdot\mathbf{r})](\mathbf{q}\cdot\nabla)^2 u(r)\right\} \tag{97a}$$

and

$$\left.\frac{d^2 F_S(q, \tau)}{d\tau^2}\right|_{\tau_B \ll \tau \ll \tau_I} = D_0^2 q^4\left[1 + \frac{\langle C \rangle}{(k_B T)q^4}\int d\mathbf{r}\; g(r)(\mathbf{q}\cdot\nabla)^2 u(r)\right] \tag{97b}$$

It is interesting to compare the results (86), (95) with those obtained by de Gennes[58] for a simple fluid, the motion of whose constituent particles is described by Liouville's rather than Smoluchowski's equation. Explicitly, he obtained

$$\left.\frac{dF(q, \tau)}{d\tau}\right|_{\tau=0} = 0 \tag{98a}$$

$$\left.\frac{d^2F(q, \tau)}{d\tau^2}\right|_{\tau=0} = \frac{k_B T}{M} q^2 \tag{98b}$$

$$\left.\frac{d^3F(q, \tau)}{d\tau^3}\right|_{\tau=0} = 0 \tag{98c}$$

$$\left.\frac{d^4F(q, \tau)}{d\tau^4}\right|_{\tau=0} = \left(\frac{k_B T}{M}\right)^2 \left[3q^4 + \frac{1}{k_B TN} \times \sum_{i,j} \langle \exp[i\mathbf{q} \cdot (\mathbf{r}_i - \mathbf{r}_j)](\mathbf{q} \cdot \nabla_i)(\mathbf{q} \cdot \nabla_j) U(r^N)\rangle\right] \tag{98d}$$

Liouville's equation is time reversal invariant, the underlying particle motions being governed by Newton's Law

$$\frac{d}{dt} \mathbf{V}_i = -\frac{1}{M} \nabla_i U(\mathbf{r}^N, t)$$

Consequently the odd-order derivatives of $F(q, \tau)$ vanish at $\tau = 0$. The Smoluchowski equation describes time irreversible diffusional motions, the systematic parts of which are driven by the interparticle forces:

$$\mathbf{V}_{Ii} = -\frac{1}{\zeta_0} \nabla_i U(\mathbf{r}^N)$$

Consequently, all orders of derivative of $F(q, \tau)$ obtained from the Smoluchowski equation will be nonzero at "$\tau = 0$" ($\tau_B \ll \tau \ll \tau_I$). The close resemblance between (95) and (98d) and the roles played by $D_0 = kT/\zeta_0$ and kT/M, respectively, reflects the formal similarity between the results [see equation (25)]

$$\frac{d^2F(q, 0)}{d\tau^2} = -\langle \dot{a}_{-\mathbf{q}} \dot{a}_{\mathbf{q}} \rangle$$

$$\frac{d^4F(q, 0)}{d\tau^4} = \langle \ddot{a}_{-\mathbf{q}} \ddot{a}_{\mathbf{q}} \rangle$$

and the expressions for $\dot{\mathbf{V}}_i$ and $\mathbf{V}_{Ii}$. Both the Liouville and Smoluchowski cumulants provide rigorous sum rules which can be used to check the consistency of model expressions for $F(q, \tau)$.

Cumulants have been evaluated from the general expressions (92) and (97) for simple systems including a one-dimensional harmonic model and a dilute suspension of hard spheres.[12, 13] In the latter case, for which $g(r) = 0$ when $r < 2a$ and $g(r) = 1$ when $r > 2a$, (92a) has the small-q limit

$$\begin{aligned} K_1 &= \frac{D_0 q^2}{1 - 8\phi}(1 - 6\phi + \phi - 1.73\phi + 0.29\phi) \\ &= D_{\text{eff}}\, q^2 \end{aligned} \tag{99}$$

where $D_{\text{eff}} = D_0(1 + 1.56\phi)$ has been identified as an effective diffusion coefficient through (81) and ϕ is the volume fraction

$$\phi = \tfrac{4}{3}\pi a^3 \langle C \rangle \tag{100}$$

Ackerson[13] has also defined an effective diffusion coefficient through

$$D'_{\text{eff}} = \frac{K_1 S(q)}{q^2}$$

which, in his Oseen approximation, gives $D'_{\text{eff}} = D_0(1 - 6\phi)$. The difference between this result and (99) is purely one of definition, but has led to some confusion.

In the first derivation of K_1 (in the absence of hydrodynamic interactions) given above, the structure factor $S(q)$ enters the result (87) as a simple normalization of (86). This has led to a physical interpretation[54] of the first cumulant result as being a consequence of free diffusion of the particle system at short times $\tau \ll \tau_I$ (see Section 4.4.2) with a "trivial normalization" by $S(q)$. Ackerson[59] has questioned this viewpoint, emphasizing that in his derivation of K_1 from the generalized Smoluchowski equation[12, 13] the short-time correlations in the drift velocity components V_I play an explicit role though they do not appear in the final result due to a formal cancellation. Thus, though equation (87) remains well established theoretically (through both derivations), its physical content is still in question.

4.3.6. Projection Operator Analysis

To investigate the behavior of $F(q, \tau)$ over times comparable with or greater than τ_I we must first cast its equation of motion into a more useful form. Projection operator methods[60–62] have been used to analyze the dynamics of systems described by the GSE by several authors, their earliest application being in the study of polymer dynamics.[63, 64] Subsequently, Wolynes and Deutch[65] studied the rotational and translational correlations of interacting Brownian particles by this method, which has also been used by Ackerson[14] in a more detailed analysis of the behavior of the dynamic structure factor. An independent derivation of many of his results was given by Dieterich and Peschel,[66] who did not, however, consider the effects of hydrodynamic interactions between the particles. Equivalent results have also been obtained in the more general treatment of Hess and Klein.[39, 67]

The correlation function of two dynamical variables can be written in the form

$$\langle A(0)B(\tau)\rangle = \frac{1}{Z}\int d\mathbf{r}^N\, d\mathbf{r}_0^N\, A(\mathbf{r}_0^N)B(\mathbf{r}^N)e^{0\tau}\delta(\mathbf{r}^N - \mathbf{r}_0^N)\exp\left[\frac{-U(\mathbf{r}_0^N)}{k_B T}\right] \tag{101}$$

analogous to (90). This reduces to

$$\langle A(0)B(\tau)\rangle = \frac{1}{Z}\int d\mathbf{r}^N \exp\left[\frac{-U(\mathbf{r}^N)}{k_B T}\right] A(\mathbf{r}^N)e^{0\tau}B(\mathbf{r}^N) \tag{102}$$

the adjoint

$$\tilde{0} = \exp\left[\frac{1}{k_B T}U(\mathbf{r}^N)\right]\nabla_j \cdot \left(\mathbf{D}_{ji}\exp\left[-\frac{1}{k_B T}U(\mathbf{r}^N)\right]\cdot \nabla_i\right) \tag{103}$$

of the Smoluchowski operator (55) acting as a time evolution operator. (An alternative formalism in which the time evolution operator for the system is formally self-adjoint has been discussed by Zwanzig[64].) In the system of Brownian particles we have only one conserved variable, the particle density, as a particle's momentum and energy are freely exchanged with the solvent. Consequently, in applying the Mori formalism we define a projector onto a Fourier component of this one "slow" variable:

$$\mathscr{P}A = \frac{\langle a_{-\mathbf{q}} A(\mathbf{r}^N)\rangle}{\langle a_{-\mathbf{q}} a_{\mathbf{q}}\rangle} a_q \tag{104}$$

where [see (26) and (27)]

$$a_{\mathbf{q}} = \frac{1}{\sqrt{N}}\sum_i e^{-i\mathbf{q}\cdot\mathbf{r}_i} \tag{105}$$

By applying the identity

$$e^{\tilde{\mathbb{O}}\tau} = e^{(1-\mathscr{P})\tilde{\mathbb{O}}\tau} + \int_0^\tau dt\; e^{\tilde{\mathbb{O}}(\tau-t)}\mathscr{P}\tilde{\mathbb{O}}e^{(1-\mathscr{P})\tilde{\mathbb{O}}t} \tag{106}$$

to the equation of motion

$$\frac{dF(q,\tau)}{d\tau} = \langle a_{-\mathbf{q}}e^{\tilde{\mathbb{O}}\tau}\tilde{\mathbb{O}}a_{\mathbf{q}}\rangle$$

we find that

$$\frac{dF(q,\tau)}{d\tau} = -K_1F(q,\tau) + \int_0^\tau dt\; M(q,\tau-t)F(q,t)S(q)^{-1} \tag{107}$$

Here K_1 is the first cumulant (80b); $M(q, t)$ is a memory function which can be written as the autocorrelation function of a random force,

$$M(q,t) = \langle f(\mathbf{q},0)^* f(\mathbf{q},t)\rangle$$

where (108)

$$f(\mathbf{q},t) = e^{(1-\mathscr{P})\tilde{\mathbb{O}}t}(1-\mathscr{P})\tilde{\mathbb{O}}a_{\mathbf{q}}$$

The equivalent equation of motion for the self-dynamic structure factor is

$$\frac{dF_S(q,t)}{d\tau} = -K_{S1}F_S(q,\tau) + \int_0^\tau dt\; M_S(q,\tau-t)F_S(q,t) \tag{109}$$

where K_{S1} is the first cumulant and the memory function $M_S(q, t)$ is given by

$$M_S(q,t) = \langle f_S(\mathbf{q},0)^* f_S(\mathbf{q},t)\rangle$$

where

$$f_S(\mathbf{q},t) = e^{(1-\mathscr{P}_S)\tilde{\mathbb{O}}t}(1-\mathscr{P}_S)\tilde{\mathbb{O}}e^{-i\mathbf{q}\cdot\mathbf{r}_1} \tag{110}$$

and

$$\mathscr{P}_S A = \langle e^{i\mathbf{q}\cdot\mathbf{r}_1}A(\mathbf{r}^N)\rangle e^{-i\mathbf{q}\cdot\mathbf{r}_1}$$

The short-time cumulant results can be recovered from (107) and (109) straight away. On setting $\tau = 0$ we obtain the first cumulants; a short-time expansion of the memory function gives

$$K_2 = M(q, 0)S(q)^{-1} \tag{111a}$$

$$K_{S2} = M_S(q, 0) \tag{111b}$$

These results are, however, of formal significance only and do nothing to simplify the evaluation of these or higher-order cumulants.

The equation of motion (107) differs from that obtained from the analysis of a single "slow" variable in a time reversal invariant system in that the "frequency" term[60] $-K_1 F(q, t)$ does not vanish. Furthermore the memory function can be expected to decay on a time scale comparable with τ_I and the familiar Markov assumption[60] cannot be made except in the discussion of very long time ($\tau \gg \tau_I$), small-q behavior, in which $F(q, \tau)$ decays only negligibly over τ_I (cf. Section 4.3.2). Before we can discuss the behavior of the dynamic structure factor at shorter times or wavelengths we must first look at the behavior of the memory functions in a little more detail.

The expressions (108) and (110) are so complicated that little hope can be held out in general for their direct evaluation. However, it is possible to get some insight into their small-q behavior from a relatively simple analysis.[14] Laplace transformation of (108) gives us

$$\begin{aligned} \tilde{M}(q, s) &= \int_0^\infty e^{-st} M(q, t)\, dt \\ &= \langle f(\mathbf{q}, 0)^* [s - (1 - \mathscr{P})\tilde{\mathrm{O}}]^{-1} f(\mathbf{q}, 0)\rangle \end{aligned} \tag{112}$$

while subsequent iteration of the identity

$$[s - (1 - \mathscr{P})\tilde{\mathrm{O}}]^{-1} = (s - \tilde{\mathrm{O}})^{-1} - (s - \tilde{\mathrm{O}})^{-1}\mathscr{P}\tilde{\mathrm{O}}[s - (1 - \mathscr{P})\tilde{\mathrm{O}}]^{-1}$$

and a resummation give

$$\begin{aligned} \langle f(\mathbf{q}, 0)^*[s - (1 - \mathscr{P})\tilde{\mathrm{O}}]^{-1} f(\mathbf{q}, 0)\rangle &= \langle f(\mathbf{q}, 0)^*(s - \tilde{\mathrm{O}})^{-1} f(\mathbf{q}, 0)\rangle \\ &\quad - \frac{\langle f(\mathbf{q}, 0)^*(s - \tilde{\mathrm{O}})^{-1} a_{\mathbf{q}}\rangle\langle a_{-\mathbf{q}}\tilde{\mathrm{O}}(s - \tilde{\mathrm{O}})^{-1} f(\mathbf{q}, 0)\rangle}{S(\mathbf{q}) + \langle a_{-\mathbf{q}}\tilde{\mathrm{O}}(s - \tilde{\mathrm{O}})^{-1} a_{\mathbf{q}}\rangle} \end{aligned} \tag{113}$$

Direct evaluation of the random force $f(\mathbf{q}, 0)$ gives [see (108)]

$$f(\mathbf{q}, 0) = i\mathbf{q} \cdot \mathbf{L}(\mathbf{q}, 0) + (K_1 - D_0 q^2) a_{\mathbf{q}} \tag{114}$$

where

$$\mathbf{L}(\mathbf{q}, 0) = \frac{1}{k_B T \sqrt{N}} \sum_{i,j} \mathbf{D}_{ij} \cdot \nabla_i U(\mathbf{r}^N) e^{-i\mathbf{q}\cdot\mathbf{r}_j}$$

From these results we see that to first order in q^2 we have

$$\tilde{M}(q, s) = \mathbf{qq} : \langle \mathbf{L}(0, 0)(s - \tilde{0})^{-1} \mathbf{L}(0, 0) \rangle \tag{115}$$

In the absence of hydrodynamic interactions $\mathbf{D}_{ij} = D_0 \delta_{ij} \mathbf{1}$ and $\mathbf{L}(0, 0)$ essentially reduces to the total force on all the particles due to their mutual interaction, which is zero. Consequently $M(q, s)$ must fall off more rapidly than q^2 in the small-q limit. This conclusion remains valid when only two-body hydrodynamic couplings are important, as the symmetry of the particle interactions remains the same. However when higher-order (three-body) interactions become important the argument breaks down and contributions of order q^2 could occur in $M(q, s)$. A similar argument shows that, to first order in q^2

$$\tilde{M}_S(q, s) = \int e^{-st} M_S(q, t)\, dt = \mathbf{qq} : \langle \mathbf{L}_S(0, 0)(s - \tilde{0})^{-1} \mathbf{L}_S(0, 0) \rangle \tag{116}$$

where

$$\mathbf{L}_S(\mathbf{q}, 0) = \frac{1}{k_B T} \mathbf{D}_{i1} \cdot \nabla_i U(\mathbf{r}^N) e^{-i\mathbf{q}\cdot\mathbf{r}_1}$$

As the total force on a single particle does not in general vanish there are contributions of the order q^2 to the self-dynamic structure factor memory function, even when three-body correlations can be neglected.

The simplest way to tackle the problems posed by the difficulties inherent in the general evaluation of the memory functions $M(q, t)$ and $M_S(q, t)$ is to represent them by some simple model, chosen, in part at least, for its analytical tractability. One such model, which leads to expressions for the dynamic structure factor which are useful in the analysis of experimental data, is that which represents the memory function by a sum of exponentially decaying terms.† Laplace transformation of the equation of motion (107) shows that a memory function consisting of n exponentials leads to a dynamic structure factor consisting of $(n + 1)$ exponentials. In the special case in which the memory function is represented by a single exponential

$$M(q, \tau) = M(q) e^{-\gamma\tau} \tag{117}$$

† Discussions of the representation of memory functions as sums of exponentials can be found in the work of Levesque and Verlet[68] on simple liquids and Lantelme *et al.*[69] on ionic liquids.

the corresponding dynamic structure factor is

$$F(q, \tau) = S(q)(\alpha_1 e^{-\Gamma_1 \tau} + \alpha_2 e^{-\Gamma_2 \tau}) \tag{118a}$$

where

$$\alpha_1 + \alpha_2 = 1, \; M(q) = \alpha_1 \alpha_2 S(q)(\Gamma_1 - \Gamma_2)^2 \tag{118b}$$

and

$$\gamma = \Gamma_2 \alpha_1 + \Gamma_1 \alpha_2 \tag{118c}$$

In the general case the relation between the exponents in the memory function and those in the dynamic structure factor is complex; the case of the two exponential memory function is considered in detail by Lantelme *et al.*[69] Here we merely note that for a dynamic structure factor of the form

$$F(q, \tau) = S(q) \sum_i^n \alpha_i e^{-\Gamma_i \tau} \tag{119a}$$

where

$$\sum_i^n \alpha_i = 1$$

we can define a mean relaxation time

$$\tau_F = \frac{\int_0^\infty F(q, \tau)\, d\tau}{S(q)} = \sum_i^n \frac{\alpha_i}{\Gamma_i} \tag{119b}$$

and that its first and second cumulants are given by

$$K_1 = \sum_i^n \alpha_i \Gamma_i \tag{119c}$$

$$K_2 = \sum_{i,j}^n \alpha_i (\delta_{ij} \Gamma_j^2 - \alpha_j \Gamma_i \Gamma_j) \tag{119d}$$

In modeling the memory function by a sum of exponentials we introduce parameters whose physical meaning is not clear and which are not amenable to direct *a priori* calculation. A less drastic procedure which allows us to retain much of the physics in the problem and provides a basis for more detailed calculations has been proposed recently by Hess and Klein,[11, 70] who calculate the memory function using mode–mode coupling theory.[71] As the system has only one conserved or "slow" variable (the Brownian particle concentration) the subspace of "fast" variables is taken to be spanned by the bilinear products $a_{\mathbf{q}-\mathbf{q}'}a_{\mathbf{q}'}$. The operator projecting onto this subspace of fast variables, $(1-\mathscr{P})$ in equation (108), is thus taken to be

$$(1-\mathscr{P})A \simeq \mathscr{P}_2 A = \frac{V}{2(2\pi)^3}\int d\mathbf{q}' \, \frac{\langle a_{-\mathbf{q}+\mathbf{q}'}a_{-\mathbf{q}'}A\rangle}{S(\mathbf{q}-\mathbf{q}')S(\mathbf{q}')} a_{\mathbf{q}-\mathbf{q}'}a_{\mathbf{q}'}$$

(A similar treatment of the more complicated simple fluid system has been given by Bosse *et al.*[72]) The memory function is then obtained in the form

$$M(\mathbf{q}, t) = \frac{1}{2}\langle C\rangle \frac{1}{(2\pi)^3}\int d\mathbf{q}' \, g(\mathbf{q}, \mathbf{q}')F\left(\frac{\mathbf{q}}{2}+\mathbf{q}', t\right)F\left(\frac{\mathbf{q}}{2}-\mathbf{q}', t\right)g(\mathbf{q}, \mathbf{q}') \quad (120)$$

where, if hydrodynamic interactions are neglected and a Gaussian factorization is assumed for the equal-time correlation functions

$$\begin{aligned} g(\mathbf{q}, \mathbf{q}') = D_0\mathbf{q}\cdot[&(\tfrac{1}{2}\mathbf{q}-\mathbf{q}')C_D(\tfrac{1}{2}\mathbf{q}-\mathbf{q}')/S(\tfrac{1}{2}\mathbf{q}+\mathbf{q}') \\ &+(\tfrac{1}{2}\mathbf{q}+\mathbf{q}')C_D(\tfrac{1}{2}\mathbf{q}+\mathbf{q}')/S(\tfrac{1}{2}\mathbf{q}-\mathbf{q}')] \end{aligned} \quad (121)$$

$C_D(\mathbf{q})$ being the Fourier transform of Ornstein–Zernicke direct correlation function.[19] The memory function for the self-dynamic structure factor can be obtained in a similar fashion[11] as

$$M_S(q, t) = \frac{\langle C\rangle}{(2\pi)^3}\int d\mathbf{q}' \, [D_0\mathbf{q}\cdot\mathbf{q}'C_D(\mathbf{q}')]^2 F_S(\mathbf{q}-\mathbf{q}', t)F(\mathbf{q}', t) \quad (122)$$

In the low-density, weak-coupling limits these results reduce to those obtained by Ackerson[14] and Dieterich and Peschel.[66]

4.3.7. Dynamics in the Small-q Limit—Cooperative and Self-Diffusion

The decay of correlations in long-wavelength fluctuations in Brownian particle density is described by the small-q limiting form of (107). As we have discussed in the previous section the memory function $M(q, t)$ tends to zero more rapidly than q^2 in dilute suspensions in which three-particle correlations can be neglected. Consequently, in the small-q limit, $F(q, t)$ for such suspensions decays exponentially at all times much greater than τ_B:

$$\frac{dF(q, \tau)}{d\tau} = -q^2 D_C F(q, \tau) \tag{123}$$

where [cf. equation (92)]

$$D_C = \lim_{q \to 0} \left(\frac{K_1}{q^2} \right) \tag{124}$$

The collective diffusion constant, is, in this special case, identical with the small-q limit of D_{eff} [cf. (99)] describing the initial decay of $F(q, \tau)$. This identical decay in the $\tau \ll \tau_I$ and $\tau \gg \tau_I$ limits has been described in terms of there being no separation of these time scales in collective diffusion in sufficiently dilute suspensions.[14] As the decay is exponential at all times the Stokes–Einstein result (43) can be established by less general arguments than those used in Section 4.3.2.[11] In suspensions of concentrations at which three-particle correlations are significant hydrodynamic contributions may enter in $M(q, t)$ to order q^2; the decay of $F(q, \tau)$ in the small-q limit will then be nonexponential and the long-time limiting decay rate will differ from that of the short-time decay determined by D_{eff}:

$$\left. \frac{dF(q, \tau)}{d\tau} \right|_{\substack{\tau \to \infty \\ q \to 0}} = -q^2 D'_C F(q, \tau) \tag{125}$$

$$D'_C = D_{\text{eff}} - \lim_{q \to 0} \frac{1}{q^2} \tilde{M}(q, 0) \tag{126}$$

Consequently a separation of time scales is predicted.

An alternative approach to the dynamics of long-wavelength fluctuations in systems with short-range interactions was pioneered by Altenberger and Deutch[73] (who treated the hydrodynamic coupling in the Oseen approximation) and refined in subsequent work.[18, 32, 74] A set of coupled equations for the reduced n-particle distribution functions

$$P_N^n(\mathbf{r}^n, t) = \frac{N!}{(N-n)!} \int P(\mathbf{r}^N, t)\, d\mathbf{r}_{n+1} \cdots d\mathbf{r}_N \tag{127}$$

(analogous to the BBYGK hierarchy in the theory of simple fluids[19]) is derived from the generalized Smoluchowski equation, the first of which is

$$\begin{aligned}\frac{\partial}{\partial t} P_N^1(\mathbf{r}_1, t) = {} & D_0 \nabla_1 \cdot \left[\nabla_1 P_N^1 + \frac{1}{k_B T} \int \nabla_1 u(\mathbf{r}_1, \mathbf{r}_2) P_N^2(\mathbf{r}_1, \mathbf{r}_2, t)\, d\mathbf{r}_2\right] \\ & + \nabla_1 \int \mathbf{D}_{11}(\mathbf{r}_1, \mathbf{r}_2) \\ & \cdot \left\{\nabla_1 P_N^2(\mathbf{r}_1, \mathbf{r}_2, t) + \frac{1}{k_B T}[\nabla_1 u(\mathbf{r}_1, \mathbf{r}_2)] P_N^2(\mathbf{r}_1, \mathbf{r}_2, t)\right\} d\mathbf{r}_2 \\ & + \nabla_1 \cdot \int \mathbf{D}_{12}(\mathbf{r}_1, \mathbf{r}_2) \\ & \cdot \left\{\nabla_2 P_N^2(\mathbf{r}_1, \mathbf{r}_2, t) + \frac{1}{k_B T}[\nabla_2 u(\mathbf{r}_1, \mathbf{r}_2)] P_N^2(\mathbf{r}_1, \mathbf{r}_2, t)\right\} d\mathbf{r}_2\end{aligned} \tag{128}$$

Here $u(\mathbf{r}_1, \mathbf{r}_2)$ is the pair potential [cf. (96)] while [cf. (79)]

$$\mathbf{D}_{11}(\mathbf{r}_1, \mathbf{r}_2) = k_B T \mathbf{b}_{11}(\mathbf{r}_1, \mathbf{r}_2) - D_0 \mathbf{1} \tag{129}$$

$$\mathbf{D}_{12}(\mathbf{r}_1, \mathbf{r}_2) = k_B T \mathbf{b}_{12}(\mathbf{r}_1, \mathbf{r}_2) \tag{130}$$

Equation (128) is now decoupled from the other members of the hierarchy by the closure relation[32]

$$P_N^2(\mathbf{r}_1, \mathbf{r}_2, t) = P_N^1(\mathbf{r}_1, t) g(|\mathbf{r}_1 - \mathbf{r}_2|) P_N^1(\mathbf{r}_2, t) \tag{131}$$

which is equivalent to an assumption of local equilibrium and is valid in the small-q, low-concentration limit.[18, 75] The most detailed calculation of this type is that carried out by Felderhof,[18] who obtains and discusses many earlier results[29, 48, 73–76] as special cases. By identifying $P_N^1(\mathbf{r}, t)$ with the particle density $C(\mathbf{r}, t)$, linearizing the decoupled equation (128) about equilibrium and performing the necessary integrations a diffusion equation

$$\frac{\partial C(\mathbf{r}, t)}{\partial t} = D_C \nabla^2 C(\mathbf{r}, t) \tag{132}$$

is found where

$$D_C = D_0[1 + \phi(\lambda_V + \lambda_0 + \lambda_D + \lambda_S + \lambda_A)] \tag{133}$$

ϕ being the volume fraction and

$$\lambda_V = -\frac{3}{a^3}\int_0^\infty [g(r) - 1]r^2\, dr \tag{134a}$$

$$\lambda_A = \frac{3}{a^3}\int_0^\infty \left(\frac{9}{8}\frac{a^6}{r^6} - \frac{5}{4}\frac{a^4}{r^4}\right) g(r) r^2\, dr \tag{134b}$$

$$\lambda_0 = \frac{3}{a^2}\int_0^\infty [g(r) - 1]r\, dr \tag{134c}$$

$$\lambda_S = \frac{75}{4} a^4 \int_0^\infty \frac{g(r)}{r^5}\, dr \tag{134d}$$

$$\lambda_D = 1 \tag{134e}$$

The identity between D_C of (133) and the small-q limit of D_{eff} of (92a) is immediately obvious; as we would expect from our previous discussion there is no indication of a separation of time scales in the limits in which this calculation is carried out. The hard-sphere result

$$D_C = D_0(1 + 1.56\phi) \tag{135}$$

is in good agreement with that found by Batchelor,[16] who related the collective diffusion coefficient to the sedimentation coefficient which was calculated using detailed two-body hydrodynamics[15] (see also Pusey[77] and Berne and Pecora[24]) giving

$$D_C = D_0(1 + 1.45\phi) \tag{136}$$

Felderhof's calculation has recently been extended by Jones[78] to treat the diffusive modes in a system of two different types of Brownian particle. An equivalent calculation has been performed by Batchelor.[16] Ackerson[12] has also reduced the generalized Smoluchowski equation to a diffusion equation in the particle density, using a method similar to that employed by Irving and Kirkwood[79] in their reduction of the Liouville equation. As the hydrodynamic interactions are treated only in the Oseen approximation, the equation he obtains is equivalent to that of Altenberger and Deutch.[73] Altenberger[80, 81] has attempted to extend the diffusion equation approach to describe the decay of finite wavelength fluctuations in more concentrated suspensions. A single-particle diffusion equation is constructed, the direct and hydrodynamic interparticle interactions being treated in a mean field approximation. The direct and pair correlation functions which arise are evaluated assuming local equilibrium. The validity of this local equilibrium approximation at nonzero wave numbers and moderately high particle concentrations is not obvious[18, 75] and the results of this approach should perhaps be viewed with some caution. The decay of finite wavelength fluctuations can be treated using the memory function and mode-coupling methods outlined in the previous section; however, the inclusion of three-body correlations and hydrodynamic interactions in the memory function remain as outstanding problems in this approach.

The diffusive motion of a single particle subject to interactions with others in the suspension determines the behavior of the self-dynamic structure factor (24). From (109) and (116) we see that as $M_S(q, t)/q^2$ does not vanish in the zero-q limit the decay of $F_S(\mathbf{q}, t)$ will be nonexponential except at very long times. The initial decay is characterized by the cumulant (94) and corresponds to free diffusion of a particle, with a coefficient differing from the single free-particle value as a result of its effectively instantaneous hydrodynamic interactions with the other particles in the system. This motion takes place over times $\tau_B \ll \tau \ll \tau_I$ and is described by the $\tau = 0$ limit of equation (109), which implies

$$\lim_{q \to 0} \frac{1}{q^2} \frac{dF_S(q, \tau)}{d\tau}\bigg|_{\tau_B \ll \tau \ll \tau_I} = -D_{\text{eff}}^S F_S(q, \tau) \tag{137}$$

where $D_{\text{eff}}^S = (1/q^2)K_{S1} = D_0(1 - 1.73\phi)$ for dilute hard spheres.

Motion at times comparable with τ_I will be affected by the particle's direct and hydrodynamic interactions with other particles. The limiting long time value of the self-diffusion coefficient can be found from the Laplace transform of (109) to be

$$D_S = D_{\text{eff}}^S - \lim_{q \to 0} \frac{\tilde{M}_S(q, 0)}{q^2} \tag{138}$$

[see equation (116)]. The decay of the memory function $M_S(q, \tau)$ can be related to that of the single-particle velocity autocorrelation function by noting that [cf. (41)]

$$\lim_{q \to 0} -\frac{1}{q^2} \frac{d^2 F_S(\mathbf{q}, \tau)}{d\tau^2} = \frac{1}{3} \langle \mathbf{V}(0) \cdot \mathbf{V}(\tau) \rangle \tag{139}$$

Laplace transformation leads to[11]

$$\tfrac{1}{3}\langle \mathbf{V}(0) \cdot \mathbf{V}(\tau) \rangle = 2D_{\text{eff}}^S \,\delta(\tau) - \lim_{q \to 0} \frac{M_S(q, \tau)}{q^2} \tag{140}$$

If hydrodynamic contributions to the memory function are neglected, it is possible, in the case of hard spheres, to evaluate the mode-coupling expression (112) for $\tilde{M}_S(q, 0)$ to first order in the volume fraction.[57] The long-time self-diffusion coefficient is then given by

$$\begin{aligned} D_S &= D^S_{\text{eff}} - \tfrac{4}{3} D_0 \phi \\ &= D_0(1 - 3.06\phi) \end{aligned} \tag{141}$$

Neglect of the hydrodynamic term in D^S_{eff} gives the self-diffusion coefficient of diffusing hard spheres, reduced from the single-particle value entirely by excluded volume effects.

Thus in the case of self-diffusion we see a definite difference in the low-q limit of single-particle dynamics at times much greater and much less than τ_I (i.e., a separation of time scales is observed), in marked contrast with the collective motion in dilute suspensions. It is interesting to note that calculations of the self-diffusion coefficient following Felderhof's and Batchelor's treatment of collective diffusion[16, 78] yield the short-time result D^S_{eff}, as was first pointed out by Wills.[28] The reason why these calculations do not give the long-time result is not clear, as they ostensibly treat the long-time, small-q thermodynamic regime.

Finally, it is worth pointing out that, in the absence of hydrodynamic interactions, the identity of the short- and long-time decays of $F(q, \tau)$ in the $q \to 0$ limit for pairwise additive direct interactions can be obtained directly by substituting (83) in the expression (43) for the collective diffusion coefficient D_C. For pairwise additive interactions, the sum $\sum_i \nabla_i U(\mathbf{r}^N)$ is always zero since pairs of particles exert equal and opposite forces on each other. Thus, through (84), $\sum_i V_{Ii}$ is also zero and only the term containing the v_B components survives in (43):

$$D_C = [3NS(0)]^{-1} \left[\sum_i \sum_j \int_0^\infty dt \, \langle \mathbf{V}_{Bi}(0) \cdot \mathbf{V}_{Bj}(t) \rangle \right] \tag{142}$$

Use of (85) and (36) then gives

$$D_C = D_0/S(0) \tag{143}$$

so that, under these conditions, the long-time collective diffusion coefficient D_C is the same as the short-time effective diffusion coefficient D_{eff} (equations (81) and (87) in the $q \to 0$ limit].

It should be noted, however, that terms in V_I contribute to the long-time self-diffusion coefficient

$$D_S = \frac{1}{3} \int_0^\infty dt \, \langle \mathbf{V}(0) \cdot \mathbf{V}(t) \rangle \tag{144}$$

since the interaction force on a single particle does not vanish.

4.4. CHARGED PARTICLES IN DILUTE SUSPENSION (NEGLIGIBLE HYDRODYNAMIC INTERACTIONS)

4.4.1. Introduction

The first dynamic light scattering experiment to give a marked indication of the rich variety of effects which strong direct interparticle interactions can cause appears to be that of Pusey *et al.*[82] These authors observed a strong q-dependence of the effective diffusion coefficient of the small charged spherical virus R17 in aqueous suspension. The same material was later studied in more detail by Schaefer and Berne,[83] who also gave a partial theoretical explanation of their results in terms of memory functions. However, because there is some doubt about the stability of this virus in solutions of very low ionic strength[82] we will not discuss these early measurements further.

In the past six years less ambiguous data have come from studies by a number of groups[21, 22, 84–88] of aqueous suspensions of small polystyrene spheres. These are spherical colloidal aggregates of polystyrene with radii in the range 200–500 Å. The surface of the particles carries a large number (typically 100–1000) of ionizable groups, e.g., $-HSO_4$.[84] In an aqueous environment these groups discharge a proton into solution and the particle acquires a large negative charge. If a significant concentration of other small ions is present in the water this particle charge is shielded well by a compact electrical double layer and the interaction force between two particles is only significant in close encounters. However, if excess small ions are removed the double layer becomes highly extended and a significant interaction force acts between particles separated by many particle radii. This situation is usually achieved by placing a small amount of ion exchange resin in the sample cell used in the DLS measurements. It is found that quartz cells cause less small-ion contamination than Pyrex cells; rigorous exclusion of atmospheric carbon dioxide is also necessary. Then the full structure usually develops in the suspension after about two weeks, with the removal of the small ions by the resin.

Under appropriate conditions of particle concentration and charge and solution ionic strength, interparticle forces can actually cause "crystallization" where the particles are constrained to remain in the neighborhood of sites on a regular lattice. Clark *et al.*[89] have recently grown single "colloidal crystals" of several millimeters dimension at particle volume fractions ϕ as low as 10^{-3}.

Here our interest is mainly in the "liquidlike" state which occurs when interparticle forces are not strong enough to cause crystallization.[84] In this situation considerable short-range order persists and any particle will, at a given time, be surrounded at the mean interparticle distance by a reasonably well-defined nearest-neighbor shell.

Since dynamics is our main concern in this paper we will discuss the average structure only briefly. The solid lines in Figure 1 show static structure factors $S(q)$, derived from measurements of the angular dependence of the average scattered intensity, for suspensions of polystyrene spheres of radius about 250 Å at particle concentrations between 1 and 9×10^{12} cm^{-3}.[84] As the concentration is reduced, the position q_m of the main peak in $S(q)$ moves to smaller wave numbers corresponding to larger average interparticle separations. Radial distribution functions $g(r)$, obtained by numerical transformation of these data [using the Fourier inverse of equation (10)], show distinct maxima corresponding to nearest and next-nearest-neighbor shells. Significant structure is still evident in the most dilute sample where the mean interparticle spacing is nearly 20 particle diameters and the volume fraction $\phi \approx 7 \times 10^{-5}$. More quantitative discussions of average structure are given by Brown *et al.*,[84] Schaefer,[90] and Dalberg *et al.*[86] Monte Carlo computer simulations, using a screened Coulombic effective pair potential, are described in a series of papers by van Megen and Snook.[91]

We turn now to the dynamics. Figure 2 shows plots of the logarithm of $c^{1/2}[g^{(1)}(q, \tau)]$ (see Section 4.2.1) against $q^2\tau$ for a suspension of the same 250 Å particles[21] at a somewhat higher concentration, $\langle C \rangle \approx 2.2 \times 10^{13}$ cm^{-3}, than for the data shown in Figure 1. In the absence of interactions data taken at any q should lie on the "free-particle" curve 4 for which [from equations (11), (15), and (35)]

$$\ln[g^{(1)}(q, \tau)] = -D_0 q^2 \tau \tag{145}$$

This was confirmed[21] by studying a suspension of particles where the interactions were well shielded by the addition of salt. It is immediately apparent from Figure 2 that interparticle interactions have a marked effect. The curves 1 to 3 are for different values of q relative to the position q_m of the main peak in $S(q)$. The data in Figure 2 (and those obtained at other

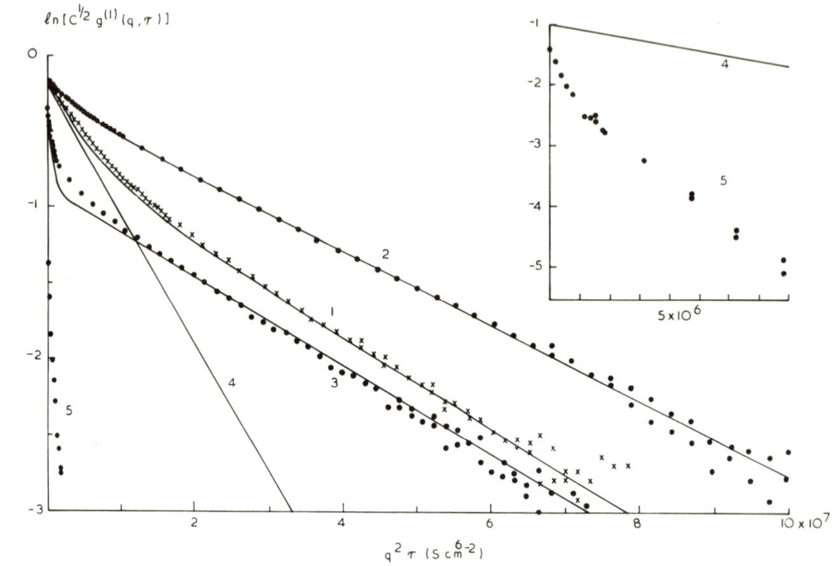

Figure 2. Semilogarithmic plots of light scattering correlation functions against $q^2\tau$ for polystyrene spheres of radius about 250 Å at number concentration $\langle C \rangle \approx 2.2 \times 10^{13}\ \mathrm{cm}^{-3}$ showing peak in $S(q)$ at $q_m \approx 2.04 \times 10^5\ \mathrm{cm}^{-1}$: 1, $q = 3.22 \times 10^5\ \mathrm{cm}^{-1}$ $(q > q_m)$; 2, $q = 2.12 \times 10^5\ \mathrm{cm}^{-1}$ $(q \approx q_m)$; 3, $q = 0.89 \times 10^5\ \mathrm{cm}^{-1}$ $(q < q_m)$; 4, "free-particle" result [equation (145)]; 5, data of curve 3 after subtraction of incoherent scattering (see Section 4.4.4). (Taken from Pusey.[21])

values of q) were analyzed by determining effective diffusion coefficients associated with the initial (D_{eff}) and final (D_L) slopes of the curves.[21] The reciprocals of these effective diffusion coefficients are plotted against q in Figure 3, where the values corresponding to the three curves of Figure 2 are indentified by arrows.

As discussed in Section 4.1, the type of information determined in a DLS experiment depends on the magnitude of q/q_m relative to one (in these highly structured systems the mean interparticle spacing L is roughly $2\pi q_m^{-1}$).[84] In the next sections we discuss the data shown in Figures 1–3 by considering separately several different regimes of q/q_m and τ/τ_I, where τ_I is the "collision time" introduced in Section 4.1.

We start in Section 4.4.2 with the regime $q > q_m$, all τ, where the measured dynamic structure factor largely reflects self diffusion (see Section 4.2.4). Section 4.4.3 deals with $\tau \ll \tau_I$, all q, where the first cumulant relationship should apply. In both cases the effect of particle polydispersity is small and, in the main, the systems can be considered as monodisperse.

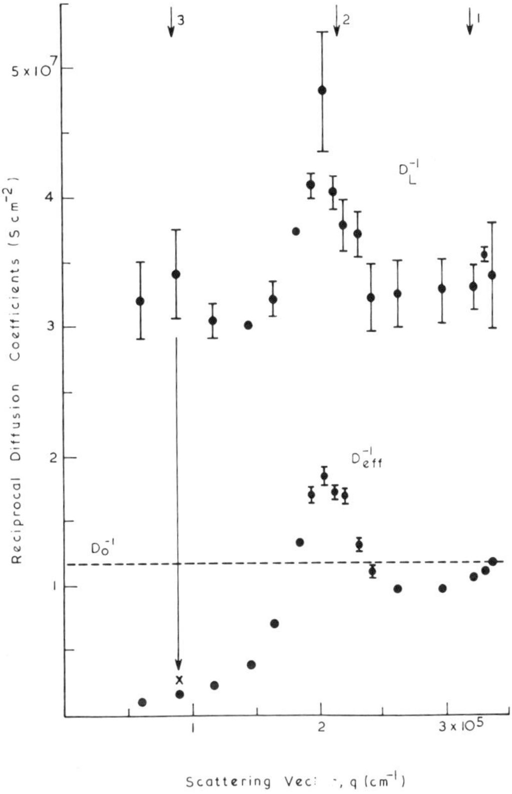

Figure 3. Reciprocal effective diffusion coefficients derived from initial (D_{eff}) and final (D_L) slopes of correlation functions obtained from the same sample as Figure 2. Arrows at the top indicate data obtained from curves 1–3 of Figure 2. × indicates D_L^{-1} obtained from curve 5 of Figure 2 (see Section 4.4.4 for further discussion). In the absence of interactions all data would lie on dashed line D_0^{-1}. (Taken from Pusey.[21])

However, in the hydrodynamic regime discussed in Section 4.4.4, where, at $q \ll q_m$, $S(q) \approx S(0) \ll 1$, incoherent scattering due to polydispersity (Section 4.2.3) becomes significant and the effects of polydispersity must be treated explicitly. Finally, in Section 4.4.5, we consider the behavior of $F(q, \tau)$ at intermediate values of q and τ. This requires the modeling of memory effects.

While we do not have the space here to discuss experimental procedures in detail, mention should be made of multiple scattering since the detection of some light which has been scattered more than once in the sample can cause distortion of the time dependence of the measured correlation functions. For small particles the intensity of multiple scattering relative to that of single scattering is proportional to $\langle C \rangle a^6$ (C is particle number density, a particle radius). In the experiments of Brown *et al.*,[84] $\langle C \rangle < 8.5 \times 10^{12}\ \text{cm}^{-3}$, $a \approx 250$ Å, multiple scattering was small. In the experiments of Pusey,[21] $\langle C \rangle \approx 2 \times 10^{13}\ \text{cm}^{-3}$, $a \approx 250$ Å, and Dalberg *et al.*[86] $\langle C \rangle < 3 \times 10^{13}\ \text{cm}^{-3}$, $a \approx 250$ Å, rough estimates indicated distortions of order 35%–50% in the initial slope of the correlation function for $q < q_m$ though smaller effects were expected at larger τ. Grüner and Lehmann[87, 88] used larger spheres, $a \approx 450$ Å, and strong multiple scattering was observed which required large corrections in some cases. These authors devised a correction procedure[92] based on the measurement of the correlation functions of both the polarized (single + multiple) and depolarized (multiple only) scattered light. While this procedure appeared to be remarkably effective,[88] the approximations involved make its absolute validity hard to assess.

Finally we should justify the claim made in Section 4.1 that hydrodynamic interactions can be neglected for these highly interacting, but dilute, systems. For short-ranged direct interactions the effect of hydrodynamic interactions on the friction coefficient ζ is less than 1% for $\phi < 10^{-3}$ (Sections 4.3.7 and 4.5.2). However, as the range of the direct interaction increases so does the effect of hydrodynamic interactions [equation (187)]. The theory outlined in Section 4.5.2 will not apply to the highly structured systems considered here. Nevertheless there is considerable evidence that the concentration dependence of the friction coefficient of an ordered array of sedimenting particles is given by[41]

$$\zeta = \zeta_0[1 + (1.6 \pm 0.3)\phi^{1/3} + \cdots] \tag{146}$$

For $\phi = 10^{-3}$, the highest concentration considered in this section, ζ will then differ from ζ_0 by about 16%. While this is not strictly negligible it is a much smaller effect than that caused by the direct interactions.

4.4.2. Single-Particle Motions ($q > q_m$, all τ)

As discussed in Section 4.2.4, for $q > q_m$, where $S(q) \approx 1$, the measured full dynamic structure factor $F(q, \tau)$ reduces approximately to the self-function $F_S(q, \tau)$, which, through equation (24), can be written

$$F_S(q, \tau) = \langle \exp[-iq\Delta x(\tau)] \rangle \tag{147}$$

where $\Delta x(\tau)$ is a single Cartesian component of the particle displacement in time τ and, without loss of generality, we have taken $\mathbf{q}$ to define the x direction. Equation (147) can be simplified as in Section 4.3.2 if Δx is assumed to be a Gaussian random variable. This assumption is likely to be valid for $\tau_B \ll \tau \ll \tau_I$, where free Brownian motion dominates (see below) and for $\tau \gg \tau_I$ where Δx comprises a many-step random walk. Its validity is questionable for $\tau \approx \tau_I$ though current data apparently do not have the precision to detect its shortcomings. With this assumption (147) becomes

$$F_S(q, \tau) = \exp\left[-\frac{q^2}{2}\langle \Delta x^2(\tau)\rangle\right] \tag{148}$$

so that a DLS measurement for $q > q_m$ gives, through equations (11), (13), (24), and (148), a direct experimental estimate of the mean-square particle displacement.

Valuable insight into particle dynamics in an interacting system can be obtained by considering a harmonically bound Brownian particle,[21, 22, 86] a model situation exactly soluble in terms of a Langevin equation.[93] This model should apply, for example, to the motion of a single particle in a colloidal crystal if the combined interparticle forces are treated in a harmonic approximation (as for simple solids). In the strongly overdamped case ($\beta/\omega \gg 1$) which applies here (see below) the velocity autocorrelation function $\Phi(\tau)$ of the particle can be written[22]

$$\begin{aligned}\Phi(\tau) &\equiv \langle V_x(0)V_x(\tau)\rangle \\ &= \frac{k_B T}{M}\left[\exp(-\beta\tau) - \left(\frac{\omega^2}{\beta^2}\right)\exp\left(\frac{-\omega^2\tau}{\beta}\right)\right]\end{aligned} \tag{149}$$

where V_x is the x component of particle velocity $\mathbf{V}$, M the particle mass,

$$\beta \equiv \zeta_0/M \qquad [= \tau_B^{-1}, \text{ equation (1)}] \tag{150}$$

and ω is the angular velocity associated with the harmonic potential. This expression leads to a mean-square displacement [see equation (32)]

$$\langle \Delta x^2(\tau)\rangle \equiv 2\int_0^\tau (\tau - t)\Phi(t)\, dt \tag{151a}$$

$$= \frac{2k_B T}{M}\left\{\frac{1}{\omega^2}\left[1 - \exp\left(\frac{-\omega^2\tau}{\beta}\right)\right] - \frac{1}{\beta^2}[1 - \exp(-\beta\tau)]\right\} \tag{151b}$$

Equations (149) and (151b) are instructive. First we note that, since $\beta/\omega \gg 1$, the two terms in $\Phi(\tau)$ decay on widely different time scales. The first large, rapidly decaying term would be present for a free Brownian particle and its decay time β^{-1} can be identified with the Brownian fluctuation time τ_B introduced in Section 4.1. The second small, but slowly decaying term is caused by the harmonic force and we make the identification

$$\tau_I = \beta/\omega^2 \tag{152}$$

where τ_I is the fluctuation time of the velocity component associated with this force (Section 4.1).

At times $\tau \ll \tau_B (\ll \tau_I)$, (151b) becomes

$$\langle \Delta x^2(\tau) \rangle \Big|_{\tau \ll \tau_B} = (k_B T/M)\tau^2 \tag{153}$$

the result for free flight of a totally unconstrained particle. Because τ_B is so small (Section 4.1) photon correlation measurements cannot be made in this limit. The first interesting limit is $\tau_B \ll \tau \ll \tau_I$, where [with use of (150) and (37)]

$$\langle \Delta x^2(\tau) \rangle \Big|_{\tau_B \ll \tau \ll \tau_I} = 2D_0 \tau(1 - \tau/2\tau_I + \cdots) \tag{154}$$

Here the particle behaves initially as a free diffuser, though, as τ becomes comparable to τ_I, the harmonic force retards further increase in $\langle \Delta x^2(\tau) \rangle$. Indeed for $\tau > \tau_I$, $\langle \Delta x^2(\tau) \rangle$ saturates at a value

$$\langle \Delta x^2(\tau) \rangle \Big|_{\tau \gg \tau_I} = 2k_B T/M\omega^2 \tag{155}$$

corresponding to the distance at which the energy stored in the "spring" is equal to the thermal energy $k_B T$. Not surprisingly this gives a long-time self-diffusion coefficient D_S of zero; here D_S is defined by

$$D_S \equiv \frac{1}{2} \frac{d}{d\tau} \langle \Delta x^2(\tau) \rangle \Big|_{\tau \gg \tau_I} \equiv \int_0^\infty \Phi(t)\, dt \tag{156}$$

While this model is strictly valid only for a particle in a colloidal crystal it seems likely that a form similar to (149), but with arbitrary parameters in the second term,[22] will hold in a suspension with liquidlike ordering:

$$\Phi(\tau) = \frac{k_B T}{M} \exp\left(\frac{-\tau}{\tau_B}\right) - A \exp\left(\frac{-\tau}{\tau_I}\right) \tag{157}$$

where we expect $A \ll k_B T/M$ and $\tau_I \gg \tau_B$. Substitution of (157) in (151a) gives

$$\langle \Delta x^2(\tau) \rangle \bigg|_{\tau \gg \tau_B} = 2D_0 \tau - 2A\tau_I[\tau + \tau_I \exp(-\tau/\tau_I) - \tau_I] \tag{158}$$

$$\langle \Delta x^2(\tau) \rangle \bigg|_{\tau_B \ll \tau \ll \tau_I} = 2D_0 \tau - A\tau^2 + \cdots \tag{159}$$

and

$$\langle \Delta x^2(\tau) \rangle \bigg|_{\tau \gg \tau_I} = 2(D_0 - A\tau_I)\tau + 2A\tau_I^2 \tag{160}$$

Free diffusion is again observed in the intermediate time regime [equation (159)], but now for $\tau \gg \tau_I$ a macroscopic self-diffusion coefficient,

$$D_S = D_0 - A\tau_I \tag{161}$$

which is smaller than D_0, can be defined.

Substitution of (158) in (148) allows evaluation of the cumulants of $F_S(q, \tau)$ [see equation (88)]:

$$K_{S1} = D_0 q^2 \tag{162a}$$

$$K_{S2}/K_{S1}^2 = A/D_0^2 q^2 \tag{162b}$$

$$K_{S3}/K_{S1}^3 = A/\tau_I D_0^3 q^4, \qquad \text{etc.} \tag{162c}$$

We note that, as q becomes large, the higher normalized cumulants tend to zero so that the measured correlation function becomes a single exponential. This is not surprising since high-q scattering probes short distances and times $\tau \ll \tau_I$ where the particle moves essentially as a free diffuser. It means, however, that if useful information about the longer-time self-diffusive motion is to be obtained, q must be chosen large enough that $S(q) \approx 1$ yet not so large that only the free diffusion is observed (see Section 4.2.4).

Figure 4 shows mean-square displacements, derived from DLS measurements,[22] for samples of polystyrene spheres of radius about 450 Å. While these data have not been analyzed quantitatively they show qualitative behavior in accord with the model outlined above. In all cases the short-time behavior ($\tau_B \ll \tau \ll \tau_I$) is close to that expected for free Brownian particles (the dashed line in Figure 4). At longer times, $\tau \approx \tau_I$, downward curvature is found. The observation of Bragg reflections showed sample 1 to be in a solidlike state. It was nevertheless possible to make measurements of

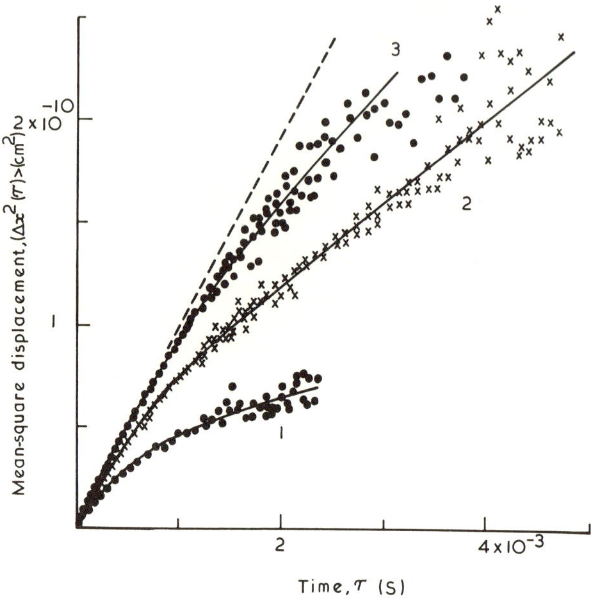

Figure 4. Mean-square displacements of interacting polystyrene spheres of radius about 450 Å. Approximate number concentrations: 1, 2.7×10^{12} cm^{-3} (sample in polycrystalline form); 2, 6.2×10^{11} cm^{-3} (sample in liquidlike state); 3, 2.4×10^{11} cm^{-3}. (Taken from Pusey.[22])

the thermal diffuse scattering at large enough q that $S(q) \approx 1$. The data for this sample show a tendency towards the saturation predicted by (155). The more dilute samples 2 and 3 were in a liquidlike state (no Bragg spots) and the free-diffusive regime extends to larger τ as expected with weakening interparticle forces. At longer times, $\tau > \tau_I$, the curves for these two samples tend to become linear. The macroscopic self-diffusion coefficients deduced from the slopes of these linear portions are markedly smaller than D_0 because of the hindering effect of the interparticle interactions.

Curve 1 of Figure 2 is in the regime $q > q_m$, $S(q) \approx 1$, and can therefore be interpreted [through equation (148)] as a plot of $(-q^2/2)\langle\Delta x^2(\tau)\rangle$. Quantitative analysis of data taken for this sample at similar values of q to curve 1 gave[21]

$$A \approx 5.5 \times 10^{-4} \text{ cm}^2 \text{ sec}^{-2}$$

and

$$\tau_I \approx 10^{-4} \text{ sec}$$

It can now be shown by direct calculation that the inequalities $A \ll k_B T/M$ and $\tau_I \gg \tau_B$ hold by several orders of magnitude, thus confirming the over-damped nature of the motion. It should be noted, however, that the product $A\tau_I \approx 5.5 \times 10^{-8}$ cm^2 sec^{-1} is not much smaller than $(k_B T/M)\tau_B \equiv D_0 \approx 8.65 \times 10^{-8}$ cm^2 sec^{-1} so that [from (161)] $D_S \approx D_0/3$, as is evident from the long-time slope of curve 1 in Figure 2.

Thus this simple intuitive model allows a plausible analysis of the experimental data. Contact can be made with the more rigorous approach of Section 4.3.7 through the following considerations. First we note that both approaches give the same first cumulant for $F_S(q, \tau)$ [equations (162a) and (137) in which $D^S_{\text{eff}} = D_0$ in the absence of hydrodynamic interactions] and first terms in the velocity autocorrelation function which are equivalent for $\tau \gg \tau_B$ [compare (157) and (140)]. Secondly, the Gaussian approximation leading to (148) is expected to be rigorously valid for $\tau \gg \tau_I$ where the particle performs a many-step diffusive random walk through the suspension. Thus we expect the long-time part of the model $F_S(q, \tau)$,

$$F_S(q, \tau)\Big|_{\tau \gg \tau_I} \propto \exp(-D_S q^2 \tau) \tag{163}$$

obtained by substituting (161) and (160) in (148), also to be rigorously valid. In terms of the approach outlined in Section 4.3.6 it is now possible to choose a two-exponential form for $F_S(q, \tau)$ which has the same short-time ($\tau_B \ll \tau \ll \tau_I$) and long-time ($\tau \gg \tau_I$) properties as that obtained from (158) and (148). This assumption implies an exponential memory function and, following the prescription of Section 4.3.6 and equation (140), it can easily be shown that a velocity autocorrelation function whose second term is identical to that of (157) is obtained.

This then establishes some consistency between the two approaches though as yet unjustified assumptions remain in each. Further support for the forms (158) and (148) for $F_S(q, \tau)$ comes from consideration of the second cumulant. Equations (84), (88), and (95b) give, for self-diffusion in the absence of hydrodynamic interactions (as an exact result within the validity of the generalized Smoluchowski equation)

$$K_{S2} = q^2 \langle V_I^2 \rangle \tag{164}$$

where V_I is the "drift velocity" component of particle velocity which results from the interparticle forces. Comparison of (164) and (162b) then suggests the identification

$$A = \langle V_I^2 \rangle \tag{165}$$

which is clearly consistent with the appearance of A in the "interaction term" of the velocity autocorrelation function (157).

It is also worth noting that, in this model, the interaction time τ_I appears first in the third cumulant (162c). This suggests that evaluation of the third cumulant of the self-dynamic structure factor, which has yet to be done, might provide a microscopic expression for τ_I which, up to now, has been introduced only as an ad hoc model parameter.

In conclusion, the simple idea of a harmonically bound Brownian particle, the more rigorous theory outlined in Section 4.3, and the experimental data all contribute to give a useful picture of single-particle motions in a strongly interacting suspension. At any instant the particle feels predominantly the effects of the strong Brownian forces. It therefore sets off on an essentially free diffusive random walk (for $\tau \ll \tau_I$). After a while ($\tau \approx \tau_I$), whatever the direction of its initial displacement, it starts to feel the repelling effect of the nearest-neighbor shell and its diffusive motion is retarded. This shell has a finite lifetime ($\sim \tau_I$) and on a long time scale ($\tau \gg \tau_I$) the particle is able to diffuse through the suspension. However, the long-time self-diffusion coefficient associated with this last motion can be much less than D_0 because of the hindering effects of the evolving nearest-neighbor shells. This picture of a particle trapped temporarily in a repulsive "cage" formed by its neighbors is similar to that frequently given for the motion of an atom in a dense atomic liquid.[19]

Brownian dynamics computer simulations by Ermak[94] and Gaylor *et al.*[95] also confirm this view of single-particle motions. The latter authors have recently obtained quantitative agreement with the DLS data[21] for both single-particle and collective motions.[96]

4.4.3 The First Cumulant ($\tau_B \ll \tau \ll \tau_I$, all q)

In the absence of hydrodynamic interactions equation (87) gives a simple result for the first cumulant K_1 of the dynamic structure factor (the initial slope of its logarithm in the limit, $\tau_B \ll \tau \ll \tau_I$):

$$K_1 = D_0 q^2 / S(q) \tag{166}$$

The "effective diffusion coefficient" associated with K_1 can thus be written (81)

$$D_{\text{eff}} = D_0 / S(q) \tag{167}$$

There have been several experimental tests of (167). Some were seriously complicated by uncertainties associated with particle stability[82, 83] and multiple scattering[85] and will not be discussed further. The data points in Figure 1 are D_0/D_{eff} for the samples described in Section 4.4.1.[84] The general form of these data follows closely that of $S(q)$ as predicted by (167). Although differences between D_0/D_{eff} and $S(q)$ are observed, they are not significantly larger than estimated experimental errors. [The data shown in Figure 1, in fact, predated the derivation of (167).] Data of similar quality have been reported by Dalberg *et al.*[86] and, after large corrections for multiple scattering, by Grüner and Lehmann.[88]

The derivations of the first cumulant relationship outlined in Section 4.3.5 are valid for identical, monodisperse particles. The derivation can easily be generalized to include the effects of polydispersity[10] with the result

$$\bar{D}_{\text{eff}} = \bar{D}_0/S^M(q) \tag{168}$$

where $\bar{D}_{\text{eff}}$ is obtained from the initial slope of the measured correlation function [equation (18)], $S^M(q)$ is the measured static structure factor [equation (17)], and $\bar{D}_0$ is the average free-particle diffusion coefficient weighted by the intensities of light scattered by each species:

$$\bar{D}_0 = \frac{\sum_i b_i^2(q) D_{0i}}{\sum_i b_i^2(q)} \tag{169}$$

$\bar{D}_0$ is the average diffusion coefficient measured in a *noninteracting* polydisperse suspension.

The particles from which the data shown in Figure 1 were obtained were significantly polydisperse (see Section 4.4.4). However, the quantities in Figure 1 labeled D_{eff}, D_0, and $S(q)$ were, in fact, $\bar{D}_{\text{eff}}$, $\bar{D}_0$, and $S^M(q)$ so that these data remain a valid test of the generalized first cumulant relation (168).

4.4.4. Low-q Limit and the Effect of Polydispersity

In the absence of hydrodynamic interactions, equations (124), (92a), and (143) predict for the long-time ($\tau \gg \tau_I$) $q \to 0$ collective diffusion coefficient,

$$D_C = D_0/S(0) \tag{170}$$

Comparison with (167) shows that, in this small-q limit, there is predicted to be no difference between the short-time $\tau \ll \tau_I$ and long-time $\tau \gg \tau_I$ behaviors. The explanation of this rather surprising result is given in Sections 4.3.6 and 4.3.7.

Table 1. Relative Strengths of the Slow Modes in $F^M(q, \tau)$: $x/S^M(q)$, due to Polydispersity [Predicted, equation (171)], and f_L from Data Analysis

q (cm^{-1})	$S^M(q)$	$x/S^M(q)$	f_L
0.87×10^5	0.16	1.19	0.57
2.12×10^5	1.81	0.10	0.87
3.22×10^5	0.89	0.21	0.61

In terms of Figure 3, the prediction of (170) implies that, as $q \to 0$, the upper curve [the reciprocal of the effective diffusion coefficient D_L derived from the *slowest* decay in $F(q, \tau)$] should drop down and ultimately coincide with the lower curve [the reciprocal effective diffusion coefficient D_{eff} obtained from the *initial* decay of $F(q, \tau)$]. Similarly in Figure 2, for $q \ll q_m$ (curve 3) the whole decay should be more rapid than for freely diffusing particles (curve 4). In neither of these figures is there any evidence of this predicted behavior. Indeed, as $q \to 0$, D_L seems to become more or less constant at a value similar to its $q > q_m$ value which has already been identified (Section 4.4.3) as the macroscopic *self-diffusion* coefficient.

The origin of these low-q, slowly decaying tails in $F(q, \tau)$, which were observed by both Pusey[21] and Dalberg *et al.*,[86] was for some time a mystery. However, the difficulty was recently resolved by Weissman,[9] who attributed them to incoherent scattering caused by "polydispersity fluctuations" as discussed in Section 4.2.3. An observation which gives strong support to this interpretation is the above-mentioned similarity between the values of D_L found at low q and the macroscopic self-diffusion coefficient measured at large q [where, since $F^I(q, \tau) \to F_S^I(q, \tau)$, there is no difference in the time dependences of coherent and incoherent scattering]: incoherent scattering is expected to show a self-diffusional decay [equation (22)].

This identification allows us to make a more quantitative analysis of the low-q data for Figures 2 and 3. For small spheres ($qa \ll 1$) the scattering amplitude b is simply proportional to a^3. Thus (23) becomes

$$x = 1 - [(\overline{a^3})^2/\overline{a^6}] \tag{171}$$

For this sample the radius moments were known from electron microscopy[84] and gave $x \approx 0.19$. Table 1 compares the strength of the incoherent scattering calculated from $x/S^M(q)$ [assuming (22) and (23) to be valid] with the amplitude f_L of the slowly decaying mode of $F^M(q, \tau)$

obtained from the numerical data analysis. For $q \approx q_m$ and $q > q_m$, $x/S^M(q)$ is considerably smaller than f_L so that most of the slow mode seen in curves 1 and 2 of Figure 2 can be attributed confidently to a genuine slow decay in the coherent scattering $F^I(q, \tau)$ due to memory effects (see Section 4.4.5). However, for $q \ll q_m$ we find the unphysical result $x/S^M(q) > 1$ which, as discussed in Section 4.2.3, almost certainly represents a breakdown of the assumption leading to (23) that number density and polydispersity fluctuations are completely uncoupled. Nevertheless it remains plausible that the slow mode for $q \ll q_m$ can be attributed *entirely* to incoherent scattering. Curve 5 of Figure 2 shows the result of subtracting $c^{1/2}xF_S^I(q, \tau)$ from the data constituting curve 3, thus providing an estimate of $c^{1/2}(1-x)F^I(q, \tau)$. $F_S^I(q, \tau)$ was calculated from equations (148) and (158) with values of A and τ_I given in Section 4.4.2; x was chosen so that the long-time decay of $c^{1/2}xF_S^I(q, \tau)$ coincided with that of curve 3. We now find, as predicted, a decay more rapid than the free diffusion result (curve 4) though non-exponentiality is still evident, caused probably by residual memory effects and indicating that the true $q \to 0$ limit has not been reached. The point marked x in Figure 3 is the reciprocal of the effective diffusion coefficient estimated from the long-time decay of curve 5. As expected, this now lies below the "free-diffusion" line D_0^{-1}.

Grüner and Lehmann[87, 88] have recently made detailed measurements on larger spheres ($a = 450$ Å) for which the relative standard deviation in radius ($\sim 6\%$) was about one third that of the spheres used in the experiments discussed above ($\sim 19\%$). The results of the data analysis for one of their samples are shown in Figure 5. The reciprocal long-time diffusion coefficient D_L^{-1} becomes small at $q \to 0$ as predicted by (170) although, again, the true $q = 0$ limit is not reached. These data show no evidence of an extra slow low-q mode, though a small effect might have been expected since the particles are still slightly polydisperse.

We digress here to give an oversimplified, yet useful, pictorial representation of the ideas discussed in this section. As described in Section 4.2.4, in the limit $q \ll q_m$ dynamic light scattering essentially observes fluctuations in a spatial Fourier component of refractive index of wavelength q^{-1} which is much greater than the mean interparticle spacing. Figure 6 depicts a possible instantaneous arrangement of particles which gives rise to such a fluctuation. We assume that there are two types of particle, strong scatterers (dots) and weak scatterers (squares) which, apart from scattering power (or refractive index), have identical properties. For clarity the particles are shown situated on lattice planes, though this need not be the case. In the high refractive index region, for this particular configuration, there are more particles altogether as well as an above-average number of strong scatterers. In the absence of particle interactions this fluctuation will decay

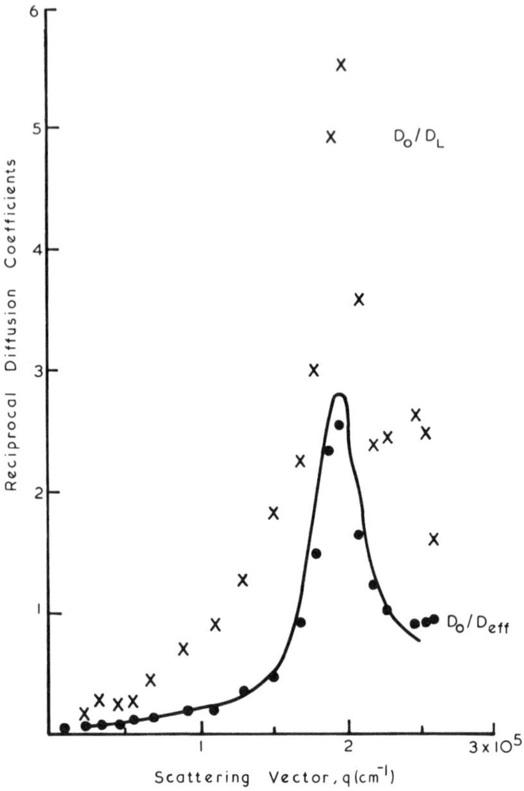

Figure 5. Reduced reciprocal diffusion coefficients derived from initial (D_0/D_{eff}) and final (D_0/D_L) slopes of correlation functions for polystyrene spheres of radius about 450 Å at number density about 1.7×10^{13} cm^{-3}. The solid line is the structure factor $S(q)$. This sample is less polydisperse than that from which Figure 3 was obtained; there is no evidence of slowly decaying modes at $q \ll q_m$. (Taken from Grüner and Lehmann.[88])

by independent freely diffusive random motions of the particles over a distance q^{-1} [with time constant $(D_0 q^2)^{-1}$, as expected]. However, in the opposite extreme of a colloidal crystal, the number density fluctuations can decay by a sort of collective "overdamped phonon" mode[84] in which lattice planes in the high-density region are pushed apart by the interparticle forces whereas planes in the low-density region are pushed closer together. This process, sketched in Figure 6, requires motion of individual particles over distances comparable to the interparticle spacing, much smaller than q^{-1}, and is therefore rapid compared to free diffusion. After this rapid decay of the number density fluctuation, the polydispersity fluctuation (the uneven spatial distribution of the two types of particle) is still

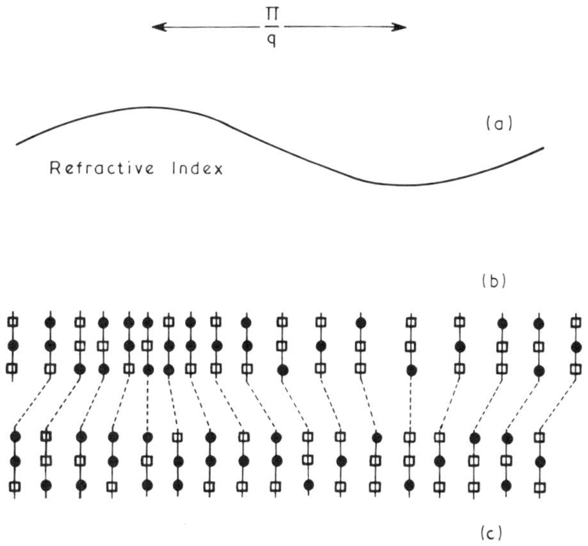

Figure 6. Simple pictorial representation of the two types of particle fluctuation in a polydisperse system. Strong scatterers (circles) and weak scatterers (squares) are shown, for simplicity, located on "lattice planes." Figure 6b shows a possible instantaneous arrangement of particles giving rise to the refractive index fluctuation shown in (a): in the high index region there is a higher density of particles as well as more strong scatterers. In (c) the number density fluctuation has decayed by relatively local particle motions. The polydispersity fluctuation remains and can only decay by the transport of particles over a distance π/q.

present and, in a colloidal crystal, will remain "frozen in." In a suspension showing strong liquidlike structure it seems reasonable that the number density fluctuations will again decay rapidly by a process similar to that described above. However, relaxation of the polydispersity fluctuation can only decay by the actual transport of particles over a distance πq^{-1}, a slow, self-diffusive process.

We conclude this section with two comments relevant to the effects of a small degree of polydispersity in a suspension with strong repulsive interactions. Firstly, we emphasize that, while incoherent scattering is present at all q, it only has important consequences at low q where the coherent scattering [amplitude $S^I(0) \ll 1$] is small and the coherent term $F^I(q, \tau)$ and incoherent term $F_S^I(q, \tau)$ decay on widely different time scales. By contrast, at $q \gtrsim q_m$, the coherent scattering is much stronger and the long-time decays of F^I and F_S^I are more similar.

Secondly, prior to the explanation of the low-q mode in terms of incoherent scattering, the similar values of D_L found at high and low q had suggested to several authors[21, 86] that the convolution approximation due to Vineyard,[97] which predicts $F^I(q, \tau) \propto F_S^I(q, \tau)$, might apply. A theoretical model for $F^I(q, \tau)$ based on this approach, was developed.[86] If the explanation, given above, of the slow mode being due to polydispersity is correct (the evidence points strongly that way), then this model must be discarded.

4.4.5. Memory Effects

In Section 4.3.6 an expression (107), exact within the range of validity of the Smoluchowski equation, was derived for the time dependence of the dynamic structure factor $F(q, \tau)$:

$$\frac{dF(q, \tau)}{d\tau} = -K_1 F(q, \tau) + \frac{1}{S(q)} \int_0^\tau dt\ M(q, \tau - t) F(q, t) \tag{172}$$

where $M(q, t)$ is the memory function and, in the absence of hydrodynamic interactions, the first cumulant K_1 is given by (166). We have mentioned before that $F(q, \tau)$ should be a single exponential both for $q \gg q_m$ (Section 4.4.2) and $q \ll q_m$ (Sections 4.3.6, 4.3.7, and 4.4.4) if, in the second limit, interactions are pairwise additive. In these limits, therefore, $M(q, \tau)$ must be zero. However, for arbitrary scattering vector a finite function will lead to a nonexponential decay in $F(q, \tau)$ (see Figure 2).

A formal expression for $M(q, t)$ is given by equation (108) but its exact evaluation has not proved possible to date. We therefore look to experiment for guidance. The Laplace transform of (172), evaluated in the zero-frequency limit, gives[11]

$$\tau_F^{-1} = K_1 - \tilde{M}(q, 0)/S(q) \tag{173}$$

where τ_F, the mean decay time of $F(q, \tau)$, is given by (119b) and

$$\tilde{M}(q, 0) = \int_0^\infty M(q, t)\, dt \tag{174}$$

Thus the difference $K_1 - \tau_F^{-1}$ gives a measure of the time integral of the memory function. Grüner and Lehmann[87, 88] have made extensive measurements of $F(q, \tau)$ for aqueous suspensions of polystyrene spheres of radius 450 Å at five different concentrations (see also Figure 5). The correlation functions were fitted to a sum of exponentials [equation (119a)]:

$$F(q, \tau) = S(q) \sum_i \alpha_i \exp(-\Gamma_i \tau) \tag{175}$$

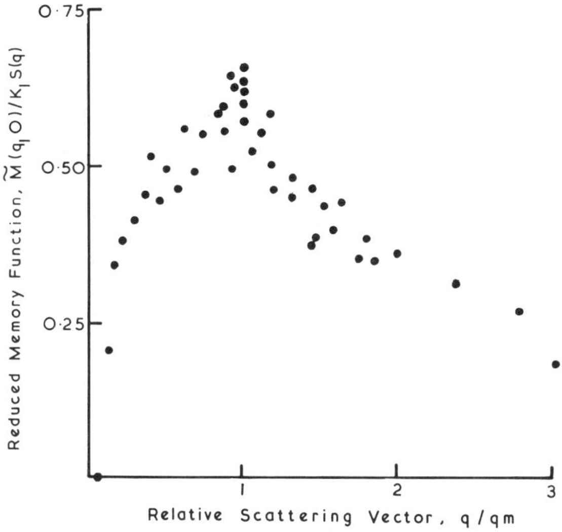

Figure 7. Reduced memory function (see Section 4.4.5) for spheres of radius about 450 Å. Five samples, ranging in number concentration from 3.6×10^{12} to 1.7×10^{13} cm^{-3}, were studied. When plotted against reduced scattering vector q/q_m the data for all samples lie roughly on a universal curve. (Taken from Grüner and Lehmann.[88])

These authors then made the surprising discovery that if the reduced memory function

$$\frac{\tilde{M}(q, 0)}{K_1 S(q)} \equiv 1 - \left(\sum_i \alpha_i / \Gamma_i \sum_i \alpha_i \Gamma_i \right)^{-1} \tag{176}$$

[see equations (173) and (119a)–(119c)] was plotted against q/q_m the results for all samples appeared to lie on a universal curve (Figure 7). As expected $\tilde{M}(q, 0) \to 0$ in the limits $q \gg q_m$ and $q \ll q_m$. It also shows a maximum at $q \approx q_m$, consistent with the slow decay of $F(q, \tau)$ observed at the peak in $S(q)$ (e.g., Figure 2, curve 2).

While it is hard to get an intuitive understanding for the detailed dependence of M on q, the slow decay of $F(q, \tau)$ at $q = q_m$ is at least reasonable since here the wavelength $2\pi q_m^{-1}$ of the spatial Fourier component of particle number density measured by light scattering is roughly equal to the mean interparticle spacing. Thus at $q = q_m$ the technique probes the system in its "preferred" state so that it is not surprising that fluctuations about this state are slow.

Recently Hess and Klein[70] calculated $\tilde{M}(q, 0)$ for an overdamped one-component-plasma model using the mode-coupling approach (Section

4.3.6). By adjusting the "plasma parameter" they obtained a reasonable description of the data of Figure 7 for $q > q_m$ though, for $q < q_m$, agreement was less good. These authors[98] have also calculated $\tilde{M}(q, 0)$ numerically in the mode-coupling approximation using a measured structure factor and no adjustable parameters. Their results are in good agreement with the data for $q > q_m$ and $q < q_m$ though there is significant disagreement in the region of the peak $q \approx q_m$. They also obtained the self-diffusional memory function $\tilde{M}_S(q, 0)$ and from it derived a long-time self-diffusion coefficient D_L^S of magnitude about $D_0/3$, in good agreement with the experimental result discussed in Section 4.4.2.

4.5. EFFECTS OF HYDRODYNAMIC INTERACTIONS

4.5.1. Introduction

In Section 4.4 we discussed systems in which, although long-ranged direct interactions caused large effects, the volume fraction ϕ was low enough for hydrodynamic interactions to be neglected. We turn now to more concentrated suspensions where these latter effects must be considered. Most of the light scattering experiments relevant to this section have been performed in the "hydrodynamic regime" $q \ll q_m$, $\tau \gg \tau_I$ (Sections 4.1 and 4.3). Here, in the absence of polydispersity, the dynamic structure factor should be a single exponential [Equation (40)]

$$F(q, \tau) = S(0)\exp(-D_C q^2 \tau) \tag{177}$$

where the collective diffusion coefficient D_C is given by the generalized Stokes–Einstein expression (43) and (48)

$$D_C = k_B T/S(0)\zeta_C \tag{178}$$

We emphasize again that D_C is expected to be the same diffusion coefficient as would be measured in a conventional "boundary spreading" measurement. $[S(0)]^{-1}$ plays the role of the "thermodynamic driving force" and is determined by the direct interactions. The collective friction coefficient ζ_C [defined by equation (48)] is determined by a combination of direct and hydrodynamic interactions and is expected to be the same as would be measured in conventional sedimentation experiments (and different, in general, from the friction coefficient describing the self-diffusion of a single particle).

While equation (177) should apply generally in the hydrodynamic regime (i.e., to arbitrarily high concentration), it must be remembered that the hydrodynamical theories discussed in Section 4.3 apply only to binary particle encounters and are therefore still limited to $\phi \ll 1$.

In the next section we consider the theoretical predictions for D_C in the $\phi \ll 1$ limit for three model situations: the hypothetical case of genuine hard spheres; longer-ranged repulsive direct interactions represented by an effective hard-sphere radius $a_{HS} > a$ (where a is the particle radius); hard spheres with a short-ranged attractive interaction represented by the addition of a Dirac delta function to $g(r)$ at $r = 2a$. We also consider briefly the effect of incoherent scattering caused by polydispersity. In Section 4.5.3 various experimental results are discussed in the light of these theoretical predictions. Section 4.5.4 is devoted to a brief discussion of the rapid growing field of dynamic light scattering studies of microemulsions. Finally in Section 4.5.5 we discuss recent DLS experiments on suspensions of large spheres at high concentration where measurements are no longer restricted to the hydrodynamic regime and theoretical predictions are few.

4.5.2. Theory of the Collective Diffusion Coefficient in the Hydrodynamic Regime

Theoretical predictions for D_C in the $q \to 0$ limit are given in Section 4.3.7 and are summarized by

$$[S(0)]^{-1} = 1 + \lambda_V \phi + O(\phi^2) \tag{179}$$

$$\zeta_C^{-1} = \zeta_0^{-1}[1 + \lambda_F \phi + O(\phi^2)] \tag{180}$$

so that, through equations (178) and (37)

$$D_C = D_0[1 + \lambda\phi + O(\phi^2)] \tag{181}$$

where

$$\lambda = \lambda_V + \lambda_F \tag{182}$$

The thermodynamic virial coefficient λ_V is given by (134a) whereas the frictional term λ_F is the sum of four terms given by (134b)–(134d).

For a genuine hard-sphere direct interaction we can take, for $\phi \ll 1$

$$\begin{aligned} g(r) &= 0, \qquad r < 2a \\ g(r) &= 1, \qquad r > 2a \end{aligned} \tag{183}$$

We then obtain, as in Section 4.3.7

$$\lambda_V = 8, \qquad \lambda_F = -6.44, \qquad \lambda = 1.56 \tag{184}$$

so that the effects of the interactions on $S(0)$ and ζ nearly cancel and

$$D_c = D_0(1 + 1.56\phi) \tag{185}$$

Some idea of the effects of longer-ranged repulsive interactions can be obtained by assuming an effective hard-sphere interaction of range a_{HS} ($>a$) for which

$$\begin{aligned} g(r) &= 0, \qquad r < 2a_{\text{HS}} \\ g(r) &= 1, \qquad r > 2a_{\text{HS}} \\ \gamma &\equiv a_{\text{HS}}/a \end{aligned} \tag{186}$$

This model has been used, for example, as a first approximation for reasonably well shielded Coulombic interactions.[99, 82, 100, 77] Evaluation of equations (134) for this $g(r)$ gives

$$\begin{aligned} \lambda_V &= 8\gamma^3 \\ \lambda_F &= -6\gamma^2 + 1 - \frac{15}{8\gamma} + \frac{9}{64\gamma^3} + \frac{75}{256\gamma^4} \end{aligned} \tag{187}$$

which naturally reduce to the result of (184) when $\gamma = 1$. In Figure 8 λ_V, λ_F, and λ [equation (182)] are plotted against γ. We see that λ_V increases with γ more rapidly than λ_F decreases so that the overall effect of this type of interaction is to increase the collective diffusion coefficient D_C. At first sight the observation that the friction coefficient ζ_C *increases* as the particles are kept further apart might seem surprising. However, this is just an expression of the fact that a closely spaced pair of particles sediments more rapidly in a constant force field than a widely spaced pair.[15]

The effects of a short-ranged attractive part to the interparticle potential can be investigated using a model suggested by Batchelor[15] and recently applied by Finsy *et al.*[101] The attraction is assumed to cause a greater probability than average that another particle will be found at distance $r = 2a$ from a given particle. This is represented by adding an extra term $g^A(r)$ to the hard-sphere radial distribution function (183):

$$g^A(r) = \frac{\alpha a}{12}\,\delta(r - 2a) \tag{188}$$

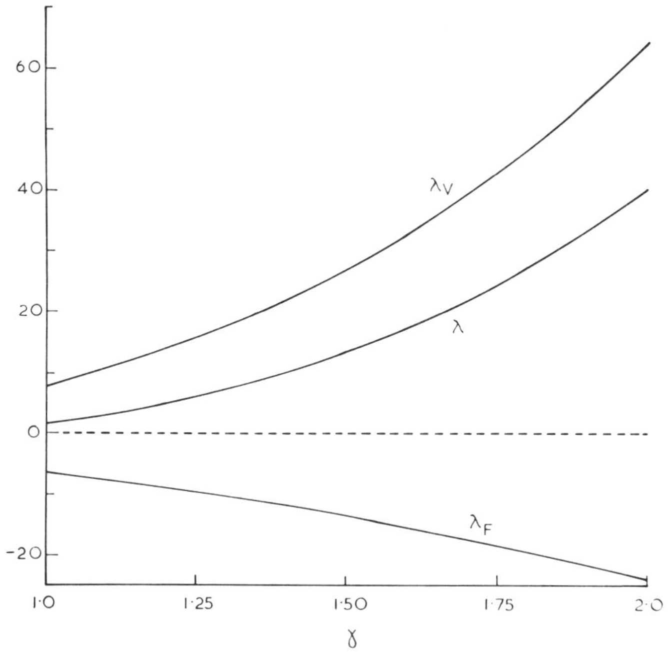

Figure 8. The effects of model repulsive interparticle interactions (see text). Coefficients λ_V, λ_F, and λ (= $\lambda_V + \lambda_F$) of the volume fraction ϕ in the first-order expansion of the thermodynamic term $S(0)$, the friction coefficient ζ, and the collective diffusion coefficient D_C are plotted against the strength of the interaction γ.

The number of particles in the nearest-neighbor shell due to the attraction is then

$$4\pi\langle C\rangle \int_0^\infty g^A(r)r^2\, dr = \alpha\phi \tag{189}$$

If, for example, $\alpha = 20$ and the volume fraction $\phi = 0.01$, this number is 0.2 so that, on average, one particle in five will have an extra particle close to it at any instant. Substitution of (183) and (188) in (134) gives

$$\begin{aligned} \lambda_V &= 8 - \alpha \\ \lambda_F &= -6.44 + 0.49\alpha \\ \lambda &= 1.56 - 0.51\alpha \end{aligned} \tag{190}$$

The slight difference between this result and that of Batchelor and Finsy *et al.* derives from the different method of evaluating the diffusion tensors (as

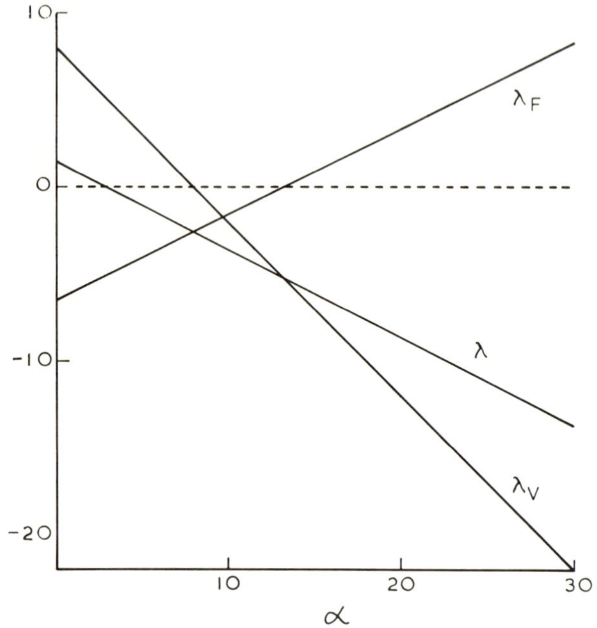

Figure 9. The effects of model attractive interactions (see text). Same data as for Figure 8, except strength of interaction is represented by α.

discussed in Section 4.3). In Figure 9, λ_V, λ_F, and λ are plotted against α. This time λ_V decreases with α more rapidly than λ_F increases and the effect of strong enough attractions is to reduce D_C below D_0.

It is thus apparent from these simple considerations that relatively small changes in the interparticle potential (i.e., small changes in γ and α) can affect the concentration dependence of D_C quite strongly and even cause it to change sign.

It should be noted that, for all the cases considered above, the "Oseen term" λ_0 [equation (134c)] makes the main contribution to λ_F. For practical purposes, therefore, we can neglect the other contributions to λ_F and obtain approximate expressions:

$$\begin{aligned} D_C &= D_0(1 + 2\phi) \qquad \text{for hard spheres} \\ D_C &= D_0[1 + 2\gamma^2(4\gamma - 3)\phi] \qquad \text{for repulsive forces} \\ D_C &= D_0[1 + \tfrac{1}{2}(4 - \alpha)\phi] \qquad \text{for attractive forces} \end{aligned} \tag{191}$$

Several early treatments of the problem did, in fact, make this approximation.

It should also be emphasized again that, while experimental data sometimes show a linear dependence of D_C on ϕ over quite a wide range of ϕ (see Section 4.5.3), the results derived in this section are only expected to be valid for small ϕ. Some idea of the expected range of validity can be obtained by quoting the ϕ^2 term in the virial expansion of $S(0)$ for hard spheres:[19]

$$[S(0)]^{-1} = 1 + 8\phi + 30\phi^2 + \cdots \tag{192}$$

Thus the requirement that the third term in (192) be less than 10% of the second term, for example, leads to the restriction $\phi < 0.027$.

Finally we discuss the effects of polydispersity. First we assume that equation (22) holds. Then in the hydrodynamic regime

$$F^M(q, \tau) = (1 - x)S^I(0)\exp(-D_C q^2\tau) + x\exp(-D_S q^2\tau) \tag{193}$$

where, for hard spheres, D_C is given by (185) and the macroscopic self-diffusion coefficient D_S by (141):

$$D_S = D_0(1 - 3.06\phi) \tag{194}$$

With use of equations (185) and (194) the cumulants of $F^M(q, \tau)$ (193) can be calculated to first order in ϕ and x with the results

$$\frac{K_1}{q^2} = D_0[1 + (1.56 - 4.62x)\phi] \tag{195}$$

$$\frac{K_2}{K_1^2} = 21.3x\phi^2 \tag{196}$$

For $x = 0.1$, say, the coefficient of ϕ in the expression (195) for the apparent diffusion coefficient is reduced to about 1.1 whereas the normalized second cumulant (196) is barely different from zero for small ϕ. Thus, within the approximation that leads to (193) (see Section 4.2.3), the effect of a small degree of polydispersity is virtually negligible at small ϕ. At larger ϕ, however, where D_S can be much less than D_0, the nonexponential nature of (193) will become more apparent (see Section 4.5.4).

4.5.3. Experimental Results

A considerable amount of data on systems for which the theory outlined in Sections 4.5.1 and 4.5.2 is expected to apply have been reported.

First we calculate typical values of q_m and τ_I for these experiments. A commonly studied material is bovine serum albumin, a small protein, usually charged, of molecular weight about 69,000 and free-particle diffusion constant $D_0 \approx 6.2 \times 10^{-7}$ cm^2 sec^{-1} (in water at 20°C).[100, 102, 103] It can be represented by a "hydrodynamic sphere" of radius about 35 Å.[102] At a weight concentration of 1%, for example, the particle number density $\langle C \rangle$ is about 9×10^{16} cm^{-3} so that the mean interparticle spacing $\langle C \rangle^{-1/3}$ is about 220 Å. For such a solution the peak in the structure factor will be at wave vector $q_m > 2\pi\langle C \rangle^{1/3} = 28 \times 10^5$ cm^{-1}. The maximum scattering vector attainable in a light-scattering experiment is about 3.5×10^5 cm^{-1} so that $q \lesssim q_m/8$. The maximum value of the particle collision time τ_I can be taken as the time needed by the particle to diffuse freely a distance equal to the interparticle spacing, i.e., $\tau_1 < \langle C \rangle^{-2/3}/6D_0 = 1.3 \times 10^{-6}$ sec. At $q = 3.5 \times 10^5$ cm^{-1} the characteristic decay time $(D_0 q^2)^{-1}$ of the free diffusional decay of the light scattering correlation function is about 13×10^{-6} sec, about $10\tau_I$. Thus, for this example, where extreme values of q_m and τ_I have been considered, we certainly have $q < q_m$ and $\tau > \tau_I$. At smaller scattering vectors the criteria $q \ll q_m$, $\tau \gg \tau_I$ for the hydrodynamic regime will be fulfilled. Nevertheless data must always be checked for departures from q^2 dependence and, if necessary, extrapolated to small q.

We now discuss the data with two main questions in mind. Firstly, is the generalized Stokes–Einstein equation (178) obeyed and, secondly, are the magnitudes of the coefficients λ, λ_V, and λ_F in agreement with the theoretical values (Section 4.5.2) at small volume fraction ϕ?

An early, comprehensive, set of data was obtained by Newman *et al.*[104] on a circular DNA which, at an ionic strength of about 0.2 *M*, assumes a compact spherical configuration with a hydrodynamic radius of about 316 Å. These authors made measurements up to effective volume fractions of about 0.06 and obtained $\lambda_V = 7.6 \pm 3.9$ from equilibrium sedimentation, $\lambda_F = -6.7 \pm 0.8$ from velocity sedimentation, and $\lambda = 1.2 \pm 0.4$ from DLS diffusion coefficient measurements. The light scattering correlation functions were fitted well by single exponentials. These results are therefore in good agreement with both the generalized Stokes–Einstein equation and the predictions (184) for hard spheres. However, several reservations should be mentioned: firstly, at $\phi = 0.06$, ϕ^2 terms may not be negligible. Secondly, the DNA is charged and, under the conditions of the experiment, the Debye–Huckel screening length is about 6 Å. If we take the effective hard-sphere radius representing the Coulombic interactions to be 10 Å greater than the hydrodynamic radius then the results should be compared with equations (187) with $\gamma \approx 1.03$. However, the changes introduced in λ and λ_F by this procedure are within the quoted experimental errors. Finally, because of the relatively large particle size, the DLS measurements were not performed fully in the hydrodynamic regime. Nevertheless, the results surprisingly showed no dependence on scattering angle in the range 35°–145°.

More recent experiments have been performed by Kops-Werkhoven and Fijnaut[105] on (uncharged) silica particles (radius ~ 230 Å), sterically stabilized by a surface layer of stearyl alcohol and suspended in cyclohexane. These authors obtained $\lambda = 1.4 \pm 0.2$ by dynamic light scattering measurements of the collective diffusion coefficient and $\lambda_F = -6 \pm 1$ by velocity sedimentation measurements, both of which are in complete agreement with the predictions for hard spheres (Section 4.5.2). Interpretation of conventional light scattering measurements was complicated by the fact that particle core and stabilizing layer have different refractive indices; nevertheless good agreement with the prediction for hard spheres is claimed. Reasonably good single exponentials were found for the light scattering correlation functions though the particles were slightly polydisperse with a standard deviation in size about 13% of the mean.

As mentioned previously there have been many DLS experiments on the protein bovine serum albumin (BSA).[100, 102, 103, 106–109] By altering the pH and the ionic strength of the solution it is possible to vary the particle charge and the degree of counterion screening over wide ranges. In qualitative agreement with the predictions of Section 4.5.2 it is found that at high charge and low ionic strength (where the Coulombic interaction becomes long ranged) the diffusion coefficient increases markedly with increasing ϕ. For example, at a weight concentration of 4%, pH 8 (where the particle charge is about $-20e$), with virtually no excess small ions, the collective diffusion coefficient can be as much as four times its free-particle value.[108] This similar to the situation for charged polystyrene spheres at small q (Section 4.4.4) where the effects of the direct interactions greatly outweigh those of the hydrodynamic interactions. On the other hand, at the isoelectric pH, where the average charge on the particle is zero, D_C can actually decrease with increasing ϕ.[100, 107]

Phillies *et al.*[106] measured the diffusion coefficients of BSA over wide ranges of pH and protein concentration and found reasonable agreement, within a rather large experimental error, with the generalized Stokes–Einstein equation. However, they took $\partial\Pi/\partial C$ [$=k_B T/S(0)$, Equation (44)] and ζ values from literature data on different samples. The frictional coefficient ζ was erroneously taken to be that for tracer (or self) diffusion from data which showed no difference between self and collective diffusion[110] and has recently been questioned.[111, 112] At reasonably high ionic strength (0.15 M) Fair *et al.*[107] found a linear dependence of D_C on ϕ up to $\phi \approx 0.24$ with D_C increasing or decreasing by some tens of percent, depending on particle charge. Weissman *et al.*[108] and more recently Weissman and Marque[109] measured both D_C and average intensity $\langle I \rangle$ [hence $S(0)$] at a range of ionic strengths. They found that, on reducing the ionic strength at constant particle concentration, D_C increased more slowly than $\langle I \rangle^{-1}$, as the Coulombic interaction increased in range. This difference is clearly due, at least in part, to increasing hydrodynamic effects on the frictional coefficient ζ_C. Reports on this work contain discussions of other possible effects which include "electrolyte friction" (see Section 4.6). The correlation functions were found to show departures from exponentiality, due possibly to polydispersity, which increased as the direct interaction became stronger.

To conclude the discussion on BSA we note that Anderson *et al.*[113] have recently measured the collective diffusion coefficient by a macroscopic method based on a principle of hydrodynamic stability. While their measurements were not compared directly with dynamic light scattering, they also found that, at low ionic strength, D_C could be as much as four times D_0. This provides some confirmation that conventional and DLS experiments, when performed under the right conditions, measure the same quantity.

Several DLS studies have been reported on various forms of hemoglobin, another compact protein of molecular weight about 64,500 and effective hydrodynamic radius about 31 Å. The most careful experiments appear to be those of Jones *et al.*,[112] who studied oxygenated normal and sickle cell heomoglobin at volume fractions up to 0.4 in 0.1 *M* KCl near their isoelectric pH's (where the average particle charge is zero). The correlation functions were close to exponential for $\phi < 0.2$ though some increase in the second cumulant was observed for $\phi > 0.2$. A linear dependence of D_C on ϕ was found over this whole concentration range though little data was obtained for $\phi < 0.05$. The value of λ describing these data was $\lambda = -0.60 \pm 0.06$. In terms of the theory of Section 4.5.2 (which should only apply at $\phi \ll 1$) this result would indicate a weak attractive part to the interparticle potential caused possibly by van der Waals forces between the uncharged particles. Similar, though less precise, results were obtained on cyanomethemoglobin by Alpert and Banks.[111] Veldkamp and Votano[114] also obtained similar results on normal hemoglobin for $\phi < 0.1$; however, their measured D_c decreased rapidly above this concentration, an observation which Jones *et al.* attribute to contaminants. In addition, Veldkamp and Votano measured the osmotic pressure of their solution. For $\phi \lesssim 0.06$ they found $\partial\Pi/\partial\phi$ to be virtually independent of ϕ. This observation is in qualitative accord with the simple theory of Section 4.5.2 for attractive forces. Note, however, that quantitative agreement is not obtained since $\lambda_V = 0$ would imply $\alpha = 8$, which would give $\lambda = -2.52$ [equation (190)] in contrast with the value $\lambda \approx -0.5$ observed.

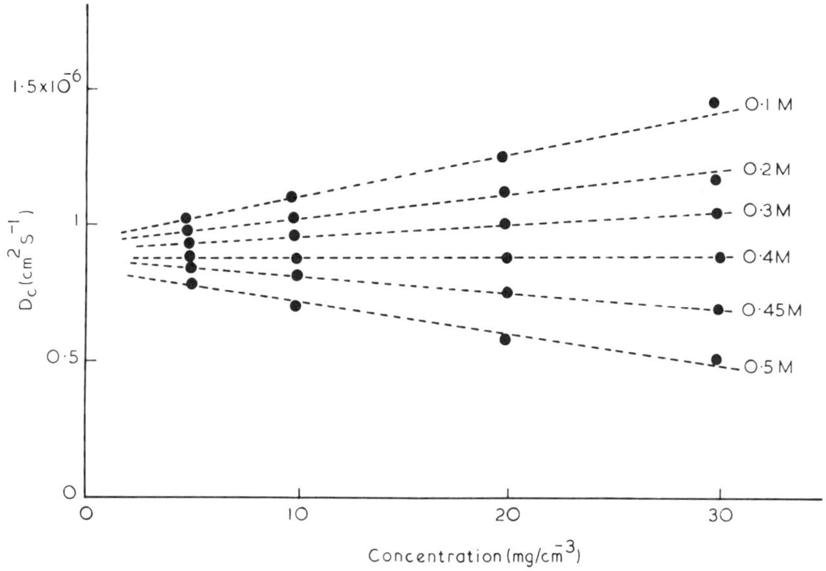

Figure 10. Dependence of the collective diffusion coefficient D_C of sodium dodecyl sulfate micelles on surfactant concentration and on the indicated molarity of added sodium chloride. The critical micelle concentration (<0.5 mg cm^{-3}) has been subtracted from the abscissa. (Taken from Corti and Degiorgio.[115])

We conclude this section with a brief discussion of recent experiments by Corti and Degiorgio[115] on aqueous solutions of sodium dodecyl sulphate (SDS) micelles. The collective diffusion coefficient was measured above the critical micelle concentration (<0.5 mg/ml) as a function of SDS and salt (NaCl) concentrations. The results are shown in Figure 10 where it is seen that relatively small changes in salt concentration can cause quite large changes in the dependence of D_C on SDS concentration. These results were analyzed using DLVO theory where the interparticle potential $u(r)$ is taken as the sum of a repulsive shielded Coulombic part and an attractive van der Waals part. Equations (134) were evaluated numerically for this potential, taking, in this low ϕ limit

$$g(r) = \exp[-u(r)/k_B T] \tag{197}$$

It was found that, with reasonable values of surface potential (determining the Coulombic interactions) and Hamaker constant (determining the van der Waals interactions), the agreement between experimental and theoretical values of λ was quite good. Difficulties with this interpretation, discussed by Corti and Degiorgio, include the possibility of changes in the particle size with solution conditions and possible aggregation (the correlation functions for salt concentrations greater than 0.4 M showed significant departures from exponentiality). Nevertheless this work shows that the point has been reached where realistic potentials (rather than the simple models of Section 4.5.2) can be used in the theory to interpret experimental results.

4.5.4. Microemulsions

Microemulsions are (usually) four-component systems comprising water, oil (a nonpolar liquid), surfactant and cosurfactant, usually an alcohol, combined in such proportions that a transparent mixture is formed. For a "water-in-oil" system, the only type considered here, the water is dispersed as small droplets of typical radius tens to hundreds of Å, surrounded by an interfacial layer of surfactant and cosurfactant, in a continuous phase consisting mainly of the oil. Under appropriate conditions the microemulsion phase of the mixture can persist to water volume fractions of 50% or more. Microemulsions are of interest both as model concentrated colloidal dispersions and because of potential industrial uses which include tertiary oil recovery.[116]

These are complicated systems described by a five-dimensional phase diagram (if temperature is included as a variable). A particular difficulty concerning their use as model colloids is found in the direct and unambiguous determination of the droplet radius since this is expected to be a strong function of position in the phase diagram. At finite droplet concentration, conventional light scattering or neutron or x-ray diffraction methods cannot be used without making assumptions about the nature of the interparticle interaction. Naturally electron microscopy cannot be used except following rapid freezing.[117] Nevertheless it appears that the dominant mechanism in determining droplet size is the tendency of the surfactant and cosurfactant to form a "monolayer" around the droplets in such a way as to keep the area per surfactant molecule roughly constant. Dilution procedures have been devised which allow for the fact that some alcohol will dissolve in the oil and can, in turn, solubilize a small amount of water in the continuous (oil) phase. These procedures can then be used to vary the droplet concentration over wide ranges with the reasonable expectation that its radius will be roughly constant.

The transparency of microemulsions derives mainly from the small droplet size. Nevertheless quite strong (single) light scattering is often observed. DLS is then an excellent technique for investigating droplet dynamics. Here we do not have the space to review in detail the rapidly

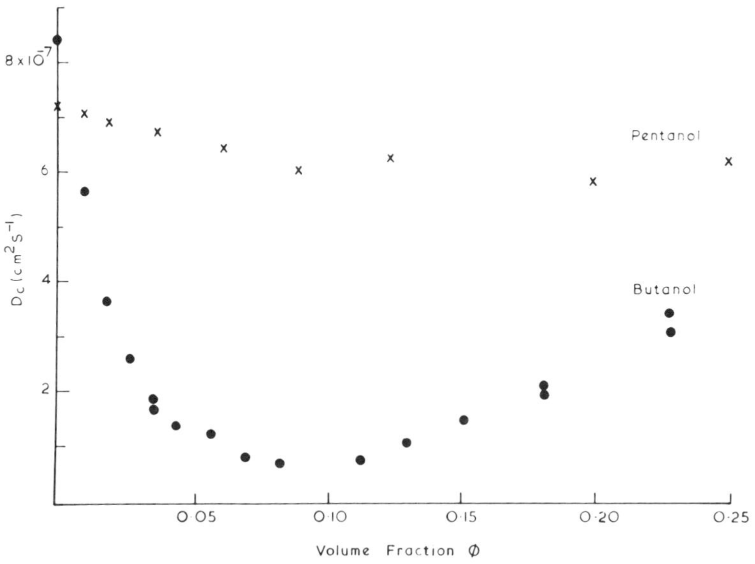

Figure 11. Dependence of the collective diffusion coefficient D_C of water-in-oil microemulsions (water, toluene, sodium dodecyl sulfate, alcohol) on water volume fraction ϕ and the nature of the cosurfactant (butanol or pentanol). (Taken from Cazabat and Langevin.[124])

growing literature[27, 101, 118–126] on this subject and will limit our comments to two studies particularly relevant to this article.

A systematic study of the systems water, cyclohexane or toluene as oil, sodium dodecyl sulphate as surfactant and butanol, pentanol or hexanol as cosurfactant has recently been reported by Cazabat and Langevin[124] (see also References 101 and 123). Figure 11 shows the concentration dependence of the collective diffusion coefficient for two series of microemulsions (with toluene as oil) which differ mainly in that series 1 has butanol as cosurfactant whereas series 2 has pentanol. These results were analyzed in terms of the theory outlined in Section 4.5.2. If it is assumed that the droplet radii do not vary with concentration, the data clearly indicate a large attractive part to the interdroplet potential for the butanol series and a smaller attractive part for the pentanol series. This work therefore suggests the interesting possibility of controlled "tailoring" of interparticle potentials in these systems by relatively minor changes of chemical composition. The correlation functions for these samples were found to be fairly good single exponentials over the whole concentration range though some peculiarities were noted for the butanol series in the region of the minimum in D_C versus ϕ which may be near a critical point.

Another set of measurements has been made recently by Cebula *et*

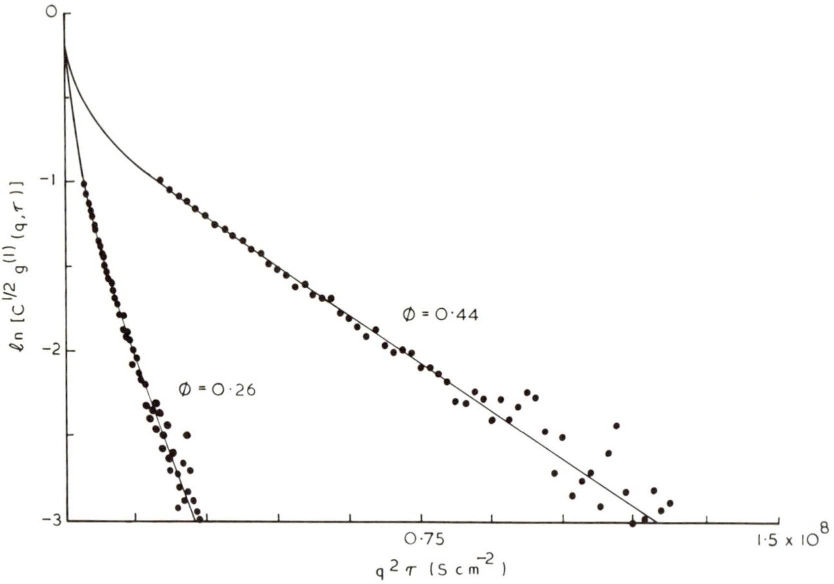

Figure 12. Semilogarithmic plots of light scattering correlation functions for water-in-oil microemulsions (water, xylene, sodium dodecyl benzene sulfonate, hexanol) at indicated water volume fractions ϕ. At short delay times τ, data points are not shown. (Taken from Cebula *et al.*[126])

al.[126] on the system water, xylene, sodium dodecyl benzene sulfonate, hexanol. Here the surfactant + cosurfactant concentration was held constant while the water volume fraction was varied in the range $0 < \phi \lesssim 0.5$. At the higher volume fractions it was possible to obtain an estimate of droplet radius by small-angle neutron diffraction, where the observed angular scattering envelope $\langle I(q) \rangle$, which showed a peak for $\phi \gtrsim 0.15$, was fitted by a simple hard-sphere model. From these measurements it was found that the number density and radius of the droplets varied with ϕ in such a way as to keep the total interfacial area between dispersed and continuous phases roughly constant (as might be expected at constant surfactant concentration).

From the point of view of dynamic light scattering a striking feature of these experiments was the highly nonexponential nature of the correlation functions observed at large ϕ (see Figure 12 for measurements at 90°). At first sight these correlation functions look similar in shape to those observed for dilute suspensions of highly charged particles (Figure 2) where curvature of the semilogarithmic plots was attributed to the transition between the short-time $\tau < \tau_I$ and long-time $\tau > \tau_I$ regimes. However, simple calculations, similar to those at the beginning of Section 4.5.3, show that, for the small-particle radius < 100 Å observed, these experiments were performed well into the hydrodynamic regime $\tau \gg \tau_I$, $q \ll q_m$. This is confirmed both by the neutron scattering experiments which gave q_m, and by the observation that data taken over a range of scattering angles superimpose on those shown in Figure 12 (when plotted as a function of $q^2\tau$).

The most likely interpretation of these results seems to be polydispersity of the droplets, which would lead to the observation of coherent (number density fluctuation) and incoherent (polydispersity fluctuation) scattering as outlined in Section 4.2.3. Then, in the hydrodynamic regime, one expects the correlation function to be the sum of two exponentials [Equation (193)], the fast one representing collective diffusion and the slow one representing self-diffusion. The data could be fitted well by the sum of two exponentials (the solid lines in Figure 12) whose decay constants were assumed to give the collective and self-diffusion coefficients D_C and D_S. In Figure 13 we plot the values so obtained as D_C/D_0 and D_S/D_0, where D_0 is the free-particle diffusion coefficient calculated, using the Stokes–Einstein equation, from the particle radii determined by small-angle neutron scattering. It is seen that, remarkably, D_C differs from D_0 by hardly more than the estimated error in each quantity up to a water volume fraction of nearly 0.5. At present no theoretical explanation of this observation can be given. By contrast, the self-diffusion coefficient D_S decreases rapidly (and roughly exponentially) with increasing ϕ in accordance with the expectation that the mobility of a single droplet will be severely hindered as the density of droplets approaches close packing. At $\phi = 0.44$ the shear viscosity η of the bulk dispersion was some 27 times greater than η_o, the viscosity of the oil. The ratio η_o/η is shown as the cross in Figure 13. We see that the increase in viscosity roughly follows the decrease in D_S. These similar trends are not surprising since both processes involve the motions of particles relative to each other.

It remains to explain why highly nonexponential correlation functions are observed in some microemulsion systems but not, at equally high volume fractions, in others. We mention two tentative explanations. Firstly the observation of a self-diffusional mode in the measured correlation function implies a significant degree of droplet polydispersity. It is possible that different systems show markedly different polydispersities, though this remains to be established. Secondly, in less viscous systems where D_S is expected to be larger, the self-diffusional mode, even when present, will have a smaller effect on the form of the correlation function.

Finally we mention briefly some intriguing results obtained by Bellocq *et al.*[125] on the system dodecane, water, potassium oleate, and hexanol. At high volume fractions these authors observed two-exponential correlation

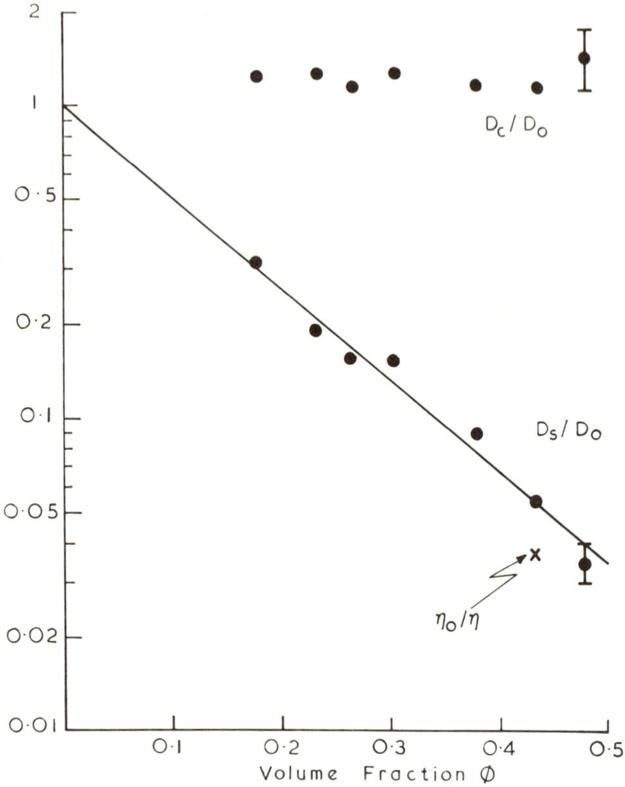

Figure 13. Dependence of collective D_C and self D_S diffusion coefficients of microemulsion droplets (see Figure 12) as a function of the water volume fraction. D_0 is the "free particle" diffusion coefficient calculated from the droplet radius determined by the analysis of small-angle neutron scattering data. The cross indicates the shear viscosity of the bulk suspension relative to that of pure xylene η_0. (Taken from Cebula *et al.*[126])

functions similar to those shown in Figure 12. However, while the diffusion coefficient associated with the slow decay was independent of q, the fast effective diffusion coefficient showed significant q dependence with a minimum at $q_m \approx 2.2 \times 10^5\ \text{cm}^{-1}$. By analogy with the charged particle systems discussed in Section 4.4 this observation would seem to imply spatial ordering on a scale q_m^{-1}. However, this spatial scale is much greater than the droplet size usually found in microemulsions and a detailed explanation of these observations has not yet been given.

4.5.5. Hydrodynamic Effects at Finite *q*

Up to now, in this section, we have considered the effects of hydrodynamic interactions only in the low-q, long-time (hydrodynamic) regime. In order to consider these effects away from $q \to 0$ we need particles of size comparable to the light wavelength so that structure will be observed in $S(q)$ and volume fractions will be large enough for hydrodynamic interactions to be appreciable. A complication with concentrated suspensions of large particles is that, because of their large light scattering cross section, they are usually quite opaque and give rise to strong multiple scattering. Fijnaut *et al.*[127] have cleverly circumvented this problem by using internally cross-linked particles of polymethylmethacrylate (PMMA) suspended in benzene, which, being a good solvent, penetrates and swells the PMMA. The resulting swollen particles are still quite spherical, with radius about 1180 Å, and have an effective refractive index close to that of the benzene. Thus fairly transparent solutions can be obtained at effective volume fractions up to about 0.40.

These samples showed typical "liquidlike" structure factors with a peak in $S(q)$ at $q_m = 2$ to 3×10^5 cm^{-1}, depending on concentration. Fijnaut *et al.* measured the dependence of the first cumulant of the dynamic structure on both q and effective volume fraction. They found that the effective diffusion coefficient D_{eff}, determined from the cumulant [equation (81)] could be described by the relation

$$D_{\text{eff}} = D_0 H(q)/S(q) \tag{198}$$

where $H(q)$ is a quantity describing the effects of hydrodynamic interactions. [In the absence of hydrodynamic interactions $H(q) = 1$ and (198) reduces to the result (87).]

In these experiments $H(q)$ was found to be always less than one so that, not surprisingly, the effect of hydrodynamic interactions was to reduce D_{eff}. For example, at a volume fraction of about 0.4, $H(q)$ increased from about 0.13 at $q \to 0$ to about 0.5 in the region of the peak in $S(q)$ at $q \approx 3 \times 10^5$ cm^{-1}. Similar q dependences were found at lower volume fractions where $H(q)$ was closer to one.

At present there is no theory which can be used for quantitative analysis of these results since, at such high volume fractions, three-body hydrodynamic effects are almost certainly important. It is worth noting, however, that, to first order in ϕ, $H(q)$ is given by equations (92). Here also, $H(q)$ starts below one for $q \to 0$ and increases towards one with increasing q.

4.6. SMALL-ION EFFECTS

So far in this chapter we have assumed that for systems of charged particles (macroions) the dominant role of the small size ions in the solution is simply to determine the form of shielded Coulombic potential between the macroions. This approach leads to an "effective two-component" (interacting particles and liquid) system. However, as mentioned previously, there have been a number of theoretical treatments[8, 102, 128, 129] which consider the small ions more explicitly. In outline Stephen's treatment[128] (specialized to spherical macromolecules) proceeds as follows: the flux of a given ionic species is written as the sum of a diffusive term and a "drift" term driven by the local electric field. These are characterized by free ion diffusion and friction coefficients. The continuity equations for each of the ionic species and the Poisson equation relating the electric field to the charge density then lead to a set of coupled differential equations describing the time evolution of the concentration of each species. These equations are linearized and solved by Fourier–Laplace transformation. In the limit of interest here, where one ionic species, the macroion, is much larger than the others (and so has a small diffusion coefficient and scatters most of the light), the decay of the light scattering correlation function is essentially that of a single exponential with an effective diffusion coefficient[24] [cf. equation (81)]

$$D(q) = D_0 \frac{q^2 + q_0^2}{q^2 + q_s^2}$$

where

$$q_0^2 = \frac{1}{k_B T \varepsilon_0 \varepsilon} \sum_{i=1}^{n} Z_i^2 \langle C_i \rangle$$

and

$$q_s^2 = \frac{1}{k_B T \varepsilon_0 \varepsilon} \sum_{i=2}^{n} Z_i^2 \langle C_i \rangle$$

are inverse square Debye shielding lengths [ε is the relative permittivity of the solution; q_0 contains the effect of the macroions (component 1) whereas q_s does not]. The related structure factor is

$$S(q) = \frac{q^2 + q_s^2}{q^2 + q_0^2}$$

and we note

$$D(q) = D_0/S(q)$$

Comparison with equation (167) shows that in this limit Stephen's theory gives results similar to those of the "effective two-component" approach we have adopted in this article.[11, 24] The description of the diffusion and drift terms making up the ionic fluxes has been refined by Friedhoff and Berne,[129] using the methods of irreversible thermodynamics. A clear and concise discussion of their analysis is given in Chapter 13 of Reference 24.

Several reservations can be stated concerning this approach. Firstly it applies only to weakly interacting systems: typically $S(q)$ shows no peak but increases monotonically from q_s^2/q_0^2 at small q to 1 at large q. Secondly, Weissman and Ware[8] have pointed out that the assumption of small concentration fluctuations (which allows linearization of the diffusion equations) effectively limits its validity to the $q \rightarrow 0$ limit. Finally, hydrodynamic interactions are not treated explicitly. To our knowledge, however, no experimental data exist at present which cannot be explained by the effective two-component approach.

It should be pointed out that there is still some doubt concerning the effect of a counterion atmosphere on the motion of an isolated macroion. Some years ago Ermak[55] performed a computer simulation showing that the additional electrostatic forces on the macroion due to fluctuations in the counterion atmosphere increased the macroion diffusion coefficient. The effect was, however, fairly small: an increase of about 10% was found for a macroion of charge $-32e$ and size 42.9 Å, and the effect decreased with increasing macroion size. Recently, however, Schurr[130] has argued on the basis of the fluctuation dissipation theorem that these forces should increase the friction coefficient of the macroion and so *reduce* its diffusion coefficient. He presents experimental results on poly-L-lysine which appear to be in excellent agreement with his analysis. Weissman and Marque[109] have invoked this "electrolyte friction" mechanism to explain recent data on BSA. However, the measurements were made on concentrated strongly interacting samples where the first-order hydrodynamic theories[131] outlined in Section 4.3.4 may not apply and higher-order effects are hard to calculate. Clearly more experiments on very dilute suspensions of highly charged particles are required.

We should mention that the existence of increased friction for macroions in a constant external force field (electric or gravitational) is well established.[132] Here, because of the different effects of the forces on the macroion and counterions, the center-of-charge of the counterion atmosphere is displaced from that of the macroion.

Finally, Phillies[133] has discussed the effects of fluctuations in the charge on a macroion; these can be significant when the mean charge is relatively small. He found that, provided the charge fluctuation time was longer than that of the scattered light, a slow self-diffusional mode entered the light scattering correlation function of an interacting suspension. Thus the effect of "charge polydispersity" is similar to that of scattering power or size polydispersity.

4.7. CONCLUSIONS

Understanding of the dynamics of interacting spherical Brownian particles has increased greatly over the last decade. A major factor contributing to this has been the data obtained from dynamic light scattering experiments on a wide variety of systems. In this last section we state briefly our view of the current status of the field and attempt to identify remaining problems.

On the theoretical side there is growing acceptance that, in the low-q, long-time hydrodynamic regime (where slow fluctuations of large spatial extent are observed), the DLS technique measures the same collective diffusion coefficient D_C as conventional macroscopic techniques. Further, D_C is given by the generalized Stokes–Einstein expression [equation (49)] which should be valid to high particle concentrations. Away from the hydrodynamic regime (i.e., at nonzero q) the theoretical framework exists to calculate [through the cumulants (Section 4.3.5)] the decay of the dynamic structure factor on a time scale short compared to the relaxation time τ_I of the local structure around a given particle. At large q, where $S(q) \approx 1$ [i.e., above any structure in $S(q)$], the decay of the dynamic structure factor is determined largely by single-particle motions which are easier to visualize and model than collective motions (e.g., Section 4.4.2).

We draw attention to two remaining areas of uncertainty in the theory. Firstly, for volume fractions ϕ greater than about 10^{-3}, hydrodynamic interactions become important. To date these have only been calculated in the binary-encounter approximation (Section 4.3.4). Thus expressions for D_C (Section 4.5.2) and the first cumulant (Section 4.3.5), based on these calculations, can only be used with confidence for $\phi \lesssim 0.05$, say. It is possible that pairwise additivity of hydrodynamic interactions remains valid to larger ϕ, but this has yet to be established. The need for a hydrodynamic theory valid to higher volume fractions, possibly along the lines of screening ideas used in the theory of polymer solutions,(134) is clear. The second area of uncertainty concerns the behavior of the dynamic structure factor at finite q and long times. In this regime, memory-function and mode-coupling approaches (Section 4.3.6) can be used but hydrodynamic interactions have not yet been incorporated into these theories. The validity of other approaches to this problem (Section 4.3.7) is still in question.

When considering the experimental work reviewed in this article it should be borne in mind that nowadays photon correlation spectoscopy is an extremely accurate technique. Lasers and detectors of high quality are available and the digital signal processing is error free. The fluctuating intensity presented to the photocathode can be analyzed to an almost arbitrarily high degree of accuracy simply by accumulating data for a long enough time. Errors in dynamic light scattering measurements arise almost entirely from uncertainties in the optics (definition of the scattering angle, stray light, etc.) and the samples (dust, multiple scattering, polydispersity, etc.). The main technical problem thus remains the production of well-characterized, dust-free samples, designed optimally for light scattering experiments.

Dilute suspensions of charged particles (Section 4.4), exhibiting long-ranged direct interactions but negligible hydrodynamic interactions, have proved to be valuable model systems whose behavior now seems to be well understood qualitatively. However, analysis of existing experimental data is complicated by the effects of polydispersity with the small particles (radius ~ 250 Å) and multiple scattering with the larger particles (~ 450 Å). There is an obvious need for stable, small, monodisperse, charged spheres which could be used to provide "definitive" data. Similarly there is still a need for definitive verification of the generalized Stokes–Einstein equation at high volume fractions. This would involve measurements of D_C by dynamic light scattering, $\partial\Pi/\partial C$ by conventional light scattering, for example, and ζ by velocity sedimentation. Preferably uncharged particles should be used so that complications associated with small ions (Section 4.6) are avoided. An intriguing, and as yet unexplained, observation in many systems is the weak and roughly linear ϕ dependence of D_C up to large ϕ. It would also be valuable to confirm, on the same sample, the identity of the collective diffusion coefficients measured conventionally and by dynamic light scattering.

The interpretation of data on several systems (e.g., Sections 4.4.4 and 4.5.4) in terms of "incoherent" scattering arising from polydispersity appears plausible. The validity of this approach could be tested further by experiments on particles which differed in refractive index (therefore scattering power) but were otherwise similar. It should also be confirmed, for example by tracer measurements, that incoherent scattering does indeed decay by self-diffusion.

On the whole, the comments made so far outline a program for the consolidation of our understanding of the relatively simple systems treated in this article and consequently may not generate great enthusiasm among more inventive experimentalists. Fortunately, one of the attractions of the study of suspensions of particles is the ease with which the complicating features of such systems can be isolated experimentally and effectively added, investigated, and understood one at a time. Early DLS experiments dealt mainly with identical, noninteracting spheres. Studies of polydisperse spheres, identical rods, and flexible coils (polymers), all in the absence of interactions, soon followed. We now have a reasonable understanding of interacting identical spheres and polymers and some idea of the effects of polydispersity in these systems. In the future we can expect to see the problems posed by still more complicated systems stimulating new experimental and theoretical developments, both drawing on and enriching our understanding not only of the suspensions themselves but also of the many other systems whose properties are determined by the interactions of their constituent particles.

4.8. ADDENDUM

Since this chapter was completed there have been many new developments, both theoretical and experimental, in the study of interacting Brownian particles. If we were to start from scratch today, the review, particulary the theory Section 4.3, would probably take a significantly different form. However, rather than attempting a complete revision, we offer this brief addendum with two main aims: (i) to draw attention to what we consider to be the most important new developments directly relevant to the subject matter of the main article (though we make no claim for completeness here); and (ii) to correct errors in and clarify misconceptions created by the article.

We start with theory. The treatment of Section 4.3 emphasizes use of the N-particle Smoluchowski equation. Recently we have developed an approach based on coupled Langevin equations used in the long-time ($\tau \gg \tau_B$) overdamped limit.[135–137] In this way we obtained formal expressions for short-time, $\tau_B \ll \tau \ll \tau_I$, expansions (to order τ^3 if hydrodynamic interactions are neglected,[135] to order τ^2 if they are included[136, 137]) of the mean-square displacement of a "test" particle and of the dynamic structure factors $F(q, \tau)$ and $F_S(q, \tau)$. From these expressions the cumulants, equations (80) and (88), may be obtained directly. Using a Smoluchowski approach but neglecting hydrodynamic interactions, Arauz–Lara and Medina–Noyola have also obtained expressions for the first three cumulants for a suspension containing two types of particle.[138]

We also recognized a direct connection between the hydrodynamic diffusion tensors $\mathbf{D}_{ij}$ ($\mathbf{D}_{ij} \equiv k_B T \mathbf{b}_{ij}$) and correlation functions of the *Brownian components* V_B of the particle velocities[135]:

$$\mathbf{D}_{ij} \equiv \int_0^\infty d\tau \, \langle \mathbf{V}_{Bi}(0) \mathbf{V}_{Bj}(\tau) \rangle_V \tag{199}$$

where the $\langle \cdots \rangle_V$ represent an average over the rapidly fluctuating Brownian velocity components but *not* over the particle positions (see also References 139 and 140). This recognition allows a simple and direct derivation[135] [along the lines of the derivation of equation (86)] of an expression for the short-time effective diffusion coefficient $D_{\text{eff}}(q)$ [equation (92a)]:

$$D_{\text{eff}}(q) = [Nq^2S(q)]^{-1} \sum_i \sum_j \langle \mathbf{q} \cdot \mathbf{D}_{ij}\{\mathbf{r}_k\} \cdot \mathbf{q} \exp\{i\mathbf{q} \cdot [\mathbf{r}_i(0) - \mathbf{r}_j(0)]\}\rangle \quad (200)$$

In the general case the diffusion tensor $\mathbf{D}_{ij}$ depends on the positions $\{\mathbf{r}_k\}$ of all the particles in the suspension (see below). Equations (92) represent a special case of (200) obtained by using the two-particle mobilities (79) for the $\mathbf{D}_{ij}$. Results equivalent to equations (92) have been obtained by a number of authors[141–144] using different approaches which are discussed in Reference 135.

It should be emphasized that it is the diffusion (or mobility) tensors defined by equation (199) which are to be identified with the hydrodynamic tensors introduced in Section 4.3.4. These differ from the tensors defined by equation (46) in that the latter contain the *full* particle velocities $\mathbf{V}_i$ and therefore include long-time ($\tau \gg \tau_I$) effects (appropriate for the description of long-time collective diffusion). As discussed in Section 4.3.6 and 4.3.7, for pairwise additive direct and hydrodynamic interactions, there is expected to be no difference between the long-time collective diffusion coefficient D_c defined by equations (48) and (49) and the short-time coefficient obtained from the $q \to 0$ limit of equation (200). However, when many-particle effects are taken into account (see below), a difference is expected.

A third theoretical approach to the dynamics of interacting Brownian particles, based on the Fokker–Planck equation, has been developed by Hess and Klein (Reference 145 and references therein). These authors view the particles in suspension as constituting an "overdamped fluid." They connect the memory function $M(q, \tau)$ [equation (108)] with a generalized longitudinal viscosity, thereby interpreting memory effects in terms of the suspension's "viscoelasticity." This approach leads to some theoretical predictions which are significantly different from those obtained by the Smoluchowski and Langevin treatments. This is a somewhat surprising finding since the Fokker–Planck, Smoluchowski, and Langevin approaches are usually thought to be equivalent in the long-time, $\tau \gg \tau_B$, overdamped limit. (The equivalence of the latter two descriptions can be shown explicitly.[31, 137])

There have been significant developments concerning the "time-scale" problem alluded to in the paragraph following equation (141). It has now become widely accepted that the theoretical approaches of Batchelor,[16] Felderhof,[18] and Jones[78] are effectively "short-time," i.e., they apply on the "plateau" time scale $\tau_B \ll \tau \ll \tau_I$ over which the spatial configuration of the particles is essentially constant.[135, 146–148] The long-time contribution [the second term in equation (138)] to the self-diffusion coefficient of hard spheres in dilute suspension has been discussed by a number of authors.[147–154] If the hydrodynamic interactions are (hypothetically) neglected, then exact calculations show this contribution to be $-2D_0\phi$,[149, 151, 152, 154] in contrast to the value $-\frac{4}{3}D_0\phi$ obtained by approximate methods [equation (141)]. However, hydrodynamic interactions act to reduce its magnitude to about $-(0.2 \pm 0.1)D_0\phi$.[148, 149, 151] Also the long-time contribution to the velocity autocorrelation function, the second term in (140), has a power-law time dependence,[150–152] a point first noted by Jacobs and Harris.[6] While this work on dilute hard-sphere suspensions has been valuable in clarifying and extending the theory, the effects predicted are small and will be difficult, if not impossible, to detect experimentally.

As mentioned above, the hydrodynamic diffusion tensors $\mathbf{D}_{ij}\{\mathbf{r}_k\}$ [equation (199)] are, in general, functions of the instantaneous positions $\{\mathbf{r}_k\}$ of all the particles in the suspension. We emphasized at several points in the main article that, in 1980, understanding of the detailed structure of these tensors was limited largely to the case of isolated pairs of particles, thereby limiting, in turn, application of the theoretical framework to relatively dilute suspensions. Important progress has been made in the last few years with the development by Mazur,[140] Mazur and van Saarloos,[155] and Muthukumar and Freed[156] of a systematic method for calculating the hydrodynamic interactions between an arbitrary number of spheres. As in the two-particle theory [equations (79)], this approach gives the $\mathbf{D}_{ij}$ as an infinite series in powers of reciprocals of the interparticle spacings, r_{ij}. However, by contrast with the two-particle theory, with increasing powers of r^{-1}, terms involving increasingly large clusters of particles contribute. For example in the cross-tensor $\mathbf{D}_{ij}$ $(i \neq j)$, in addition to the usual two-particle terms [equation (79)], three-particle contributions of order $r_{ik}{}^{-2} r_{kj}{}^{-2}$ and $r_{ik}{}^{-2} r_{kj}{}^{-4}$, a four-particle term of order $r_{ik}{}^{-2} r_{kl}{}^{-3} r_{lj}{}^{-2}$ etc. are found.

Beenakker and Mazur[157] have been able to perform a formal resummation of these many-particle hydrodynamic effects. They were then able to evaluate, albeit approximately, but, they argue, accurately, equation (200) for the q-dependent (short-time) effective diffusion coefficient $D_{\text{eff}}(q)$ for a suspension of hard spheres. Recent light scattering measurement of $D_{\text{eff}}(q)$ by van Megen *et al.*[158] for polymethyl-methacrylate spheres of radius ~ 1900 Å, suspended in an index-matched mixture of liquids (to avoid multiple scattering), show good agreement with the theoretical predictions of Beenakker and Mazur up to volume fractions of about $\phi = 0.45$.

An earlier study[159] on larger spheres of the short-time self diffusion ceoffi-cient D^S_{eff} [the high-q limit of $D_{eff}(q)$] also showed good agreement with these calculations. These experiments extend the work of Fijnaut *et al.*[127] discussed in Section 4.5.5.

We should emphasize that, at high volume fractions, the many-particle terms in $\mathbf{D}_{ij}$ cause large effects rather than minor perturbations. This is seen from evaluations[159–161] for hard spheres of the high- and low-q limits of $D_{eff}(q)$ [equation (200)] which used two-particle expressions for the $\mathbf{D}_{ij}$ but the accurate values for the radial distribution function required to evaluate the ensemble average. These calculations led to nonphysical behavior (negative diffusion coefficients) at volume fractions well below the expected onset of crystallization. However, as outlined above, when many-particle hydrodynamic effects were included, as in the calculations of Beenakker and Mazur, sensible predictions for $D_{eff}(q)$ were obtained in good agreement with the experiments.

An interesting conclusion of Beenakker and Mazur's calculation is that the $q \rightarrow 0$ limit of $D_{eff}(q)$ is relatively independent of concentration even though the hydrodynamic factor $H(0)$ and the thermodynamic factor $S(0)$ [equation (198)] individually show very strong concentration dependences. Thus, for hard spheres, there seems to be a strong, though presumably fortuitous, cancellation between the thermodynamic force driving collective (long wavelength, $q \rightarrow 0$) diffusion and the hydrodynamic drag retarding it. This goes some way towards explaining the lack of concentration dependence of the collective diffusion coefficient observed in a number of experiments and noted at various points in the main article (Sections 4.5.3, 4.5.4, and 4.7). It should be remembered, however, that Beenakker and Mazur's theory calculates the short-time collective diffusion coefficient $\lim_{q \rightarrow 0} D_{eff}(q)$ whereas most measurements provide the long-time coefficient D_c [equation (49)]. When interactions are not pairwise additive (as in a concentrated suspension) these two coefficients are expected to differ due to memory effects embodied in the second term of (126) (see the discussion in Section 4.3.7). The similarity between the predictions of the (short-time) theory and the (long-time) measurements seems to indicate that memory effects in collective diffusion may be small; this remains to be established theoretically.

The idea that size or scattering-power polydispersity causes incoherent light scattering and therefore allows the study of self-diffusion even in strongly interacting suspensions (see Section 4.2.3) has become quite widely accepted. The Utrecht group[162–164] were able to prepare small colloidal silica spheres of roughly equal size but variable refractive index providing thereby a system exhibiting "scattering-power polydispersity." Furthermore the degree of index matching with the solvent (cyclohexane or cycloheptane) could be varied by changing the temperature. Thus the relative amplitudes of the coherent and incoherent contributions to the measured correlation function [equations (22) and (23)] could be varied. By extrapolating their measurements to $q = 0$, these authors were able to study the concentration dependence of both the (long-time) collective and self-diffusion coefficients for this "hard-sphere" system over a wide range of volume fraction. Härtl and Versmold[(165)] studied a mixture of two individually quite monodisperse fractions of charged polystyrene spheres differing in mean size by about 28%. They found strong evidence for incoherent scattering well below the peak in the structure factor but, as we anticipated in Section 4.2.3, the incoherent scattering was relatively much less important at and above the peak.

As we also anticipated in Section 4.2.3 it has proved possible to extend Vrij's theoretical treatment of polydisperse hard spheres, based on the Percus–Yevick approximation, to calculate the amplitude of the incoherent scattering the $q \to 0$ limit.[(166)] It turns out that there is a strong coupling between fluctuations in the number density of particles (determined by the osmotic compressibility of the suspension) and concentration or "polydispersity" fluctuations, so that the "decoupling approximation" leading to equation (22) is poor for hard spheres. Nevertheless, for a not-too-polydisperse suspension, we still expect $F^M(q, \tau)$ (in the low-q limit) to be the sum of a collective mode [$F^I(q, \tau)$ in (22)] and an "average" self mode. However, the amplitude x of the self or incoherent mode is now a complicated function of the particle size distribution and the volume fraction; explicit expressions for x for hard spheres, calculated via the Percus–Yevick approximation, are given in Reference 166. Detailed experimental verification of these predictions is still required.

One error, which should be corrected, concerns our discussion in Section 4.4.2 of the Gaussian approximation applied to the self-dynamic structure factor. By expanding the exponential in (147), using the fact that odd moments of the displacement $\Delta x(\tau)$ are zero, and resumming we get the result, valid for arbitrary statistical properties of $\Delta x(\tau)$,[(167)]

$$F_S(q, \tau) = \exp\left\{ -\frac{q^2}{2} \langle \Delta x^2(\tau) \rangle + \frac{q^4}{24} [\langle \Delta x^4(\tau) \rangle - 3\langle \Delta x^2(\tau) \rangle^2] + \cdots \right\} \quad (201)$$

For short times $\tau_B \ll \tau \ll \tau_I$ it can be shown that

$$\langle \Delta x^2(\tau) \rangle = 2D^S_{\text{eff}}\, \tau \quad (202)$$

where D^S_{eff} is the short-time self-diffusion coefficient discussed in Section 4.3.7, and that

$$\langle \Delta x^4(\tau) \rangle - 3\langle \Delta x^2(\tau) \rangle^2 \propto \tau^\mu \quad (203)$$

where $\mu = 2$ if hydrodynamic interactions are included[136, 137] and $\mu > 3$ if they are not.[135] Thus

$$F_S(q, \tau)\Big|_{\tau_B \ll \tau \ll \tau_I} = 1 - \frac{q^2}{2} \langle \Delta x^2(\tau) \rangle \tag{204}$$

a result which is valid at all q accessible to a light scattering experiment on Brownian particles. At long times $\tau \gg \tau_I$ we have

$$\langle \Delta x^2(\tau) \rangle = 2D_S \tau \tag{205}$$

where D_S is the long-time self-diffusion coefficient discussed in Section 4.3.7 and we expect

$$\langle \Delta x^4(\tau) \rangle - 3\langle \Delta x^2(\tau) \rangle^2 \propto \tau^\nu \tag{206}$$

where $\nu < 2$ [167] (for a simple one-dimensional random walk, it can be shown that $\nu = 1$). Combination of (205) and (206) thus gives

$$\lim_{\tau \to \infty} \langle \Delta x^4(\tau) \rangle = 3\langle \Delta x^2(\tau) \rangle^2 \tag{207}$$

so that, as implied above equation (163), $\Delta x(\tau)$ becomes a Gaussian variable for $\tau \gg \tau_I$. However, substitution of (205) and (206) in (201) does *not* give (163) (unless $\nu < 1$, which is not generally the case). In other words *equation (163) is not generally true*. Gaussian statistics for $\Delta x(\tau)$ are not sufficient to give (163); the rate at which Gaussian statistics are approached as $\tau \to \infty$, as described by the parameter ν, is also important. (It is somewhat comforting to note that the same mistake was made in the early theory of neutron scattering by simple liquids![168]) Note, however, that equation (163) *is* valid in the $q \to 0$ limit [see equations (201), (205), and (206)]:

$$\lim_{\substack{q \to 0 \\ \tau \gg \tau_I}} F_S(q, \tau) = \exp(-D_S q^2 \tau) \tag{208}$$

To summarize the above, the first cumulant of $F_S(q, \tau)$ *at all* q gives the short-time self-diffusion coefficient D^S_{eff} through equations (202) and (204). The *low-q, long-time* behavior of $F_S(q, \tau)$ gives the long-time self-diffusion coefficient D_S through (208); in this limit $F_S(q, \tau)$ would have to be obtained from the incoherent scattering of a polydisperse (in size and/or scattering power) suspension. At high q, where $F_S(q, \tau)$ can be obtained as the high-q

limit of coherent scattering [equation (24)], non-Gaussian terms may be important in $F_S(q, \tau)$ at times $\tau > \tau_I$ [equation (201)] even though $\Delta x(\tau)$ itself becomes Gaussian. Thus the diffusion coefficient D_L obtained from the long-time slope of a plot of ln $F_S(q, \tau)$ against $q^2\tau$ (e.g., Figure 2) cannot, in the high-q limit, be interpreted definitively as D_S. The actual magnitude of these non-Gaussian terms remains to be established. A detailed study of the q-dependence of incoherent scattering would clearly be valuable.

In Section 4.4.4 we argued that the similarity between the long-time diffusion coefficients D_L (Figure 3) at high q [interpreting this as D_S through the incorrect equation (163)] and at low q was evidence for incoherent scattering (measuring D_S) at low q. With our current increased confidence in the existence of incoherent scattering, we could now invert this argument to claim that non-Gaussian terms in equation (201) appear to be relatively unimportant at high q (in this system, at least)! Again a more detailed study is required.

One topic which was not dealt with in the body of the chapter, although there has been considerable discussion in the literature, concerns so-called "reference-frame corrections" (see, for example, Phillies[(29)] and Schurr[(169)]). This subject has recently been discussed in detail by Kops–Werkhoven *et al.*[(170)] who conclude that there is no ambiguity in the relationship between, for example, diffusion and sedimentation. Thus our equation (49) is correct provided both the diffusion coefficient D_c and the sedimentation coefficient ζ^{-1} are referred to the volume-fixed or laboratory frame.

In Section 4.5.4 we discussed briefly water-in-oil microemulsions, describing them as droplets of water dispersed in a continuous phase of oil. While this model is probably correct for the systems considered in Section 4.5.4 (at least at not too high volume fraction), we should mention that such a description of microemulsions is not universally accepted. There is evidence that, at some points in the complicated phase diagram at least, the two phases may be distributed in a more complex bicontinuous structure.[(171)]

Finally, we list without comment a few other recent papers on interacting Brownian particles.[(172–197)]

ACKNOWLEDGMENTS

Tribute should be paid to the cooperative atmosphere prevailing among researchers in this field. We have benefited greatly from numerous discussions and from the free exchange of preprints.

REFERENCES AND NOTES

1. *Colloid Stability*, Faraday Discussions of the Chemical Society, Vol. 65, The Chemical Society, London (1978).
2. A. Bouiller, J. P. Boon, and P. Deguent, Photon correlation study of Brownian motion, *J. Phys.* (*Paris*) **39**, 159–165 (1978).
3. G. L. Paul and P. N. Pusey, Observation of a long-time tail in Brownian motion, *J. Phys. A: Math., Gen.* **14**, 3301–3327 (1981).
4. *Selected Papers on Noise and Stochastic Processes*, N. Wax, ed., Dover, New York (1954).
5. L. D. Landau and E. M. Lifshitz, *Fluid Mechanics*, Pergamon, London (1959).
6. G. Jacobs and S. Harris, Macromolecular self-diffusion and momentum autocorrelation functions in dilute solutions, *J. Chem. Phys.* **67**, 5655–5657 (1977).
7. A. R. Altenberger, The role of inertial effects in macroparticle diffusion, *J. Polym. Sci. Polym. Phys. Ed.* **17**, 1317–1324 (1979).
8. M. B. Weissman and B. R. Ware, Applications of fluctuation transport theory, *J. Chem. Phys.* **68**, 5069–5076 (1978).
9. M. B. Weissman, Quasielastic light scattering from solutions in the small wave vector limit, *J. Chem. Phys.* **72**, 231–233 (1980).
10. P. N. Pusey, Some experiments using quasi-elastic light scattering, in *Light Scattering in Liquids and Macromolecular Solutions*, (V. Degiorgio, M. Corti, and M. Giglio, eds.), Plenum, New York (1980).
11. W. Hess, Diffusion coefficients in colloidal and polymeric solutions, in *Light Scattering in Liquids and Macromolecular Solutions*, (V. Degiorgio, M. Corti, and M. Giglio, eds.) Plenum, New York (1980).
12. B. J. Ackerson, Brownian motion of interacting particles, University of Colorado, Ph.D. thesis (1976).
13. B. J. Ackerson, Correlations for interacting Brownian particles, *J. Chem. Phys.* **64**, 242–246 (1976).
14. B. J. Ackerson, Correlations for interacting Brownian particles, II, *J. Chem. Phys.* **69**, 684–690 (1978).
15. G. K. Batchelor, Sedimentation in a dilute suspension of spheres, *J. Fluid Mech.* **52**, 245–268 (1972).
16. G. K. Batchelor, Brownian diffusion of particles with hydrodynamic interaction, *J. Fluid Mech.* **74**, 1–29 (1976).
17. B. U. Felderhof, Hydrodynamic interaction between two spheres, *Physica* **89A**, 373–384 (1977).
18. B. U. Felderhof, Diffusion of interacting Brownian particles, *J. Phys. A: Math. Gen.* **11**, 929–937 (1978).
19. J. P. Hansen and I. R. McDonald, *Theory of Simple Liquids*, Academic, London (1976).
20. J. C. Brown, P. N. Pusey, and R. Dietz, Photon correlation study of polydisperse samples of polystyrene in cyclohexane, *J. Chem. Phys.* **62**, 1136–1144 (1975).
21. P. N. Pusey, Intensity fluctuation spectroscopy of charged Brownian particles: the coherent scattering function, *J. Phys. A: Math. Gen.* **11**, 119–135 (1978).
22. P. N. Pusey, The study of Brownian motion by intensity fluctuation spectroscopy, *Phil. Trans. R. Soc. London* **A293**, 429–439 (1979).
23. A. Vrij, Influence of polydispersity on the light scattering of concentrated suspensions of spherical particles, *Chem. Phys. Lett.* **53**, 144–147 (1978); Light scattering of a concentrated multicomponent system of hard spheres in the Percus–Yevick approximation, *J. Chem. Phys.* **69**, 1742–1747 (1978).
24. B. J. Berne and R. Pecora, *Dynamic Light Scattering*, Wiley, New York (1976).

25. R. Zwanzig, Elementary derivation of time-correlation formulas for transport coefficients, *J. Chem. Phys.* **40**, 2527–2533 (1964).
26. A. R. Altenberger, On the theory of generalized diffusion processes, *Acta Phys. Pol.* **A46**, 661–666 (1974).
27. A. Vrij, E. A. Nieuwenhuis, H. M. Fijnaut, and W. G. M. Agterof, Application of modern concepts in liquid state theory to concentrated particle dispersions, in *Colloid Stability*, Faraday Discussions of the Chemical Society, Vol. 65, The Chemical Society, London (1978).
28. P. R. Wills, Isothermal diffusion and quasielastic light-scattering of macromolecular solutes at finite concentration, *J. Chem. Phys.* **70**, 5865–5874 (1979).
29. G. D. J. Phillies, Effects of intermacromolecular interactions on diffusion, I, *J. Chem. Phys.* **60**, 976–982 (1974).
30. A. Einstein, *Investigations on the Theory of the Brownian Movement*, Dover, New York (1956).
31. R. Zwanzig, Langevin theory of polymer dynamics in dilute solutions. *Adv. Chem. Phys.* **15**, 325–333 (1969).
32. S. Harris, Perturbation solution of the one particle generalized Smoluchowski equation, *J. Chem. Phys.* **65**, 5408–5412 (1976).
33. S. Chandrasekhar, Stochastic processes in physics and astronomy, *Rev. Mod. Phys.* **15**, 1–89 (1943); reprinted in reference 4.
34. P. Mazur and I. Oppenheim, Molecular theory of Brownian motion, *Physica* **50**, 241–258 (1970).
35. J. Albers, J. M. Deutch, and I. Oppenheim, Generalized Langevin equations, *J. Chem. Phys.* **54**, 3541–3546 (1971).
36. J. M. Deutch and I. Oppenheim, Molecular theory of Brownian motion of several particles, *J. Chem. Phys.* **54**, 3547–3555 (1971).
37. T. J. Murphy and J. L. Aguirre, Brownian motion of interacting particles I, *J. Chem. Phys.* **57**, 2098 (1972).
38. G. Wilemski, On the derivation of Smoluchowski equations with corrections in the classical theory on Brownian motion, *J. Stat. Phys.* **14**, 153–169 (1976).
39. W. Hess and R. Klein, Dynamical properties of colloidal systems I, *Physica* **94A**, 71–90 (1978).
40. A. R. Altenberger, Molecular derivation of the generalized Kirkwood–Riseman equation for a dilute polymer solution, *Acta Phys. Pol.* **A47**, 861–865 (1974).
41. J. Happel and H. Brenner, *Low Reynolds Number Hydrodynamics*, 2nd Edition Noordhoff, Leyden (1973).
42. R. Kubo, The fluctuation–dissipation theorem, *Rep. Prog. Phys.* **29**, 255–284 (1966).
43. R. Zwanzig, Hydrodynamic fluctuations and Stokes' law friction, *J. Res. N.B.S.* **68B**, 143–145 (1964).
44. C. W. Oseen, *Neuere Methoden und Ergebnisse in der Hydrodynamik*, Akademische Verlagsgellschaft, Leipzig (1927).
45. B. U. Felderhof and J. M. Deutch, Frictional properties of dilute polymer solutions I, *J. Chem. Phys.* **62**, 2391–2397 (1975).
46. M. Stimson and G. B. Jeffrey, The motion of two spheres in a viscous fluid, *Proc. R. Soc. London* **A111**, 110–116 (1926).
47. H. Brenner, Stokes resistance of an arbitrary particle II; An extension, *Chem. Eng. Sci.* **19**, 599–629 (1964).
48. J. L. Aguirre and T. J. Murphy, Brownian motion of N interacting particles II, *J. Chem. Phys.* **59**, 1833–1840 (1973).
49. B. U. Felderhof, Force density induced on a sphere in linear hydrodynamics, I, *Physica* **84A**, 557–568 (1976); Force density induced on a sphere in linear hydrodynamics, II, *ibid.*

84A, 569–576 (1976); R. Schmitz and B. U. Felderhof, Creeping flow about a sphere, *ibid.* **92A**, 423–437 (1978).
50. R. B. Jones, Hydrodynamic interaction of two permeable spheres I, *Physica* **92A**, 545–556 (1978); Hydrodynamic interaction of two permeable spheres II, *ibid.* **92A**, 557–570 (1978); Hydrodynamic interactions of two permeable spheres III, *ibid.* **92A**, 571–583 (1978).
51. B. U. Felderhof and R. B. Jones, Faxen theorems for spherically symmetric polymers in solution, *Physica* **93A**, 457–464 (1978).
52. P. Reuland, B. U. Felderhof, and R. B. Jones, Hydrodynamic interaction of two spherically symmetric polymers, *Physica* **93A**, 465–475 (1978).
53. R. C. Ball and P. Richmond, Dynamics of colloidal dispersions, *Phys. Chem. Liq.* **9**, 99–116 (1980).
54. P. N. Pusey, Dynamics of interacting Brownian particles, *J. Phys. A: Math. Gen.* **8**, 1433–1439 (1975).
55. D. L. Ermak, A computer simulation of charged particles in solution, I: Technique and equilibrium properties, *J. Chem. Phys.* **62**, 4189–4196 (1975); D. L. Ermak, A computer simulation of charged particles in solution II: Polyion diffusion coefficient, *J. Chem. Phys.* **62**, 4197–4203 (1975).
56. S. A. Allison, E. L. Chang, and J. M. Schurr, The effects of direct and hydrodynamic forces on macromolecular diffusion, *Chem. Phys.* **38**, 29–41 (1979).
57. R. J. A. Tough, unpublished calculations.
58. P. G. deGennes, Liquid dynamics and inelastic scattering of neutrons, *Physica* **25**, 825–839 (1959).
59. B. J. Ackerson, private communication.
60. D. Forster, *Hydrodynamic Fluctuations, Broken Symmetry, and Correlation Functions*, W. A. Benjamin, Reading, Massachusetts (1975).
61. H. Mori, Transport, collective motion and Brownian motion, *Prog. Theor. Phys.* **33**, 423–455 (1965).
62. B. J. Berne, J. P. Boon, and S. A. Rice, On the calculation of autocorrelation functions of dynamical variables, *J. Chem. Phys.* **45**, 1086–1096 (1966).
63. M. Bixon, Dynamics of polymer molecules in dilute solution, *J. Chem. Phys.* **58**, 1459–1466 (1973).
64. R. Zwanzig, Theoretical basis for the Rouse-Zimm model in polymer solution dynamics, *J. Chem. Phys.* **60**, 2717–2720 (1974).
65. P. G. Wolynes and J. M. Deutch, Dynamical orientation correlations in solution, *J. Chem. Phys.* **67**, 733–741 (1977).
66. W. Dieterich and I. Peschel, Memory function approach to the dynamics of interacting Brownian particles, *Physica* **95A**, 208–224 (1979).
67. W. Hess and R. Klein, Dynamical properties of colloidal systems II, *Physica* **99A**, 463–493 (1979).
68. D. Levesque and L. Verlet, Computer "experiments" on classical fluids III, *Phys. Rev.* **A2**, 2514–2528 (1970).
69. F. Lantelme, P. Turq, and P. Schofield, On the use of memory functions in the study of the dynamical properties of ionic liquids, *J. Chem. Phys.* **71**, 2507–2513 (1979).
70. W. Hess and R. Klein, Long-time versus short-time behavior of a system of interacting Brownian particles, *J. Phys. A: Math. Gen.* **13**, L5–L10 (1980).
71. T. Keyes in *Statistical Mechanics: Time-Dependent Processes*, (B. J. Berne, ed.), Plenum, New York (1977), pp. 259–309.
72. J. Bosse, W. Götze, and M. Lücke, Mode-coupling theory of simple classical liquids, *Phys. Rev.* **A17**, 434–454 (1978).
73. A. R. Altenberger and J. M. Deutch, Light scattering from dilute macromolecular solutions, *J. Chem. Phys.* **59**, 894–898 (1973).

74. S. Harris, Diffusion effects in solutions of Brownian particles, *J. Phys. A: Math. Gen.* **11**, 1895–1898 (1976).
75. W. Hess and R. Klein, Theory of light scattering from a system of interacting Brownian particles, *Physica* **85A**, 509–527 (1976).
76. J. L. Anderson and C. C. Reed, Diffusion of spherical macromolecules at finite concentration, *J. Chem. Phys.* **64**, 3240–3250 (1976).
77. P. N. Pusey, in *Photon Correlation and Light Beating Spectroscopy*, (H. Z. Cummins and E. R. Pike, eds.) Plenum, New York (1974), pp. 387–429.
78. R. B. Jones, Diffusion of tagged interacting spherical polymers, *Physica* **97A**, 113–126 (1979).
79. J. M. Irving and J. G. Kirkwood, The statistical mechanical theory of transport processes IV: The Equations of hydrodynamics, *J. Chem. Phys.* **18**, 817–829 (1950).
80. A. R. Altenberger, Generalized diffusion processes and light scattering from a moderately concentrated solution of spherical particles, *Chem. Phys.* **15**, 269–277 (1976).
81. A. R. Altenberger, On the wave vector dependent mutual diffusion of interacting Brownian particles, *J. Chem. Phys.* **70**, 1994–2002 (1979).
82. P. N. Pusey, D. W. Schaefer, D. E. Koppel, R. D. Camerini-Otero, and R. M. Franklin, A study of the diffusion properties of R17 virus by time-dependent light scattering, *J. Phys. (Paris)* **33**, C1-163–168 (1972).
83. D. W. Schaefer and B. J. Berne, Dynamics of charged macromolecules in solution, *Phys. Rev. Lett.* **32**, 1110–1113 (1974).
84. J. C. Brown, P. N. Pusey, J. W. Goodwin, and R. H. Ottewill, Light scattering study of dynamic and time-averaged correlations in dispersions of charged particles, *J. Phys. A: Math. Gen.* **8**, 664–682 (1975).
85. D. W. Schaefer and B. J. Ackerson, Melting of colloidal crystals, *Phys. Rev. Lett.* **35**, 1448–1451 (1975).
86. P. S. Dalberg, A. Bøe, K. A. Strand, and T. Sikkeland, Quasielastic light scattering study of charged polystyrene particles in water, *J. Chem. Phys.* **69**, 5473–5478 (1978).
87. F. Grüner and W. Lehmann, The *k* dependence of the long-time diffusion in systems of interacting Brownian particles, *J. Phys. A: Math. Gen.* **12**, L303–L307 (1979).
88. F. Grüner and W. Lehmann, On the long time diffusion of interacting Brownian particles, in *Light Scattering in Fluids and Macromolecular Solutions*, (V. Degiorgio, M. Corti, and M. Giglio, eds, Plenum, New York (1980).
89. N. A. Clark, A. J. Hurd, and B. J. Ackerson, Single colloidal crystals, *Nature* **281**, 57–60 (1979).
90. D. W. Schaefer, Colloidal suspensions as soft core liquids, *J. Chem. Phys.* **66**, 3980–3984 (1977).
91. K. Gaylor, W. van Megen, and I. Snook, Structure of dispersions of small, strongly interacting particles, *J. Chem. Soc. Farad. Trans. II* **75**, 451–455 (1979), and references therein.
92. F. Grüner and W. Lehmann, Multiple scattering of light in a system of interacting Brownian particles, *J. Phys. A: Math. Gen.* **13**, 2155–2170 (1980).
93. M. C. Wang and G. E. Uhlenbeck, On the theory of Brownian motion II, *Rev. Mod. Phys.* **17**, 323–342 (1945), reprinted in reference 4.
94. D. L. Ermak, Polyion–polyion interaction effects in a model polyelectrolyte solution, *Automedica (GB)* **3**, 39–45 (1979).
95. K. J. Gaylor, I. K. Snook, W. van Megen, and R. O. Watts, Brownian dynamics studies of dilute dispersions, *Chem. Phys.* **43**, 233–239 (1979).
96. K. J. Gaylor, I. K. Snook, and W. van Megen, Comparison of Brownian dynamics with photon correlation spectroscopy of strongly interacting colloidal particles, *J. Chem. Phys.* **75**, 1682–1689 (1981).

97. G. H. Vineyard, Scattering of slow neutrons by a liquid, *Phys. Rev.* **110**, 999–1010 (1958).
98. W. Hess and R. Klein, Dynamical properties of colloidal systems III: Collective and self diffusion of interacting charged particles, *Physica* **105A**, 552–576 (1981).
99. P. Doty and R. F. Steiner, Macro-ions I: Light scattering theory and experiments with bovine serum albumin, *J. Chem. Phys.* **20**, 85–94 (1952).
100. T. Raj and W. H. Flygare, Diffusion studies on bovine serum albumin by quasielastic light scattering, *Biochemistry*, **13**, 3336–3340 (1974).
101. R. Finsy, A. Devriese, and H. Lekkerkerker, Light scattering study of the diffusion of interacting particles, *J. Chem. Soc. Farad. Trans. II* **76**, 767–775 (1980).
102. P. Doherty and G. B. Benedek, The effect of electric charge on the diffusion of macromolecules, *J. Chem. Phys.* **61**, 5426–5434 (1974).
103. J. D. Harvey, R. Geddes, and P. R. Wills, Conformational studies of BSA using laser light scattering, *Biopolymers* **18**, 2249–2260 (1979).
104. J. Newman, H. L. Swinney, S. A. Berkowitz, and L. A. Day, Hydrodynamic properties and molecular weight of FD bacteriophage DNA, *Biochemistry* **13**, 4832–4838 (1974).
105. M. M. Kops-Werkhoven and H. M. Fijnaut, Light scattering from sterically stabilized silica particles, in *Light Scattering in Fluids and Macromolecular Solutions*, (V. Degiorgio, M. Corti, and M. Giglio, eds.), Plenum, London (1980); Light scattering and sedimentation experiments on silica dispersions at finite concentrations, *J. Chem. Phys.* **74**, 1618–1625 (1981).
106. G. D. J. Phillies, G. B. Benedek, and N. A. Mazer, Diffusion in protein solutions at high concentrations: a study by quasielastic light scattering, *J. Chem. Phys.* **65**, 1883–1892 (1976).
107. B. D. Fair, D. Y. Chao, and A. M. Jamieson, Mutual translational diffusion coefficients in bovine serum albumin solutions measured by quasielastic laser light scattering, *J. Colloid Interface Sci.* **66**, 323–330 (1978).
108. M. B. Weissman, R. C. Pan, and B. R. Ware, Electrostatic contributions to the viscosities and diffusion coefficients of macroion solutions, *J. Chem. Phys.* **70**, 2897–2903 (1979).
109. M. B. Weissman and J. Marque, Small-ion friction, hydrodynamic interaction and charge fluctuation effects in solutions of a globular macroion, *J. Chem. Phys.* **73**, 3999–4004 (1980).
110. K. H. Keller, E. R. Canales and S. I. Yum, Tracer and mutual diffusion coefficients of proteins, *J. Phys. Chem.* **75**, 379–387 (1971).
111. S. S. Alpert and G. Banks, The concentration dependence of the hemoglobin mutual diffusion coefficient, *Biophys. Chem.* **4**, 287–296 (1976).
112. C. R. Jones, C. S. Johnson, and J. T. Penniston, Photon correlation spectroscopy of hemoglobin: diffusion of oxy-HbA and oxy-HbS, *Biopolymers* **17**, 1581–1593 (1978).
113. J. L. Anderson, F. Rauh, and A. Morales, Particle diffusion as a function of concentration and ionic strength, *J. Phys. Chem.* **82**, 608–616 (1978).
114. W. B. Veldkamp and J. R. Votano, Effects of intermolecular interaction on protein diffusion in solution, *J. Phys. Chem.* **80**, 2794–2801 (1976).
115. M. Corti and V. Degiorgio, Quasielastic light scattering study of intermicellar interactions, in *Light Scattering in Liquids and Macromolecular Solutions*, (V. Degiorgio, M. Corti, and M. Giglio, eds.), Plenum, New York (1980).
116. D. O. Shah and R. S. Schecter, *Improved Oil Recovery by Surfactant and Polymer Flooding*, Academic, New York (1977).
117. A. M. Bellocq, J. Biais, B. Clin, P. Lalanne, and B. Lemanceau, Study of dynamical and structural properties of microemulsions by chemical physics methods, *J. Colloid Interface Sci.* **70**, 524–536 (1979).
118. A Graciaa, J. Lachaise, P. Chabrat, L. Letamendia, J. Rouch, C. Vaucamps, M. Bourrel,

and C. Chambu, Light beating spectroscopy measurements of microemulsion diffusion coefficient, *J. Phys. (Paris) Lett.* **38**, 253–257 (1977).
119. A. Graciaa, J. Lachaise, P. Chabrat, L. Letamendia, J. Rouch, and C. Vaucamps, Light beating spectroscopy measurements of micelles mutual diffusion coefficient within oil in water microemulsions in the presence of sodium chloride, *J. Phys. (Paris) Lett.* **39**, 235–238 (1978).
120. R. A. Day, B. H. Robinson, J. H. R. Clarke, and J. V. Doherty, Characterization of water-containing reversed micelles by viscosity and dynamic light scattering methods, *J. Chem. Soc. Faraday Trans. II* **75**, 132–139 (1979).
121. M. Zulauf and H. Eicke, Inverted micelles and microemulsions in the ternary system H_2O/aerosol–OT/isooctane as studied by photon correlation spectroscopy, *J. Phys. Chem.* **83**, 480–486 (1979).
122. E. Sein, J. R. Lalanne, J. Buchert, and S. Kielich, Dynamic aspect of Rayleigh scattering and viscosity of ternary system aerosol-OT/water/carbon tetrachloride, *J. Colloid Interface Sci.* **72**, 363–366 (1979).
123. A. M. Cazabat, D. Langevin, and A. Pouchelon, Light-scattering study of water-in-oil microemulsions, *J. Coll. Interface Sci.* **73**, 1–12 (1980).
124. A. M. Cazabat and D. Langevin, Diffusion of interacting particles: light scattering study of microemulsions, *J. Chem. Phys.* **74**, 3148–3158 (1981).
125. A. M. Bellocq, G. Fourche, P. Chabrat, L. Letamendia, J. Rouch, and C. Vaucamps, Dynamic light scattering study of concentrated W/O microemulsions, *Opt. Acta* **27**, 1629–1639 (1980).
126. D. J. Cebula, R. H. Ottewill, J. Ralston, and P. N. Pusey, Investigations of microemulsions by light scattering and neutron scattering, *J. Chem. Soc. Faraday Trans. I*, **77**, 2585–2612 (1981).
127. H. M. Fijnaut, C. Pathmamanoharan, E. A. Nieuwenhuis, and A. Vrij, Dynamic light scattering from concentrated colloidal dispersions, *Chem. Phys. Lett.* **59**, 351–355 (1978).
128. M. J. Stephen, Spectrum of light scattered from charged macromolecules in solution, *J. Chem. Phys.* **55**, 3878–3883 (1971).
129. L. Friedhoff and B. J. Berne, Irreversible thermodynamic analysis of electrophoretic light scattering experiments, *Biopolymers* **15**, 21–28 (1976).
130. J. M. Schurr, A theory of electrolyte friction on translating polyelectrolytes, *Chem. Phys.* **45**, 119–132 (1980).
131. M. B. Weissman, Hydrodynamic interaction effects on spherical macroion diffusion, *J. Chem. Phys.* **73**, 3997–3998 (1980).
132. Z. Alexandrowicz and E. Daniel, Sedimentation and diffusion of polyelectrolytes. Part 1: Theoretical description, *Biopolymers* **1**, 447–471 (1963).
133. G. D. J. Phillies, Contribution of slow charge fluctuations to light scattering from a monodisperse solution of macromolecules, *Macromolecules* **9**, 447–450 (1976).
134. K. Freed, Polymer dynamics and the hydrodynamics of polymer solutions, in *Progress in Liquid Physics*, (C. Croxton, ed.), Wiley, New York (1978), 343–390.
135. P. N. Pusey and R. J. A. Tough, Langevin approach to the dynamics of interacting Brownian particles, *J. Phys. A: Math. Gen.* **15**, 1291–1308 (1982); Corrigendum **16**, 2889 (1983).
136. P. N. Pusey and R. J. A. Tough, Hydrodynamic interactions and diffusion in concentrated particle suspensions, *Faraday Discuss. Chem. Soc.* **76**, 123–136 (1983).
137. R. J. A. Tough and P. N. Pusey, Short-time dynamics of hydrodynamically-interacting Brownian particles, to be published.
138. J. L. Arauz-Lara and M. Medina-Noyola, Sum rules for $S_{\alpha\beta}(k, \omega)$ for two diffusing species, *Physica* **122A**, 547–562 (1983).

139. D. L. Ermak and J. A. McCammon, Brownian dynamics with hydrodynamic interactions, *J. Chem. Phys.* **69**, 1352–1360 (1978).
140. P. Mazur, On the motion and Brownian motion of *n* spheres in a viscous fluid, *Physica* **110A**, 128–146 (1982).
141. W. B. Russel and A. B. Glendinning, The effective diffusion coefficient detected by dynamic light scattering, *J. Chem. Phys.* **74**, 948–952 (1981).
142. H. M. Fijnaut, Wave vector dependence of the effective diffusion coefficient of Brownian particles, *J. Chem. Phys.* **74**, 6857–6863 (1981).
143. G. D. J. Phillies and P. R. Wills, Light scattering spectrum of a suspension of interacting Brownian macromolecules, *J. Chem. Phys.* **75**, 508–514 (1981).
144. P. R. Wills, Wave vector dependence of the effective diffusion coefficient for solutions of macromolecules, *J. Phys. A: Math. Gen.* **14**, 3093–3099 (1981).
145. For a comprehensive review see W. Hess and R. Klein, Generalized hydrodynamics of systems of Brownian particles, *Adv. Phys.* **32**, 173–283 (1983).
146. P. N. Pusey and R. J. A. Tough, Dynamic light scattering: a probe of Brownian particle dynamics, *Adv. Colloid Interface Sci.* **16**, 143–159 (1982).
147. J. A. Marqusee and J. M. Deutch, Concentration dependence of the self-diffusion coefficient, *J. Chem. Phys.* **73**, 5396–5397 (1980).
148. G. K. Batchelor, Sedimentation in a dilute polydisperse system of interacting spheres. Part 1: General formulae, *J. Fluid Mech.* **119**, 379–408 (1982); Diffusion in a dilute polydisperse system of interacting spheres, *J. Fluid Mech.* **131**, 155–175 (1983).
149. S. Hanna, W. Hess, and R. Klein, Self-diffusion of spherical Brownian particles with hard-core interaction, *Physica* **111A**, 181–199 (1982).
150. S. Hanna, W. Hess, and R. Klein, The velocity autocorrelation function of an overdamped Brownian system with hard-core interaction, *J. Phys. A: Math. Gen.* **12**, L493–498 (1981).
151. R. B. Jones and G. S. Burfield, Memory effects in the diffusion of an interacting polydisperse suspension I: Projection formalism at long wavelength; II: Hard spheres at low density, *Physica* **111A**, 562–576 and 577–590 (1982).
152. B. J. Ackerson and L. Fleishman, Correlations for dilute hard core suspensions, *J. Chem. Phys.* **76**, 2675–2679 (1982).
153. R. J. A. Tough, Self-diffusion in a suspension of interacting Brownian particles, *Mol. Phys.* **46**, 465–474 (1982).
154. H. N. W. Lekkerkerker and J. K. G. Dhont, On the calculation of the self-diffusion coefficient of interacting Brownian particles, *J. Chem. Phys.* **80**, 5790–5792 (1984).
155. P. Mazur and W. van Saarloos, Many-sphere hydrodynamic interactions and mobilities in a suspension, *Physica* **115A**, 21–57 (1982).
156. M. Muthukumar and K. Freed, Cluster expansion for concentration dependence of self-friction coefficients for suspensions of interacting spheres; Cluster expansion for concentration dependence of cooperative friction coefficients for suspensions of interacting spheres, *J. Chem. Phys.* **78**, 497–510 and 511–519 (1983).
157. C. W. J. Beenakker and P. Mazur, Self-diffusion of spheres in a concentrated suspension, *Physica* **120A**, 388–410 (1983); Diffusion of spheres in a concentrated suspension: resummation of many-body hydrodynamic interactions, *Phys. Lett.* **98A**, 22–24 (1983); Diffusion of spheres in a concentrated suspension II, in press.
158. W. van Megen, R. H. Ottewill, S. M. Owens, and P. N. Pusey, Measurement of the wave-vector dependent diffusion coefficient in concentrated particle dispersions, *J. Chem. Phys.*, in press.
159. P. N. Pusey and W. van Megen, Measurement of the short-time self-mobility of particles in concentrated suspension. Evidence for many-particle hydrodynamic interactions, *J. Phys. (Paris)* **44**, 285–291 (1983).

160. A. B. Glendinning and W. B. Russel, A pairwise additive description of sedimentation and diffusion in concentrated suspensions of hard spheres, *J. Colloid Interface Sci.* **89**, 124–143 (1982).
161. W. van Megen, I. Snook, and P. N. Pusey, Diffusion in concentrated hard sphere dispersions: effects of two and three particle mobilities, *J. Chem. Phys.* **78**, 931–936 (1983).
162. M. M. Kops-Werkhoven, H. J. Mos, P. N. Pusey, and H. M. Fijnaut, Dynamic light scattering and optical contrast in concentrated silica dispersions, *Chem. Phys. Lett.* **81**, 365–370 (1981).
163. M. M. Kops-Werkhoven and H. M. Fijnaut, Dynamic behavior of silica dispersions studied near the optical matching point, *J. Chem. Phys.* **77**, 2242–2253 (1982).
164. M. M. Kops-Werkhoven, C. Pathmamanoharan, A. Vrij, and H. M. Fijnaut, Concentration dependence of the self diffusion coefficient of hard spherical particles measured with photon correlation spectroscopy, *J. Chem. Phys.* **77**, 5913–5922 (1982).
165. W. Härtl and H. Versmold, An experimental verification of incoherent light scattering, *J. Chem. Phys.* **80**, 1387–1389 (1984).
166. P. N. Pusey, H. M. Fijnaut, and A. Vrij, Mode amplitudes in dynamic light scattering by concentrated liquid suspensions of polydisperse hard spheres, *J. Chem. Phys.* **77**, 4270–4281 (1982).
167. J. P. Boon and S. Yip, *Molecular Hydrodynamics*, McGraw-Hill, New York (1980); A. Rahman, K. S. Singwi, and A. Sjölander, Theory of slow neutron scattering by liquids. I, *Phys. Rev.* **126**, 986–996 (1962).
168. *Inelastic Scattering of Neutrons in Solids and Liquids*, International Atomic Energy Agency, Vienna (1961).
169. J. M. Schurr, The thermodynamic driving force in mutual diffusion of hard spheres, *Chem. Phys.* **65**, 217–223 (1982).
170. M. M. Kops-Werkhoven, A. Vrij, and H. N. W. Lekkerkerker, On the relation between diffusion, sedimentation and friction, *J. Chem. Phys.* **78**, 2760–2763 (1983).
171. See, for example, *Physics of Amphiphiles: Micelles, Vesicles and Microemulsions*, International School of Physics "Enrico Fermi," V. Degiorgio and M. Corti, eds., North-Holland, Amsterdam, in press.
172. A. R. Altenberger, On the Rayleigh light scattering from dilute solutions of charged spherical macromolecules, *Opt. Acta* **27**, 345–352 (1980).
173. R. S. Hall and C. S. Johnson, Experimental evidence that mutual and tracer diffusion coefficients for hemoglobin are not equal, *J. Chem. Phys.* **72**, 4251–4253 (1980).
174. R. S. Hall, Y. S. Oh, and C. S. Johnson, Photon correlation spectroscopy in strongly absorbing and concentrated samples with applications to unliganded hemoglobin, *J. Phys. Chem.* **84**, 756–767 (1980).
175. Y. S. Oh and C. S. Johnson, The wave vector dependence of diffusion coefficients in photon correlation spectroscopy of protein solutions, *J. Chem. Phys.* **74**, 2717–2720 (1981).
176. D. R. Bauer, *J. Phys. Chem.* **84**, 1592–1598 (1980); also in *Polymer Colloids II*, (R. M. Fitch, ed.), Plenum, New York (1980).
177. R. Giordano, A. Salleo, S. Salleo, F. Mallamace, and F. Wanderlingh, Diffusion coefficient of lysozyme in water, *Opt. Acta* **27**, 1465–1472 (1980).
178. C. M. Trotter and D. N. Pinder, Laser light scattering from concentrated solutions of polystyrene latex spheres: a comparative study. *J. Chem. Phys.* **75**, 118–127 (1981).
179. A. M. Cazabat, D. Chatenay, D. Langevin, and A. Pouchelon, Light scattering study of microemulsions and its relation to percolation phenomena, *J. Phys. Lett.* (*Paris*) **41**, L441–L445 (1980).
180. C. Van den Broeck, F. Lostak and H. N. W. Lekkerkerker, The effect of direct interactions on Brownian diffusion, *J. Chem. Phys.* **74**, 2006–2010 (1981).

181. G. D. J. Phillies, Experimental demonstration of multiple scattering suppression in quasielastic-light-scattering spectroscopy by homodyne coincidence techniques, *Phys. Rev.* **A24**, 1939–1943 (1981).
182. G. D. J. Phillies, Non-hydrodynamic contribution to the concentration dependence of the self-diffusion of interacting Brownian particles, *Chem. Phys.* **74**, 197–203 (1981).
183. T. Tsang and H. T. Tang, Light scattering and dynamics of interacting Brownian particles, *J. Chem. Phys.* **76**, 3873–3876 (1982).
184. J. M. Schurr, The fluctuating-force formalism of friction drag coefficients, *Chem. Phys.* **71**, 101–104 (1982).
185. T. Ohtsuki and K. Okano, Diffusion coefficients of interacting Brownian particles, *J. Chem. Phys.* **73**, 1443–1450 (1982).
186. T. Ohtsuki, Dynamical properties of strongly interacting Brownian particles, *Physica* **110A**, 606–616 (1982).
187. T. Ohtsuki, Generalized diffusion equation of interacting Brownian particles, *Chem. Phys. Lett.* **98**, 121–124 (1983).
188. T. Ohtsuki, Dynamical properties of strongly interacting Brownian particles. III: Binary mixtures, *Physica* **122A**, 212–230 (1983).
189. G. T. Evans and C. P. James. A caiculation of the self-diffusion coefficient for a dilute solution of Brownian particles, *J. Chem. Phys.* **79**, 5553–5557 (1983).
190. B. U. Felderhof and R. B. Jones, Linear response theory of the sedimentation and diffusion in a suspension of spherical particles, *Physica* **119A**, 591–608 (1983).
191. B. U. Felderhof and R. B. Jones, Cluster expansion of the diffusion kernel of a suspension of interacting Brownian particles, *Physica* **121A**, 239–244 (1983).
192. B. U. Felderhof and R. Jones, Diffusion in hard sphere suspensions, *Physica* **122A**, 89–104 (1983).
193. J. K. G. Dhont, Multiple Rayleigh–Gans–Debye scattering in colloidal systems—general theory and static light scattering, *Physica* **120A**, 238–262 (1983).
194. R. I. Cukier, Diffusion of interacting Brownian particles in a fluid with fixed macroparticles, *J. Chem. Phys.* **79**, 3911–3920 (1983).
195. N. Yoshida, Contribution of electrostatic interactions to the concentration dependence of the self-diffusion coefficient of Brownian particles, *Chem. Phys. Lett.* **102**, 83–87 (1983).
196. C. Andries, W. Guedens, and J. Clauwaert, Photon and fluorescence correlation spectroscopy and light scattering of eye-lens proteins at moderate concentrations, *Biophys. J.* **43**, 345–354 (1983).
197. B. Nystrom and R. M. Johnsen, Effect of concentration and ionic strength of lysozyme, *Chem. Scr.* **22**, 82–84 (1983).

5

Quasielastic Light Scattering from Dilute and Semidilute Polymer Solutions†

D. W. Schaefer

Sandia National Laboratories
Albuquerque, New Mexico 87185

C. C. Han

National Bureau of Standards
Washington, D.C. 20234

5.1. INTRODUCTION

Application of photon correlation spectroscopy (PCS) to polymer dynamics dates back to the inception of the technique in the early 1960s. Early work demonstrated that the self-diffusion constant of large molecules could be obtained by application of PCS to dilute solutions. In addition, some early work dealt with the initial concentration dependence of the diffusion constant. Nevertheless, major advances in the understanding of the dynamics of chain molecules has only occurred in the last few years. In this period there has been a dramatic increase in he amount and significance of PCS activity in polymer physics. The purpose of this review is to summarize and hopefully justify this more recent work.

Two significant factors led to the upsurge of interest by the light scattering community in polymer dynamics. One factor is the application of new theoretical techniques based on scaling methods, renormalization concepts, and linear response techniques to polymer systems. These methods not only brought simple interpretation of known results, but also led to new insights and novel predictions.

† Prepared by Sandia National Laboratories, for the United States Department of Energy under Contract No. DE-AC04-76DP00789

The development of small angle x-ray and neutron techniques has also contributed to the increased interest in polymer physics. These methods are ideally suited to polymers since they are sensitive to spatial dimensions characteristic of chain molecules (100 Å). In addition, neutron scattering coupled with deuterium tagging permits unique experimental measurements such as the properties of a single tagged chain in a matrix of chemically identical untagged chains. Small angle techniques have primarily elucidated static properties since flux problems have precluded quasielastic studies of polymer dynamics. Except for limited experiments based on the newly developed spin-echo technique, PCS has provided the only microscopic probe of polymer dynamics.

The goal of this chapter is not to review the PCS literature but rather to demonstrate how PCS has enhanced understanding of the statistical properties of chain molecules in solution. Since polymer gels, polymer glasses, and phase phenomena are treated elsewhere in this volume, these subjects are not considered. Basic understanding of PCS techniques and polymer fundamentals is assumed. In addition, lengthy derivations are eliminated in favor of ultimate results and interpretation.

For simplicity the single chain problem is treated before adding the complicating effects of interchain interactions and entanglements. The dynamics of a single polymer chain are complex, with many possible approaches to the problem. Therefore, this subject is first treated in a simplified manner employing scaling ideas and later in more detail and rigor using the Zwanzig–Mori formalism. In both cases the renormalization or "blob" concept is introduced to interpret subtle aspects of the molecular wright dependence of the dynamics of a single chain. After the single chain problem, the virial regime is briefly reviewed. The discussion of the virial regime naturally leads to consideration of semidilute solutions. In contrast to the dilute case, the semidilute system is sufficiently complex that a rigorous treatment, even with severe approximations, is not possible. Fortunately, however, dynamical scaling yields results which now seem to be confirmed by PCS measurements in good solvents. Little consideration is given to the concentrated regime since an adequate model is lacking and few experimental data are available.

5.2. THE SINGLE CHAIN

5.2.1. Basic Polymer Statistics

Understanding of the statistical properties of chain molecules centers around two fundamental concepts: the ideal or random-flight chain and

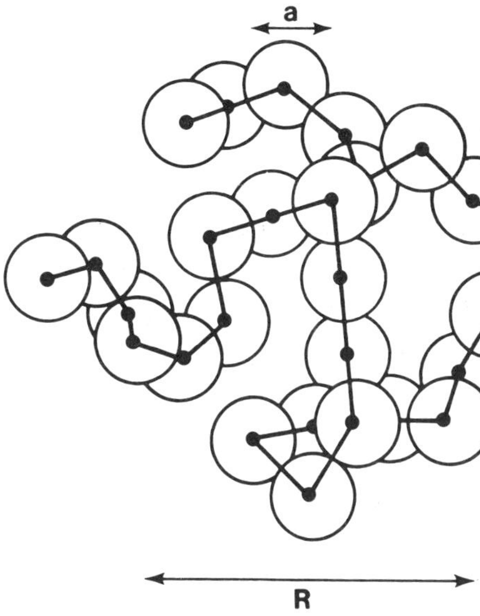

Figure 1. Schematic picture of a polymer. a^3 is the volume associated with a monomer.

chain expansion due to excluded volume. Review of these ideas will lay the groundwork for the discussion of the dynamical properties both in dilute and semidilute solution. The next few sections will attempt to develop a simple and somewhat schematic view of polymers and their dynamical motions.

The simplest model of a polymer is that of a series of connected, noninteracting balls (monomers) of diameter a as illustrated in Figure 1. If the connections between the monomers are stiff universal joints, then the sequence of vectors connecting them defines a random-flight trajectory. In the limit of many joints the radius R of such a sequence is well known[1]

$$R \equiv \langle R^2 \rangle^{1/2} = n^{1/2} a N^{1/2} \tag{1}$$

where N is the number of steps in the sequence or the degree of polymerization, n is a constant, and a is the segment length. The constant n is related to the chain stiffness and can be calculated from the characteristic ratio C_∞ of the chain.[1] At this point it is not necessary to precisely define the radius R, which may be the end-to-end distance, the radius of gyration R_g, or the radius of hydration R_h. These various lengths differ through the factor n. In consideration of dynamic properties R will generally be associated with the hydrodynamic radius R_h. In polymer solutions, ideal or

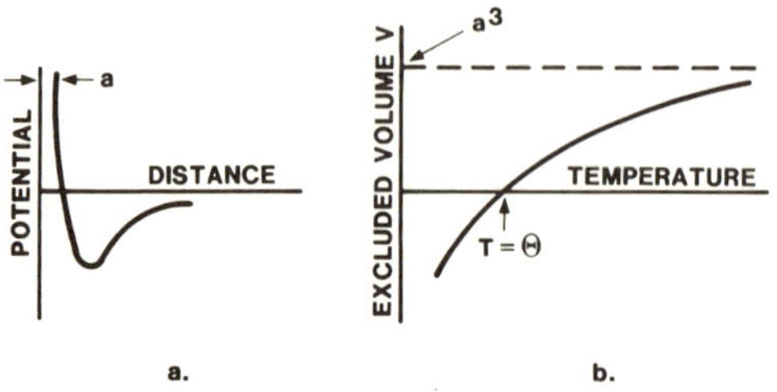

Figure 2. The effective potential between monomers is similar to that between any molecules: repulsive at short range and attractive at larger distances. Figure 2b shows the temperature dependence of the effective interaction between monomers, which is called the excluded volume v. At high temperature, v approaches a^3.

Gaussian behavior described by equation (1), is observed in some cases. The particular temperature at which equation (1) is observed is called the theta temperature ($T = \theta$) and at this point the effective two-body interaction between monomers (excluded volume) vanishes.

Away from the theta point, chains may be swollen or collapsed due to the repulsive or attractive forces between monomers. The forces between monomers are conveniently characterized by an excluded volume v, a concept borrowed from the virial theory of dilute gases.[2] If the temperature is high, monomer interaction is dominated by the repulsive part of the pair potential illustrated in Figure 2a. At high T then the effective excluded volume is just the hard core volume, $v \cong a^3$. At lower temperatures the attractive part of the pair potential dominates and the excluded volume becomes negative, indicative of a net attraction. The point at which $v = 0$, illustrated in Figure 2b, is the theta temperature where the excluded volume or effective two-body interaction vanishes. In this chapter only the regime $v \geq 0$ will be considered. The collapsed state ($v \leq 0$) is intimately connected with phase separation and therefore is not treated here.

The properties of polymers in the swollen state ($T > \theta$) were originally worked out in the mean-field approximation by Flory.[3] Basically, Flory considers the free energy of the chain to consist of an entropic or elastic contribution plus an enthalpic contribution due to the excluded volume interaction. Minimization of the free energy then leads to the so-called Flory law

$$R = a\left(\frac{v}{a^3}\right)^{1/5} n^{1/5} N^{3/5} \tag{2}$$

Although the Flory approach has well-known shortcomings,[4] the 3/5 exponent in the molecular weight dependence of R is well established experimentally for polymers dissolved in good solvents. Renormalization-group calculations yield an exponent 2% smaller than Flory's calculation, but the difference is beyond experimental resolution at this time.

In solvents of intermediate quality, excluded volume effects are weak and scaling exponents between 1/2 and 3/5 are often observed experimentally. This intermediate regime can be analyzed by perturbation theory,[5] where R is expanded about $T = \theta$ in terms of an expansion factor $Z = va^{-3}N^{1/2}$. Perturbation theory is only expected to be valid near the theta point where v is approximately linear in the temperature increment $\tau = (T - \theta)/T$

$$\frac{v}{a^3} = 1 - 2\chi = (1 - 2\chi_s)\tau, \qquad \tau \ll 1 \tag{3}$$

where the Flory[6] reduced residual chemical potential χ and reduced residual partial-molar entropy χ_s have been introduced. The validity of perturbation theory requires $Z \ll 1$ or $\tau \ll N^{-1/2}$. In addition, the so-called two-parameter perturbation approach also requires that two-body interactions dominate three-body interactions. In virial language this condition becomes

$$v > w\rho \tag{4}$$

where w is the three-body excluded volume or third-virial coefficient. By analogy with simple fluids,[7] $w \sim a^6$ so that condition (4) reduces to $\tau \gg N^{-1/2}$ since the density ρ inside the chain is proportional to $N^{-1/2}$.† Except for unusual circumstances, therefore, it is unlikely to have weak two-body interactions which still dominate three-body effects.

Recently, Farnoux *et al.*[8] developed an interesting alternative to perturbation theory for the intermediate regime. In this so-called "blob" approach short sequences are approximated as ideal ($R \sim N^{1/2}$) whereas long sequences are considered fully swollen ($R \sim N^{3/5}$). In spite of its simplicity, this model has been very successful in explaining both static and dynamic properties of dilute polymers. The blob concept will play a key role in the analysis of chain dynamics which follows.

† The symbol $\sim$ is used to define power-law dependences and does not imply that both sides of the equation have the same dimensions.

5.2.2. Dynamical Regimes

In a photon correlation experiment dynamical information is extracted from the intensity correlation function $\langle I(q, t)I(q, 0)\rangle$, which is directly related to the intermediate scattering function $S(q, t)$

$$\langle I(q, t)I(q, 0)\rangle - \langle I^2\rangle = \beta S^2(q, t), \qquad t > 0 \tag{5a}$$

$$S(q, t) = \frac{1}{V}\sum_{ij} \exp\{i\mathbf{q}\cdot[\mathbf{r}_i(t) - \mathbf{r}_j(0)]\} \tag{5b}$$

where V is the scattering volume, $\mathbf{q}$ is the scattering vector, and $\mathbf{r}_i(t)$ is the position of the monomer i at time t. β is a constant which depends on geometrical factors. Interpretation of light scattering experiments then requires a model for $S(q, t)$ which is consistent with the structure of the polymer and the laws of physics. Since $S(q, t)$ has been calculated exactly only for models with severe approximations, the complete interpretation of scattering experiments is therefore not possible at this time.

Fortunately, considerable progress in the interpretation of PCS experiments can be made on the basis of rather simple arguments. For example, in the limit $qR \ll 1$, $S(q, t)$ is sensitive to fluctuations whose Fourier spatial wavelength q^{-1} is large compared to the size of a single chain. In this regime it follows immediately that the dominant relaxation process is center-of-mass (CM) diffusion. By analogy to hard-sphere systems $S(q, t)$ is exponential with a characteristic decay rate $\Omega = Dq^2$. By contrast, when $1 \ll qR \ll qa$, only internal chain distortions are important and CM diffusion can be neglected. Finally, when $qa \simeq 1$ the motion of single monomers becomes important and the characteristic time of $S(q, t)$ is related to monomer mobility.

5.2.2.1. The Fundamental Relaxation Time. The scaling approach to polymer dynamics rests on the assumption that there is only one fundamental relaxation rate.(9) If the fundamental relaxation rate can be identified then scaling permits characterization of the relaxation processes in dynamical regimes where the fundamental process is no longer dominant. Support for the idea of a single relaxation parameter comes from the static properties of chain molecules. Here a single parameter R_g characterizes the monomer pair distribution function for both ideal and swollen chains. This single characterstic length underlies static scaling methods.

By analogy to the static situation it is reasonable to believe that the fundamental chain relaxation process will dominate when $qR = 1$. That is, the fundamental rate Ω_0 is associated with the fundamental length R. Since $qR = 1$ defines the crossover from CM to internal motion either of these processes could be used to define Ω_0. From the CM diffusion side, Ω_0 is the rate associated with diffusion over a distance R

$$\Omega_0 = \frac{D}{R^2} \tag{6}$$

Ω_0 could equally well be derived from the tumbling time of a Brownian sphere of radius R

$$\Omega_0 = \frac{kT}{\zeta_{ch} R^2} \tag{7}$$

where ζ_{ch} is the overall chain friction constant. Finally, if there is indeed but one characteristic rate at $qR = 1$ then chain distortions must also display consistent dynamics.

The dynamics of the distortional breathing mode can be established by balance of elastic forces with viscous forces as the chain expands or contracts from its equilibrium distribution.[9] The elastic force can be obtained from linear response theory by imagining that the chain ends are subjected to a fictitious force **f**. Since the energy associated with the resulting distortion $\delta\mathbf{r}$ is $\mathbf{f} \cdot \delta\mathbf{r}$, then the mean distortion is

$$\langle \delta\mathbf{r} \rangle = \frac{1}{Q} \int \delta\mathbf{r} \exp(-\beta H_0 + \mathbf{f} \cdot \delta\mathbf{r}) \tag{8}$$

where Q is the partition function and H_0 is the energy in the absence of f. For small distortions equation (8) can be linearized

$$\langle |\delta\mathbf{r}| \rangle = \frac{R^2}{kT} f \tag{9}$$

Inversion of (9) gives the elastic force associated with a distortion δr. When this force is balanced by the viscous force, we find

$$\frac{kT}{R^2} \langle |\delta\mathbf{r}| \rangle = \zeta_{ch}(\delta\dot{r}) \tag{10}$$

where ζ_{ch} is still the friction constant associated with the entire chain. Equation (10) implies an exponential time correlation function $S(q, t)$ with decay rate equivalent to equation (7). All three possible relaxation processes, therefore, give the same relaxation time since (6) and (7) are consistent with the Einstein formula for D

$$D \sim \frac{kT}{\zeta_{\text{ch}}} \tag{11}$$

5.2.2.2. Models of Chain Friction. The fundamental relaxation rate Ω_0 depends on the self-diffusion constant D (or equivalently on ζ_{ch}). It is D that contains model specific information. Because of the connectedness of the chain, the motions of the individual monomers are coupled by elastic as well as hydrodynamic interactions. Within certain approximations, however, the coupling can be handled and a simple picture of polymer dynamics emerges.

A reasonable starting point for calculation of D is the Kubo formula[4]

$$D = \frac{kT}{\zeta_{\text{ch}}} = \frac{1}{3}\frac{1}{N^2}\int_0^\infty \left\langle \sum_n \mathbf{V}_n(0) \cdot \sum_m \mathbf{V}_m(t) \right\rangle dt \tag{12}$$

where $\mathbf{V}_n(t)$ is the instantaneous velocity of monomer n at time t. The monomer velocities are correlated because the motion of one monomer necessarily leads to a velocity field in the solvent. This solvent motion then leads to a hydrodynamic force which moves other monomers. Solution of equation (12) requires detailed knowledge of the velocity correlations between monomers, and different levels of approximation have been considered to treat this problem.

Within what is known as the Rouse model, hydrodynamic interactions between monomers are completely ignored so that $\langle \mathbf{V}_n(0) \cdot \mathbf{V}_m(t)\rangle = \delta_{nm}\langle \mathbf{V}_n(0) \cdot \mathbf{V}_m(t)\rangle$ and equation (12) reduces to

$$D \cong \frac{D_m}{N} \tag{13}$$

where D_m is the monomer diffusion constant and η is the solvent viscosity,

$$D_m = \frac{kT}{\zeta_m} = \frac{1}{3}\int_0^\infty \langle \mathbf{V}_1(0) \cdot \mathbf{V}_1(t)\rangle \, dt \tag{14}$$

In the Rouse model then

$$\Omega_0 = \frac{D_m}{R^2 N} \tag{15}$$

Although the Rouse model is generally associated with ideal chains, equation (15) depends only on the lack of hydrodynamic interactions and is not dependent on ideal statistics. In spite of the drastic approximations involved in the Rouse model, Rouse-type dynamics may be expected under certain conditions. At extremely short times, for example, $S(q, t)$ may not reflect hydrodynamic interactions due to the finite relaxation time associated with the solvent viscosity. Also, in concentrated solution and melts, where hydrodynamic interactions are screened by other chains, Rouse behavior is expected. The latter case is just now under study by quasielastic neutron scattering.[69]

Solution of equation (12) with hydrodynamic interactions was originally obtained by Kirkwood and Riseman[10] (KR), and their calculation has been summarized in the present context by de Gennes.[4] The central approximation in the KR approach is that temporal and spatial correlations contained in equation (12) are independent so that a given monomer is assumed to feel the average rather than instantaneous flow field of all the other monomers. The result of the KR approach is relatively simple,

$$D = \frac{kT}{6\pi\eta}\left\langle\frac{1}{r}\right\rangle \tag{16}$$

where $\langle 1/r \rangle$ is an average over the static monomer pair correlation function $g(r)$. In the infinite chain limit there is only one length scale and $\langle 1/r \rangle$ is proportional to R^{-1} so that

$$D \sim \frac{kT}{\eta R} \tag{17}$$

and the fundamental relaxation time follows from equation (17)

$$\Omega_0 \sim \frac{kT}{\eta R^3} \tag{18}$$

when equation (18) is specialized to theta conditions ($R \sim N^{1/2}$), the result is generally associated with Zimm,[11] who first calculated the mode structure for the case of an ideal chain with hydrodynamic interactions. Equation (18) is more general, however, and should be valid for swollen chains as well.

The form of equation (18) arises because of the form of the flow field which results from monomer motion. This field, which is described by the Oseen tensor, is long range (decaying as $1/r$) so that monomers substantially removed from each other become hydrodynamically coupled. As a result, the effective friction constant in the KR approximation is greatly reduced compared to the Rouse case. The particular average in equation (16) is called the hydrodynamic radius R_h,

$$R_h = \left\langle \frac{1}{r} \right\rangle^{-1} \tag{19}$$

5.2.2.3. Scaling[4]. Having established the rate of the fundamental mode at $qR = 1$, it is now possible to proceed to the neighboring limiting regimes using scaling ideas.[12] In particular, for $qR \neq 1$ the relaxation rate is presumed to follow power law behavior,

$$\Omega = \Omega_0(qR)^x \tag{20}$$

where the exponent x is to be determined. For example, when $qR \ll 1$ then equation (20) should reduce to a diffusional form, i.e., $\Omega \sim q^2$, so that $x = 2$, and using (18),

$$\Omega = Dq^2, \qquad qR \ll 1 \tag{21}$$

as expected. In the opposite limit, $qR \gg 1$, the relaxation processes of interest are distortions internal to the chain so Ω should be independent of R. Therefore, from (18) and (20), $x = 3$ and

$$\Omega \sim \frac{kTq^3}{\eta}, \qquad qa \ll 1 \ll qR \tag{22a}$$

Equation (22a) predicts the q dependence of the relaxation rate for a Zimm chain with hydrodynamic interactions. If the same scaling argument is used for a Rouse chain (no hydrodynamic coupling) one finds

$$\Omega \sim \frac{kTq^4}{\eta} \tag{22b}$$

Some evidence exists for Rouse behavior in melts.[69]

In the limit $qa \cong 1$ a new relaxation process associated with the monomer mobility comes into the problem so that scaling is certain to fail. Nevertheless it is reasonable to expect that monomer translational diffusion will dominate and that

$$\Omega \sim D_m q^2 \tag{23}$$

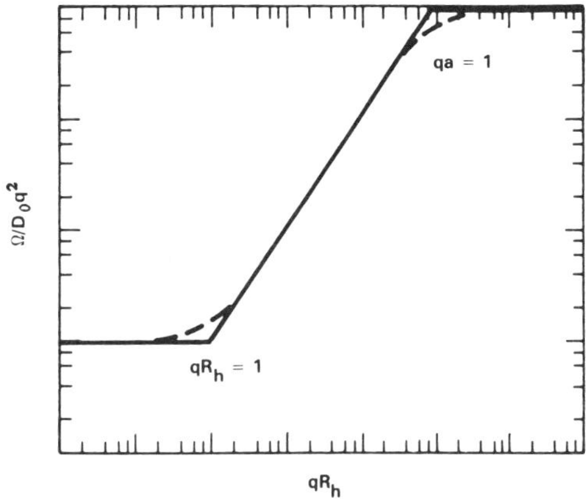

Figure 3. Schematic picture of the normalized relaxation rate $\Omega/D_0 q^2$ associated with $S(q, t)$. Crossover is observed at $qR_h = 1$ and $qa = 1$.

This regime is not accessible by light scattering but is a subject of investigation by quasielastic neutron scattering.[13]

5.2.2.4. Summary and Experimental Data. The above analysis indicates that three dynamical regimes will be observed depending on the value of qR. The three regimes are indicated schematically in Figure 3, where the reduced relaxation rate $\Omega/D_0 q^2$ is plotted versus log qR_h. In the regime $qR_h \ll 1$, $\Omega \sim D_0 q^2$ as expected for center-of-mass diffusion of a chain whose diffusion constant is D_0 at infinite dilution. At $qR_h = 1$, a crossover is observed to the regime where $\Omega \sim q^3$. The crossover is expected to occur at $qR_h = 1$ since R_h is the length scale associated with the fundamental relaxation rate Ω_0. In the regime $qR_h \gg 1$ scaling predicts q^3 dependence regardless of solvent quality. In this intermediate regime $qa \ll 1 \ll qR$, Ω is sensitive to internal dynamics or Rouse–Zimm modes. Finally, when $qa = 1$, a crossover is expected to diffusional motion of individual monomers with monomer diffusion constant D_m.

The general features of Figure 3 are reasonably confirmed by experimental data. Figure 4 shows measured reduced relaxation rates for polystyrene in several solvents ranging from a theta solvent (CH = cyclohexane) to a marginal solvent (EA = ethyl acetate) to good solvents (TOL = toluene, BZ = benzene, and THF = tetrahydrofuran). The mean relaxation rate Ω was obtained from the initial slope of the intensity correlation function defined in equation (5a):

$$\Omega(q) = -\lim_{t \to 0} \frac{d \ln[S(q, t)/S(q, 0)]}{dt} \tag{24a}$$

$$= D_0 q^2, \qquad qR \ll 1, \qquad \phi \to 0 \tag{24b}$$

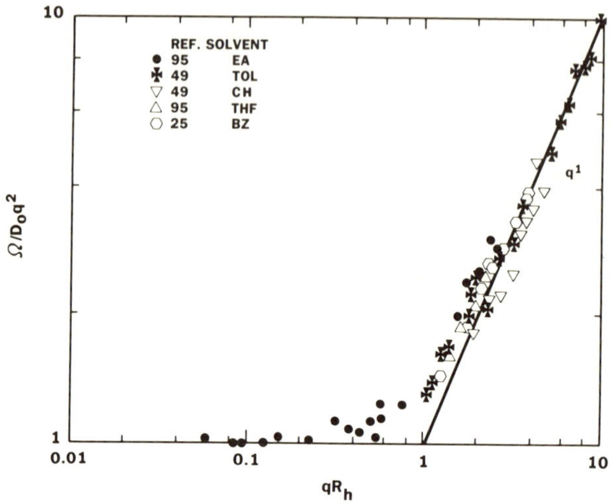

Figure 4. Normalized relaxation rate for PS in several solvents. A universal curve is seen with $\Omega \sim q^3$ in the intermediate region regardless of solvent quality. The log–log plot obscures slight differences in the relaxation rate in theta solvents as opposed to good solvents.

D_0 is the diffusion constant at infinite dilution and ϕ is monomer volume fraction.

The data in Figure 4 demonstrate both the beauty and limitations of the scaling approach. In the regime $qR \ll 1$, diffusional behavior is observed and the relation $\Omega = D_0 q^2$ has been found in many systems. The precise meaning of D_0, however, is not defined by scaling, and in fact questions still exist concerning the dependence of D_0 on the molecular weight. In the following sections a more detailed analysis of D_0 is presented.

Figure 4 shows that the crossover regime spans about one order in q. This crossover regime, which is not subject to scaling analysis, covers many experimental situations. Fortunately, the crossover can be treated by methods outlined in Section 5.2.4.

In the intermediate region, $qa \ll 1 \ll qR$, the data are consistent with scaling predictions, namely, $\Omega \sim q^3$. The scatter in the data, however, obscures any differences which might exist between the data sets in theta vs. good solvents. In fact, slight differences are predicted by linear response theory and these differences are analyzed in more detail in Section 5.2.4.

The high-q regime, where monomer diffusion dominates, is not subject to study by light scattering since the conditions $qa \simeq 1$ cannot be realized. Nevertheless, some results from quasielastic neutron scattering are available, and these data are consistent with the existence of crossover to $\Omega \sim D_m q^2$ at high q. Figure 5 shows data of Nicholson *et al.*[13] taken by

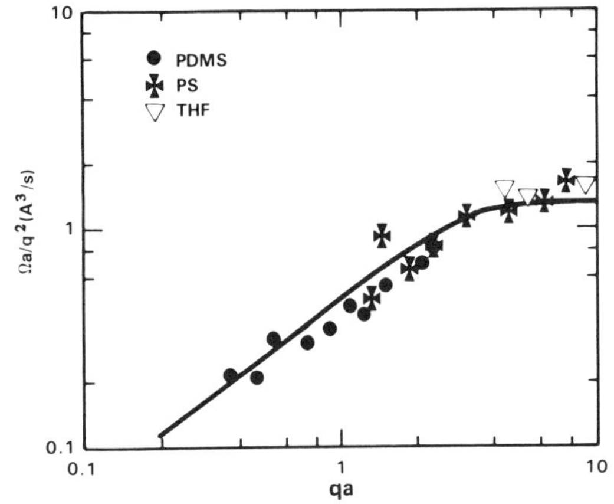

Figure 5. Normalized relaxation rate in the regime $qa \simeq 1$. These quasielastic neutron data[13] are consistent with a crossover to diffusive behavior near $qa = 1$. For $qa > 1$, the relaxation rate is diffusive, the diffusion constant being that of monomer or segment.

neutron spin-echo techniques. The line in the figure is calculated on the basis of the theory outlined in the following section. These data suggest a crossover to q^2 behavior at high q, but they are certainly not definitive.

5.2.3. Center-of-Mass Diffusion ($qR \ll 1$)

Almost all photon correlation experiments performed to date in dilute solution have investigated the regime $qR \ll 1$ where the decay constant Ω is proportional to the diffusion constant D of the chain, equation (21). The profusion of diffusional data is due to the relatively long wavelength of visible light as well as the difficulty of preparing very high molecular weight chains. In theta systems, interpretation of D_0 (D at infinite dilution) is relatively straightforward since D_0 is inversely proportional to the hydrodynamic radius $R_h = \langle r^{-1} \rangle^{-1} \sim D^{-1}$, which in turn scales as $N^{1/2}$ by equation (1). In better solvents, however, interpretation of D_0 is not trivial since D_0 is found to scale with molecular weight to a power intermediate between the ideal value of 1/2 and the Flory exponent 3/5 [equation (2)]. Much of the rest of this section is devoted to an analysis of these intermediate exponents in terms of the so-called "Blob" model, which is a simple conceptualization of the renormalization concept as applied to polymers.

5.2.3.1. Theta Systems. The first systematic study of D_0 in theta systems by PCS is the work of King *et al.*[14] on polystyrene (PS) in cyclohexane (CH). The molecular weight dependence of D_0 for PS in CH at the theta temperature is shown in Figure 6. Equation (17) suggests that D_0

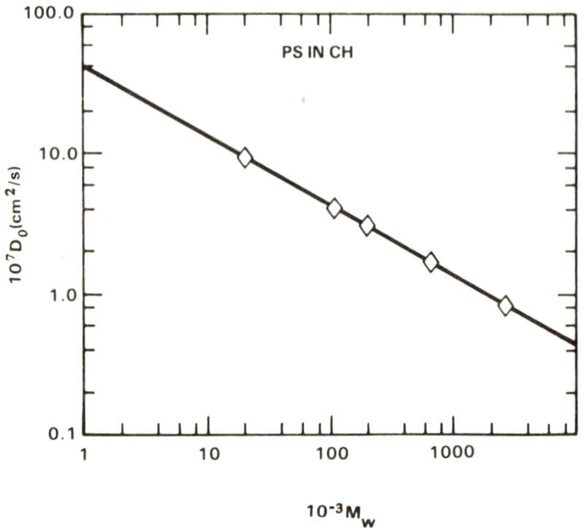

Figure 6. Molecular weight dependence of the diffusion constant in dilute theta solution.[14] D_0 is obtained from the relaxation rate of $S(q, t)$ at $qR \ll 1$ and infinite dilution. The line is of slope $-1/2$ as expected for an ideal chain. $T = 35°C$, PS in cyclohexane.

should scale as $N^{-1/2}$ in this system since the chains are ideal. The line in Figure 6 shows this power law dependence.

5.2.3.2. Marginal and Good Solvents. (a) *Molecular Weight Dependence.* In good solvents, when the chains are highly swollen, both scaling arguments and linear response theory predict $\Omega \sim \langle r^{-1} \rangle \sim N^{-\nu_h}$, where $\nu_h \cong 3/5$ is the Flory exponent. All recent light scattering experiments, however, show weaker power law dependence. Table 1 summarizes the measured exponents ν_h found for polystyrene (PS) and poly-α-methyl styrene (PAMS) in several solvents. A reasonable explanation[15, 16] for the anomalous exponents is that the hydrodynamic radius $R_h = \langle r^{-1} \rangle$ is sensitive to short sequences within the chain and therefore R_H does not reach its asymptotic limit ($N \to \infty$) within the range of experimental values of the molecular weight.

It has been known for many years that short chains and short sequences within chains show less swelling than longer chains.[3] This scale dependence of swelling coupled with the sensitivity of Ω to short distances explains the observed intermediate exponents. Basically the probability of intrachain contact increases with molecular weight so short chains with few intrachain contacts are nearly ideal.

The ideality of short sequences can be crudely modeled by considering the chain to be a sequence of " blobs " or renormalized monomers as shown

Table 1. Parameters for Polymer Systems[a]

System	T (°C)	Θ (°K)	τ	$10^2\eta$	ν_h	L (Å)	N_τ	χ_{pcs}	χ	χ_s
PS in										
CH[14]	35	307[6]	0	0.762	0.491	8.19	—	0.5	0.5[6]	0.2[6]
EA[22]	25	229[28]	0.232	0.426	0.515	8.66	2620	0.480	0.49[3]	0.47[6]
MEK[21]	25	0[6]	1	0.38	0.525	8.78	656	0.458	0.494[6]	0.50[6]
BZ[25]	20	(100)[6]	0.66)	0.649	0.537	9.73	482	0.451	0.44[3]	0.41[3]
THF[23, 68, 87]	25	—	—	0.46	0.559	10.13	80.0	0.379	—	—
TOL[24, 86]	20	(160)[6]	(0.46)	0.589	0.567	8.03	49.8	0.347	0.36[71]	0.39[3]
EBZ	25	—	—	—	—	—	—	—	0.40[6]	0.42[6]
PAMS in										
TOL[27]	25	—	—	0.552	0.534	7.92	193	0.411	0.36[70]	—
BZ[27]	30	—	—	0.561	0.536	7.87	163	0.403	—	—
PDMS in										
TOL	20	—	—	—	—	—	—	—	0.43[89]	—

[a] T is the temperature at which diffusion measurements were made. In some cases different temperatures were reduced to a common temperature, assuming $D_0 \sim T/\eta$. η is the viscosity used in fitting D_0 vs. M to equation (27) with L and N_τ being the parameters resulting from such a fit. X_{pcs} is calculated using equation (25b), assuming $n = 1.67$ for PS and $n = 1.84$ for PAMS. v_h is the apparent hydrodynamic exponent found from the slope of log D_0 vs. log M. τ is calculated using equation (68). Numbers in parentheses are very approximate.

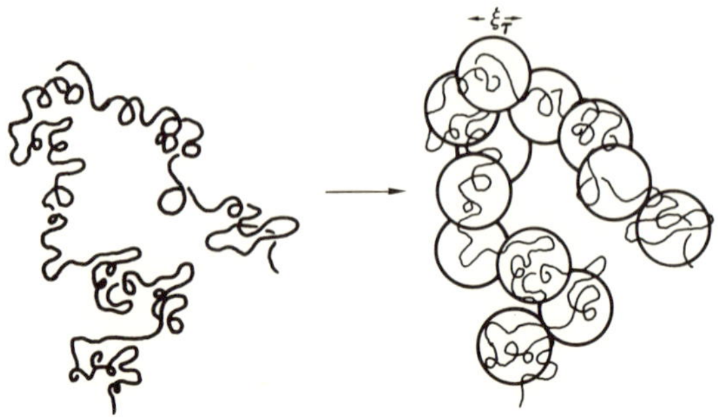

Figure 7. A schematic picture of the thermal blob model of polymers.[8] The actual chain on the left is divided into blobs on the right whose temperature-dependent radius ξ_τ is the mean distance smaller than the radius at which the chain is ideal. Inside the blob the chain is assumed to be ideal, whereas the renormalized chain, consisting of a sequence of blobs of radius ξ_τ, is swollen due to excluded volume. The blob model correctly predicts many polymer properties but it is a drastic oversimplification.[17] Most global properties are not strongly affected, however, by correcting[18] the deficiencies of the blob model.

in Figure 7. The actual chain on the left is divided into a sequence of blobs of radius ξ_τ. Within the blob the chain is ideal so that

$$\xi_\tau = (nN_\tau)^{1/2}a \tag{25a}$$

where N_τ is the temperature-dependent cutoff for ideal behavior. N_τ can be obtained by equating equations (1) and (2) with the result $[v = a^3(1 - 2\chi)]$

$$N_\tau = \frac{n^3}{(1 - 2\chi)^2} \tag{25b}$$

The renormalized chain or sequence of blobs shows swelling so that by analogy with equation (2)

$$R \sim \left(\frac{N}{N_\tau}\right)^{3/5} \xi_\tau \tag{26}$$

A qualitative explanation for the intermediate exponents ν_h in Table 1 follows from equations (25) and (26). That is, since D_0 is proportional to the first reciprocal moment of the monomer pair distribution function, it is sensitive to sequences short compared to ξ_τ. Since short sequences are nearly ideal, exponents closer to 1/2 than 3/5 are reasonable even in relatively good solvents where the radius of gyration R_g scales with molecular weight dependence near $N^{3/5}$.

Akcasu and Han[16] have worked out a detailed expression for D_0

within the blob model. Their result depends on $x = N_\tau/N$, the ratio of the chain to the blob molecular weight

$$D_0 = \frac{kT}{6\pi\eta}\frac{1}{LN^{1/2}}\frac{12}{(6\pi)^{1/2}}x^{1/2}2\left(1-\frac{x}{3}\right) + \frac{1}{1-\nu}(x^{\nu-1}-1) - \frac{1}{2-\nu}(x^{\nu-1}-x) \quad (27)$$

where $L \simeq na$ is an unknown length scale which is equal to a for a completely flexible chain. Equation (27) shows the expected crossover from ideal to swollen behavior as x increases:

$$D_0 = \frac{kT}{6\pi\eta L}\frac{12}{(6\pi)^{1/2}}\frac{1}{(1-\nu)(2-\nu)N^{\nu}} \sim N^{-\nu}, \qquad x \ll 1 \quad (28)$$

$$D_0 = \frac{kT}{6\pi\eta L}\frac{16}{(6\pi)^{1/2}N^{1/2}} \sim N^{-1/2}, \qquad x = 1 \quad (29)$$

Figures 8–12 show the existing literature data for the molecular weight dependence of the diffusion constant of PS in various solvents. In all cases the line is a best fit to the data using equation (27) with L and N_τ as

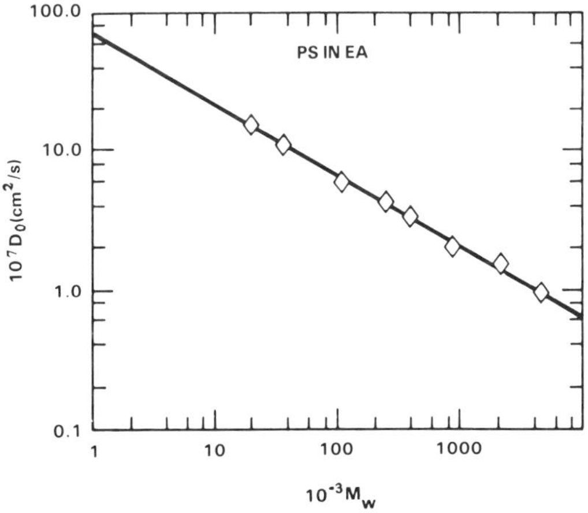

Figure 8. Molecular weight dependence of D_0 for PS in ethyl acetate[22] which is a near theta solvent. The line is a fit to equation (27) with L and N_τ as parameters. These parameters are listed in Table 1. For this system, $N_\tau \cong 2600$, which means a sequence must contain 2600 monomers before showing appreciable swelling due to excluded volume interactions. $T = 25°C$.

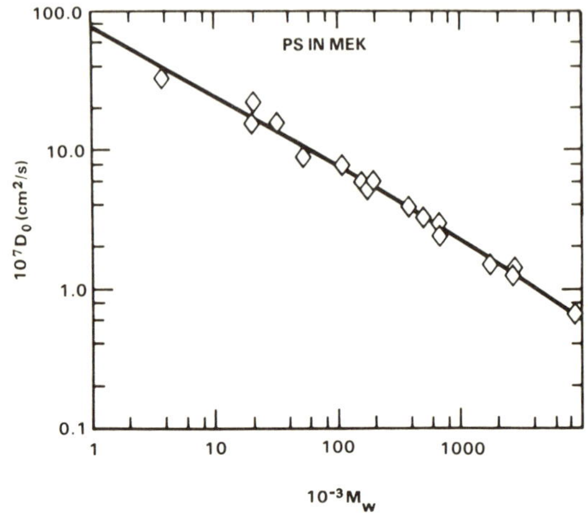

Figure 9. Molecular weight dependence of D_0 for PS in methyl ethyl ketone,[21] a marginal solvent. $T = 25°C$.

parameters. The values of these parameters are collected in Table 1 for these data and others for poly-α-methyl styrene (PAMS).[27] Examination of N_τ in Table 1 indicates that N_τ is a measure of solvent quality. For good solvents such as THF and TOL, N_τ is of the order of 100 monomers whereas for marginal solvents such as EA, $N_\tau \sim 10^3$. Increasing N_τ correlates directly with decreasing values of the apparent exponent ν_h. The Flory

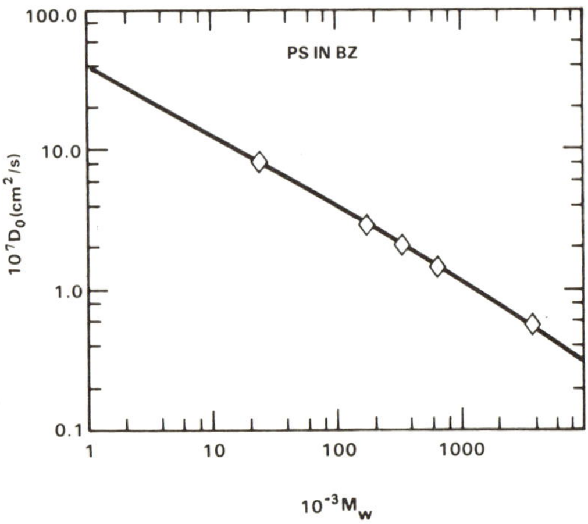

Figure 10. Molecular weight dependence of D_0 for PS in benzene.[26] Benzene is probably best classified as a marginal solvent. $T = 20°C$.

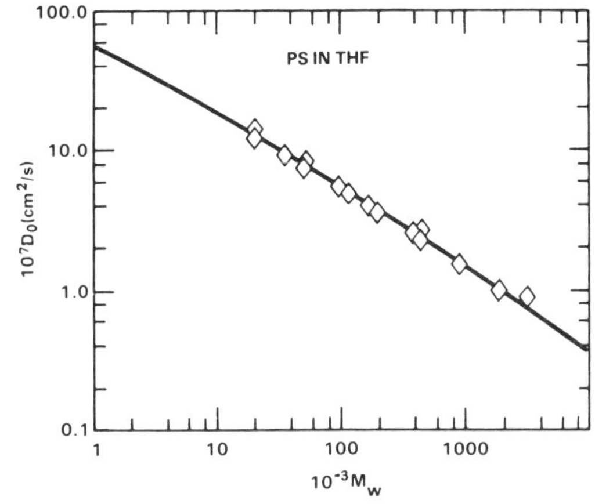

Figure 11. Molecular weight dependence of D_0 for PS in THF.[23, 68, 87] The data are plotted for $T = 25°C$. Data of References 68 and 87 have been reduced to 25°C assuming $D \sim T/\eta$.

reduced residual free energy χ can be calculated from N_τ using equation (25b) and the results are shown in Table 1 as χ_{pcs}. For comparison, χ measured by other means is also shown and the agreement is surprisingly good.

The blob model is known to be a drastic oversimplification since short sequences can deviate substantially from ideality.[17, 18] Nevertheless, the blob concept provides a simple framework to understand polymer statistics

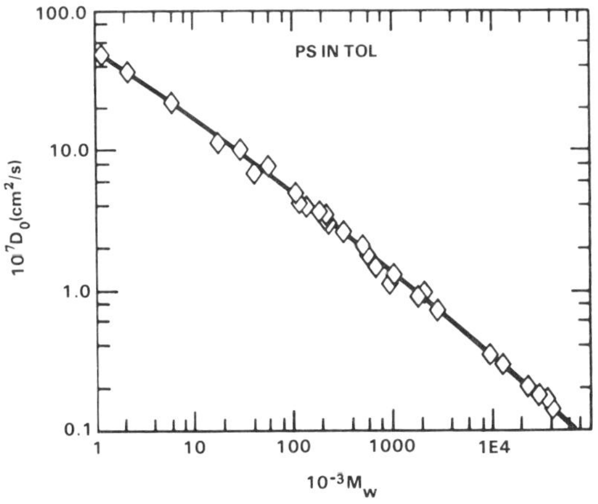

Figure 12. Molecular weight dependence of D_0 for PS in toluene. Toluene is a good solvent so the data approached a slope of $-3/5$ at high molecular weight.

in the intermediate molecular weight range. In this regime no completely satisfactory theory is available.

(b) *Temperature Dependence.* The temperature dependence of chain swelling can also be treated within the blob model. Unfortunately there is very little experimental data on the temperature dependence of the hydrodynamic radius, probably because most organic solvents boil at a relatively low temperature.

The temperature dependence of D_0 follows from equation (27) if the temperature dependence of N_τ is known. Near $\tau = \theta$ the excluded volume is expected to be approximately linear in the temperature increment $\tau = (T - \theta)/T$, so that

$$N_\tau \sim \left(\frac{a^3}{v}\right)^2 n^3 \sim \frac{n^3}{(1 - 2\chi_s)^2 \tau^2} \tag{30}$$

assuming n is temperature independent. Unfortunately, existing data are insufficient to confirm this temperature dependence for N_τ.[16]

5.2.4. Internal Dynamics and the Dynamic Structure Factor

5.2.4.1. Preliminary Considerations. The scaling analysis outlined in Section 5.2.2.3 shows that the basic form of the dynamic structure factor can be obtained by simple scaling analysis. In particular, three regimes separated by $qR_h = 1$ and $qa = 1$ were identified and the functional dependence of the relaxation rate on temperature, molecular weight, viscosity, and q were established. In spite of the success of this analysis, many shortcomings are present in the scaling approach. In the present section, more detailed analysis of $S(q, t)$ is reviewed for several models of chain molecules. This analysis both justifies the scaling approach and fills in details which cannot be obtained by scaling.

The limitations of scaling analysis include the following problems. First, no underlying justification for scaling or power laws was given. Second, the shape of $S(q, t)$ and the relationship between the characteristic frequency Ω and $S(q, t)$ is not obvious. In addition, scaling yields power law exponents but not coefficients. Information on chain statistics and monomer interactions is available from coefficients as shown below. Finally, scaling does not treat the transition regions $qa = 1$ and $qR = 1$. These regions are important experimentally and fortunately $S(q, t)$ is available in the crossover regimes for certain simplified models of polymer chains.

Ideally, interpretation of dynamic scattering experiments requires a theory that can predict $S(q, t)$ under actual experimental conditions which are characterized by temperature, concentration, and a chain model consistent with the chemical structure of the polymer. Unfortunately, exact expressions for $S(q, t)$ are available at present only for a single unperturbed (Θ-condition) Gaussian chain without hydrodynamic interaction (Rouse model), and in the infinite chain limit, with hydrodynamic interaction and preaveraged Oseen tensor (Rouse–Zimm model). In this sense, a complete interpretation of dynamic scattering experiments on polymer solutions is an unsolved problem. Nevertheless the interpretation of scattering experiments can proceed in terms of the initial slope $\Omega(q)$ of the normalized intermediate scattering function $S(q, t)$. The initial slope is defined by equation (24a) and should reduce to q^2D as $q \rightarrow 0$. $\Omega(q)$ turns out to be identical to the relaxation rate Ω which was derived by scaling analysis in Section 5.2.2.3. By using linear response theory, $\Omega(q)$ can be calculated for many different models. For example, $\Omega(q)$ as a function of temperature and concentration has been calculated in terms of the "blob" model.[19]

No attempt will be made in this section to derive $S(q, t)$ and $\Omega(q)$ for the various models. Rather, the results of these various calculations will be analyzed graphically to demonstrate the effect of various experimental parameters such as hydrodynamic coupling, monomer excluded volume, temperature, and momentum transfer q. Ideally one should start with a theory for $S(q, t)$ which somehow incorporates hydrodynamic coupling and a temperature-dependent distribution function in a form which does not require approximations such as preaveraging of the Oseen tenor, moment expansions, or asymptotic limits. Unfortunately, such a theory is not available so it is necessary to resort to various levels of approximation to study the influence of various factors on the dynamics.

Table 2 lists the models which are currently available with the appropriate theoretical and experimental references. Akcasu, Benmouna, and Han[20] have treated most of these models starting from a general formalism based on the eigenfunction method and linear response theory. For this review only selected models are discussed to illustrate trends. For example, the q dependence of $S(q, t)$ is analyzed for a Rouse–Zimm ring (model 3) which is the only realistic model for which $S(q, t)$ is available for $qR_g \sim 1$. Data on polystyrene in cyclohexane and toluene are compared with this theory. The dependence of $S(q, t)$ on hydrodynamic coupling is analyzed using models 1 and 4 which are valid only for $qR \gg 1$. The effects of excluded volume and hydrodynamic preaveraging are discussed only with respect to the characteristic frequency $\Omega(q)$ since full calculations of $S(q, t)$ are not available. In all cases, comparison with experimental data is shown when data are available.

In 1965, Pecora showed[29] for the first time that the intermediate

Table 2. Models for Chain Dynamics

Name, model	Function calculated	Regime	Statistics	Hydrodynamics	Monomer interaction	Geometry	Refs.
1. Rouse	S, Ω	All	Gaussian	None	None	Linear	20, 29, 30
2. Rouse	S, Ω	All	Gaussian	None	None	Ring	20
3. Rouse–Zimm ring	S, Ω	All	Gaussian	Preav.	None	Ring	20
4. Rouse–Zimm ∞ chain	S, Ω	$KR_g \gg 1$	Gaussian	Preav.	None	Ring or linear	20, 31, 32, 34
5. (unnamed)	Ω	All	Quasi-Gaussian	Preav.	Blob approx.	Linear	19, 34
6. (unnamed)	Ω	All	Quasi-Gaussian	Nonpreav.	Blob approx.	Linear	19, 34
7. (unnamed)	Ω	All	Gaussian	Nonpreav.	None	Ring	40
8. Freely jointed chain	Ω	All	Freely jointed	Preav.	None	Linear	41
9. Sliding rod	Ω	All	Sliding rod	Preav.	None	Linear	42
10. Harris–Herst	Ω	All	Wormlike	Preav.	None	Linear	44

scattering function $S(q, t)$ for a flexible polymer chain can be calculated by using a Green's function solution of the Fokker–Plank equation for the bead-and-spring model. Later, de Gennes and Debois-Violette[30, 31] calculated $S(q, t)$ in the intermediate q region, that is, for $qR_g \gg 1 \gg qa$, where R_g is the radius of gyration of the polymer and a is the segment length of the polymer. They demonstrated the q^3 and q^4 dependence of relaxation rate in the cases with and without hydrodynamic interaction for a Gaussian chain. They also showed that $S(q, t)$ will follow a universal shape in this intermediate region if time is scaled by a characteristic frequency.

If one follows the approach of Zwanzig and Fixman[20, 32–34] by assuming that the distribution function, ψ, of the monomers is a function which satisfies a dynamical equation

$$\frac{\partial \psi}{\partial t} = D\psi \tag{31}$$

with D as a linear, time-independent diffusion operator, then, one can define a self-adjoint operator $\mathscr{L}$ through

$$D(\psi A) \equiv -\psi \mathscr{L} A \tag{32}$$

where A is an arbitrary dynamical variable such as Fourier transform of the monomer density $\rho(q, t) \equiv \sum_j e^{iq \cdot R_j}$.

The intermediate scattering function can be written as

$$\begin{aligned} S(q, t) &= \langle \rho(q, 0)\rho(q, t) \rangle \\ &\equiv \langle \rho, e^{-t\mathscr{L}} \rho \rangle \\ &\equiv \langle \rho, \rho(t) \rangle \end{aligned} \tag{33}$$

For an eigenvalue problem[5, 35]

$$\mathscr{L} v_n = \omega_n v_n \tag{34}$$

Equation (33) can be written as

$$\begin{aligned} S(q, t) &= \sum_n \langle \rho, v_n \rangle \langle v_n, e^{-t\mathscr{L}} \rho \rangle \\ &= \sum_n e^{-\omega_n t} \, | \langle v_n, \rho \rangle |^2 \end{aligned} \tag{35}$$

Also the characteristic frequency $\Omega(q)$ can be obtained through differentiation as

$$\Omega(q) = \sum_n \omega_n |\langle v_n, \rho \rangle|^2 / S(q, 0)$$
$$= \langle \rho, \mathscr{L}\rho \rangle / \langle \rho, \rho \rangle \tag{36}$$

Whenever a model polymer can be solved as an eigenvalue problem, *in principle* the corresponding scattering problem is solved. This principle can be seen from cases 1 through 4 in Table 2.

In the case in which $S(q, t)$ cannot be solved, progress may still be possible through the calculation of the characteristic frequency $\Omega(q)$. This procedure is discussed later. One should observe from equation (36) that the initial slope of $S(q, t)$ or characteristic frequency $\Omega(q)$ is actually a weighted average of all the relaxation times (or eigenvalues). $S(q, t)$ always decays exponentially at short time, but this short-time decay rate, which is the characteristic frequency $\Omega(q)$, bears no simple relationship with the shortest or longest (terminal) relaxation times.

5.2.4.2. Intermediate Scattering Function, $S(q, t)$. (*a*) *q dependence.* In the generalized Kirkwood diffusion equation,[36] a linear flexible polymer is approximated by a bead-and-spring model. All forces are concentrated on these beads. The frictional force on the *j*th bead is

$$f = -\zeta(v_j - v'_j) \tag{37}$$

where ζ is the frictional coefficient, v_j is the velocity of the *j*th bead, and v'_j is the velocity of the solvent at the *j*th bead's position produced by all other beads through hydrodynamic interaction which is again approximated by the Oseen tensor as

$$v'_j = \sum_l T_{jl}\zeta(v_l - v'_l) \tag{38}$$

with

$$T_{jl} = \frac{1}{8\pi\eta}\left(\frac{1}{R_{jl}^3}\right)[R_{jl}^2\mathbf{1} + \mathbf{R}_{jl} : \mathbf{R}_{jl}]$$

and $T_{jj} = 0$. $\zeta(V_l - V'_l)$ is the force exerted on the fluid by the *l*th bead.

The eigenvalues have been calculated by B. Zimm[11] in 1956 for a preaveraged Oseen tensor. The corresponding case with no hydrodynamic interaction was solved in 1953 by Rouse.[37] In the ring polymer case,[5, 38, 39] because of the cyclic boundary condition, both cases with and without hydrodynamic interactions can be calculated analytically at any chain length N [see Equations (17)–(72) of Reference 20].

Figure 13 shows the variation of $S(q, t)$ as a function of $\Omega(q)t$. For a ring polymer the initial slopes of all curves are equal to -1. It is observed that the shape function tends to a straight line when $qR_g \leq 0.87$ or $qa \geq 10$. The curves also cluster around the curve corresponding to $qa = 1.5$ when

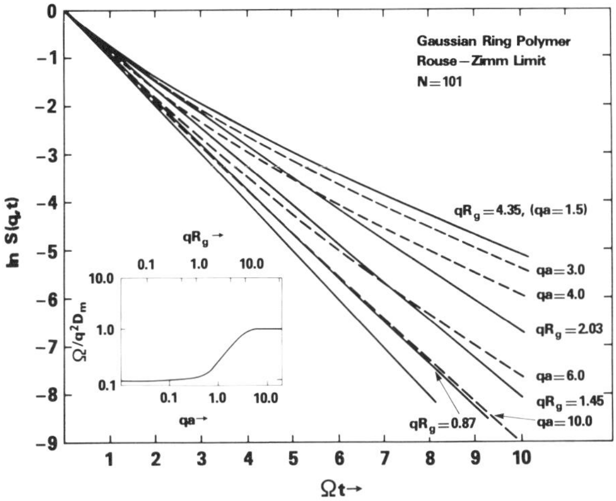

Figure 13. The variation of the shape function ln $S(q, t)$ with normalized time Ωt for various values of q in the presence of hydrodynamic interaction ($B = 0.38$). Also shown is the variation of the initial slope $\Omega(q)$ with (qa).

$qR_g \gg 1$ but $qa \leq 2$. These tendencies become more apparent as N increases. At the same time, the characteristic frequency $\Omega(q)$ approaches a q^3 dependence (see insert of Figure 13).

Since the open and closed chains are identical as chain length $N \to \infty$, the normalized intermediate scattering function $S(q, t)$ with preaveraged Oseen tensor can be found in this limit as

$$S(q, t) = \left(e^{-\alpha\phi_0(t)} + 2\sum_{s=1}^{\infty} e^{-\alpha\phi_s(t)}\right)\left(1 + 2\sum_{s=1}^{\infty} e^{-\alpha s}\right)^{-1} \tag{39}$$

where

$$\phi_s(t) \equiv s + \frac{1}{2\pi}\int_{-\pi}^{\pi} dp\, \frac{1 - \exp(-\alpha_p t)}{1 - \cos p} \cos ps$$

$$\alpha_p \equiv 2W(1 - \cos p)[1 + 2BZ(p)]$$

$$Z(p) \equiv \sum_{n=1}^{\infty} \frac{\cos pn}{(n)^{1/2}}$$

$$\alpha = \frac{q^2 a^2}{6}$$

$$W \equiv \left(\frac{3}{a^2}\right) D_m$$

where B is the draining parameter defined below.

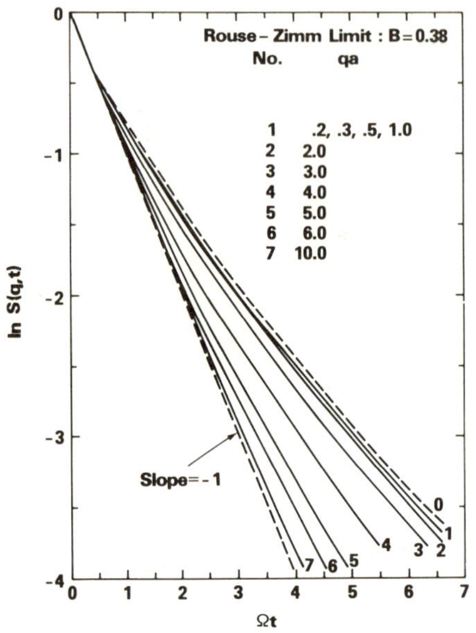

Figure 14. The variation of the shape function ln $S(q, t)$ with normalized time Ωt for various values of qa in the case of an infinite chain with hydrodynamic interaction.

In Figure 14, $S(q, t)$ at various qa values is plotted for the Rouse–Zimm case. In the Rouse case, $S(q, t)$ can be recovered from equation (39) by letting $B = 0$. The Rouse case is displayed in Figure 15 for various qa values.

The form of the normalized $S(q, t)$ in the intermediate q region where $qa \ll 1$ and $qR_g \gg 1$ can be obtained from equation (39) in the limit of $qa \to 0$. The $qR_g \gg 1$ has already been taken into account through $N \to \infty$. The general expression is

$$S(q, t) = \int_0^\infty du \exp\left[-u - J\left(u, \Omega t, \frac{B}{\sqrt{\alpha}}\right)\right] \tag{40a}$$

where

$$J\left(u, Qt, \frac{B}{\sqrt{\alpha}}\right) \equiv \frac{2}{\pi}\int_0^\infty dx \frac{\cos xu}{x^2} \times \left\{1 - \exp\left[-(\Omega t)x^2 \frac{1 + B(2\pi/x\alpha)^{1/2}}{1 + 2B(\pi/\alpha)^{1/2}}\right]\right\} \tag{40b}$$

With

$$\Omega(q) = \frac{1}{12}\left(\frac{kT}{\zeta}\right)q^4a^2\left[1 + \left(\frac{2\sqrt{\pi}B}{\sqrt{\alpha}}\right)\right] \tag{41}$$

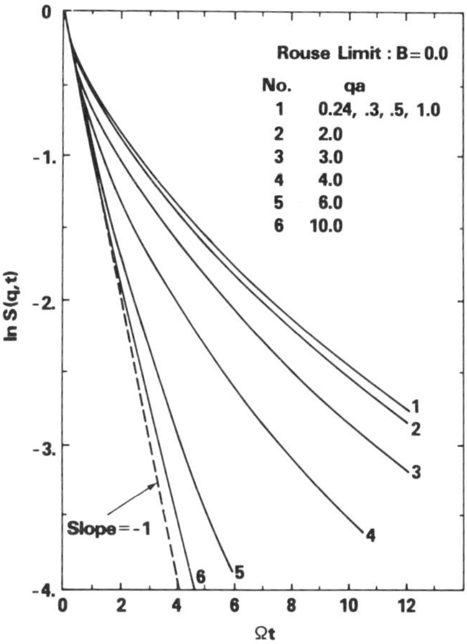

Figure 15. Variation of the shape function ln $S(q, t)$ with normalized time, Ωt, for a Rouse chain.

where B is the draining parameter

$$B \equiv \frac{\zeta/\eta a}{\pi\sqrt{6\pi}} \tag{42}$$

The Rouse model is recaptured with $B = 0$ as

$$S(q, t) = \int_0^\infty du \exp\left\{-u - (\Omega t)^{1/2} g\left[\frac{u}{(\Omega t)^{1/2}}\right]\right\} \tag{43a}$$

where

$$g(u) \equiv \frac{2}{\pi}\int_0^\infty dx \frac{\cos xu}{x^2}[1 - \exp(-x^2)] \tag{43b}$$

and the characteristic frequency

$$\Omega(q) \equiv \frac{1}{12}\left(\frac{kT}{\zeta}\right)q^4 a^2 \tag{44}$$

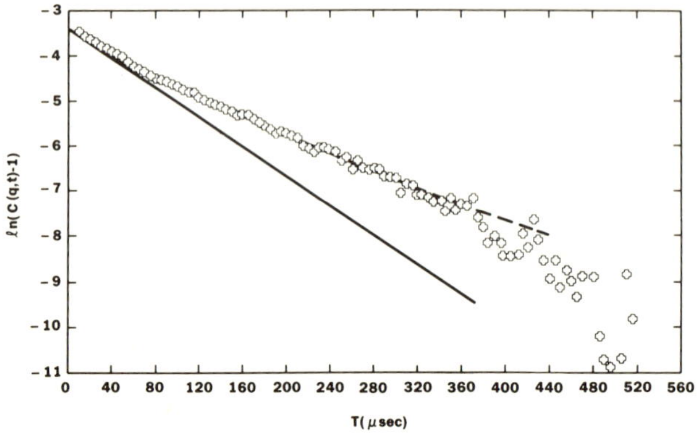

Figure 16. Experimental correlation data of PS ($M_w = 44 \times 10^6$) in cyclohexane at a scattering angle $\theta = 120^\circ$ with a delay time of 5 μsec. The dashed curve is the best fit using the asymptotic shape function, and the solid line is the corresponding initial slope. $T = 35^\circ$C.

On the other hand, the Rouse–Zimm limit can be reached as $qa \ll 2B(6\pi)^{1/2}$ as

$$S(q,\,t) = \int_0^\infty du\,\exp\{-u - (Qt)^{2/3}h[u(Qt)^{-2/3}]\} \tag{45a}$$

where

$$h(u) \equiv \frac{2}{\pi}\int_0^\infty dx\,\frac{\cos xu}{x^2}\left[1 - \exp\left(\frac{-x^{3/2}}{\sqrt{2}}\right)\right] \tag{45b}$$

The initial slope follows as:

$$\Omega(q) = \frac{1}{6\pi}\left(\frac{kT}{\eta}\right)q^3 \tag{46}$$

The following points should be noticed:

(i) $S(q,\,t)$ is expressed in the intermediate q region as a function of a single variable, $\Omega(q)t$, which combines both q and t. When time is scaled as $\tau = \Omega(q)t$, a universal curve as a function of τ, is obtained for all q values in this q region. This conclusion justifies the interpretation of $\Omega(q)$ as a "characteristic frequency" in the sense of dynamic scaling.

(ii) $\Omega(q)$ is the initial slope of $S(q,\,t)$ at $t = 0$.

(iii) The q dependence of $\Omega(q)$ follows a power law in the intermediate q region.

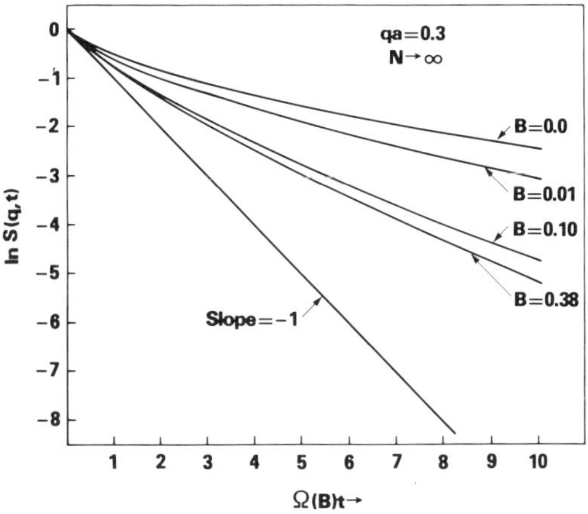

Figure 17. The effect of the draining parameter B on the shape function $\ln S(q, t)$.

(iv) For large times $S(q, t) \to \exp[-(2/\sqrt{\pi})(\Omega t)^{1/2}]$ in the Rouse case and $S(q, t) \to \exp[-1.35(\Omega t)^{2/3}]$ in the Rouse–Zimm case, although this asymptotic region may not be reached in the realistic experimental time range, i.e., $\Omega t \leq 10$.

In Figure 16, experimental data of $\ln[C(q, t) - 1] = \ln[\beta S^2(q, t)]$ for polystyrene with molecular weight of 44×10^6, $c = 0.11$ mg/ml, and $q = 3.2 \times 10^5$ cm^{-1} is plotted vs. time to demonstrate the asymptotic shape function. Here $C(q, t) = \langle I(q, t)I(q, 0)\rangle / \langle I^2 \rangle$ is defined in equation (5). The dotted line is the best fit to equation (45). The solid line is the initial slope of the asymptotic function fit. It is clear that the experimental range is controlled by the signal-to-noise ratio at large times. In this case, correlation data are truncated at 400 μsec in the analysis.

Hydrodynamic Interaction Strength. The notation $B \equiv \zeta_0/\eta a \pi \sqrt{6\pi}$ is used to represent the hydrodynamic interaction strength. It should be noted that in order to keep the dynamical operator as positive definite and avoid degeneracy, B has to be smaller than ~ 0.6.[20, 46–48] In Figure 17, the asymptotic shape function is plotted with several B values to demonstrate the change from the Rouse–Zimm limit (a Flory value of $B = 0.38$ is used[1]) to Rouse limit of $B = 0$.

In general, the effect of B on the shape of $S(q, \Omega t)$ is such that $S(q, \Omega t)$ increases with decreasing B at fixed Ωt and qa (see Figure 17) from the Zimm value given by equation (45) to the Rouse value given by equation (43). It has been conjectured in the literature that the Rouse limit may be followed as polymer concentration is close to 1.

5.2.4.3. Characteristic Frequency $\Omega(q)$. *Hydrodynamic Interaction Strength B.* In cases where $S(q, t)$ can be calculated, $\Omega(q)$ can easily be

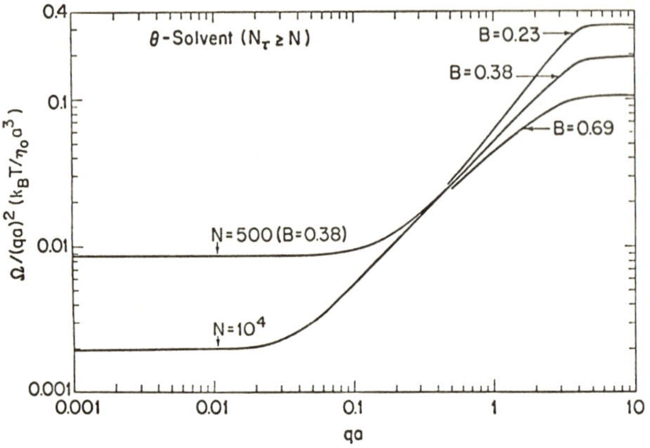

Figure 18. The variation of the initial slope $\Omega(q)$ with (qa) under Θ conditions for various values of the draining parameter B and two values of N. Note that the curves for different B's and a fixed N coincide at small values of qa.

obtained through differentiation as shown in the previous section. When $S(q, t)$ cannot be calculated, progress may still be possible by using the procedure developed by Akcasu and Gurol[34] in 1976 to calculate the characteristic frequency $\Omega(q)$. For example: the effect of preaveraging Oseen tensor can be compared for both Gaussian (or Θ) chain and excluded volume chain at different hydrodynamic interaction strength.

In Figure 18, the variation of $\Omega(q)$ with qa for a Θ chain of two different chain lengths, N, and various values of B are displayed. Notice that the curves for different B's coincide for a given N at small values of qa. The curves are plotted following equation (17) of Reference 20, are calculated with a preaveraged Oseen tensor, but the effect of preaveraging the Oseen tensor can also be studied through the Akcasu–Gurol formalism.

Preaveraging. With the preaveraged Oseen tensor, $\Omega(q)$ can be expressed in the small qa region as

$$\frac{\Omega(q)}{q^2 D_v} = \frac{(1-v)(2-v)}{2} \kappa_v \frac{\gamma((1-v)/2v, \kappa_v^2) - \kappa_v^{-1/v}\gamma((2-v)/2v, \kappa_v^2)}{\gamma(1/2v, \kappa_v^2) - \kappa_v^{-1/v}\gamma(1/v, \kappa_v^2)} \tag{47}$$

where

$$\kappa_v^2 \equiv (qR_g)^2(1+v)(1+2v)/3$$

$$R_g \equiv lN^v[2(1+v)(1+2v)]^{-1/2}$$

$$D_v = \frac{kT[3\pi(1+v)(1+2v)]^{-1/2}}{\eta R_g \pi(1-v)(2-v)}$$

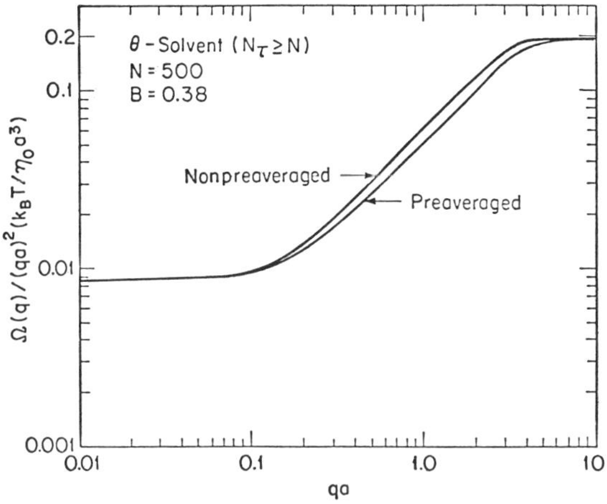

Figure 19. Variation of $\Omega(q)/q^2(kT/\eta)$ with (qa) under Θ conditions with a preaveraged and nonpreaveraged Oseen tensor. Note that the maximum difference (15%) is reached at $qa \approx 1$.

and where

$$\gamma(\mu, x) \equiv \int_0^x dt\, t^{\mu-1} \exp(-t)$$

is the incomplete gamma function. This expression yields $\Omega(q)$ in the Θ-condition with $\nu = 1/2$, and in a good solvent with $\nu = 3/5$.

The expression for $\Omega(g)$ has also been obtained without the approximation of preaveraging the Oseen tensor. Again, in the small qa region, it can be reduced as

$$\frac{\Omega(q)}{q^2 D_\nu} = \frac{3(1-\nu)(2-\nu)}{16\nu} \kappa_\nu^{(-1-1/\nu)} \left[\int_0^1 du\,(1-u)e^{-u^{2\nu}\kappa_\nu^2}\right]^{-1}$$
$$\times \int_0^{\kappa_\nu{}^2} du\left(1 - \frac{u^{1/2\nu}}{\kappa_\nu^{1/\nu}}\right) u^{1/2\nu-2}$$
$$\times \left[-u^{-1/2} + (2+u^{-1})e^{-u}\int_0^{u^{1/2}} dt\, e^{t^2}\right] \tag{48}$$

where D_ν, κ_ν, and R_g have the same meaning as in equation (47). This result is valid for both theta and good solvents with $\nu = 1/2$ and $\nu = 3/5$, respectively.

In Figure 19, the effect of preaveraging the Oseen tensor is demonstrated for a Θ chain. The important feature is in the intermediate region

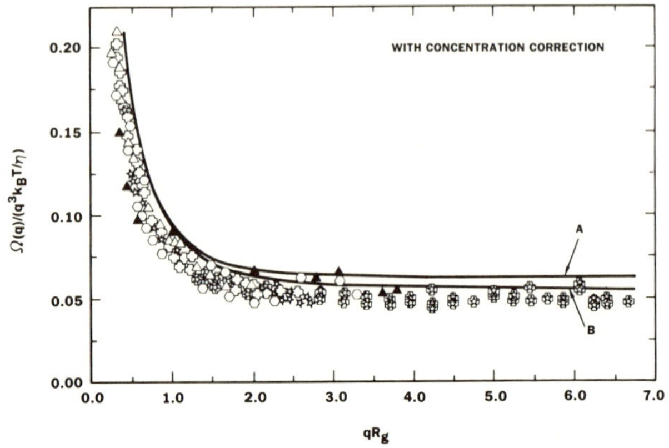

Figure 20. Variation of the reduced first cumulant Ω as a function of qR_g at the Θ condition with concentration correction.

($qa \ll 1$ and $gR_g \gg 1$). In both cases—with and without preaveraging—$\Omega(q)$ approaches q^3 dependence but with different magnitude. Actually, the same effect also appears in the excluded volume chain case.

In this intermediate q region, this effect can be summarized as follows: For the Θ chain:

$$\Omega(q) = 0.053 \frac{kT}{\eta} q^3 \qquad \text{with preaveraged Oseen tensor}$$

$$= 0.0625 \frac{kT}{\eta} q^3 \qquad \text{without preaveraged Oseen tensor}$$

For the excluded volume chain:

$$\Omega(q) = 0.071 \frac{kT}{\eta} q^3 \qquad \text{with preaveraged Oseen tensor}$$

$$= 0.079 \frac{kT}{\eta} q^3 \qquad \text{without preaveraged Oseen tensor}$$

In Figures 20 and 21,[49] $\Omega(q)/[(kT/\eta)q^3]$ is plotted vs. qR_g for various molecular weights of polystyrene in Θ (CH, 35°C) and good (TOL and THF) solvents, respectively. Also the calculated curves according to equations (47) and (48) were included. It was concluded[49] that the dynamic light scattering experiments in the diffusion, transition, as well as in the intermediate q regions can be interpretated on the basis of the characteristic

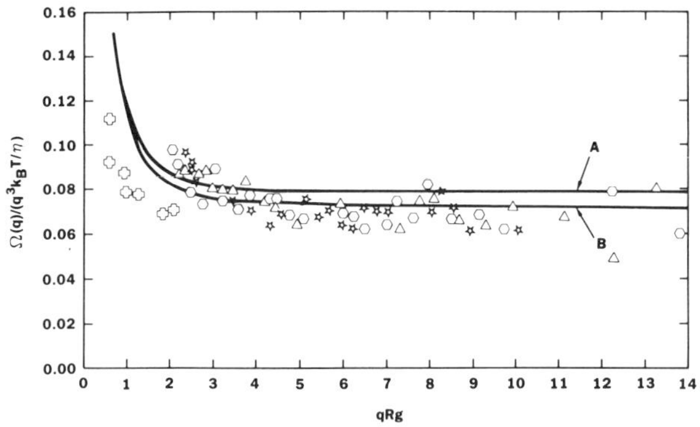

Figure 21. Variation of the reduced first cumulant Ω as a function of qR_g in good solvents. Curves A and B are the same as in Figure 20, except $v = 0.6$.

frequency $\Omega(q)$ as far as trends are concerned. In both Θ and excluded volume chain cases, experimental data agree better with the preaveraged Oseen tensor calculation. It is suggested that the Oseen tensor approximation needs improvement, but the preaveraged Oseen tensor calculation may be interpretated as a better model. This result, however, cannot eliminate the possibility that the equilibrium distribution function itself needs improvement, especially at short distances.

Effect of Temperature (Excluded Volume). In the previous section, we discussed two extreme cases of excluded volume effect: the Θ-chain case, which has no excluded volume at all, and the excluded-volume chain case, which has excluded volume effect everywhere along the chain. In this section, we use the blob model to demonstrate $\Omega(q)$ as a function of temperature, although we should keep in mind that $\Omega(q)$ can be calculated for a better model through the Akcasu–Gurol formalism when it is available.

In the blob model, chain statistics of mean-square distance are expressed as[8, 51, 25]

$$\langle R_{nm}^2 \rangle = |n - m| a^2 \qquad \text{if } |n - m| < N\tau \tag{49a}$$

and

$$\langle R_{nm}^2 \rangle = \left(\frac{|n - m|}{N_\tau} \right)^{2v} \xi_\tau^2 \qquad \text{if } |n - m| > N_\tau \tag{49b}$$

where N_τ is defined in equation (25b)

$$N_\tau \sim \tau^{-2} \tag{49c}$$

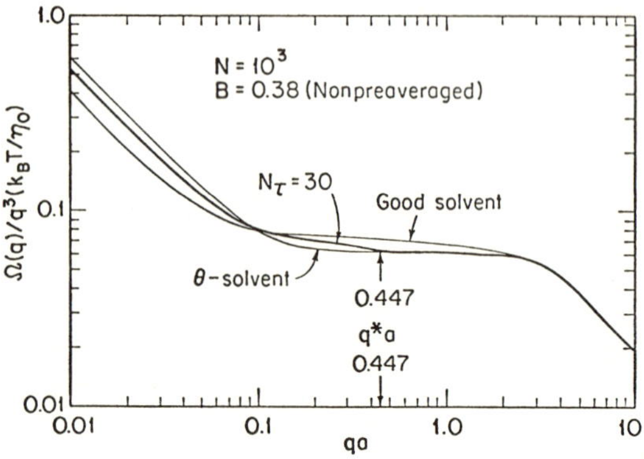

Figure 22. Variation of $\Omega(q)/q^3(kT/\eta)$ with (qa) using a nonpreaveraged Oseen tensor, in good solvent, in Θ solvent, and for an intermediate temperature corresponding to $N_\tau = 30$. This figure also illustrates the tendency to a crossover from good to Θ conditions when q is increased from $q < q^*$ to $q > q^*$.

and

$$\xi_\tau^2 = N_\tau a^2 \tag{49d}$$

The characteristic frequency $\Omega(q)$ has been calculated with[19] and without[44] preaveraging the Oseen tensor as a function of temperature $\tau \equiv (T - \Theta)/T$. In Figure 22, the effect of temperature crossover according to the blob model is illustrated. The plateau corresponding to $\Omega(q) \sim q^3$ behavior is replaced by a smooth transition from 0.0790 to 0.0625. The transition occurs at around q^*, which is defined as $q^* \equiv \sqrt{6}/\xi_\tau$.

A transition to theta behavior at high q, similar to Figure 22, has been reported by Richter *et al.*[82] using quasielastic neutron scattering. The observed crossover, however, occurs very near the temperature where PS shows a change in flexibility,[83] so it is possible that the transition represents a change in the effective segment length na rather than the crossover predicted by the blob model. For the static structure factor, no transition to theta behavior is observed[18] at high q, so it would be surprising if the dynamics reflected such a crossover. The blob model represents a severe approximation and it seems unlikely that crossovers as abrupt as Figure 22 will ever be observed.[17, 18]

5.3. VIRIAL REGIME

In this section we will discuss various predictions of the concentration dependence of diffusion constant, D, in the dilute region as a function of temperature and molecular weights. Some experimental results are included for comparison.

The concentration dependence of D can be examined through the Gibbs–Duhem expression[5] as

$$D = \frac{kT}{\zeta_{\text{ch}}}\left(1 - \frac{N_A V_1}{M} c\right)(1 + 2A_2 Mc + \cdots) \tag{50}$$

where T is the temperature of the solution and ζ_{ch} is the frictional coefficient of the polymer molecule in solution. A_2 is the second virial coefficient of the osmotic pressure, V_1 is the partial specific volume of polymer with molecular weight M, and N_A is Avogadro's number. In this expression the concentration dependence of D is separated into two parts: The first contribution is from the chemical potential, which involves the virial coefficients. The second contribution is due to the hydrodynamic interaction, which is included in the frictional coefficient ζ_{ch}. One can immediately draw several conclusions: (i) The concentration dependence of D can be nonzero at Θ temperature, because ζ_{ch} may still be concentration dependent even at $T = \Theta$. (ii) If we write $D(c) = D(0) + Sc$, the sign and magnitude of the initial slope S of D as a function of concentration at a given temperature and molecular weight will depend on the relative magnitudes of A_2 and the concentration coefficient of ζ_{ch}.

In the theoretical calculations, the concentration dependence of D is normally expressed as

$$D(c) = D(0)(1 + k_D c_v) \tag{51}$$

with concentration coefficient k_D and concentration c_v which is the volume fraction occupied by polymer chains and defined as

$$c_v \equiv cN_A v_h/M \tag{52}$$

when v_h is the hydrodynamic volume of the polymer molecule.

In the theoretical calculation of k_D, it is the calculation of ζ_{ch}[52–57] which depends on the model and approximations used. In Figure 23, several theoretical calculations are included. Curve 5 represents the calculation of Yamakawa[5, 52] by using the bead-and-spring model with the drift velocity correction. Curve 6 is calculated by Pyun and Fixman[54] using the interpenetrable sphere model. Curve 7 is the calculation of Altenberger and Deutch[55] for hard spheres. Akcasu and Benmouna[56] and Akcasu[57] have also calculated k_D through the intermediate scattering function $S(q, t)$,

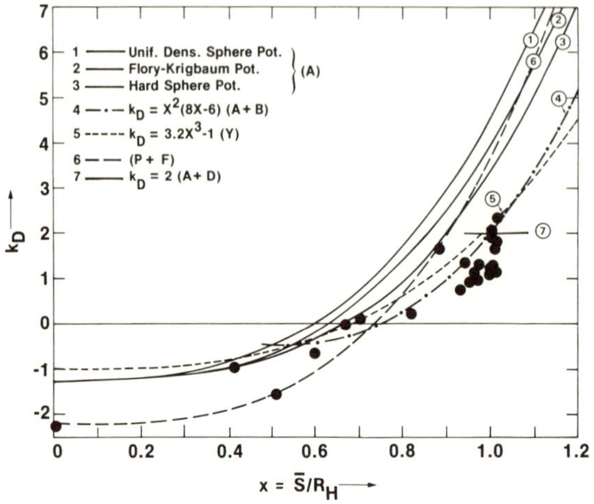

Figure 23. Second virial coefficient of the diffusion constant as a function of the ratio of the hard sphere radius to the hydrodynamic radius.

rather than using the Gibbs–Duhem formulation. In their calculation, various models and potentials (curves 1–4) were used.

Included in Figure 23 are also some experimental results for polystyrene with various molecular weights, temperature, and solvents.[23, 27, 58] All models and experimental results show similar trends as regards the temperature dependence of k_D except for the hard sphere model, which does not apply at lower temperature. In all cases, k_D vanishes at a temperature corresponding to $\bar{S}/R_h \cong 0.72$ with $\bar{S}$ defined as $(3M^2A_2/16\pi N_A)^{1/3}$. This value of $\bar{S}/R_h$ marks the transition from theta to good solvent behavior. In the good solvent region $k_D > 0$ and the diffusion coefficient increases with concentration. In this region, Yamakawa's result (curve 5) and that of Akcasu *et al.*[56] (curve 4) are in good agreement with experimental results. The assumption of replacing the distance between two monomers belonging to two different molecules by their center-of-mass separation[56] is justified probably because the molecular interpenetration is not serious. This assumption could be the cause of $k_D = 0$ at Θ condition in the calculation of Akcasu *et al.* where the molecular interpenetration could be significant.

In the poor solvent region, $k_D < 0$ and the theoretical predictions are very sensitive to the models used to describe the translational diffusion of a pair of interacting molecules. Calculations of Pyun and Fixman,[54] Yamakawa,[52] and Akcasu[57] all give correct negative k_D. But it seems that more refined models and more precise measurements are still needed in this region.

5.4. SEMIDILUTE SOLUTIONS

5.4.1. Introduction

As the concentration of a polymer solution is increased, eventually the virial expansion [equation (51)] is inadequate. The virial method would certainly be inadequate when the chains begin to overlap and entangle because new dynamical processes involving interchain interactions and disentanglement come into the problem. The number concentration at the overlap is approximately

$$\rho^* \simeq \frac{3N}{4\pi R_g^3} \tag{53}$$

and virial expansions certainly fail for $\rho > \rho^*$. For typical polymers ($N \sim 1000$), chain overlap occurs at concentrations of about 1% by volume. In a sense then the solution is still dilute since a particular monomer is still surrounded overwhelmingly by solvent molecules and seldom is in contact with monomers. The regime $\rho^* \ll \rho \ll a^{-3}$ is called the semidilute regime. The system is dilute in terms of overall monomer density but the system still displays strong dynamical effects due to the interchain contacts.

The transition from dilute to semidilute is illustrated in Figures 24 and 25, which show the mean relaxation rate Ω/q^2 for polystyrene of various

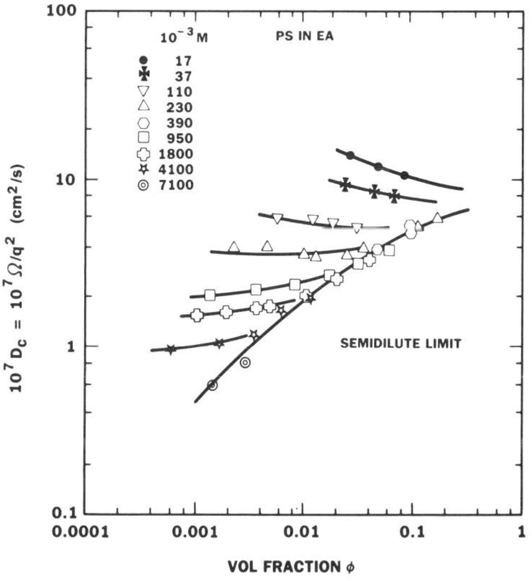

Figure 24. Concentration dependence of the relaxation rate for PS in EA[22] in the dilute and semidilute region at 25°C. ρ^* is defined as the concentration where the molecular-weight-dependent dilute curves intersect the semidilute limit. EA is a near theta solvent. At low concentration the ordinate is the self-diffusion constant and at high concentration it is the collective diffusion constant.

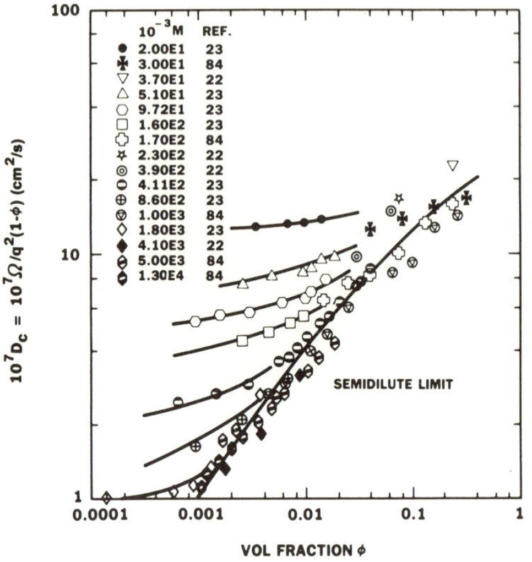

Figure 25. Concentration dependence of the relaxation rate for PS in THF. All the data have been reduced to 25°C. Substantial scatter between different groups is found, particularly at high concentration.

molecular weights in ethyl acetate (EA) and tetrahydrofuran (THF). Several observations can be made concerning these data, the most important of which is that the Ω becomes independent of molecular weight at concentrations near ρ^*. The regime where Ω is independent of molecular weight is called semidilute and the molecular-weight independence results from the fact that concentration fluctuations relax without overall chain motion. It should also be noted that for a given molecular weight the semidilute regime occurs at higher concentrations in marginal solvents (EA) than good solvents (THF). This difference is a consequence of the greater degree of swelling in good solvents.

In contrast to dilute systems, the dynamics of semidilute solutions are not firmly established either theoretically or experimentally. On the theoretical side, the results of the self-consistent field calculation of Freed and Edwards[59] contrasts with the scaling approach of de Gennes.[12] Contradictory experiments also exist with inconsistent conclusions reported by different groups working on the same system. In some sense then, it is premature to review semidilute systems at this time. Nevertheless, this report is more optimistic and an attempt is made to resolve the body of conflicting information through a careful identification of the conditions under which theoretical assumptions are satisfied. The general approach is that of Reference 22, although the details of the model are somewhat refined.

5.4.1.1. Hydrodynamic Screening. The concept of hydrodynamic screening is central to the understanding of semidilute polymer solutions.

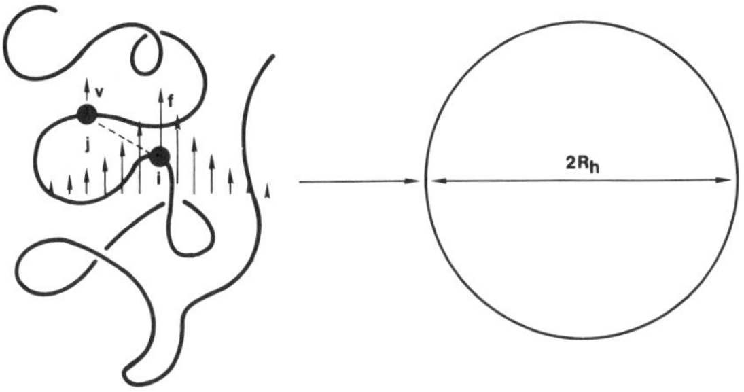

Figure 26. Schematic picture of the hydrodynamics of a single chain. A force f is exerted on the solvent by monomer i. This force leads to a solvent flow field decaying as $1/r$. This flow field leads to a solvent velocity at monomer j which couples the motion of the two monomers. If the flow fields of all monomers are summed self-consistently, then the chain friction constant is the same as that of a sphere of radius $R_h = \langle 1/r \rangle^{-1}$, where r is the intersegment distance in the chain.

Screening can be understood by first recalling the situation for a single chain in dilute solution. Here, the Kirkwood–Riseman calculation[10] shows that the solvent flow caused by the motion of any monomer leads to a solvent-mediated coupling with other monomers in the chain. In fact, in the regime of $qR \ll 1$, the motion of the chain is equivalent to that of a hard sphere of the same hydrodynamic radius. This situation is represented schematically in Figure 26 where the flow field around a single moving monomer is shown as well as the equivalent hard sphere.

In semidilute solution, the single-chain picture remains correct for short distances. That is, short sequences are coupled hydrodynamically and tend to move in a correlated manner. The situation for widely separated monomers on the same chain, however, is quite different because a distant monomer also feels the flow field caused by monomers on other chains. The flow fields from the neighboring chains are not correlated with that arising from monomers on the same chain so that the motion of widely separated monomers on the same chain becomes uncorrelated. This situation is shown schematically in Figure 27, which shows the equivalent hard sphere picture of a semidilute solution. Conceptually the system is replaced by a space-filling (or in some cases overlapping) group of spheres (concentration blobs of radius ξ_ρ) whose motion is uncorrelated. The radius of the spheres, ξ_ρ, is roughly the distance between interchain contacts and is proportional to the range of the static pair correlation function. Experiments sensitive to distances small compared to ξ_ρ ($q\xi_\rho \gg 1$) should yield results similar to a single chain, whereas in the opposite limit ($q\xi_\rho \ll 1$) dynamics characteristic of the diffusion of sphere of radius ξ_ρ are observed.

The characteristic length ξ_ρ is called a screening length. In fact, the

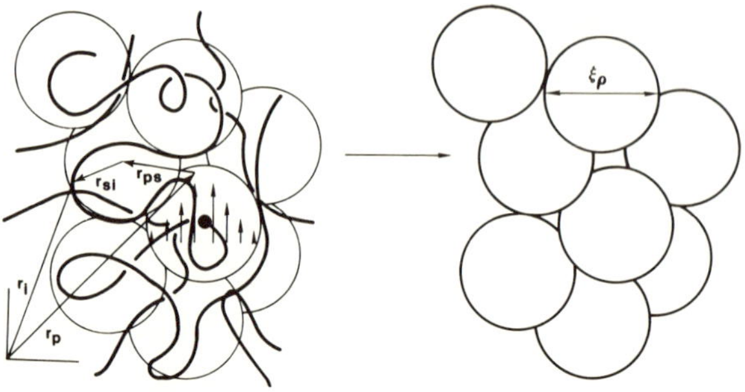

Figure 27. Schematic picture of screening in semidilute solution. In contrast to Figure 26, the flow-induced coupling is shorter ranged since distant monomers feel the random fields induced by the uncorrelated motion of monomers on other chains. The blob picture on the right shows the size of the effective Rouse blob in semidilute solution. Hydrodynamic coupling exists within the blob, but the motion of different blobs is uncorrelated even though they may share a common chain.

radial dependence of flow field (screened Oseen tensor) in semidilute solution has the classic Debye–Huckle form[59]

$$G(r) \sim \frac{1}{r} e^{-r/\xi_\rho} \tag{54}$$

and thus the identification of ξ_ρ as a screening length. Equation (54) contrasts with the flow field in dilute solution which is described by the Oseen tensor whose radial dependence decays as $1/r$. A reasonable body of evidence[22] shows that the screening length ξ_ρ is proportional to the range of the pair correlation function. This proportionality is reasonable since in the Kirkwood–Riseman picture, interchain contacts determine both the range of the pair correlation function as well as the range of hydrodynamic interaction.[4]

An alternative but equivalent description[12] of semidilute polymers can be obtained from a balance of elastic and frictional forces. The elastic force arises from the osmotic rigidity of the transient network formed by interchain contacts and the frictional resistance is that of a blob described above. The term "pseudogel" is often used to describe semidilute solutions when $q\xi_\rho \ll 1 \ll qR_g$. This term reflects the similarities between the transient network in semidilute solution and the permanent network in cross-linked swollen gels.

5.4.1.2. Reptation. In addition to screening and concentration blobs, one further concept, reptation, is necessary to appreciate semidilute-polymer dynamics. Reptation is the snakelike motion by which a chain completely renews its configuration.[4, 66] In congested systems, an individual chain is severely restricted by neighboring chains and effectively moves in a "tube" created by surrounding chains.[4] The reptation time T_R is the time required for a chain to completely escape a given tube and thereby renew its configuration. When the relaxation time of concentration blobs (described above) becomes comparable to T_R, then analogy to gels breaks down and a new dynamical process involving overall chain motion is important. Later we show that the reptation becomes the dominant process for $g^2 R_g^2 \ll \rho^*/\rho$.

5.4.2. Dynamical Regimes

The aspects of polymer dynamics which are probed by PCS are contained in the intermediate scattering function $S(q, t)$ in equation (5). A complete theory for $S(q, t, \rho)$ is obviously a formidable many-body problem so it is not surprising that simplified models are necessary. Fortunately, armed with the concepts of hydrodynamic screening, concentration blobs, pseudogels, and reptation, it is possible to predict the general properties of $S(q, t, \rho)$. A developing body of experimental evidence confirms many of these predictions.

Following Figure 27, $S(q, t)$ can be broken down into contributions due to chains p, concentration blobs or "strands" s within chains, and monomers i, j within strands. If $\mathbf{r}_i = \mathbf{r}_p + \mathbf{r}_{ps} + \mathbf{r}_{si}$ is the position of monomer i, then

$$\begin{aligned} S(q, t) = \frac{1}{V} \Big\langle \sum_p \exp\{i\mathbf{q} \cdot [\mathbf{r}_p(0) - \mathbf{r}_p(\mathrm{t})]\} \sum_s \exp\{i\mathbf{q} \cdot [\mathbf{r}_{ps}(0) - \mathbf{r}_{ps}(t)]\} \\ \times \sum_{ij} \exp\{i\mathbf{q} \cdot [\mathbf{r}_{si}(0) - \mathbf{r}_{sj}(t)]\} \\ + \sum_p \exp\{i\mathbf{q} \cdot [\mathbf{r}_p(0) - \mathbf{r}_p(t)]\} \sum_{s \neq s'} \exp\{i\mathbf{q} \cdot [\mathbf{r}_{ps}(0) - \mathbf{r}_{ps'}(t)]\} \\ \times \sum_{ij} \exp\{i\mathbf{q} \cdot [\mathbf{r}_{si}(0) - \mathbf{r}_{s'j}(t)]\} \Big\rangle \end{aligned} \tag{55}$$

Equation (55) presumes that the center-of-mass $\mathbf{r}_p$ of different chains is uncorrelated. This assumption is true in dilute solution and is also true within the reptation model of semidilute solutions. In order to simplify $S(q, t)$, it is convenient to specialize to three dynamical regimes and to

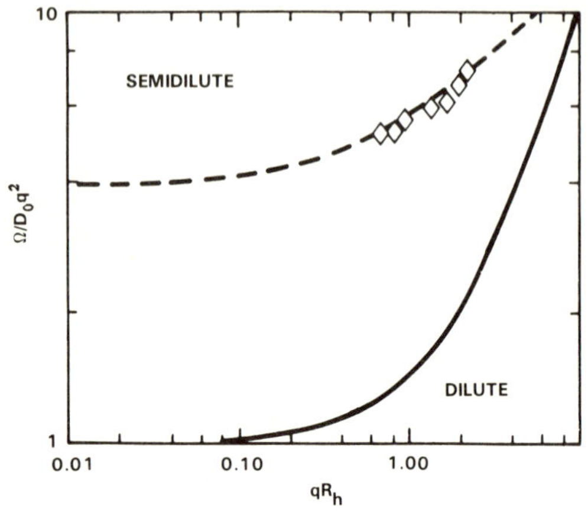

Figure 28. Comparison of the q dependence of the relaxation rate in dilute and semidilute solutions. The solid line is a best fit to the data in Figure 4, and the points are for a semidilute solution of PS in THF ($M = 4.1 \times 10^6$, $c = 0.01$ g/ml). The dotted line is to guide the eye. The blob model implies that the dotted line would intersect the solid line near $q\xi_\rho = 1$ and that the asymptote to the ordinate occurs at $\Omega/D_0q^2 = R_h/\xi_\rho$.

examine the concentration dependence of the relaxation processes within each dynamical regime separately.

5.4.2.1. The Single-Chain Limit: $q\xi_\rho \gg 1$. In the regime $q\xi_\rho \gg 1$, fast relaxation processes characteristic of monomer motion within strands dominate. In this limit $\mathbf{r}_p(0) = \mathbf{r}_p(t)$ and $\mathbf{r}_{ps}(0) = \mathbf{r}_{ps}(t)$ so equation (55) becomes

$$S(q, t) = \rho g_\rho \langle \exp\{i\mathbf{q} \cdot [\mathbf{r}_{si}(0) - \mathbf{r}_{sj}(t)]\} \rangle \tag{56}$$

where the correlation function in equation (56) has been examined in Section 5.2.2.3 by scaling and in Section 5.2.4 by linear response theory. Basically single-chain dynamics are recovered with the mean relaxation rate Ω scaling as q^3 as in equation (22).

Figure 28 shows data in the regime $q\xi_\rho \gtrsim 1$ for PS in semidilute solution. The line in the figure is the line which fit a similar plot (Figure 4) for polystyrene in dilute solution. The data in Figure 28 are not extensive, but it suggests that the crossover to q^3 scaling occurs over a very broad range in q. It is also possible that the transition to q^3 behavior occurs at $qR_h = 1$ rather than at $q\xi_\rho = 1$ as expected from the blob model. Clearly, more experimental data are needed.

5.2.2.2. The Pseudogel Regime: $q\xi_\rho \ll 1 \ll qR_g$. When $q\xi_\rho \ll 1 \ll qR_g$, equation (55) becomes

$$S(q, t) = \frac{\rho}{g_\rho} \langle g_\rho^2 \exp\{i\mathbf{q} \cdot [\mathbf{r}_{ps}(0) - \mathbf{r}_{ps}(t)]\}\rangle \tag{57}$$

where ρ is the monomer concentration and g_ρ is the number of monomers in a blob of radius ξ_ρ. Equation (57) assumes different blobs are uncorrelated due to screening as discussed above. In this equation we are concerned with the motion of the center of mass of a concentration blob whose dynamical properties have been obtained by de Gennes through analogy to cross-linked gels.[12] Equation (57) requires a Fokker–Plank equation for the time-dependent distribution function $P(\Delta r_{ps}; t)$ for the center of mass $\mathbf{r}_{ps}$ of a piece of chain between interchain contacts. The Fourier components of this distribution function, however, should follow the same dynamical equation as the Fourier components of the displacement $\mathbf{r} = \mathbf{r}_{ps}$ as Tanaka[72] assumed for his calculation of $S(q, t)$ for cross-linked gels. The equation of motion for $P(\Delta r; t)$ is determined by a balance of osmotic and frictional forces which lead to a Fourier-transformed dynamical equation of this form[63, 65].

$$-E_0 q^2 P(q; t) = \frac{\rho}{g_\rho} \zeta_\rho (1 - \phi)^{-1} \dot{P}(q; t) \tag{58}$$

where P is the distribution function hidden in equation (57), $E_0 = c\, \partial\pi/\partial c$ is the osmotic rigidity, $\zeta_\rho \simeq 6\pi\eta\xi_\rho$ is the blob friction constant, and ρ/g_ρ is the density of blobs. Solving for P in equation (58) and using (57), we obtain

$$S(q, t) = \frac{\rho}{g_\rho} \langle g_\rho^2 \exp[-D_c(1 - \phi)q^2 t]\rangle_{\xi_\rho} \simeq \rho g_\rho e^{-\Omega t} \tag{59}$$

$$\frac{\Omega}{(1 - \phi)q^2} = D_c \cong \frac{E_0 g_\rho}{\rho 6\pi\eta\xi_\rho} \tag{60}$$

Here D_c is the cooperative diffusion constant. The average over ξ_ρ in equation (59) arises because there is a distribution in blob radii ξ_ρ which is determined by the monomer pair correlation function. Experiments in cross-linked gels show that measured correlation functions are nearly exponential, so little generality is lost if the final average is ignored and ξ_ρ is identified with the range of the pair correlation function.[22] The various parameters in equation (60) depend on solvent quality (i.e., temperature), so to predict the concentration dependence of the decay rate g_ρ, ξ_ρ, and E_0 must be determined. With the exception of theta systems all these parameters depend on the probability of binary contacts.

Good Solvents. In good solvents the osmotic rigidity E_0 follows from scaling analysis.[73] In the semidilute regime the osmotic pressure is presumed to follow the power law

$$\pi \sim \frac{\rho}{N}\left(\frac{\rho}{\rho^*}\right)^x; \qquad \rho \gg \rho^* \tag{61}$$

where $\rho^* \sim N^{-4/5}$ [see equation (54)] and the exponent is fixed by the requirement that π be independent of N for $\rho > \rho^*$. This requirement leads to $x = 5/4$ so

$$\pi \sim \rho^{9/4} \tag{62}$$

Equation (62) is now well confirmed for polymers in good solvents.[70]

The osmotic rigidity $E_0 \sim \rho\, \partial\pi/\partial\rho$ immediately follows from (62):

$$E_0 \sim \rho^{9/4} \tag{63}$$

Since π is proportional to the number of binary contacts between chains, the number of monomers between binary contacts g_2 also follows. That is, since ρ/N is the number of chains per unit volume, $\rho^{9/4}(N/\rho) = N\rho^{5/4}$ is the number of binary contacts per chain. Therefore,

$$\frac{N}{g_2} \sim N\rho^{5/4} \tag{64}$$

or

$$g_2 \sim \rho^{-5/4} \tag{65}$$

In good solvents $g_\rho = g_2$ since concentration blobs are identified with strands between binary contacts.

The distance $\xi_\rho \sim \xi_2$ between contacts follows

$$\xi_\rho \sim g_2^{3/5} a \sim \rho^{-3/4} \tag{66a}$$

$$\simeq a\phi^{-3/4} n^{-1/4}(1 - 2\chi)^{-1/4} \tag{66b}$$

using equation (2). ϕ is the monomer volume fraction. Equation (66b) does not follow directly from the simple arguments presented here.[22,74]

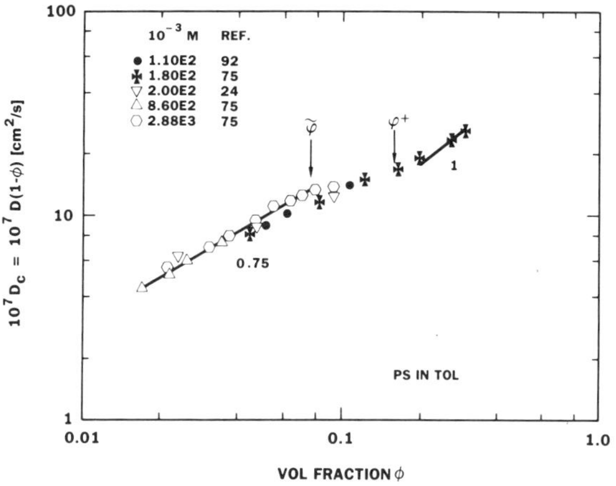

Figure 29. Concentration dependence of the collective diffusion constant D_c for polystyrene in semidilute solution in toluene. Toluene is a good solvent and scaling is observed below $\phi = 0.08$. This data may show a small marginal regime between $\tilde{\phi}$ and $\phi^{\dagger}$. All the data have been corrected to 25°C. Most of these data were measured by gradient diffusion. The data for the 1.8×10^6 sample due to Rehage and Ernst and are recorded in Reference 75.

In semidilute good solvents, density dependence of the relaxation rate Ω follows from equations (60), (63), (65), and (66):

$$D_c = \frac{\Omega}{q^2(1-\phi)} \sim \rho^{3/4} \sim \phi^{3/4} \tag{67}$$

For reasons which should become clear later, the scaling law equation (67) is seldom observed experimentally. Although scaling behavior has been claimed for many systems, it appears that only the PS/TOL and PDMS/TOL systems show unequivocal scaling exponents. From Table 1, these two systems represent a very good solvent (N_τ small for PS in TOL) and a very flexible molecule ($n \cong 1$ for PDMS). In addition to these two systems, the work of Yu *et al.*[68] shows a clear scaling regime for PS in THF at 30°C. These data, however, seem inconsistent with other data on PS in THF at 25°C, so they will be analyzed later.

Figure 29 shows data on semidilute PS in TOL obtained both by PCS and by gradient diffusion. This latter method is expected to yield the cooperative diffusion constant D_c when the gradient is small.[50] The data in Figure 29 are consistent with equation (67) for $\phi < 0.07$. Above $\phi = 0.07$ there appears to be two transitions which are discussed below.

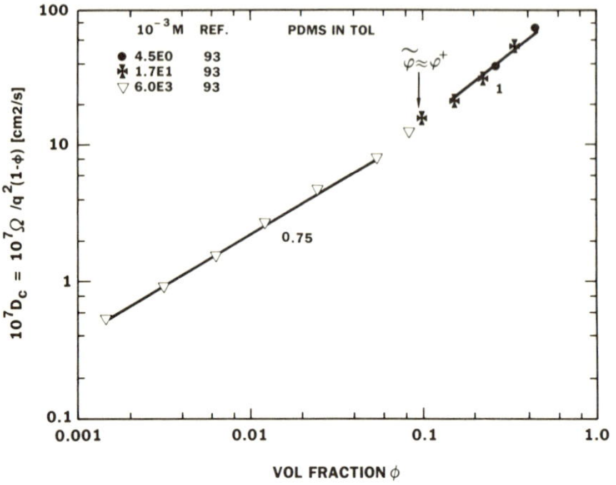

Figure 30. Concentration dependence of the collective diffusion constant for polydimethylsiloxane (PDMS) in toluene. PDMS is a very flexible chain, so no marginal regime is observed. $T = 20°C$.

PDMS is a very flexible molecule with $n = C_\infty/6 \simeq 1$. This system also shows a large region in semidilute solution where the scaling law is obeyed. The data of Munch *et al.*[92] for this system are shown in Figure 30 with a line of slope = 0.75 drawn through the data below $\phi = 0.1$. Above $\phi = 0.1$ a transition to a larger slope is observed. Below we show that such a transition is expected for a very flexible chain.

The breakdown of scaling at high concentration is due to the fact that the chains are ideal at small length scales in contrast to the assumption of equation (67), which depends on the fact that the chains are swollen: i.e., $\xi_\rho \sim g_\rho^{3/5}$. We know from Section 5.2.3 that swelling is found only for distances greater than ξ_τ, the temperature-dependent length which characterizes the statistics of the single chain. The volume fraction $\tilde{\phi}$ where equation (67) breaks down is found from equating ξ_ρ in equation (66) with ξ_τ in equation (25) or

$$\tilde{\phi} \cong \frac{3(1-2\chi)}{4\pi n^3} \tag{68}$$

$\tilde{\phi}$ is tabulated in Table 4 for several systems. Note that the predicted cutoff for scaling behavior is generally below $\phi = 0.05$.

Marginal Solvents. Above $\tilde{\phi}$ the chains are nearly ideal on all length scales and binary interactions are weak. In such a system, binary contacts occur at random with a probability proportional to $\phi^2 \sim \rho^2$. Following the

Table 3. Exponents for the Density Dependence of Various Quantities[a]

Quantity	Good	Marginal	Theta	Quantity	Good	Marginal	Theta
g_ρ	$-5/4$	-1	-2	D_c	$3/4$	$1/2$	1
ξ_ρ	$-3/4$	$-1/2$	-1	T_R/N^3	$3/2$	$5/2$	3
g_2	$-5/4$	-1	-1	E_0	$9/4$	2	3
ξ_2	$-3/4$	$-1/2$	$-1/2$	E_{gel}	$9/4$	2	2
g_3	—	-2	-2	N^2D_s	$-7/4$	$-5/2$	-3
ξ_3	—	-1	-1	η_r/N^3	$15/4$	$9/2$	5
g_t	$-5/4$	-2	-2	s	$-1/2$	$-1/2$	-1
ξ_t	$-3/4$	-1	-1				

[a] ξ_ρ is the range of the pair correlation function and is proportional to the hydrodynamic screening length. ξ_2 and ξ_3 are the distance between binary and ternary contacts. ξ_t is the radius in the repetation model. g_ρ, g_2, g_3, g_t are the number of monomers associated with the above lengths. T_R is the tube renewal time. E^0 and E_{gel} are the osmotic and elastic moduli. D_s is the self-diffusion coefficient. η is the viscosity, and s is the sedimentation constant.

logic of equations (64)–(66) then, one finds

$$\pi \sim E_0 \sim \rho^2 \tag{69}$$

$$g_\rho \sim g_2 \sim \rho^{-1} \tag{70}$$

$$\xi_\rho \sim \xi_2 \sim g_2^{1/2} a \sim \rho^{-1/2} \tag{71}$$

$$\cong \frac{n^{3/2} a}{(va^{-3})^{1/2}} \phi^{-1/2} \tag{72}$$

or using equation (60)

$$D_c = \frac{\Omega}{q^2(1-\phi)} \sim \phi^{1/2} \tag{73}$$

Once again, the coefficients in equation (72) do not follow directly,[22] but this equation will be useful later.

Experimental results consistent with equation (73) are observed in a large number of systems.[22, 64, 75, 78] Figures 31 and 32, for example, show PS in EA and MEK. Both systems show over a decade in concentration where $D_c \sim \phi^{1/2}$ is predicted by equation (73). Table 1 shows that these systems are marginal in the sense that excluded volume effects are weak but not absent ($N_\tau \simeq 10^3$). These solvents are generally called marginal solvents

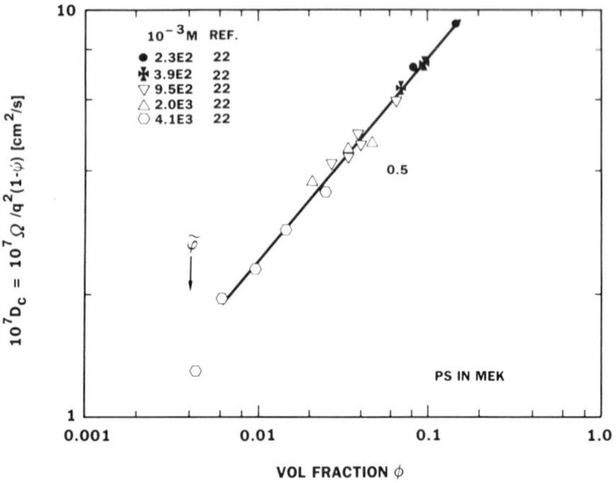

Figure 31. Concentration dependence of the collective diffusion constant for PS in MEK at 25°C.

and the regime above $\tilde{\phi}$ is called marginal semidilute. In the PS/MEK system there is a hint of a transition to steeper slope or scaling behavior below $\tilde{\phi} \cong 0.004$. In the PS/EA system, on the other hand, there appears to be a transition to steeper slope above $\phi = 0.08$. These transitions are probably manifestation of similar behavior found in PS/TOL in Figure 29. Other systems have been studied which may show marginal behavior such as PS/BZ and PS/THF. Interpretation of these systems is more complex, however, so discussion is deferred until the end of this section. PAMS dissolved in toluene(78) and PS in ethylbenzene and chlorobenzene(75) also show marginal behavior for $\phi > 0.01$. These observations are reasonably consistent with $\tilde{\phi}$ calculated from equation (68) using N_τ from Table 1.

Marginal systems described by equation (73) are usually associated with Edwards, who first obtained equation (73) in the mean-field approximation. Because of this approximation this regime is also called the mean-field regime. To confuse the situation further, British scientists often call $\phi > \tilde{\phi}$ the semiconcentrated regime. Continental scientists generally treat the marginal regime with patched-up scaling laws.(45, 95)

Theta Systems. As the concentration is increased or the binary interaction parameter, $va^{-3} = 1 - 2\chi$, is decreased and eventually three-body interactions dominate two-body effects. This situation can be expected either near the theta temperature or at high concentration. These systems, which are generically referred to as theta systems, cannot be described by

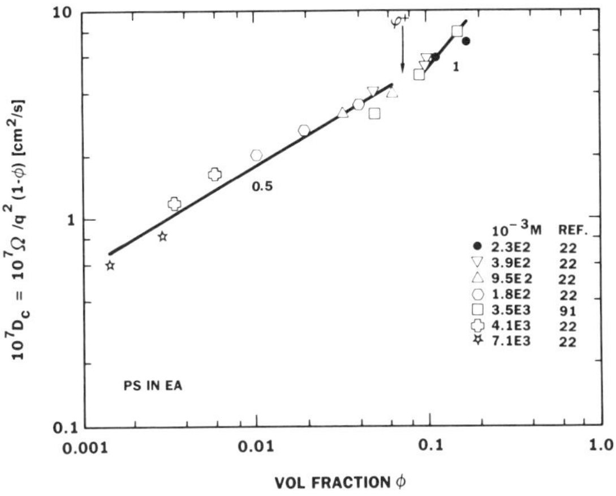

Figure 32. Concentration dependence of the collective diffusion constant for PS in ethyl acetate at 25°C.

the preceding theory where two-body interactions determine E_0, g_ρ, and ξ_ρ. In theta systems, both the osmotic pressure and osmotic rigidity E_0 are proportional to the probability of ternary contacts:

$$E_0 \sim E_3 \sim \rho^3 \tag{74}$$

Therefore, the blob friction constant in equations (58) and (60) must be calculated for a strand spanning ternary contacts that is $\xi_\rho \sim \xi_3$ and $g_\rho \sim g_3$,

$$g_\rho \sim g_3 \sim \rho^{-2} \tag{75}$$

$$\xi_\rho \sim \xi_3 \sim g_3^{1/2} a \sim \rho^{-1} \tag{76}$$

$$\simeq \frac{an^2}{(wa^{-6})^{1/2}} \phi^{-1} \tag{77}$$

where w is the three-body excluded volume ($w \simeq a^6 n_3$). Equation (77) was first obtained by Moore for the static correlation range[76] and was applied by Schaefer *et al.*[22] to dynamics. Application of equations (74)–(77) to equation (60) yields

$$D_c = \frac{\Omega}{q^2(1-\phi)} \sim \rho \sim \phi \tag{78}$$

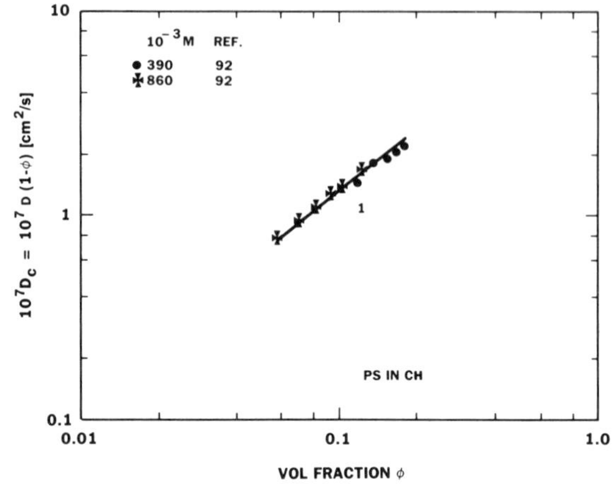

Figure 33. Concentration dependence of the collective diffusion constant of PS in cyclohexane at the theta temperature. Data were obtained by gradient diffusion.

Equation (78) is expected to apply to theta solvents for all concentrations above ρ^* and to other systems where the ρ is high enough that three-body effects dominate. This concentration, $\rho^\dagger$, can be found by equating equation (77) with equation (72) assuming $w = n^3a^6$,

$$\phi^\dagger = \rho^\dagger a^3 \simeq (1 - 2\chi) \tag{79}$$

This equation is highly approximate since the assumption $w = a^6n^3$ is unverified. For a few systems,[77] v and w are known from static properties and it appears that equation (79) is correct within a factor of 3.

Unfortunately, experimental data on cooperative diffusion in theta solvents are very limited. This scarcity is due to the fact that very high molecular weights are required to achieve chain overlap below $\phi = 0.1$. In theta solvents, however, high molecular weight chains are heavily self-entangled, making it difficult to achieve equilibrium systems. In fact, months or even years may be required to prepare such systems.[93]

Although PCS data is lacking on theta systems, both gradient diffusion[88] and sedimentation[75] results exist which are consistent with equation (78). Figure 33, for example, shows Roots and Nystrom's gradient diffusion data[88] for PS in CH at the theta temperature. Although the concentration range covered is limited, the data are consistent with a power law exponent of 1.0.

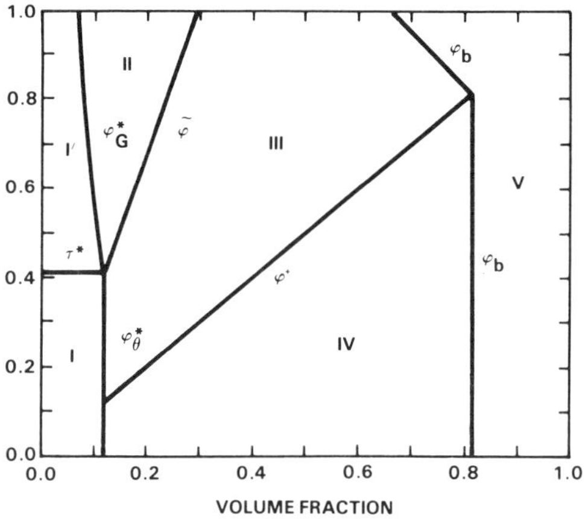

Figure 34. T–C diagram for a 20-unit chain with $n = 1.5$. The excluded volume parameter v is assumed to be linear in the temperature increment $\tau = (T - \Theta)/T$.

Thetalike behavior is also observed in good and marginal solvents at high concentration. The systems PS/TOL, PS/EA, and PDMS/TOL in Figures 29–31 all show power-law exponents near unity at high concentration. There are numerous potential experimental problems at high concentration, so these Θ-like exponents above $\phi = 0.1$ may be questionable. In fact, several workers have observed concentration independence or negative exponents at high concentration.[22, 60]

T–C Diagrams. The various pseudogel regimes described above can be collected graphically through a temperature-concentration diagram (T–C diagram) as proposed by Daoud and Jannink.[80] Such a diagram is shown in Figure 34 for a 20-unit chain with $n = 1.5$ and $(1 - 2\chi) = \tau$. The various lines in the figure represent the approximate transition regions where the nature of polymer dynamics changes. The line τ^*, for example, separates dilute theta systems (region I) from dilute swollen systems (region I′). The equation for this line follows from equation (25b) with $N_\tau = N$. Lines ϕ^*_Θ and ϕ^*_G represent the crossover to semidilute solution and follow from equation (53) for theta and good solvents.

As discussed above, the semidilute regime is subdivided into good (II), marginal (III), and theta (IV) domains with the crossovers $\tilde{\phi}$ and $\phi^\dagger$ defined by equations (68) and (79). Finally, the crossovers ϕ_b to the concentrated regime V are determined by the condition $\xi_\rho = na$ and follow directly from equations (72) and (77).

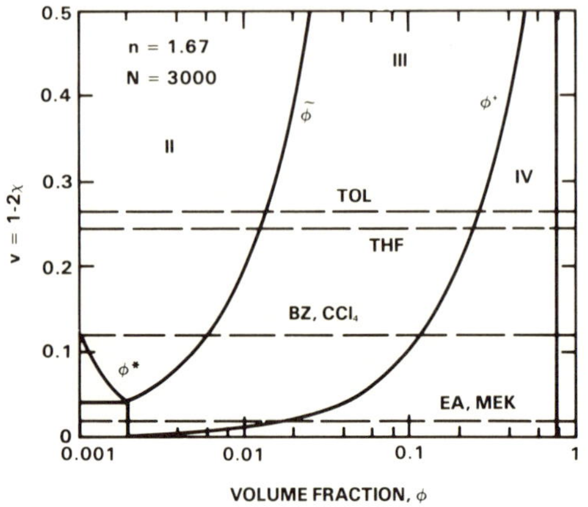

Figure 35. T–C diagram for PS. The ordinate is the coupling constant $(1 - 2\chi) = va^{-3}$ and various solvents are represented by the horizontal lines.

A more universal form of the T–C diagram can be constructed by use of the parameter $(1 - 2\chi) = va^{-3}$ instead of temperature on the ordinate. The semilog version of such a plot is shown in Figure 35 for $N = 3000$ and $n = 1.67$. This value of n is chosen for PS where $n = C_\infty/6 = 1.67$ when equation (1) is applied to the radius of gyration. The behavior of PS in various solvents can be predicted from horizontal lines at the appropriate values of $(1 - 2\chi)$ obtained from Table 1. Lines appropriate to several solvents are included. Plots like Figure 35 should accurately reflect trends with temperature, chain rigidity, and solvent quality, but caution should be exercised in quantitative predictions based on the diagrams. After all, the assumptions leading to the diagram are crude, and although power law exponents are probably accurate, the coefficients required for quantitative analysis are largely unknown.

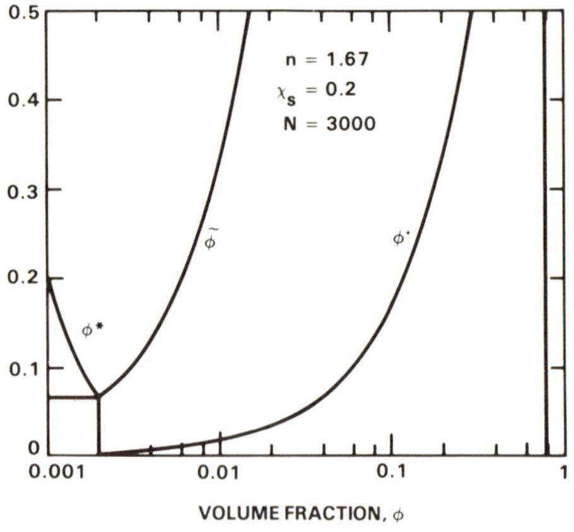

Figure 36. T–C diagram for PS in cyclohexane. This diagram assumes $\chi_s = 0.2$ and that $(1 - 2\chi)$ is linear in the temperature increment $\tau = (T - \Theta)/T$.

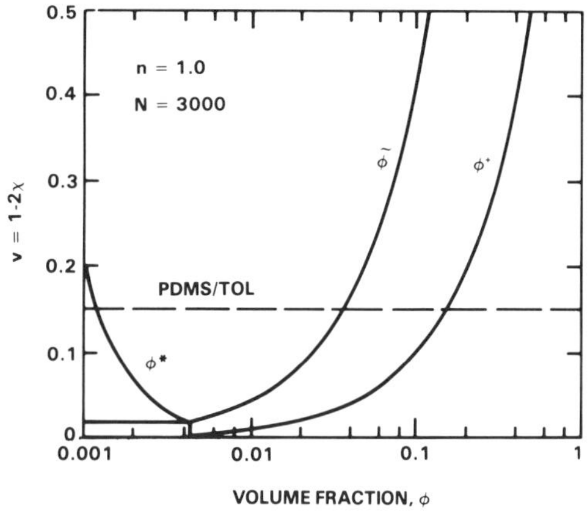

Figure 37. *T–C* diagram for a completely flexible chain. This diagram should be appropriate for PDMS. $1 - 2\chi$ for PDMS in toluene is shown as a horizontal line.

For systems like PS in CH where va^{-3} is a strong function of temperature, it is possible to construct a temperature–concentration diagram as shown in Figure 36. This diagram assumes $\chi_s = 0.2$ and should be reasonably accurate for PS in CH and probably is fairly close for PS in cyclopentane as well.

The marginal regime (III) is found only in stiff polymer chains. Figure 37 shows the T–C diagram appropriate to a highly flexible chain such as PDMS where $n = C_\infty/6 = 1.0$. For $n \leqq 1$ region III is small and crossover may occur directly from good-solvent scaling behavior to theta behavior. Such a crossover is consistent with the PDMS data in Figure 30, where a distinct transition to higher slope is observed near $\tilde{\phi} \simeq \phi^\dagger = 0.1$.

Other Systems. Conflicting data and/or interpretations exist for cooperative diffusion of PS in BZ and THF. In BZ, for example, Adam and Delsanti[25] and Munch *et al.*[90] report exponents consistent with scaling laws, whereas Schaefer[61] finds an exponent of 0.5 consistent with marginal behavior. Similar discrepancies exist for PS in THF.

Figure 38 shows existing data for PS in BZ. These data can be fitted nicely with the scaling exponent of 0.75, confirming the observation of Adam and Delsanti[25] and Munch *et al.*[90] Scaling behavior at these high concentrations, however, is inconsistent with the T–C diagram in Figure 35 and also conflicts with viscosity[94] and sedimentation data.[75] The sedimentation data, in particular, are consistent with $\phi^\dagger \simeq 0.03$. In other words, it appears that most of the data in Figure 38 are in the crossover region between marginal and theta power laws as evidenced by the upward curvature of the data and the limiting slopes shown on the figure. The apparent exponent near 0.75 is a consequence of an attempt to fit the data in a crossover region and is not necessarily indicative of scaling.

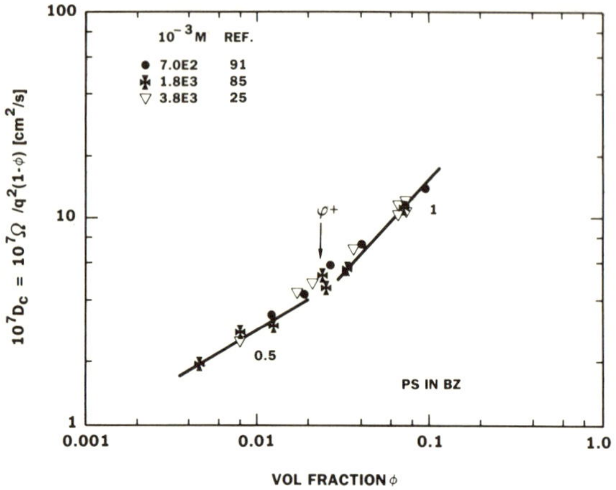

Figure 38. Collective diffusion constant for PS in benzene. The limiting slopes of 0.5 and 1.0 suggest that BZ is a marginal solvent. $T = 20°C$. The data of Reference 85 were taken at 25°C and reduced to 20°C assuming $D_c \sim T/\eta$.

The PS/THF system has been studied by at least four groups with different results and conclusions. Figure 39 shows the data of Schaefer *et al.*,[22] Mandema *et al.*,[23] and Amis and Han.[84] Although the scatter is substantial below $\phi = 0.05$, the data seem to follow power law behavior with a slope of ~ 0.6. From Table 1, this system should show scaling behavior below $\tilde{\phi} \simeq 0.05$ so we attribute the intermediate exponent to mean that the data are in the region-I–region-II crossover. The data of Yu *et al.*[68] are

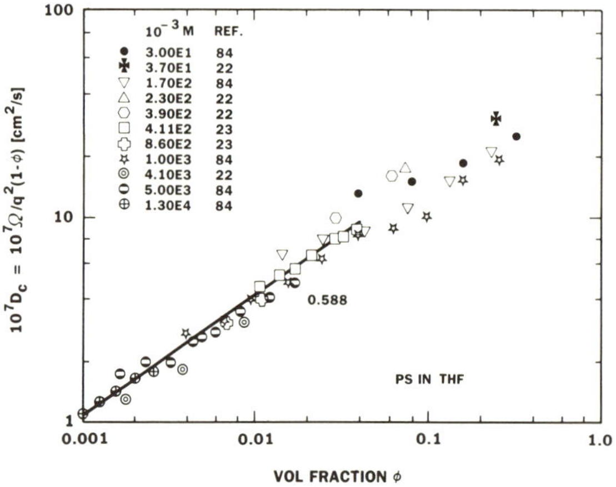

Figure 39. Collective diffusion constant for PS in tetrahedrofuran (THF) at 25°C.

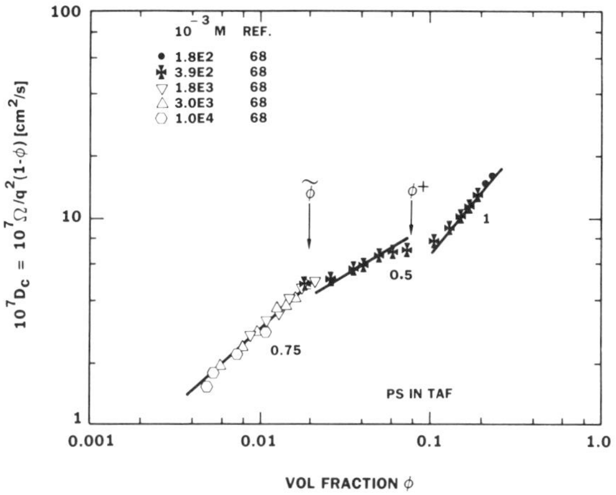

Figure 40. Collective diffusion constant of PS in THF measured at 30°C and reduced to 25°C. These are the only data that show two distinct transitions. The theory outlined here implies that the slope should be 0.5 between $\tilde{\phi}$ and $\phi^{\dagger}$. The data, however, show a substantially smaller slope.

not numerically consistent with Figure 39, but these data do show the expected crossovers. These data are plotted in Figure 40 where the ordinate is reduced to 25°C assuming $D_c \sim T/\eta$. These data show scaling below $\phi = 0.02$ and a thetalike power law above $\phi = 0.1$. In the marginal regime between 0.02 and 0.1, however, the power law exponent appears to be smaller than 0.5. It seems that in spite of the abundance of data, the PS/THF system is not well understood.

Table 4 tabulates the various crossovers both inferred from the model presented here and from the experimental data. Although some of the

Table 4. Crossover Points

System	$\tilde{\phi}$ Equation (68)	$\tilde{\phi}$ Observed	$\phi^{\dagger}$ Equation (79)	$\phi^{\dagger}$ Observed
PS in				
EA	0.002	—	0.04	0.08
MEK	0.004	(0.004)	0.08	—
BZ	0.004	—	0.1	0.03
THF	0.01	0.02	0.2	0.1
TOL	0.02	0.07	0.3	0.2
PDMS in				
TOL	—	—	0.1	0.1

observed crossovers differ from the predicted values by as much as a factor of 5, there is a reasonably consistent trend with solvent quality. In view of the simplicity of the model used, even accuracy within a factor of 5 is encouraging.

5.4.2.3. Reptation: $qR_g \lesssim 1$. As q is decreased, one probes concentration fluctuations of longer Fourier spatial wavelength. Eventually, the condition is approached where the collective relaxation time $(D_c q^2)^{-1}$ becomes comparable to the reptation time T_R. At this point, reptation becomes the dominant relaxation mechanism and Ω is expected to reflect the chain self-diffusion constant D_s. Both T_R and D_t can be found from the simple reptation model described above.

The reptation time and self-diffusion constant can be obtained by a model based on one-dimensional diffusion of blobs in a tube.[4] That is, a polymer is pictured as a Rouse chain of renormalized monomers (blobs) of radius ξ_t, where the Rouse monomer radius or tube radius ξ_t is determined by the space-filling condition:

$$\rho = \frac{g_t}{\xi_t^3} \tag{80}$$

Here g_t is the number of monomers in a "tube blob." Equation (80) coupled with the appropriate chain statistics

$$\xi_t \sim g_t^\nu a, \qquad \tfrac{1}{2} < \nu < \tfrac{3}{5} \tag{81}$$

yields the tube radius as listed in Table 3 for the various concentration regimes described above. Note that in marginal systems the tube radius is not proportional to $\xi_\rho \sim \xi_2$. The multiplicity of length scales in marginal systems has led to considerable confusion in the interpretation of experiments which are sensitive to reptation. The model proposed here is not without conceptual problems, so reptation in marginal systems should still be considered an open question. In good solvents problems do not arise since all lengths (ξ_ρ, ξ_2, ξ_t) have the same concentration dependence.

The friction constant associated with the Rouse blobs in the tube is not necessarily proportional to ξ_t^{-1} because screening tells us that monomers are hydrodynamically coupled over distances comparable to the range ξ_ρ of the pair correlation function, not the tube radius ξ_t. Therefore, the friction constant of a tube blob ζ is

$$\zeta \sim 6\pi\eta\xi_\rho \tag{82}$$

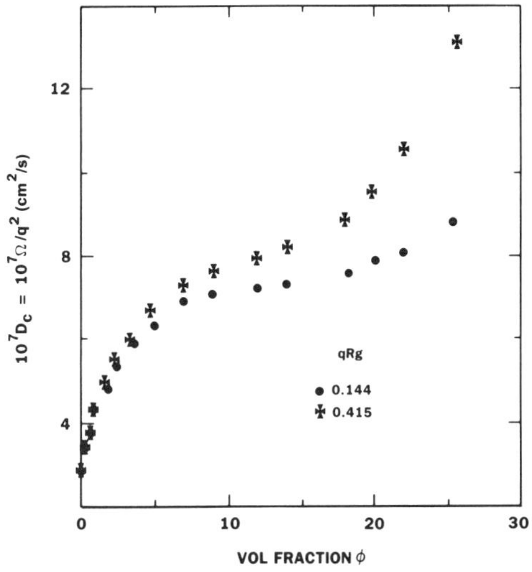

Figure 41. Concentration dependence of the initial slope of $S(q, t)$ for two values of qR_g.[68] For $qR_g \ll 1$ the mean relaxation rate is substantially reduced due to a reptation-type diffusion process. For $qR_g \ll 1$, collective diffusion is not possible.

Using $D_1 = g_\rho kT/N\zeta$ it is possible to calculate the time for the chain to escape its tube by *one-dimensional* diffusion[10]

$$T_R \sim \frac{(\text{tube length})^2}{D_1} \sim \frac{N^3 \xi_\rho \xi_t^2 \eta}{g_t^2 g_\rho kT}, \qquad \varphi \ll 1 \tag{83}$$

The reptation time calculated by equation (83) is listed in Table 3.

The crossover point for transition from cooperative diffusion to reptation can be found by equating the cooperative relaxation time $(D_c q^2)^{-1}$ with T_R, with the final result

$$q^2 R_g^2 = \left(\frac{\rho^*}{\rho}\right)^{9/4} \tag{84}$$

in good solvents. For $q^2 R_g^2 < (\rho^*/\rho)^{9/4}$, reptation is the dominant relaxation process. In marginal and theta systems, equation (84) is slightly modified $[q^2 R_g^2 = 0.44(\rho^*/\rho)^3]$.

The data of Yu *et al.*[68] shown in Figure 41 illustrate the transition to slower reptation dynamics at small q. These data show that as concentration increases, the correlation function is increasingly dominated by a slow relaxation process when $qR_g \ll 1$. The strong q dependence of Ω/q^2 at high concentration may explain the discrepancies in values of D_c reported by different groups working on the same system such as the PS/THF system described above.

The form of $S(q, t)$ in the reptation regime can be obtained[67] by evaluating equation (55) for $q\xi_\rho \ll 1$, $qR_g \ll 1$

$$S(q, t) = \rho g_\rho \left\{ \exp[-q^2(D_s + D_c)(1 - \phi)t] + \left[\left(\frac{N}{g_\rho} \right) - 1 \right] \exp[-q^2 D_s(1 - \phi)t] \right\} \quad (85)$$

where we have assumed that the chain center of mass relaxes by self diffusion, D_s being the three-dimensional self-diffusion constant. Equation (85) predicts a bimodal decay, and this fact accounts for the nonexponential correlation functions which are generally reported in a semidilute solution.[61] In cross-linked gels, by contrast, the second term of equation (85) is absent and the observed correlation functions are exponential.

The amplitude A of the collective and self terms in equation (85) shows that

$$\frac{A_s}{A_c} = \left(\frac{N}{g_\rho} \right) - 1 \cong \left(\frac{\rho}{\rho^*} \right)^y, \qquad \begin{array}{l} y = 5/4 \text{ (good)} \\ y = 1 \text{ (marginal)} \\ y = 2 \text{ (theta)} \end{array} \quad (86)$$

In other words, reptative self-diffusion becomes more important at high concentration. This behavior is seen not only in Figure 41 but also in the bimodal correlation functions of Jamieson *et al.*[81] shown in Figure 42. Here a slow relaxation process becomes increasingly important at high concentration for $qR_g < 1$.

In the limit $qR_g \ll 1$ or $t \gg (D_c q^2)^{-1}$, equation (85) simplifies to

$$S(q, t) \cong \rho N \exp[-q^2 D_s(1 - \phi)t] \quad (87)$$

This equation implies that the reptative diffusion constant D_t can be measured by PCS. The concentration dependence of the relaxation rate $\Gamma = D_s q^2(1 - \phi)$ can be obtained through the tube model. That is, since a time T_R is required for the chain to escape its tube whose trajectory is of dimension R_t

$$D_s \sim \frac{R_t^2}{T_R} = \left[\left(\frac{N}{g_t} \right)^{1/2} \xi_t \right]^2 T_R^{-1} \quad (88a)$$

$$\sim kTN^{-2} g_t g_\rho \xi_\rho^{-1} \eta^{-1}, \qquad \varphi \ll 1 \quad (88b)$$

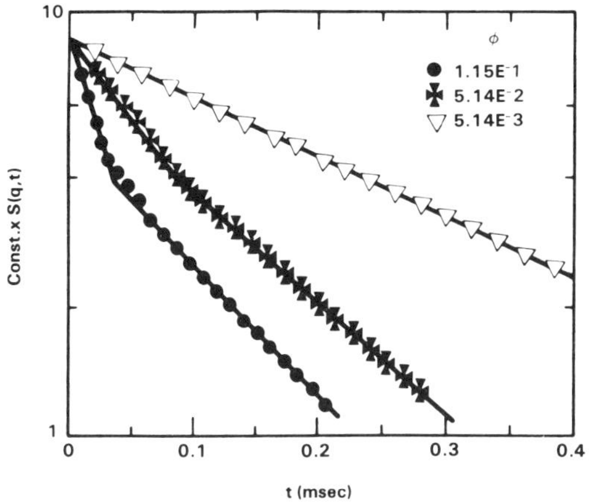

Figure 42. Measured correlation function for PS in THF at three concentrations.[81] At high concentration $S(q, t)$ is bimodal, reflecting both a collective and reptation relaxation mode. The data are for $M = 3.9 \times 10^5$ at 30°C. The upper curve is in the dilute regime.

The concentration dependence obtained from equation (88b) is listed in Table 3. For convenience, other quantities[4] are also listed, such as the elastic modulus E_{gel} and the viscosity $\eta_r = E_{\text{gel}} T_R$ and the sedimentation coefficient $s = g_\rho \xi_\rho^{-1}$.

Very little experimental work has been reported on the slowly decaying translational mode in semidilute solution, although forced Rayleigh scattering has been reported.[62] Amis and Han,[84] however, have studied this mode for PS in THF, and their data are shown in Figure 43. Here $N^2\Gamma/q^2(1 - \phi)$ is plotted vs. volume fraction ϕ and lines are drawn with the appropriate slopes for good marginal and theta behavior. In the figure, Γ is the relaxation rate of $S(q, t)$ for $t \gg (D_c q^2)^{-1}$. $N^2\Gamma/q^2(1 - \phi)$ should equal D_s in equation (88) if the model used here is correct. Here $(1 - \phi)$ is a backflow correction. The data are reasonably consistent with equation (88b) with the crossover $\tilde{\phi}$ occurring at 0.01, which is close to $\tilde{\phi} \cong 0.02$ found in Figure 40 for the same system. At the highest concentrations, the curve drops precipitously and is not consistent with the ϕ^{-3} behavior predicted above $\phi^\dagger$. It must be realized that the precision of the data is poor for these very long correlation times, so ϕ^{-3} behavior is not necessarily out of the question. These data are also at the upper limit of semidilute solutions and very likely interchain interactions limit the validity of the friction constants used.

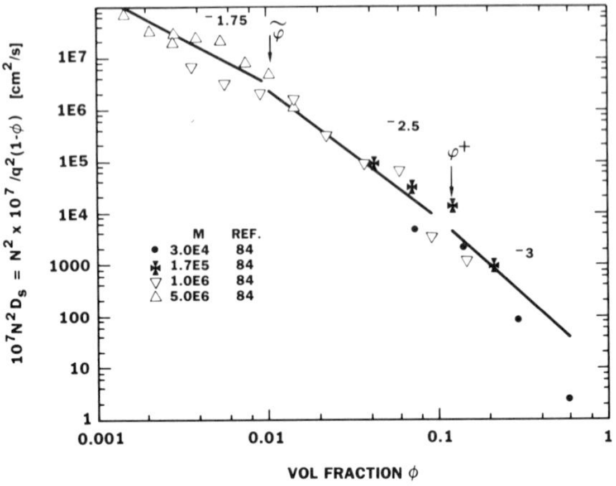

Figure 43. Self-diffusion constant taken from the measured relaxation rate Γ of the slowly decaying mode of $S(q, t)$. The ordinate should be N^2D_s and should follow the power laws indicated by the lines in the figure.

5.4.3. Conclusions

In this discussion of semidilute systems we have attempted to develop a framework that is useful to understand a large body of data that appear contradictory. In many cases, our interpretation conflicts with that of the experimenters themselves. Unfortunately, we have been forced, because of space limitations, to ignore these credible alternative interpretations. The model presented here, however, makes a broad range of predictions which have been subject to only limited experimental study.

It is interesting to note that in this entire review the concept of an "entanglement" has been completely ignored. Although entanglements may be important to certain viscoelastic properties, there is no evidence as yet that they play a significant role in $S(q, t, \rho)$ in the domains easily accessible to PCS measurements.

If rubberlike elasticity due to entanglements were playing a role in $S(q, t)$, then D_c would be proportional to E_{gel} rather than E_0.[63] In good and marginal solvents $E_{gel} \sim E_0$, so the observed concentration dependence of Ω would not change. In theta solvents ($T = \Theta$ not $\phi > \phi^\dagger$), however, D_c would be concentration independent in contradiction to Figure 33. Thus, based on limited data available, entanglements do not seem to affect cooperative diffusion.

If entanglements played a significant role in the reptation process, then D_s should show a dependence on N to a power larger than 2 and should also display a distinct change in behavior below the critical molecular weight for entanglements.[66] To our knowledge, neither expectation is realized. Recent experiments by Selser[78] and theory by Ronca[79] do suggest that the detailed shape and q dependence of $S(q, t)$ reflect entanglements. This effect, however, appears as a small quartic term in the reptation part of $S(q, t)$. It appears, then, that although the broad features of semidilute dynamics do not reflect entanglements, undoubtedly they will be necessary to explain increasingly detailed aspects of polymers at high concentration.

REFERENCES

1. P. J. Flory, *Statistical Mechanics of Chain Molecules*, Interscience, New York (1969).
2. C. Croxton, *Introduction to Liquid State Physics*, Wiley, New York (1975).
3. P. J. Flory, *Principles of Polymer Chemistry*, Cornell University, Ithaca, New York (1953), Chap. 14.
4. P. G. de Gennes, *Scaling Concepts in Polymer Physics*, Cornell University, Ithaca, New York (1979).
5. H. Yamakawa, *Modern Theory of Polymer Solutions*, Harper and Row, New York (1971).
6. P. J. Flory, *Discuss. Faraday Soc.* **49**, 7 (1970).
7. Reference 2, Chap. 2.
8. B. Farnoux, F. Boue, J. P. Cotton, M. Daoud, G. Jannink, M. Nierlich, and P. G. de Gennes, *J. Phys. (Paris)* **39**, 77 (1978).
9. Reference 4, Chap. 6.
10. J. Kirkwood and J. Riseman, *J. Chem. Phys.* **16**, 565 (1948).
11. B. Zimm, *J. Chem. Phys.* **24**, 269 (1956).
12. P. G. de Gennes, *Macromolecules* **9**, 587, 594 (1976).
13. L. K. Nicholson, J. S. Higgins, and J. B. Hayter, *Macromolecules* **14**, 836 (1981).
14. T. A. King, A. Knox, W. I. Lee, and J. D. G. McAdam, *Polymer* **14**, 151 (1973).
15. J. des Cloizeaux and G. Weill, *J. Phys. (Paris)* **40**, 99 (1979).
16. A. Z. Akcasu and C. C. Han, *Macromolecules* **12**, 276 (1979).
17. D. W. Schaefer and J. G. Curro, *Ferroelectrics* **30**, 49 (1980).
18. J. G. Curro and D. W. Schaefer, *Macromolecules* **13**, 1199 (1980).
19. M. Benmouna and A. Z. Akcasu, *Macromolecules* **11**, 1187, 1193 (1978).
20. A. Z. Akcasu, M. Benmouna, and C. C. Han, *Polymer* **21**, 866 (1980).
21. (a) T. A. King, A. Knox, and J. D. G. McAdam, *Polymer* **14**, 293 (1973); (b) N. C. Ford, F. E. Karaz, and J. E. M. Owen, *Discuss. Faraday Soc.* **49**, 228 (1970).
22. D. W. Schaefer, J. F. Joanny, and P. Pincus, *Macromolecules* **13**, 1280 (1980).
23. W. Mandema and Z. Zeldenrust, *Polymer* **18**, 835 (1977).
24. P. N. Pusey, J. M. Vaughn, and V. G. Williams, *J. Chem. Soc., Trans. II* **70**, 1696 (1974).
25. M. Adam and M. Delsanti, *Macromolecules* **10**, 1229 (1977).
26. *Polymer Handbook*, 2nd Ed., J. Brandrup and E. H. Immergut, eds., Interscience, New York (1975).
27. J. Selser, *Macromolecules* **14**, 346 (1981).
28. S. Saeki, S. Konno, N. Kuwahara, M. Nakata, and M. Kaneko, *Macromolecules* **7**, 521 (1974).
29. R. Pecora, *J. Chem. Phys.* **43**, 1562 (1965).
30. P. G. de Gennes, *Physics*, **3**, 37 (1967).
31. E. Dubois-Violette and P. G. de Gennes, *Physics* **3**, 181 (1967).

32. R. Zwanzig, *J. Chem. Phys.* **60**, 2717 (1974).
33. (a) M. Fixman, *J. Chem. Phys.* **42**, 3831 (1965); (b) C. W. Pyun and M. Fixman, *J. Chem. Phys.* **42**, 3838 (1965).
34. A. Z. Akcasu and H. Gurol, *J. Polym. Sci., Polym. Phys. Ed.* **14**, 1 (1976).
35. J. des Cloizeaux, CEN (Saclay) Reports (1976).
36. J. Riseman and J. G. Kirkwood, in *Rheology, Theory and Applications*, F. R. Eirich, ed., Adademic Press, New York (1956).
37. P. E. Rouse, Jr., *J. Chem. Phys.* **21**, 1272 (1953).
38. J. E. Shore and R. Zwanzig, *J. Chem. Phys.* **63**, 5445 (1975).
39. R. Bellman, *Introduction to Matrix Analysis*, 2nd Ed., McGraw-Hill, New York (1970).
40. W. Burchard and M. Schmidt, *Polymer* **21**, 745 (1980).
41. A. Z. Akcasu and J. S. Higgins, *J. Polym. Sci.* (*Polym. Phys. Ed.*) **15**, 1745 (1977).
42. M. Benmouna, A. Z. Akcasu, and M. Daoud, *Macromolecules* **13**, 1703 (1980).
43. A. Perico, P. Piaggio, and C. Cuniberti, *J. Chem. Phys.* **62**, 2690 (1975).
44. M. Schmidt and W. Stockmayer (private communication).
45. G. Pouyet, J. Francois, J. Dayantis, and G. Weill, *Macromolecules* **13**, 176 (1980).
46. R. Zwanzig *et al.*, *Proc. Nat. Acad. Sci. USA* **60**, 381 (1981).
47. R. Ullman, *Macromolecules* **7**, 300 (1974).
48. J. T. Fong and A. Peterlin, *J. Res. Nat. Bur. Stand. Sect. B, Math. Sci.* **80B**, 273 (1976).
49. C. C. Han and A. Z. Akcasu, *Macromolecules* **14**, 1080 (1981).
50. M. Adam, M. Delsanti, and G. Pouyet, *J. Phys. Lett.* (*Paris*) **40**, L-435 (1979).
51. M. Daoud, thesis, Universite de Paris (1977).
52. H. Yamakawa, *J. Chem. Phys.* **36**, 2995 (1962).
53. S. Imai, *J. Chem. Phys.* **50**, 2116 (1969).
54. C. W. Pyun and M. Fixman, *J. Chem. Phys.* **41**, 957 (1964).
55. A. R. Altenberger and J. M. Deutch, *J. Chem. Phys.* **59**, 894 (1973).
56. A. Z. Akcasu and M. Benmouna, *Macromolecules* **11**, 1193 (1978).
57. A. Z. Akcasu, *Polymer* **22**, 1169 (1981).
58. C. C. Han and A. Z. Akcasu, *Polymer* **22**, 1165 (1981).
59. K. F. Freed and S. F. Edwards, *J. Chem. Phys.* **61**, 1189, 3626 (1974).
60. G. D. Patterson, J. R. Stevens, G. R. Alms, and C. P. Lindsey, *Macromolecules* **12**, 661 (1979).
61. D. W. Schaefer, *Polym. Prepr.* **19**, 452 (1978).
62. H. Hervet, L. Leger, and F. Rondelez, *Phys. Rev. Lett.* **42**, 1681 (1979).
63. F. Brouchard and P. G. de Gennes, *Macromolecules* **10**, 1157 (1977).
64. D. W. Schaefer, *Polym. Prepr.* **19**, 733 (1978).
65. E. Geissler and A. M. Hecht, *J. Phys. Lett.* (*Paris*) **40**, L-173 (1979).
66. J. Klein, *Macromolecules* **11**, 852 (1978).
67. E. J. Amis, C. C. Han, and Y. Matsushita, *Polymer* **25**, 650 (1984).
68. T. L. Yu, H. Reihanian, and A. M. Jamieson, *Macromolecules* **13**, 1590 (1980).
69. J. S. Higgins, L. K. Nicholson, and J. B. Hayter, *Polym. Prepr.* **22**, 86 (1981).
70. I. Noda, N. Kato, T. Kitano, and M. Nagasawa, *Macromolecules* **14**, 668 (1981).
71. C. E. H. Bawn, R. F. J. Freeman, and A. R. Kamaliddin, *Trans. Faraday Soc.* **46**, 677 (1950).
72. T. Tanaka, L. O. Hocker, and G. B. Benedek, *J. Chem. Phys.* **59**, 5151 (1973).
73. M. Daoud, J. P. Cotton, B. Farnoux, G. Jannink, G. Sarma, H. Benoit, R. Duplessix, R. Picot, and P. G. de Gennes, *Macromolecules* **8**, 804 (1975).
74. A. R. Khokhlov, *Polym. Sci. USSR* **20**, 3087 (1978).
75. B. Nyström and J. Roots, *J. Macromol. Sci. Rev. Macromol. Chem.* **C19**, 35 (1980).
76. M. A. Moore, *J. Phys.* (*Paris*) **38**, 265 (1977).

77. K. Okano, E. Wada, K. Kurita, and H. Fukuro, *J. Appl. Crystallogr.* **11**, 507 (1978).
78. J. Selser, *J. Chem. Phys.* **79**, 1044 (1983).
79. G. Ronca, to be published.
80. M. Daoud and G. Jannink, *J. Phys.* (*Paris*) **37**, 973 (1976).
81. A. M. Jamieson, H. Reihanian, J. G. Southwick, T. L. Yu and J. Blackwell, *Ferroelectrics* **30**, 267 (1980).
82. D. Richter, B. Ewen, and J. B. Hayter, *Phys. Rev. Lett.* **45**, 2121 (1980).
83. C. Reiss and H. Benoit, *C.R. Acad. Sci. Paris* **253**, 268 (1961).
84. E. J. Amis and C. C. Han, *Polymer* **23**, 1403 (1982).
85. D. W. Schaefer, unpublished.
86. B. Appelt and G. Meyerhoff, *Macromolecules* **13**, 657 (1980).
87. M. E. McDonnell and A. M. Jamieson, *J. Macromol. Sci. Phys.* **B13**, 67 (1977).
88. J. Roots and B. Nyström, *Macromolecules* **13**, 1595 (1980).
89. R. S. Chahal, W. P. Kao, and D. J. Patterson, *Chem. Soc., Faraday Trans.* **69**, 1834 (1973).
90. J. P. Munch, S. Candau, J. Hertz, and G. J. Hild, *J. Phys.* (*Paris*) **38**, 971 (1977).
91. J. Roots, B. Nyström, B. Porsch, and L. O. Sundelöf, *Polymer* **20**, 337 (1979).
92. J. P. Munch, P. L. Lemarichal, and S. Candau, *J. Phys.* (*Paris*) **38**, 1499 (1977).
93. P. Mathiez, C. Mouttet, and G. Weisbuch, *J. Phys.* (*Paris*) **41**, 519 (1980).
94. M. Adam and M. Delsanti, *Polymer. Prepr.* **22**, 104 (1981).
95. M. Daoud and G. Jannink, *J. Phys. Lett.* (*Paris*) **41**, L-217 (1980).

6

Dynamic Light Scattering in Bulk Polymers

G. D. Patterson

Department of Chemistry
Carnegie–Mellon University
Pittsburgh, Pennsylvania 15213

6.1. INTRODUCTION

Light scattering in bulk polymers is due to fluctuations in the local dielectric tensor ε of the medium.[1, 2] In this chapter, the types of fluctuations that occur in polymers and their manifestation in the light scattering will be presented. The features which are characteristic of polymers will be elucidated as well as the general behavior of light scattering from liquids and amorphous solids. Most of the published work on dynamic light scattering in bulk polymers will be reviewed in the course of the chapter.

6.2. LIGHT SCATTERING

In a typical light scattering experiment, incident light of well-defined frequency, intensity, and direction is scattered by a fluctuation in the sample through an angle θ in the scattering plane. The magnitude of the momentum transfer vector for the scattering process is given by

$$q = \frac{4\pi n}{\lambda} \sin \frac{\theta}{2} \tag{1}$$

where λ is the wavelength of the incident light in a vacuum and n is the refractive index of the medium. The value of q determines the length scale over which the fluctuations must persist in order to scatter light at angle θ according to $L = 2\pi/q$. If the fluctuations are dynamic the scattered intensity will be a function of time. The dynamics of the fluctuations can be determined either by measuring the intensity correlation function

$$C(t) = \frac{\langle I(t)I(0)\rangle}{\langle I\rangle^2} \tag{2}$$

where the angle brackets denote a time average, or by obtaining the frequency spectrum of the scattered light $S(q, \omega)$. If the fluctuations are sufficiently rapid (frequencies above 10 MHz), the frequency spectrum can be determined with a Fabry–Perot interferometer. If the characteristic relaxation time is longer than 10^{-6} sec, then a digital photon correlator is most useful to obtain the intensity correlation function. In the ideal case $C(t)$ is related to the relaxation function for the fluctuations by

$$C(t) = 1 + \phi^2(t) \tag{3}$$

where

$$\phi(t) = \frac{\langle \Delta\varepsilon(t)\Delta\varepsilon(0)\rangle}{\langle \Delta\varepsilon^2\rangle} \tag{4}$$

and $\Delta\varepsilon(t)$ is the magnitude of the dielectric tensor fluctuation of the correct symmetry to give rise to the scattering.

The incident light is typically polarized either vertically or horizontally with respect to the scattering plane. If both the incident and the scattered light are polarized in the vertical direction (VV scattering), the symmetry of the fluctuations is longitudinal. The HV scattering has transverse symmetry. Only these two cases will be treated in this chapter.

6.3. SOURCES OF LIGHT SCATTERING

Polymers are a well-known class of photoelastic materials.[3] This means that the elements of the macroscopic dielectric tensor can be related to the state of stress or strain in the material. Another important property of polymers is that they are viscoelastic.[4] This means that the strain in a sample is a function not just of the instantaneous stress, but of the stress history. Light scattering from polymers will have some of the characteristics normally associated with liquids and some features typical of amorphous solids.

Like all liquids and amorphous solids, polymers are compressible. The dielectric tensor is a function of the density ρ. Since density is a scalar, only the diagonal components of ε will be affected and light scattering due to density fluctuations will be observed in the VV scattering geometry. This means that the symmetry of the density fluctuations is longitudinal.

Because polymers are viscoelastic, the derivative of the dielectric tensor with respect to the density may be a function of the time over which it is measured. It takes time for the internal structure of the fluid to reach a new equilibrium state after the imposition of a new density. However, at present there is no good evidence for dispersion in this quantity in polymers. Nor is the temperature of the liquid expected to affect ε appreciably at constant density.

The other major source of light scattering is microscopic birefringence. The local dielectric tensor can have an anisotropy. Any arrangement of atoms with lower than T_d symmetry will lead to such an effect. Polymer fluids are inherently anisotropic because of the shape of the chain molecules. Local birefringence can also arise from the way the molecules pack in the amorphous state. If a chain molecule fluid is cross-linked and stretched, it becomes birefringent due to the orientation of the chains. If the same fluid is strained in the glassy state, the magnitude and even the sign of the birefringence may be different.(5) This corresponds to an intermolecular effect due to the change in local structure upon shear under conditions where the chains are not free to adopt new conformations. The importance of these two competing effects will be considered below.

The shear-optic coefficient is inherently a function of the time scale over which it is measured in polymer fluids. The instantaneous value will reflect the glassy state birefringence. As the local fluid structure relaxes, the chains will be free to orient and the birefringence will change accordingly. Finally the shear stress will be relieved by chain disentanglement and the birefringence will disappear.

6.4. THEORY

A full theory of light scattering from a viscoelastic medium has been presented by Rytov.(6) All of the qualitative features to be discussed in this chapter can be understood in terms of this theory. All the dielectric derivatives and mechanical and thermal properties of the materials are assumed to depend on the time scale of measurement. The theory will be applied to the special case of polymer fluids.

Relaxation behavior in polymers can also be expressed in terms of the frequency dependence of the compression and shear modulus. At high frequencies polymer fluids behave as hard elastic solids. There will be a limiting high-frequency modulus of compression K_∞ and a corresponding shear modulus G_∞. At zero frequency the modulus of compression K_0 is equal to $1/\beta_t$, the inverse of the isothermal compressibility, and the shear modulus is zero. At intermediate frequencies, the moduli can be expressed as complex numbers where the real part reflects the elastic character and the imaginary part the viscous response of the material. There will in general be a number of processes which contribute to the relaxation of the moduli at intermediate frequencies.

In all fluids it takes time for the atoms to slip by one another in response to an applied shear stress. This corresponds to rearrangement of the local fluid structure. Actual exchange of particles will take a much longer time. The importance of structural relaxation in light scattering will be stressed below. Polymer molecules will tend to orient and change their conformation under a shear stress. And finally the molecules will disentangle from one another. If we associate a relaxation time τ_k with each of these processes and a relaxation strength N_k^2, then the shear modulus can be expressed as

$$G = G_\infty - \frac{1}{2}\sum_k \frac{N_k^2}{1 + i\omega\tau_k} \tag{5}$$

The modulus of compression has the same relaxational form. An important quantity for light scattering is the longitudinal modulus $M = K + 4/3G$.

As a liquid is cooled the relaxation times increase and eventually become comparable to normal experimental times (10^2 sec). This phenomenon is called the glass transition. If the liquid is cooled faster than it can maintain internal equilibrium, it will be trapped in a nonequilibrium state and will be a glass. Any attempt to define a glass transition temperature T_g must involve the time scale of the experiment. While it is convenient to speak of T_g as if it were a fixed temperature, it should be understood that no phase transition is involved and that equilibrium liquids can easily exist at temperatures below T_g if sufficient time is allowed.

Adiabatic density fluctuations usually involve a change in temperature. As a result light scattering will depend on the specific heat. The specific heat also has a relaxational form because it takes time for all the processes which contribute to C_v to come to equilibrium after an imposed temperature jump. The specific heat will be smallest at high frequency and will increase to its thermodynamic value at zero frequency. At intermediate frequencies it will be complex and have a relaxational form similar to the mechanical moduli.

The shear-optic coefficient X will also have a relaxational form which reflects the same dynamics as the shear modulus, but with different weights depending on the coupling between the shear and the dielectric tensor for each mode. It can be expressed formally as

$$X = X_\infty - \sum_k \frac{n_k N_k}{1 + i\omega\tau_k} \tag{6}$$

where the n_k reflect the coupling of each mode of relaxation to the dielectric tensor.

The spectrum due to light scattering by longitudinal density fluctuations can be calculated if the frequency dependences of the material properties are specified. The basic equation is

$$S(q, \omega) = \frac{-k_b T}{2\pi i\omega}\left(\frac{Y^2 C q^2}{\Delta} - \text{c.c.}\right) \tag{7}$$

where k_B is Boltzmann's constant, T is the absolute temperature, $Y = \partial\varepsilon/\partial\rho$, $C = \rho C_v/T + \kappa q^2/i\omega T$, κ is the thermal conductivity, $\Delta = (Mq^2 - \rho\omega^2)C + K^2\alpha^2 q^2$, α is the thermal expansion coefficient, and c.c. denotes the complex conjugate. The spectral features are determined by the roots of the equation $\Delta = 0$.

Although the polymer case is very much more complicated, it is instructive to consider the spectrum for a simple compressible, viscous fluid with finite thermal conductivity and expansivity. The modulus of compression will be set equal to K_0 and the shear modulus will be representative of a purely viscous fluid without relaxation of the viscosity, $G = i\omega\eta_s$, where η_s is the shear viscosity. Under these conditions there will be three roots of the dispersion equation $\Delta = 0$. The first root is imaginary and corresponds to a peak centered at the incident frequency with Lorentzian linewidth

$$\Gamma_c \approx \frac{\kappa q^2}{\rho C_v}\left(1 + \frac{K_0 \alpha^2 T}{\rho C_v}\right)^{-1} \tag{8}$$

The intensity of this peak depends on the thermal expansivity and the dynamics of the fluctuations are determined by the direct conduction of heat. This central peak is usually called the Rayleigh peak. There is nothing characteristic of polymer dynamics in this feature. The other two roots are complex and correspond to frequency shifted peaks. They were predicted by Brillouin in 1922[7] and are due to light scattering by thermal sound waves in the liquid. The peak frequency shift is given by $\pm qV_l$, where V_l is the longitudinal velocity of sound for waves of wave-vector magnitude q. The

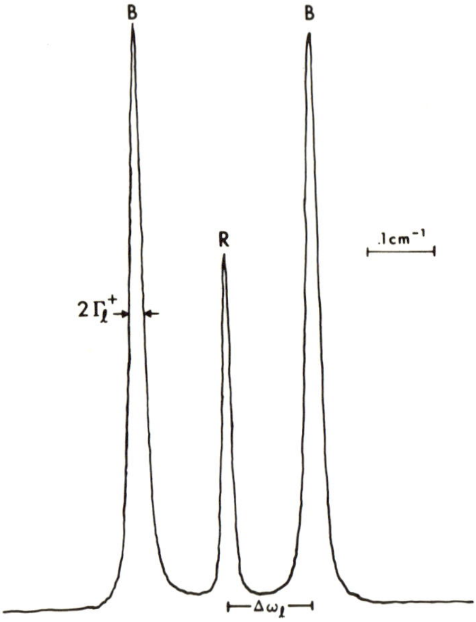

Figure 1. Typical Rayleigh–Brillouin spectrum showing a central Rayleigh peak (R) and two shifted Brillouin peaks (B) with splitting $\Delta\omega_l$ and linewidth Γ_l.

sound velocity is a function of q in a viscoleastic liquid. The value of V_l predicted by the above conditions is given by

$$V_l \approx \left[\frac{K_0}{\rho}\left(1 + \frac{K_0 \alpha^2 T}{\rho C_v}\right) - \frac{\Gamma_l^2}{q^2}\right]^{1/2} \tag{9}$$

where Γ_l is the predicted linewidth of the shifted peaks. The linewidth is due to the attentuation of the sound waves in the liquid. The predicted value is

$$\Gamma_l \approx \frac{2q^2\eta_s}{3\rho} + \frac{\kappa q^2}{2\rho C_p}(\gamma - 1) \tag{10}$$

where C_p is the specific heat at constant pressure and $\gamma = C_p/C_v$.

This overall three-peak structure is characteristic of light scattering due to longitudinal density fluctuations. A typical Rayleigh–Brillouin spectrum is shown in Figure 1. While the simplest assumptions lead to a three-peak structure, there are features that are qualitatively wrong about the predictions for the sound velocity and the Brillouin linewidth. The linewidth is predicted to increase without bound as the shear viscosity increases. For polymers this would lead to peaks that would be too broad to observe. The actual Brillouin peaks in polymer fluids are well defined, even near the glass transition where the viscosity becomes very large. The sound velocity is predicted to decrease as the linewidth increases, but this is opposite to the actual effect. The source of the discrepancy is the assumption that the shear modulus is purely viscous with a frequency-independent shear viscosity.

The thermal conductivity of polymer fluids is typical of organic liquids and is very low. The effect of thermal conductivity on the Brillouin linewidth is not measurable for polymers. Thus, terms involving κ will be ignored in the treatment of Γ_l. However, the low value of κ means that the sound waves are adiabatic and hence there will be a temperature change associated with the waves. The relaxational character of the specific heat will have an effect on the Brillouin linewidth.

In order to illustrate the effect of relaxing moduli on the predicted spectrum, let the longitudinal modulus be given by $M = M_\infty - M_r/1 + i\omega\tau$. There will now be four roots of the dispersion equation, due to the introduction of another frequency term in the expression for the modulus. Two of the roots correspond to the Brillouin peaks. The character of the other two roots depends on the value of τ, but they both correspond to central peaks. The prediction of dynamic central peaks associated with relaxation in the moduli is very important for the understanding of light scattering in polymers. The first treatment of this problem for liquids in general is generally attributed to Mountain.[8] If Γ_C is arbitrarily chosen as the smallest root, it will be given by

$$\Gamma_C \approx \frac{K_0 \kappa q^4/\rho^2 C_v}{K_0 q^2/\rho(1 + K_0 \alpha^2 T/\rho C_v) + M_\infty \kappa q^4 \tau/\rho^2 C_v} \tag{11}$$

If τ is negligible the result given in equation (8) is recovered. For the large values of τ expected near the glass transition, the smallest root becomes proportional to τ^{-1}. Thus a study of the narrow central line near T_g will probe the relaxation behavior of the moduli directly. The other central line will have a linewidth given by

$$\Gamma_2 = \frac{1}{\tau} + \frac{\kappa q^2}{\rho C_v} - \Gamma_C - 2\Gamma_l \tag{12}$$

When the inverse relaxation time is large relative to the thermal conductivity term, the second central peak reflects the slowly relaxing adiabatic density fluctuations. When the relaxation time becomes long, the narrow central peak is due to isothermal density fluctuations, and the second central peak now is determined by the thermal conductivity. A detailed discussion of this simple case has been presented by Allain-Demoulin *et al.*[9]

The use of a relaxing modulus also produces the correct qualitative behavior for the Brillouin splitting and linewidth. If terms involving the thermal conductivity can be neglected, the linewidth is predicted to be

$$\Gamma_l \approx \frac{M_R q^2/2\rho\tau}{1/\tau^2 + M_\infty q^2/\rho + K_0^2 \alpha^2 T q^2/\rho^2 C_v} \tag{13}$$

The Brillouin linewidth is predicted to go through a maximum as τ increases. The sound velocity under the same conditions is given by

$$V_l \approx \left(\frac{M_\infty}{\rho} + \frac{K_0^2\alpha^2 T}{\rho^2 C_v} - \frac{\Gamma_l^2}{q^2} - \frac{2\Gamma_2\Gamma_l}{q^2}\right)^{1/2} \tag{14}$$

For small values of τ this reduces to

$$V_l \approx \left[\frac{K_0}{\rho}\left(1 + \frac{K_0\alpha^2 T}{\rho C_v}\right) + \frac{3\Gamma_l^2}{q^2}\right]^{1/2} \tag{15}$$

For large values of τ the sound velocity reaches its elastic limit given by the q independent terms in equation (14). The increase in V_l as the relaxation time or q increases is correctly predicted. An example of the Rayleigh–Brillouin spectrum of a polymer when the relaxation time that determines Γ_l is short and when the linewidth is near its maximum value is shown in Figure 2. The central peak due to relaxation is clearly visible in the second case.

Any attempt to interpret the Rayleigh–Brillouin spectrum in a quantitative manner must deal with the full frequency dependence of K, G, and C_v. Direct attempts to solve the dispersion equation are no longer helpful because the algebra becomes intractable. However, certain features of the spectrum can be discussed in detail. There will always be a central peak whose intensity depends on the thermal expansivity and whose linewidth reflects the thermal conductivity. This feature has not been studied in detail for polymers and no further mention will be made of it. There will always be two Brillouin peaks with finite width. All sources of hypersonic attenuation will contribute to this width.

All the dynamic processes which relax K, G, and C_v will make a contribution to Γ_l, but only some of them will be significant. In order to

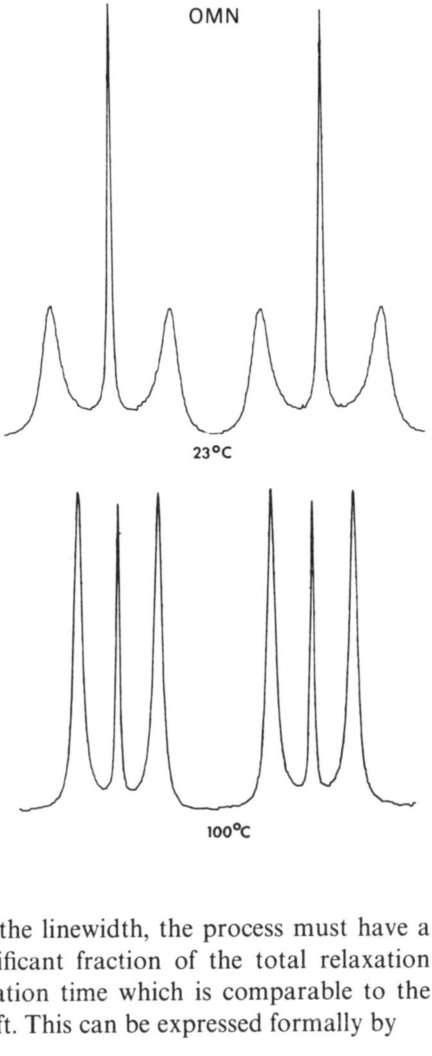

Figure 2. Rayleigh–Brillouin spectrum of octamethyl nonane showing two Fabry–Perot orders at 100°C, where the relaxation time is short, and two at 23°C, where Γ_l is near its maximum value and the central peak due to relaxation is visible.

make a measurable contribution to the linewidth, the process must have a relaxation strength which is a significant fraction of the total relaxation modulus. It must also have a relaxation time which is comparable to the inverse of the Brillouin frequency shift. This can be expressed formally by

$$\Gamma_l = \frac{q^2}{2\rho} \sum_k \frac{M_{Rk}\tau_k}{1 + (qV_l\tau_k)^2} \tag{16}$$

This means that the interpretation of the Brillouin linewidth for an arbitrary fluid can be extremely complicated, but some qualitative progress has been made.

One of the most important consequences of the use of relaxing moduli is the prediction of additional central peaks related directly to the dynamics of the relaxing processes. The intensity of these peaks will reflect the relative relaxation strengths of each process. Near the glass transition some of these processes will relax slowly enough to be observed using photon correlation spectroscopy. Thus, Brillouin scattering and photon correlation spectroscopy are complementary probes of molecular motion in polymers.

Although the use of a simple relaxation form with discrete relaxation times illustrates qualitatively the correct form of the spectrum, real moduli are usually[4] represented in terms of a continuous distribution of relaxation times. In principle the nature of the distribution could be detected by measuring Γ_l as a function of frequency (angle), but in practice the useful range of frequencies is too small. One of the main advantages of studying the dynamic central peaks using photon correlation spectroscopy is that a very large range of times can be studied in the relaxation function. This makes the detection and characterization of a distribution of relaxation times much easier.

Anisotropy fluctuations can also have longitudinal symmetry. And they can couple to longitudinal strain fluctuations. The resulting spectrum will have a central peak due to direct relaxation of the anisotropy plus features characteristic of the longitudinal density fluctuations. Experimentally, the anisotropy fluctuations contribute only a very small fraction of the intensity of the longitudinal Brillouin peaks. The presence of another strong central peak in the VV spectrum can complicate the analysis of the data.

The spectrum due to transverse fluctuations in the dielectric tensor is determined by the dynamics of the anisotropy fluctuations and their coupling to shear strains. In the simplest model the spectrum consists of a single central peak with a width determined by the collective reorientation time of the molecules. For polymer fluids, the dynamics will be much more complicated. Both local changes in conformation and overall reorientation of the molecules are expected to contribute to the spectrum. However, there is usually a dominant mode of anisotropy fluctuation which determines the spectrum and the results may appear more simple than at first expected. When the relaxation time for the shear modulus is short, the transverse sound velocity is imaginary and no shifted Brillouin peaks occur. But there will be coupling between the shear fluctuations and the orientation of the molecules. Under these conditions the VH spectrum is predicted to be[10]

$$S(q, \omega) \propto \frac{\Gamma_{\text{or}}}{\Gamma_{\text{or}}^2 + \omega^2} \sin^2 \frac{\theta}{2} + \Gamma_{\text{or}} \cos^2 \frac{\theta}{2} \times \frac{q^4\eta_s^2/\rho^2(1 - R) + \omega^2}{[\Gamma_{\text{or}}(q^2\eta_s/\rho) - \omega^2]^2 + \omega^2[\Gamma_{\text{or}} + (q^2\eta_s/\rho)(1 - R)]^2} \tag{17}$$

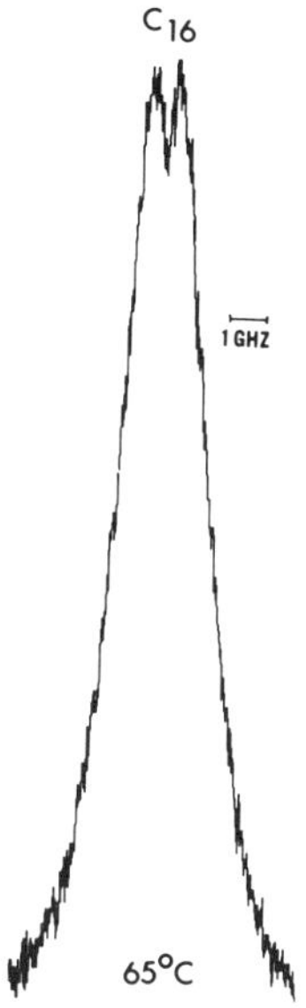

Figure 3. Depolarized (VH) spectrum of *n*-hexadecane at 65°C showing the central dip feature.

where Γ_{or} is determined by the overall collective reorientation time for the molecules and R is a parameter that is equal to the fraction of the shear viscosity that is due to coupling to molecular reorientation. When the quantity $(q^2\eta_s)/(\rho\Gamma_{\text{or}})$ is less than a critical value (approximately 0.7), the spectrum is predicted to display a central dip feature. The spectrum of *n*-hexadecane at 65°C is shown in Figure 3 as an example of this type of spectrum in polymers.

The spectral form of the Rytov theory for VH scattering is

$$I_{\text{VH}} = -\frac{k_B T}{2\pi i\omega}\left(\frac{X^2 q_2^2}{4A} + \frac{1}{2}\sum_k \frac{n_k^2}{1 + i\omega\tau_k} - \text{c.c.}\right) \tag{18}$$

where $q_2 = -(2\pi n/\lambda)\sin\theta$ and $A = Gq^2 - \rho\omega^2$. In addition to the central peaks determined by the τ_k, the spectral features are determined by the roots of the equation $A = 0$, and the relaxational character of X. The Rytov theory also predicts a central dip when the relaxation times are in the appropriate range. When the shear relaxation time becomes longer, propagating shear waves can give rise to shifted Brillouin peaks. A spectrum showing both the longitudinal and transverse peaks for a bulk polymer is presented in Figure 4.

Near the glass transition the width of the central peaks will become narrow enough to be studied with photon correlation spectroscopy. Under these conditions the dynamics of the intramolecular orientation fluctuations

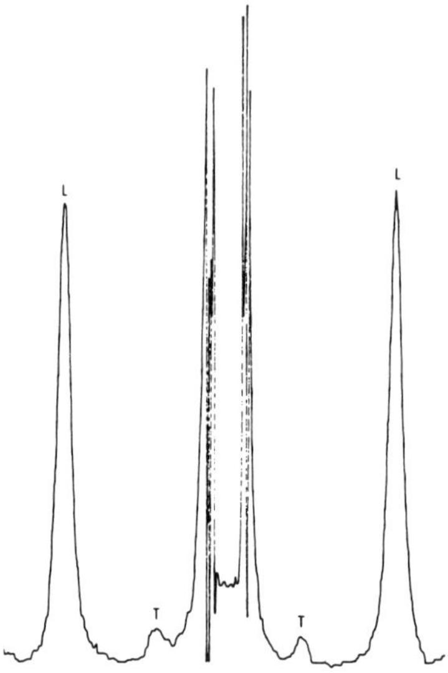

Figure 4. Rayleigh–Brillouin spectrum (VV + VH) of bisphenol-A polycarbonate showing both longitudinal (L) and transverse (T) Brillouin peaks.

will be dominated by the rate at which the fluid structure changes. There may also be direct scattering by anisotropic fluctuations in the fluid structure. The resulting relaxation function is expected to reflect the complex dynamics observed near T_g and to display a distribution of relaxation times.

6.5. APPLICATIONS

5.1.6. Brillouin Spectroscopy

The measurement of Brillouin linewidths in polymer fluids has recently been thoroughly reviewed.[11] Typical linewidths are in the range 100–1000 MHz and typical frequency shifts in the range 2–20 GHz. The spectra are obtained with a Fabry–Perot interferometer. A detailed discussion of the experimental procedures has been previously presented.[12] The most important results of these studies will be given here.

The *n*-alkanes are an ideal system with which to illustrate many of the effects that are common to polymer fluids. The observed linewidths[13, 14] decrease monotonically with temperature from values above 300 MHz near the melting point to less than 200 MHz at higher temperatures. The linewidth increases with chain length at any particular temperature, but the linewidths are similar when compared at constant density. In fact, the longitudinal sound velocities determined using Brillouin spectroscopy in the *n*-alkanes[15] have been shown to depend only on the density for all the homologues including polyethylene. Also, the measured values of V_l are within experimental error of the low-frequency speed of sound in these liquids.

The total shear viscosity is a strong function of chain length in polymer fluids. Above a critical molecular weight, the viscosity is determined by the presence of an entanglement network. However, the value of the shear modulus associated with the network is low in comparison to K_0 (so that V_l will not be affected) and the relaxation time is very long relative to the inverse of the Brillouin frequency (so that Γ_l will not reflect this motion). At lower molecular weights the viscosity is determined primarily by the rate of overall molecular reorientation. But again these motions will not measurably affect the Brillouin spectrum. The rates of intramolecular *trans–gauche* isomerization are also too slow to contribute to the linewidth. The conclusion is that for polymer fluids, only the local fluid structural rearrangements are rapid enough to contribute significantly to Γ_l. For polymer fluids local translational and rotational motions are highly coupled, but a full conformational transition is not required to relax most of the local stress.

The dynamic contributions to the Brillouin linewidth can be grouped into a frequency-dependent shear viscosity and into a volume viscosity which depends on both K and C_v. If the hypersonic shear viscosity could be obtained, then the contribution of the volume viscosity could be calculated from the Brillouin linewidth. For the shorter *n*-alkanes this has actually been carried out.[14] The total shear viscosity must be reduced by the fraction due to coupling to overall reorientation, because the orientational relaxation time is long relative to the inverse of the Brillouin frequency. This fraction can be obtained by measuring the depolarized Rayleigh spectrum under conditions where the central dip is observed. Approximately one third of the total shear viscosity is due to coupling to reorientation in *n*-hexadecane at 65°C. When this correction is made, the hypersonic volume and shear viscosities turn out to be comparable in the *n*-alkanes. This result may not be completely general, but it does suggest that the structural relaxation contributions may be comparable for both volume and shear viscosities. The relaxation in the specific heat may be very different for each polymer.

The longitudinal Brillouin linewidth depends on both the relaxation strength and the relaxation time for each dynamic process. This complicates the calculation of a relaxation time from measurements of Γ_l alone. Also, both quantities are a function of temperature so that the times cannot be inferred just from the temperature dependence of the linewidth. Analysis of Brillouin linewidth data in polymers has been carried out in a manner similar to that used in mechanical and dielectric relaxation studies. An

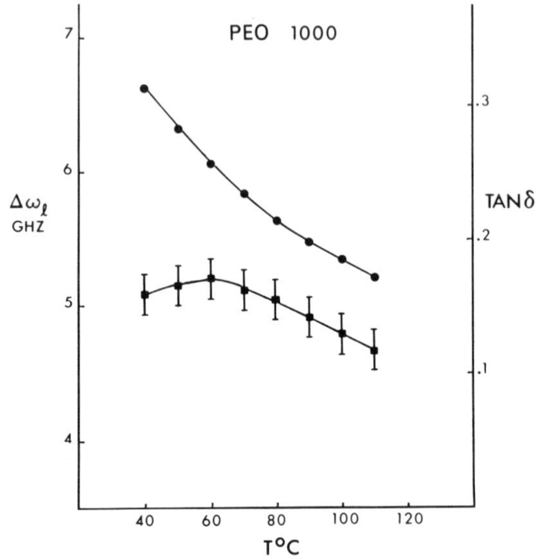

Figure 5. Brillouin frequency (●) and tan δ (■) plotted versus temperature for poly(ethylene oxide). A maximum in tan δ is observed at 60°C.

attempt is made to observe a maximum in the linewidth, and the frequency measured at this maximum is then compared with the frequencies determined by other techniques. The logarithm of the frequency of maximum loss is plotted against the reciprocal of the temperature at which the maximum was observed. Many of these types of plots have been constructed for polymers.[16]

In the *n*-alkanes the liquid crystallizes before Γ_l reaches a maximum value so that this type of analysis was not carried out. In poly(ethylene oxide) (PEO) the melting point for 1000 molecular weight material is low enough to allow the observation of a maximum in the Brillouin linewidth. Mechanical relaxation data are often expressed in terms of the loss tangent, the ratio of the imaginary part of the modulus to the real part. In Brillouin scattering this quantity is given by

$$\tan \delta \approx 2\Gamma_l/\Delta\omega_l \tag{19}$$

The Brillouin frequency and tan δ are plotted *versus* temperature for PEO in Figure 5. The maximum loss is observed at 60°C and a frequency of 6.06 GHz.[17] The melting point of high molecular weight PEO is also 60°C, so that the use of lower molecular weight polymer was essential in order to detect the maximum. The Brillouin result is compared with high-frequency

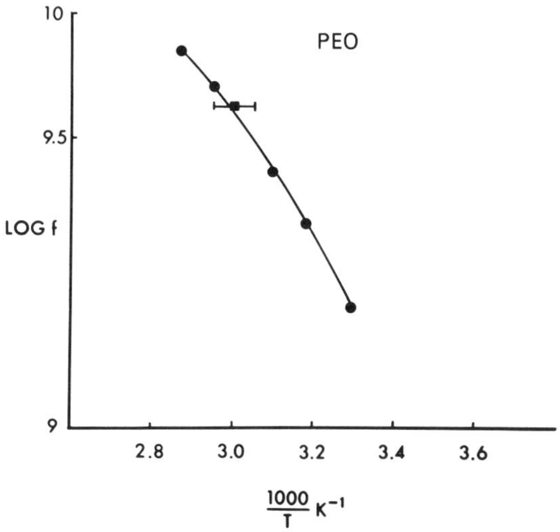

Figure 6. Comparison of dielectric relaxation data in the GHz range for PEO with the Brillouin result. The data are presented as a transition map where the logarithm of the frequency is plotted versus the reciprocal of the temperature of maximum loss.

dielectric relaxation data[18] in Figure 6. The agreement is quite remarkable and suggests that there is something very universal about the glass transition. The similarity of dynamics is based on the dependence on the local fluid structural relaxations discussed above.

In order to use lower molecular weight polymers to determine the characteristic temperatures of maximum loss, it is necessary to know the chain length dependence of the linewidths. Wang and Huang[19] found essentially no molecular weight dependence in poly(propylene glycol) (PPG). However, the presence of hydrogen bonding end groups in PPG will tend to obscure the intrinsic chain length dependence of Γ_l. For the *n*-alkanes mentioned above there is a definite chain length dependence to Γ_l but no maximum was obtained. A study of polystyrene[20] revealed that the loss maximum in this polymer reaches its limiting value very quickly with molecular weight. Approximately 50 monomer units would be required. The limiting molecular weight for a particular polymer will depend on the flexibility of the chain and on the packing of the fluid, but rather low molecular weights can usually be employed. This is a great help in sample preparation and for crystallizable polymer allows the measurements to be carried out in the necessary temperature range.

There is very little complementary data at hypersonic frequencies. At lower frequencies more than one temperature of maximum loss is commonly observed. These data tend to group into discrete transition lines

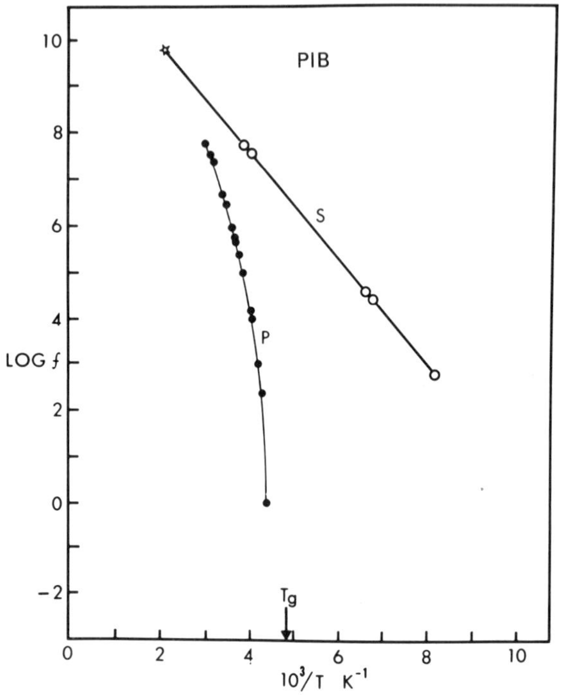

Figure 7. Transition map for PIB. The Brillouin result falls on an extrapolation of the secondary relaxation line (S) above the temperature where the primary (P) and secondary lines appear to merge.

when plotted as the logarithm of the frequency versus the reciprocal of the temperature of maximum loss. One of these transition lines can be described by the relation

$$\log f = A + \frac{B}{T - T_0} \tag{20}$$

and is referred to as the primary glass relaxation. There is also at least one other transition line which displays an Arrhenius temperature dependence and which tends to merge with the primary line at higher frequencies. This is called the secondary glass relaxation. There is always a good correlation between the extrapolation of the secondary glass relaxation line and the temperature of maximum loss determined by Brillouin scattering in polymers. A transition map plot for polyisobutylene (PIB)[21] is shown in Figure 7. It is typical of the results for many polymers.

In some cases there are also two maxima in the loss tangent determined by Brillouin scattering. This has been shown to be the case in

Table 1. Hypersonic Loss Maxima

Polymer	T_{max}(°C)	f_{max} (GHz)
Polystyrene 600 MW	120	5.8
Polystyrene 2,100 MW	220	5.1
Polystyrene 20,000 MW	240	5.5
Poly(propylene glycol)	50	5.43
	100	4.4
Poly(ethylene oxide)	60	6.06
Polyisobutylene	200	4.95
Poly(dimethyl siloxane)	−30	5.0
	40	3.8
Polypropylene	180	4.11
Poly(vinyl acetate)	200	4.5
Bisphenol-A Polycarbonate	280	5.43
Poly(methyl acrylate)	170	4.3
Poly(butadiene)	125	5.0
Poly(isoprene)	100	5.9

poly(propylene glycol)[22] and in poly(dimethyl siloxane).[23] In the case of PPG the higher temperature loss agrees very well with dielectric relaxation data at high frequencies[24] and the lower temperature loss falls on an extrapolation of the secondary transition line. This is the only clear evidence that the two processes may not have merged at hypersonic frequencies. The two characteristic motions can be understood as the most local structural rearrangements involving only nearest neighbors, and a cooperative restructuring including all neighbors out to some correlation length.

Many polymers have now been studied by Brillouin scattering. One limitation is that the temperature of maximum loss may fall above the decomposition temperature of the polymer. The highest measured loss maximum[25] is at 280°C for bisphenol-A polycarbonate. A list of temperatures of maximum hypersonic loss for polymers is given in Table 1.

As polymer fluids are cooled towards the glass transition temperature, T_g, the Brillouin linewidths become narrower. This is quite reasonable since the characteristic relaxation times are becoming much longer and the hypersonic viscosity should be reduced. However, the values of Γ_l near T_g remain too large to be accounted for by the structural rearrangements associated with the glass transition. One significant source of Brillouin line broadening near T_g is relaxation of the vibrational specific heat. It has also been proposed that there is another class of mechanical processes which would only become slow at very low temperatures. The narrowing of the Brillouin lines at very low temperatures has been observed by Vacher *et al.*[26] This has been interpreted as evidence for these very local mechanical relaxation processes in polymers at low temperatures.

Brillouin spectroscopy has established that the local fluctuations in polymer fluids are much like those in any viscoelastic liquid. Although the macroscopic viscosity can be very high, the local viscosity is determined by the normal structural rearrangements found in typical liquids.

6.5.2. Dynamic Central Peaks

When the Brillouin linewidth is near its maximum value a dynamic central peak due to relaxation in the longitudinal modulus is clearly evident. No systematic study of this feature seems to have been carried out in polymers in the high-temperature range where the density fluctuations are adiabatic. As the liquid is cooled towards T_g this central peak becomes narrow enough to be studied using photon correlation spectroscopy. Under these conditions the density fluctuations being observed are isothermal since they are much slower than the rate at which temperature can equilibrate on a length scale of $2\pi/q$. Light scattering by thermal density fluctuations is much weaker than the usual concentration fluctuation scattering observed using photon correlation spectroscopy in polymer solutions. And only part of the total intensity is due to slowly relaxing fluctuations. Thus, high incident power and long averaging times are required to obtain the data. In many polymers, there will also be a strong central peak due to anisotropy fluctuations, but in this part of the chapter only weakly anisotropic systems will be considered. It is important to eliminate purely elastic sources of scattering if the pure homodyne correlation function is to be obtained. This has only been realized for a few polymers. The technique has been to prepare the polymer by thermal polymerization of the monomer under very clean vacuum conditions.[27, 28]

A single relaxation time model for the isothermal density fluctuations would predict that the relaxation function observed by light scattering was a simple exponential decay. Mechanical relaxation in polymers is usually characterized by a distribution of relaxation times, especially near the glass transition. The relaxation functions observed by photon correlation spectroscopy near T_g are in fact highly nonexponential. This raises many questions about the collection, analysis, and interpretation of the data. The present state of understanding of this problem will be discussed below.

The actual measurements of photon correlation spectroscopy in bulk polymers are carried out with a digital correlator. The number of photons n in a finite sampling interval T is counted and the discrete function $C(iT)$ is calculated as

$$C(iT) = \frac{\langle n(iT)n(0)\rangle}{\langle n\rangle^2} \tag{21}$$

where i is an integer equal to the delay channel number. Often the clipped correlation function is actually calculated

$$C_k(iT) = \frac{\langle n(iT)n_k(0)\rangle}{\langle n\rangle\langle n_k\rangle} \tag{22}$$

where $n_k = 1$ if $n > k$ and 0 otherwise. The integer k is called the clipping level. For a Gaussian random process no distortions in the relaxation function are introduced by using clipping. The clipped correlation function is related to the relaxation function for the fluctuations by[29]

$$C_k(iT) = 1 + \frac{1 + k}{1 + \langle n\rangle} f(A)g(T)\phi^2(iT) \tag{23}$$

where $f(A)$ accounts for the loss of correlation due to the finite coherence area observed in the scattering, and $g(T)$ corrects for the distortion of the data due to the finite sampling interval.

The discrete correlation function is typically collected for two decades of time. This would be sufficient to characterize a single exponential decay. But the relaxation functions observed near the glass transition decay too slowly and data collected using several sample times must be combined to give the full $\phi^2(iT)$. In order to accurately combine such data, absolute values of the relaxation function must be obtained for each sample time. The clipping level k and the average number of counts per sampling interval $\langle n\rangle$ are accurately known. The coherence factor $f(A)$ can be obtained experimentally by measuring the intercept of the relaxation function under conditions where T is sufficiently short and the true intercept for the slowly relaxing part of $\phi^2(iT)$ is known. There are always processes which lead to light scattering which relax faster than those observed by photon correlation spectroscopy. However, the fraction of the scattered light due to slowly relaxing fluctuations can be accurately determined for polymer solutions. The calculated intercept was always realized for sufficiently small coherence area A.

The distortion of the data due to the finite sampling interval is more difficult to deal with for nonexponential decays. If we represent the relaxation function as a sum of exponentials, then components with relaxation times comparable to or shorter than T will make no contribution to $C_k(iT)$ after the first few intervals. If the first few points in each run are neglected the effect of time distortion will be minimized in the final composite relaxation function.

There are always small discrepancies between the absolute values of $\phi^2(iT)$ calculated for different runs at the same sample time. Small vertical

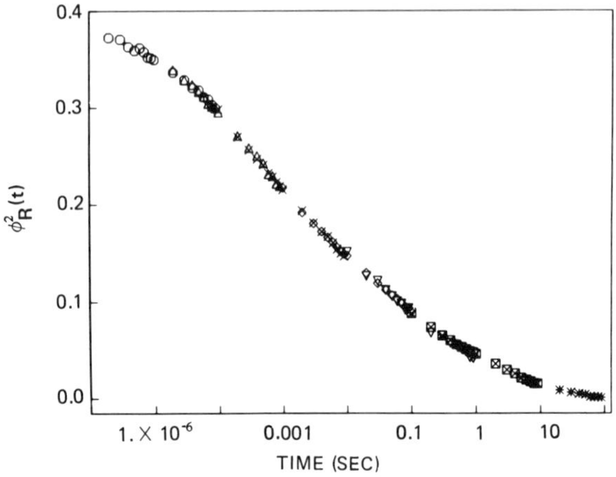

Figure 8. Composite relaxation function for poly(ethyl methacrylate) (PEMA) at 70°C plotted as $\phi^2(t)$ versus log t. There is significant change in the relaxation function over eight decades in time.

adjustments of the data are necessary to finally match up the runs at different sample times. A composite relaxation function chosen to show the large range of relaxation times apparent in polymer fluids near T_g is shown in Figure 8. The most convenient way to plot these data is $\phi^2(t)$ against log t.

A single exponential decay can be characterized by the decay time τ. A nonexponential decay could be analyzed in terms of a distribution of relaxation times $\rho(\tau)$. However, the inverse Laplace transform of the real data is notoriously unreliable. In practice, it is found that the data can be described very well by the empirical Williams–Watts[30] relaxation function

$$\phi(t) = \exp[-(t/\tau)^{\beta}] \tag{24}$$

where $0 < \beta \leq 1$. The parameter β is a measure of the width of the underlying distribution of relaxation times implied by the nonexponential character of the relaxation function. A full discussion of this distribution has been presented by Lindsey and Patterson.[31] The average relaxation time for the Williams–Watts function is given by

$$\langle \tau \rangle = \frac{\tau}{\beta}\,\Gamma(1/\beta) \tag{25}$$

where $\Gamma(x)$ is the gamma function. This allows the data to be meaningfully characterized by two parameters. Attempts to obtain more parameters with which to characterize the data might be meaningful in some cases, but if the data fit very well to the empirical form, then only two parameters are probably justified.

The first experimental study of photon correlation spectroscopy in bulk polymers was reported by Jackson *et al.*[32] They chose poly(methyl methacrylate) (PMMA) as the sample and examined it below the glass transition. A slowly relaxing component was detected. While this paper pointed the way to further work, it also illustrated some of the problems associated with light scattering from bulk polymers. It was correctly noted that only a small fraction of the scattered light was associated with the desired fluctuations. Most of the scattering could be attributed to static impurities. This leads to a heterodyne spectrum with a very low intrinsic intercept and poor precision in the data.

The observed relaxation function in PMMA was nonexponential, but it was arbitrarily analyzed as the sum of two exponential decays. This was largely motivated by the concept of two characteristic relaxation frequency regions being associated with the glass transition phenomenon. However, the interpretation of data taken below T_g is clouded by uncertainty as to the state of the sample. One of the fundamental assumptions usually made in photon correlation spectroscopy is that the scattering is due to equilibrium fluctuations that occur about a stationary state. Polymer glasses are not at equilibrium nor are they even in a stationary state near T_g. The sample will be undergoing macroscopic relaxation during the experiment and the local fluctuations will change accordingly.

Two other studies of PMMA have also been reported.[33, 34] The agreement between the three reports is poor and probably reflects the different samples used as well as the different ways of analyzing the nonexponential decays. Above T_g there was a slowly relaxing component whose relaxation time decreased very rapidly with temperature as the sample was heated. This is qualitatively correct and the results were correlated with other measurements of relaxation associated with the primary glass transition. However, the data were analyzed as a single exponential decay and the effect of scattering by static impurities was ignored. Until a sample of PMMA can be prepared with only intrinsic scattering due to thermal density fluctuations, it will be difficult to resolve the confusion created by the earliest papers in this field. Even if the elastic scattering could be eliminated above T_g, part of the intensity observed below the glass transition would be due to fluctuations whose relaxation time was too long to measure and the correlation function would be partially heterodyned.

In order to carry out a quantitative study of photon correlation spectroscopy in a weakly anisotropic polymer near T_g a very pure sample of

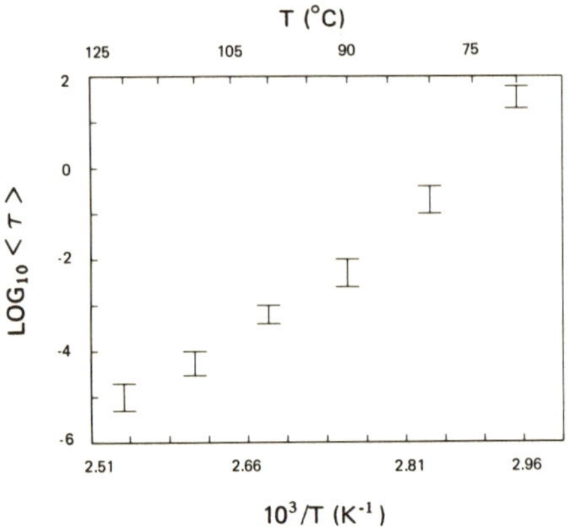

Figure 9. Logarithm of the average relaxation time $\langle\tau\rangle$ versus $1/T$ for PEMA near the glass transition.

PEMA was prepared by thermal polymerization from the monomer.[28] Almost all the scattered intensity from this sample could be attributed to thermal density fluctuations, and the fraction of slowly relaxing fluctuations was estimated to be nearly 75%. This meant that the value of $\phi^2(t)$ at the shortest times was large since most of the scattered light was due to this effect.

The average relaxation time was determined from 70°C (T_g = 65°C) to 130°C, where the relaxation time became too short to accurately measure.[35] A plot of $\log\langle\tau\rangle$ versus $1/T$ is shown in Figure 9. The average relaxation time changes by a decade with a 5°C change in temperature near T_g. This is characteristic of the primary glass transition.

Normal mechanical relaxation spectroscopy is difficult for unsupported liquids above T_g due to clamping problems. Light scattering shares with dielectric relaxation the advantage that measurements are easily carried out above the glass transition. In PEMA the dielectric relaxation is dominated by a secondary relaxation process which has a different temperature dependence than the primary glass transition.[36] Although there is no clear demarcation of two decays in the measured light scattering relaxation function for PEMA, the shape of the decay as characterized by the parameter β changes dramatically as the sample is cooled. At 120°C the value of β is near 0.4. This is typical of many materials near T_g. But the value of β drops to 0.16 at 70°C. This great increase in the breadth of the distribution of relaxation times is probably due to the presence of both the primary and secondary relaxation processes. At higher temperatures these processes will tend to merge, but as T_g is approached they separate rapidly. The data in PEMA reinforce the idea that these two processes are really just two regions of a continuous distribution of relaxation times connected with structural relaxation in liquids.

6.5.3. Depolarized Rayleigh Scattering

In small molecule fluids, the overall width of the depolarized Rayleigh peak is interpreted in terms of the collective reorientation time of the molecules[37]

$$\tau_{ls} = \frac{g_2}{J_2}\left(\frac{C'\eta_s}{T} + \tau_0\right) \tag{26}$$

where g_2 is the static pair orientation correlation parameter, $J_2 \approx 1$ is the corresponding dynamic quantity, C' depends only on the size and shape of the molecule, and τ_0 is an inertial limit for the single-particle reorientation time. This same formalism can still be useful for polymer fluids, but a new understanding of C' will result. The reorienting entity will be only a part of the molecule. Also, the macroscopic viscosity will not be the appropriate quantity which determines the linewidth, but a local viscosity must be invoked.

A systematic study of the depolarized Rayleigh linewidths in the *n*-alkanes has been carried out.[38] It illustrates the crossover from the small molecule regime to the polymer case. Although there are many motions which could contribute to the depolarized Rayleigh spectrum, the actual data can be fitted successfully with a single Lorentzian peak with half-width at half-height $\Gamma = (2\pi\tau_{ls})^{-1}$ under conditions where the central dip is not resolved. A typical result is shown in Figure 10. This suggests that there is a dominant mode of anisotropy relaxation which determines the linewidth.

For the *n*-alkanes with $n \leq 28$, the value of C' and τ_{ls} increases with chain length at constant η_s/T. This is consistent with overall reorientation of the molecule as the dominant mode of anisotropy relaxation in the shorter *n*-alkanes. For the longer chains, C' and τ_{ls} decrease as n increases at constant η_s/T. This means that a more local mode of motion is becoming the dominant one. Interestingly, the temperature dependence of τ_{ls} is similar for all the homologues. The value is comparable to the barrier for *trans–gauche* isomerization, but also similar to the apparent activation energy for the shear viscosity itself. The asymptotic limit where τ_{ls} was independent of chain length was not achieved at 1000 molecular weight polyethylene. Depolarized Rayleigh scattering in pure PPG has been studied by Jones and Wang.[39] They also associated their observed linewidths with a local mode of relaxation.

Under the right conditions a central dip can be observed in the depolarized Rayleigh spectrum of the *n*-alkanes.[40] From this spectrum the value of the coupling parameter R can be determined. For *n*-hexadecane the value of R was found to be 0.33 and for *n*-docosane $R = 0.38$. As the chain length increases, a greater fraction of the total shear viscosity is due to

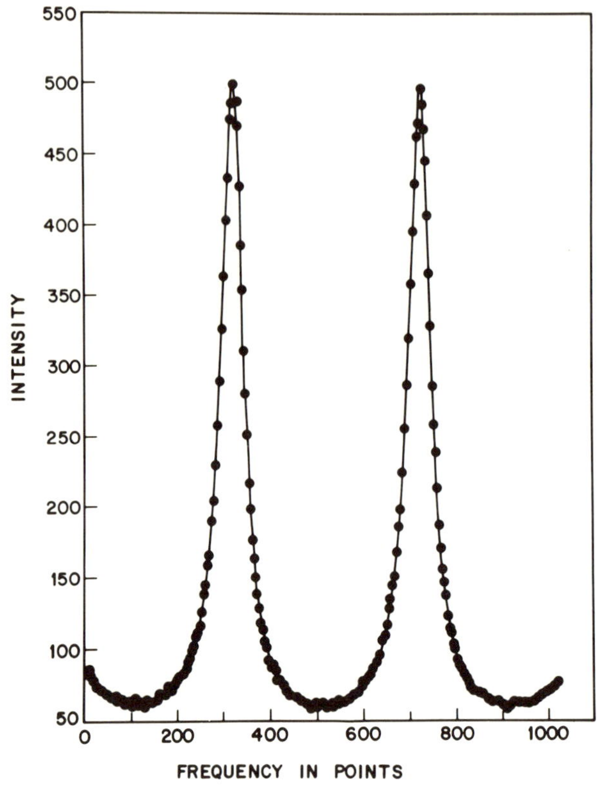

Figure 10. Depolarized Rayleigh spectrum of *n*-hexadecane at 20°C showing the computer fit to two Fabry–Perot orders. The solid line is the calculated spectrum and every fourth data point is shown by a •. The free spectral range is 7.41 GHz.

coupling to molecular reorientation. The temperature region over which the dip is observed is narrow in the *n*-alkanes. The dip was clearly observed in *n*-docosane near 110°C. For longer chains the viscosity will remain high at higher temperatures and the dip feature would no longer be observable.

When the chains exceed a certain length, the polymer fluid displays a low-frequency elasticity characteristic of a rubber network. It has not yet been established whether the existence of this entanglement network can be detected in the depolarized Rayleigh spectrum. If the intramolecular correlations in orientation are short ranged, then the existence of long-ranged restrictions on large-scale motion of the polymer will have no measurable effect on the orientation fluctuations which determine the scattered intensity. Experiments to study this regime are presently in progress.

There are only a few polymers for which transverse Brillouin peaks have been observed. Only a qualitative study[41] of the transverse linewidths in bisphenol-A polycarbonate has been carried out. At low temperatures the lines are very narrow. As the polymer is heated above the glass transition the lines broaden and eventually become overdamped at temperatures near the hypersonic loss maximum for the longitudinal peaks. One reason why the transverse Brillouin intensity is so weak in most polymers is that the shear optic coefficient can contain contributions of opposite sign which will lead to very small overall birefringence at hypersonic frequencies.

As the liquid is cooled towards T_g, the macroscopic viscosity increases dramatically. This is due to a decrease in the local mobility as the liquid packs into increasingly tight arrangements. Under these conditions, the depolarized Rayleigh peak becomes narrow enough to be analyzed using photon correlation spectroscopy. The dynamics of the relaxation of the chain orientation will be determined by the fluid structural rearrangements more than any internal barriers to reorientation.

Optically clean polystyrene can be prepared by thermal polymerization of pure styrene in the scattering cell.[42] As in the case of PEMA[35] the observed relaxation functions near T_g are nonexponential.[43] They can be described very well by the empirical Williams–Watts function with a value of $\beta = 0.4$. A typical composite relaxation function for polystyrene showing the fit to the empirical function is shown in Figure 11. In contrast to PEMA the shape of the relaxation function in polystyrene does not change over the temperature interval 100–130°C ($T_g \approx 100$°C). The logarithm of the average relaxation time determined from τ and β is plotted against the reciprocal of the temperature in Figure 12. The dramatic change in $\langle\tau\rangle$ with temperature near T_g is characteristic of the primary glass transition.

The slowly relaxing part of the polarized scattering in polystyrene is also dominated by the anisotropy fluctuations and the average relaxation times were found to be the same.[44] However, the value of β was 0.34 for the longitudinal fluctuations, including both density and anisotropy. A relaxation frequency can be calculated as $f = (2\pi\langle\tau\rangle)^{-1}$. The results for polystyrene using light scattering are compared to the frequencies of maximum dielectric or mechanical loss on a transition map in Figure 13. The light scattering frequencies are two orders of magnitude lower than the dielectric ones, but the temperature dependence is the same. Apparently the anisotropy fluctuations are determined by a much larger molecular subunit than the dielectric relaxation. The two techniques are complementary and probe different parts of the spectrum of relaxation times in the polymer fluid.

There is also a quantitative difference between the relaxation frequency calculated above and a frequency of maximum loss when the relaxation function is not a single exponential.[45] If we assume a relaxation function of

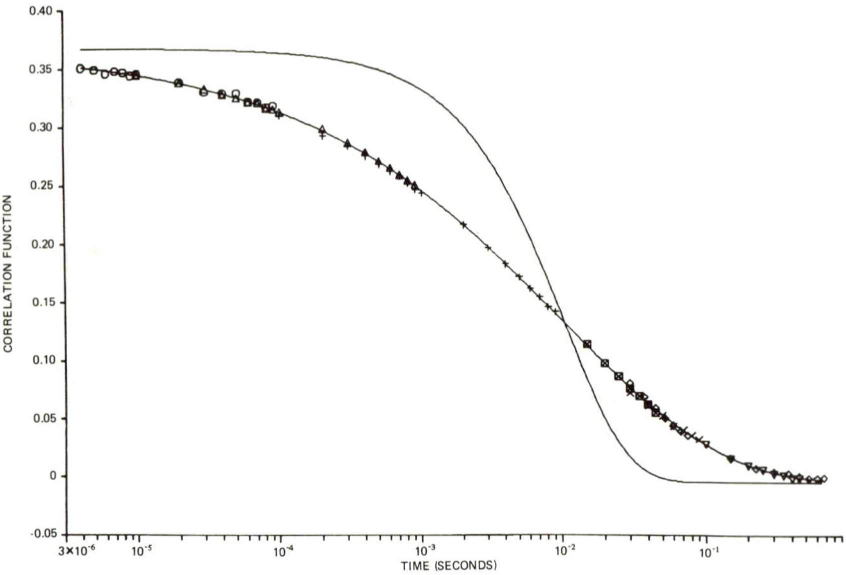

Figure 11. Composite relaxation function for the depolarized Rayleigh scattering from polystyrene at 114.5°C. The line through the data is the best fit to the Williams–Watts function. A single exponential decay is also shown for comparison.

the Williams–Watts form, a corresponding susceptibility can be calculated. The frequency at which the maximum value of the imaginary part of the susceptibility occurs can then be calculated and compared to $1/2\pi\langle\tau\rangle$ as a function of β. For values of β typical of polystyrene, the light scattering relaxation frequency is lower than the frequency of maximum loss by a factor between 2 and 3. Only for a single exponential decay would they be equal. The same type of quantitative difference occurs if the half-width at half-height (HWHH) is used to characterize the relaxation frequency of the spectrum for a system with a nonexponential relaxation function. The HWHH is a much lower frequency than f calculated above for $\beta = 0.35$.

Because of the breadth of the relaxation function, there has been an attempt[(46)] to extract more information from the data in polystyrene than just $\langle\tau\rangle$ and β. This approach used a graphical method of detecting deviations from a pure Williams–Watts form. However, there are numerous pitfalls in the use of this method which should be avoided. The combined data may have certain distortions due to the use of different sampling intervals. These could be mistaken for an intrinsic relaxation process. In order to use the graphical method, the observed relaxation function must be renormalized so that the zero time intercept is 1. This requires a very accurate knowledge of the apparent intercept of the slowly relaxing part of

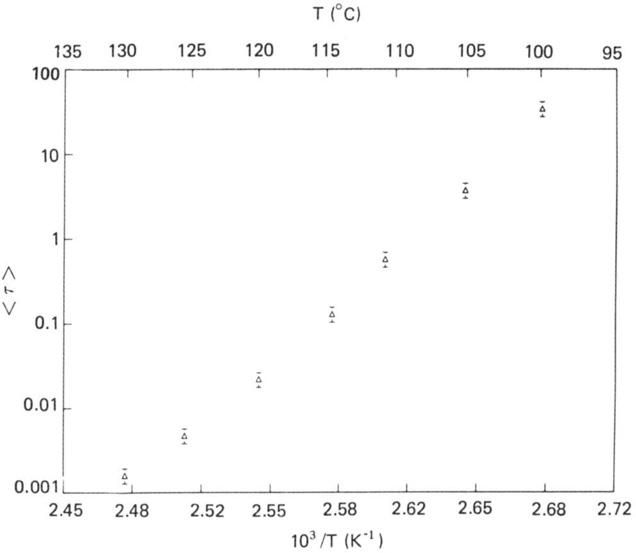

Figure 12. Logarithm of $\langle\tau\rangle$ versus $1/T$ for anisotropy fluctuations in polystyrene.

the light scattering relaxation function. Small errors in the renormalization can lead to significant distortions in the plotted data at short times. The renormalized relaxation function is plotted in the form $\log[-1/2 \ln \phi_N^2(t)]$ versus $\log t$. For a pure Williams–Watts function, this would yield a straight line with slope equal to β and intercept equal to $\beta \log \tau$. Treatment of the polystyrene data in this manner[46] produced overall agreement with the previous results,[43] but also revealed deviations from the pure

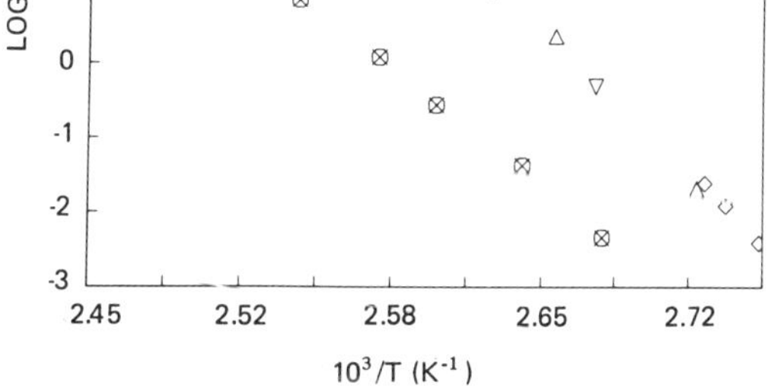

Figure 13. Transition map for polystyrene showing results from light scattering (⊗), dielectric (▽, △), and mechanical (◇) relaxation.

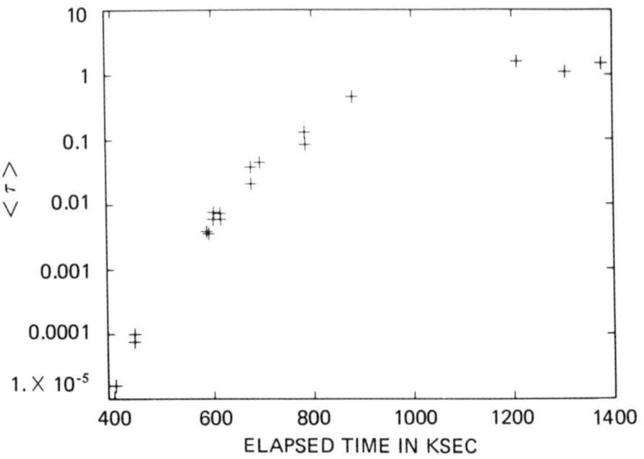

Figure 14. Average relaxation times versus elapsed time for depolarized Rayleigh scattering near the end of the thermal polymerization of styrene at 90°C.

Williams–Watts form at short time. In the case of the polarized scattering from polystyrene studied by Lee *et al.*, there may indeed be a difference between the relaxation functions for the density and anisotropy fluctuations, and the graphical technique could reveal this difference. But when the actual data can be fitted very well with the two-parameter Williams–Watts function, it may not be meaningful to try to extract more parameters.

Heating the polymer fluid near T_g produces more mobility and $\langle \tau \rangle$ becomes shorter. Mobility can also be increased by adding a small molecule liquid to act as a plasticizer. This effect has been studied[(47)] during the final stages of the thermal polymerization of styrene at 90°C. The value of β was again found to be 0.4 for the depolarized scattering. The average relaxation time is plotted versus elapsed time in Figure 14. The average relaxation time changes over five decades as the last few percent of styrene is converted to polystyrene and the system approaches its end point at 90°C. These results suggest that the nonexponential relaxation function is a characteristic of the fluctuations near the glass transition, whether the mobility is induced by temperature or dilution.

6.6. CONCLUSIONS

Light scattering from bulk polymers is a promising new application of photon correlation spectroscopy. When it is combined with Rayleigh–Brillouin spectroscopy in the frequency domain, a reasonably complete picture of the dynamics of the glass transition in polymers emerges. There have only been a very few successful studies of photon correlation spectroscopy in bulk polymers so far, but the results indicate that this will be a fertile area for future research. The main difficulty is in the preparation of samples with high optical purity. Many more polymers are currently under investigation.

REFERENCES

1. A. Einstein, Theorie der Opaleszenz von homogenen Flussigkeiten und Flussigkeitgemischen in der Nahe des kritischen Zustandes, *Ann. Phys. (Leipzig)* **33**, 1275–1298 (1910).
2. M. Smoluchowski, Molekular-kinetische Theorie der Opaleszenz von Gasen im kritischen Zustande, sowie einiger verwandter Erscheinungen, *Ann. Phys. (Leipzig)* **25**, 205–226 (1908).
3. J. Kestens, *The Photoelastic Effect and Its Applications*, Springer-Verlag, New York (1975).
4. J. Ferry, *Viscoelastic Properties of Polymers*, Wiley, New York (1970).
5. R. D. Andrews and J. F. Rudd, Photoelastic properties of polystyrene in the glassy state. I. Effect of molecular orientation, *J. Appl. Phys.* **28**, 1091–1095 (1957).
6. S. M. Rytov, Relaxation theory of Rayleigh scattering, *Sov. Phys. JETP* **31**, 1163–1171 (1970).
7. L. Brillouin, Diffusion de la lumiere et des rayons X par un corps transparent homogene, influence de l'agitation thermique, *Ann. Phys. (Paris)* **17**, 88–122 (1922).
8. R. D. Mountain, Thermal relaxation and Brillouin scattering in liquids, *J. Res. Nat. Bur. Stand.* **70A**, 207–220 (1966).
9. C. Allain-Demoulin, P. Lallemand, and N. Ostrowsky, Theoretical study of light scattering spectrum of a pure relaxing fluid. Application to viscous fluids at low temperature, *Mol. Phys.* **31**, 581–601 (1976).
10. G. R. Alms, D. R. Bauer, J. I. Brauman, and R. Pecora, Generalized hydrodynamics and the depolarized Rayleigh doublet in anisaldehyde, *J. Chem Phys.* **59**, 5304–5309 (1973).
11. G. D. Patterson, Hypersonic relaxation in amorphous polymers, *CRC Crit. Rev. Solid State Mater. Sci.* **9**, 373–397 (1980).
12. G. D. Patterson, in *Methods of Experimental Physics*, R. Fava, ed., Vol. 16A, pp. 170–204, Academic, New York (1980).
13. J. V. Champion and D. A. Jackson, Brillouin scattering from liquid *n*-alkanes, *Mol. Phys.* **31**, 1159–1168 (1976).
14. G. D. Patterson and C. P. Lindsey, Hypersonic attenuation in the *n*-alkanes, *J. Appl. Phys.* **49**, 5039–5041 (1978).
15. G. D. Patterson and J. P. Latham, Brillouin scattering from molten linear polyethylene, *Macromolecules* **10**, 736–738 (1977).
16. N. G. McCrum, B. E. Read, and G. Williams, *Anelastic and Dielectric Effects in Polymeric Solids*, Wiley, New York (1967).
17. G. D. Patterson and J. P. Latham, Hypersonic relaxation in poly(ethylene oxide), *Macromolecules* **10**, 1414 (1977).
18. C. H. Porter and R. H. Boyd, A dielectric study of the effects of melting on molecular relaxation in poly(ethylene oxide) and polyoxymethylene, *Macromolecules* **4**, 589–594 (1971).
19. C. H. Wang and Y. Y. Huang, Brillouin–Rayleigh scattering studies of polypropylene glycol. III, *J. Chem. Phys.* **64**, 4847–4852 (1976).

20. G. D. Patterson, Chain length dependence of hypersonic relaxation in amorphous polymers, *J. Polym. Sci. Polym. Phys. Ed.* **15**, 579–582 (1977).
21. G. D. Patterson, Brillouin scattering and hypersonic relaxation in amorphous polymers, *J. Polym. Sci. Polym. Phys. Ed.* **15**, 455–464 (1977).
22. G. D. Patterson, D. C. Douglass, and J. P. Latham, Hypersonic relaxation and the glass–rubber relaxation in poly(propylene glycol), *Macromolecules* **11**, 263–265 (1978).
23. A. Adshead and S. M. Lindsay, The hypersonic loss processes in polydimethyl siloxane and the effects of crosslinking, *Polymer* **21**, 1355–1356 (1980).
24. S. Yano, R. R. Rahalkar, S. P. Hunter, C. H. Wang, and R. H. Boyd, Studies of molecular relaxation of poly(propylene oxide) solutions by dielectric relaxation and Brillouin scattering, *J. Polym. Sci. Polym. Phys. Ed.* **14**, 1877–1890 (1976).
25. G. D. Patterson, Brillouin scattering and hypersonic relaxation in amorphous polymers, *J. Macromol. Sci. Phys.* **B13**, 647–664 (1977).
26. R. Vacher and J. Pelous, Hypersonic properties of PMMA between 4 and 300 K, *Phys. Lett.* **58A**, 139–140 (1976).
27. R. W. Coakley, R. S. Mitchell, J. R. Stevens, and J. L. Hunt, Rayleigh–Brillouin light-scattering studies on atactic polystyrene, *J. Appl. Phys.* **47**, 4271–4277 (1976).
28. D. S. Mahler, R. W. Coakley, J. R. Stevens, and J. L. Hunt, Rayleigh–Brillouin light-scattering study of atactic polyethylmethacrylate (PEMA), *J. Appl. Phys.* **49**, 5029–5031 (1978).
29. E. Jakeman, C. J. Oliver, and E. R. Pike, Clipped correlation of integrated intensity fluctuations of Gaussian light, *J. Phys. A* **4**, 827–835 (1971).
30. G. Williams and D. C. Watts, Nonsymmetrical dielectric relaxation behaviour arising from a simple empirical decay function, *Trans. Faraday Soc.* **66**, 80–85 (1970).
31. C. P. Lindsey and G. D. Patterson, Detailed comparison of the Williams–Watts and Cole–Davidson functions, *J. Chem. Phys.* **73**, 3348–3357 (1980).
32. D. A. Jackson, E. R. Pike, J. G. Powles, and J. M. Vaughan, On the possibility of detecting slow molecular reorientation in polymers by photon correlation spectroscopy, *J. Phys. C: Solid State Phys.* **6**, L55–L58 (1973).
33. T. A. King and M. F. Treadaway, Molecular motion in solid polymers detected by photon correlation spectroscopy, *Chem. Phys. Lett.* **50**, 494–496 (1977).
34. C. Cohen, V. Sankur, and C. J. Pings, Laser correlation spectroscopy of amorphous polymethylmethacrylate, *J. Chem. Phys.* **67**, 1436–1441 (1977).
35. G. D. Patterson, J. R. Stevens, and C. P. Lindsey, Photon correlation spectroscopy of poly(ethyl methacrylate) near the glass transition, *J. Macromol. Sci. Phys.* **B18**, 641–648 (1980).
36. Y. Ishida and K. Yamafuji, Studies on dielectric behaviors in a series of polyalkylmethacrylates, *Kolloid-Z.* **177**, 97–116 (1961).
37. G. R. Alms, T. D. Gierke, and G. D. Patterson, Observation and analysis of the depolarized Rayleigh doublet in isotropic MBBA and measurement of the de Gennes viscosity coefficients, *J. Chem. Phys.* **67**, 5779–5787 (1977).
38. G. D. Patterson, C. P. Lindsey, and G. R. Alms, Depolarized Rayleigh spectroscopy in the *n*-alkanes, *J. Chem. Phys.* **69**, 3250–3253 (1978).
39. D. R. Jones and C. H. Wang, Depolarized Rayleigh scattering and backbone motion of polypropylene glycol, *J. Chem. Phys.* **65**, 1835–1840 (1976).
40. G. D. Patterson and G. R. Alms, Depolarized Rayleigh dip spectra in the *n*-alkanes, *Macromolecules* **10**, 1237–1239 (1977).
41. G. D. Patterson, Brillouin scattering and polymer science, *Adv. Chem.* **174**, 141–162 (1979).
42. G. R. Alms, G. D. Patterson, and J. R. Stevens, A study of the thermal polymerization of styrene by depolarized Rayleigh light scattering spectroscopy, *J. Chem. Phys.* **70**, 2145–2154 (1979).

43. G. D. Patterson, C. P. Lindsey, and J. R. Stevens, Depolarized Rayleigh spectroscopy of polystyrene near the glass–rubber relaxation, *J. Chem. Phys.* **70**, 643–645 (1979).
44. C. P. Lindsey, G. D. Patterson, and J. R. Stevens, Photon correlation spectroscopy of polystyrene near the glass–rubber relaxation, *J. Polym. Sci. Polym. Phys. Ed.* **17**, 1547–1555 (1979).
45. G. D. Patterson and C. P. Lindsey, Photon correlation spectroscopy near the glass transition, *Macromolecules* **14**, 83–86 (1980).
46. H. Lee, A. M. Jamieson, and R. Simha, Photon correlation spectroscopy of polystyrene in the glass transition region, *Macromolecules* **12**, 329–332 (1979).
47. G. D. Patterson, J. R. Stevens, G. R. Alms, and C. P. Lindsey, Depolarized Rayleigh spectroscopy of concentrated solutions of polystyrene in styrene, *Macromolecules* **12**, 661–662 (1979).

7

Critical Phenomena

B. Chu

Chemistry Department
State University of New York at Stony Brook
Stony Brook, New York 11794–3400

7.1. INTRODUCTION

This chapter is concerned with the application of laser light scattering, including photon correlation spectroscopy, to critical opalescence in real fluid systems.[1] We begin with a brief discussion of critical phenomena and then provide an outline of the subjects covered. The aim is not to treat the subject exhaustively, but rather to point out topics of current interest, and to provide a guide to the vast literature on critical opalescence and to possible future directions of research for the noninitiate.

The field of critical phenomena has witnessed noteworthy achievements over the last decade or so by both theorists and experimentalists. Many physical systems have phase transitions terminating in a critical point, such as the gas–liquid critical point of a fluid and the critical consolute point of a binary mixture. The first striking fact is the similarity of critical behavior in seemingly dissimilar systems. Aside from numerous but often erroneous reports of preliminary experimental observations on minor details of different quantitative behavior of a system near its critical point, the concept of universality for real systems has started as a fascinating hypothesis. Its subsequent elucidation and demonstration using experiments with a high degree of precision represent a truly dramatic achievement in condensed matter physics. As the physical behavior of critical systems depends only on general geometrical properties and is largely independent of detailed microscopic interactions, the modern approach explains critical phenomena in terms of simple statistical models and the renormalization-group picture of university. There exist many recent reviews of critical phenomena. For example, a very good book[2] by Eugene Stanley providing a general review of the subject is being updated; The Domb–Green series[3] contains detailed reviews. Elementary reviews on critical-point universality and fluids,[4] liquid–liquid critical phenomena,[5] critical exponents for binary fluid mixtures,[6] and critical phenomena in fluids[7] represent good starting material where many additional pertinent references have been listed. The renormalization-group method[8] has been extended to dynamic critical phenomena.[9] Detailed equations on scaling laws including scaled equation of state, corrections to scaling, correlation scaling function, and gravity effects[10] have been summarized by the Sengers,[11] while results of dynamical scaling, mode coupling theory, and the dynamic renormalization group approach have been outlined by Kawasaki and Gunton.[12]

In comparing experimental results with theoretical predictions, the discussions usually center on critical exponents which are the consequences of scaling, universality, and the renormalization group. Until recently, insufficient attention has been given to the range of asymptotic scaling behavior. As the structure of the corrections to the asymptotic behavior depend on the thermodynamic properties, it becomes essential that we know the effects of correction terms to simple power laws. By using those simple power laws, one often obtains only effective exponents if experimental data were fitted in the intermediate neighborhood of the critical point $\{10^{-4} \le \varepsilon[\equiv(T - T_C)/T_C] \le 10^{-3}\}$. A correct comparison requires an analysis of data very close to the critical point ($\varepsilon \le 10^{-4}$) where correction terms are much less important. Unfortunately, measurements in the very immediate neighborhood of the critical point are difficult to perform and the results are subject to other experimental errors such as those related to gravity and multiple scattering. Thus, an alternative approach using data over a wide range with extended scaling[13–15] is perhaps a better compromise. However, the Wagner expansion has a poor convergence because of the small value of the correction-to-scaling exponent Δ_1 and its range of applicability is limited. A more promising approach might be a crossover description from Ising-like behavior close to the critical point to mean-field behavior further away from the critical point.[14] For details, the reader is referred to the chapter by Sengers and Sengers[16] for the question "How close is close to the critical point?" Suffice it to say that simple power laws are applicable only in the asymptotic region not easily accessible experimentally. Therefore, they are not generally as useful in engineering problems. Extended scaling has illustrated the limitations on simple power laws; but the poor convergence again limits its usage over a broad temperature range. The crossover approach looks promising; but much work remains to be done.

Table 1 lists critical power laws and their exponent values. A compari-

Table 1. Critical Power Laws and Exponent Values

Quantity	Thermodynamic path	Power law[a]	Exponent: Mean field	Exponent: Renormalization group[b] ($d = 3$)
Coexistence curve	$\rho = \rho_{c \times c}$, $\varepsilon < 0$	$\Delta\tilde{\rho} = \pm B \lvert\varepsilon\rvert^{\beta}$	$\beta = 1/2$	$\beta = 0.325 \pm 0.001$
Susceptibility	$\rho = \rho_c$, $\varepsilon > 0$	$\tilde{\chi} = C_0^+ \lvert\varepsilon\rvert^{-\gamma}$	$\gamma = 1$	$\gamma = 1.240 \pm 0.002$
	$\rho = \rho_{c \times c}$, $\varepsilon < 0$	$\tilde{\chi} = C_0^- \lvert\varepsilon\rvert^{-\gamma'}$	$\gamma' = 1$	$(\gamma = \gamma')$
Critical isotherm	$\varepsilon = 0$	$\Delta\tilde{\mu} = \pm D \lvert\Delta\tilde{\rho}\rvert^{\delta}$	$\delta = 3$	$\delta = 4.82 \pm 0.02$
Specific heat	$\rho = \rho_c$, $\varepsilon > 0$	$\tilde{c}_v = (A^+/\alpha)\lvert\varepsilon\rvert^{-\alpha}$	$\tilde{c}_v = c_0$	$\alpha = 0.109 \pm 0.004$
	$\rho = \rho_{c \times c}$, $\varepsilon < 0$	$\tilde{c}_v = (A^-/\alpha)\lvert\varepsilon\rvert^{-\alpha'}$	$\tilde{c}_v = c_0 + \Delta c_v$	$(\alpha = \alpha')$
Correlations	$\varepsilon = 0$	$\tilde{\chi}(q) = C_1/q^{2-\eta}$	$\eta = 0$	$\eta = 0.0315 \pm 0.0025$
	$\rho = \rho_c$, $\varepsilon > 0$	$\xi = \xi_0^+ \lvert\varepsilon\rvert^{-\nu}$	$\nu = 1/2$	$\nu = 0.630 \pm 0.001$
	$\rho = \rho_{c \times c}$, $\varepsilon < 0$	$\xi = \xi_0^- \lvert\varepsilon\rvert^{-\nu'}$	$\nu' = 1/2$	$(\nu = \nu')$

[a] $\tilde{\chi} \equiv (\partial\tilde{\rho}/\partial\tilde{\mu})_T$, where $\tilde{\rho} = \rho/\rho_c$, $\tilde{\mu} = (\mu)(\rho_c/P_c)$; $\Delta\tilde{\mu} = [\mu(\rho, T) - \mu(\rho_c, T)]\rho_c/P_c$, where $\mu(\rho_c, T)$ is the chemical potential on the critical isotherm at temperature T; $\tilde{c}_v = (c_v/v)(T_c/P_c)$; $q = (4\pi/\lambda)\sin(\theta/2)$.

[b] References 17, 18.

son of recent experimental results on both exponents and amplitude ratios shows general excellent agreement with the predictions of the renormalization group method. In particular, the values of α, β, γ, ν, and η have been verified quantitatively by a small number of systems[5, 6, 14, 15, 19–22] even though it cannot be said that universality has been established conclusively in fluid systems. With the values of critical exponents more firmly established, future studies are being focused on amplitude ratios[23–26] and on numerical scale factors which depend upon structure. It is also important to note that polymers of very narrow molecular weight distributions in theta solvents behave like binary critical fluid mixtures.[19, 27] Finally, the 1980 Cargése Summer Institute on Phase Transitions (July 16–30, 1980, Corsica, France) has reported on the recent experimental and theoretical status of critical phenomena.[28]

The brief outline above represents mainly a summary of equilibrium critical phenomena in fluids. In this chapter we shall restrict ourselves to discussions on critical opalescence with emphasis on critical dynamics, even though applications of dynamic light scattering on critical fluid systems have been rather limited in scope in recent years.

7.2. CRITICAL FLUCTUATIONS

A comparison of the results of the approximate mode-coupling calculations to the critical behavior of dynamical properties of fluids by means of quasielastic light scattering has been reviewed by Swinney and Henry[29] and others.[9] As the static critical behavior is so closely related to critical dynamics, we shall provide an outline of the pertinent equations for both the static intensity and the dynamical (linewidth) critical properties of fluid systems.

7.2.1. Static Critical Behavior

7.2.1.1. Light Scattering. At the critical density of fluids very near the gas–liquid critical point, or at the critical solution concentration of binary liquids very near its critical mixing point, the structure factor $\tilde{\chi}(q)$ has the form[21]

$$\tilde{\chi}(q) = C_0 \varepsilon^{-\gamma} g(q\xi) \tag{1}$$

where $g(x \equiv q\xi)$ is the correlation scaling function and the amplitude C_0 is defined such that $\lim_{x \to 0} g(x) = 1$. For $x \ll 1$,

$$g(x) = 1/(1 + x^2) \tag{2}$$

which is the Ornstein–Zernike correlation function. For $x \gg 1$,

$$g(x) = \frac{C_1}{x^{2-\eta}}\left(1 + \frac{C_2}{x^{(1-\alpha)/\nu}} + \frac{C_3}{x^{1/\nu}}\right) \tag{3}$$

which is often referred to as the Fisher–Langer expansion.[30] Earlier light scattering intensity data have often been interpreted in terms of the Fisher approximant

$$g_F(x) = 1/(1 + x^2)^{1-\eta/2} \tag{4}$$

which cannot accommodate simultaneously the correct amplitudes of the leading terms in the small and large x expansions. The Fisher–Burford[31] scaling function has the form

$$g_{\mathrm{FB}}(x) = \frac{(1 + \psi^2 x^2)^{\eta/2}}{1 + (1 + \psi^2\eta/2)x^2} \tag{5}$$

which yields $C_1 = \psi^\eta/(1 + \psi^2\eta/2)$ in the asymptotic expansion of equation (3) for large x. Unfortunately, equation (5) does not reproduce accurately the exactly known correlation function of the two-dimensional Ising model over an appropriate range of x.[32] The correlation scaling function of Ferrell and Scalapino[33] has a similar defect.[21b] Thus, Tracy and McCoy[32] have questioned the reliability of any of the correlation scaling functions [equations (2)–(5)] for determining the exponent η from experimental scattering data.

Values of C_1, C_2, and C_3 in equation (3) have been calculated by Chang *et al.*,[21b] who adopted the sum rule $C_2 + C_3 = -0.9$ obtained from an ε expansion to third order[33] and followed essentially Bray's procedure.[34] The results are tabulated in Table 2. Bray could reproduce the correlation scaling function of the two-dimensional Ising model to within 0.03% of any value of x. Furthermore, his scaling function agrees well with the theoretical values near four dimensions calculated using ε expansion techniques.[35] Therefore, the correlation scaling function proposed by Bray is assumed to be an adequate representation for three-dimensional Ising-like systems.

In practice, the light scattering intensity has to be corrected for background and turbidity as well as the effects of multiple scattering and of gravitation, if any, and the temperature dependence of $(\partial n/\partial c)^2_{T,P}$ before the excess scattered intensity can be related to $\tilde{\chi}(q)$ of equation (1). The pro-

Table 2. Values of C_1, C_2, C_3 in the Correlation Scaling Function $g(x)$ [Equation (3)] for Various Values of η and ν

η	1/54	0.0315	0.0410
ν	5/8	0.6300	0.6380
C_1	0.9529	0.9268	0.9089
C_2	1.733	2.403	3.591
C_3	−2.745	−3.303	−4.491

cedures vary and the corrections can be substantial, depending upon the nature of the system and experimental conditions. One detailed description by Chang *et al.*[21b] represents the best careful consideration on linearized turbidity and double scattering corrections so far. However, their approach is specifically designed for only 90° scattering angle.

By using the Fisher relation

$$\gamma = (2 - \eta)\nu \tag{6}$$

and by assuming that in the experimental temperature range $\varepsilon \leq 2.7 \times 10^{-3}$, corrections to scaling can be neglected, Chang *et al.*[21b] obtained for the 3-methylpentane-nitroethane system

$$\eta = 0.017 \pm 0.015, \qquad \nu = 0.625 \pm 0.006$$

$$\xi_0 = (2.29 \pm 0.10)\ \text{Å}, \qquad \gamma = 1.240 \pm 0.017$$

$$C_1 = 0.96 \pm 0.04$$

The results are insensitive to small changes in the definition of the order parameter. The correlation scaling functions proposed by Fisher and Burford [equation (5)] and by Ferrell and Scalapino gave equally satisfactory representations of the experimental scattering data leading to similar values for the critical exponents.

Chang *et al.*[21b] also considered corrections to scaling using

$$g_c(x) = g(x)[(\text{equation (3)}] + g_1(x)\Gamma_1\varepsilon^{\Delta_1} \tag{7}$$

where the scaling variable x is now defined as

$$x = q\xi_0\varepsilon^{-\nu} \tag{8}$$

Γ_1 and $\Delta_1[=0.5]$ are, respectively, the correction-to-scaling amplitude and the correction-to-scaling exponent, and $g_1(x)$ has the form

$$g_1(x) \simeq 1/(1 + x^2)^{(2-\eta-\Delta_1/\nu)/2} \tag{9}$$

By taking the proportionality constant between the scattered intensity and $\tilde{\chi}(q\xi)$ $(\tilde{I}_0)$, the background scattered intensity (ΔI), η, ν, ξ_0, and Γ_1 as the six adjustable parameters and by using the same linearized approximation with $\eta = 1/54$ and $\nu = 5/8$, they got

$$\eta = 0.016 \pm 0.016, \qquad \nu = 0.628 \pm 0.013$$
$$\xi_0 = (2.23 \pm 0.25)\ \text{Å}, \qquad \Gamma_1 = 1.1 \pm 3.5$$
$$\gamma = 1.246 \pm 0.033, \qquad C_1 = 0.96 \pm 0.04$$

Therefore, the correction-to-scaling term does not appear to be significant in this case.

With the critical exponents η, γ, and ν fixed at values listed in Table 1 according to the renormalization-group theory, and again with the appropriate spectral function to evaluate $g(x)$ [equation (3)] and the remaining four parameters $\tilde{I}_0$, ΔI, ξ_0, and Γ_1 as adjustable parameters, Chang *et al.*[21b] found that the experimental data were consistent with the predictions from the renormalization-group theory for Ising-like systems. Values of the critical exponents based on theoretical estimates from numerical analysis of high-temperature series expansions[36] with $\eta = 0.041$, $\nu = 0.638$, and $\gamma = 1.250$ have generally been recognized as being biased.[28]

One lesson to be learned from the above discussion is that the determination of η is in the fringe of experimental error limits. The measured scattered intensity must be corrected for turbidity and background. The effects of multiple scattering and gravity have to be considered. After all these corrections and considerations, the scattering data are fitted to a correlation scaling function using *six* adjustable parameters. While the values of the critical exponents γ and ν can be determined by this procedure, the small magnitude of η has made its precise characterization difficult. It is doubtful that we can push such analysis much further using scattering data obtained at one scattering angle. If we plan to vary $x(=q\xi)$ by fixing ξ and varying only q, the range of $q\xi$ which can be covered by means of light scattering is usually limited to the domain where the value of η is relatively insensitive to changes of x in the small $q\xi$ range. Efforts[37] have been made to increase ξ by making measurements in the very immediate neighborhood of the critical point and by varying the wavelength of visible light. Unfortunately, we still have to contend with the effects of multiple scattering and of gravity.

7.2.1.2. Small-Angle X-Ray Scattering (SAXS) and Small-Angle Neutron Scattering (SANS). An alternative approach is to vary q over a much larger range by means of small-angle x-ray scattering (SAXS) or small-angle neutron scattering (SANS). The condition $q\xi \gg 1$ where the scattering is proportional to $q^{\eta-2}$ can be met easily. However, it is essential that the condition $qa \ll 1$ is also satisfied, where a is the average intermolecular distance.

In an investigation of small-angle x-ray scattering of argon, Schmidt and his co-workers found $\eta = 0.07 \pm 0.01$[38] without correction to scaling and $\eta = 0.03 \pm 0.02$[39] if allowance is made for correction to scaling. The equations used in the fitting procedure were different from those used by Chang *et al.*[21b] Schmidt and his co-workers first took equations (1), (2), and (6) to obtain

$$[X(q)]^{-1} = \frac{\varepsilon^{\gamma}}{C_0} + \left[\frac{\xi_0^2 \varepsilon^{-\eta\nu}}{C_0}\right] q^2 \tag{10}$$

and tried to determine the slope $S[=\xi_0^2 \varepsilon^{-\eta\nu}/C_0]$ as a function of ε. η was determined by assuming $\xi = \xi_0 \varepsilon^{-\nu}$ and $\nu = 0.63$. Therefore, equation (3) was never used. In his correction to scaling, Schmidt and his co-workers[39] extended equation (10) by using an expression similar to equation (7).

$$g_c(x) = g(x)\ [\text{equation (10)}] + g_2(x)\Gamma_1 \varepsilon^{\Delta_1} \tag{11}$$

where

$$g_2(x) = 1/(1 + D_0 x^2) \tag{12}$$

with D_0 being another universal constant for all systems of the same universality class. Schmidt and his co-workers concluded that corrections to scaling were needed for argon near the critical point because the experimental η value is in agreement with the theoretical calculations *only* when allowance is made for corrections to scaling.

By combining light scattering[40] and small-angle neutron scattering[41] data, it is possible to determine the value of critical exponent η directly. The temperature dependence of the correlation length ξ and the susceptibility $\chi(0)$ of an isobutyric acid/D_2O mixture of critical composition [mass fraction of isobutyric acid $(y) = 0.36$; phase separation temperature $(T_p) =$ 44.910°C] was studied by measuring the angular dependence of scattered light intensity as a function of the temperature difference $(T - T_c)$ in the temperature range 1.10^{-2} $K \leq (T - T_c) \leq 3.0$ K. The correlation length $\xi(=\xi_0 \varepsilon^{-\nu})$ was computed using equation (4) from scattering data measured at two fixed scattering angles ($\theta_1 + \theta_2 = 180°$) and the critical exponent γ was determined using equations (1) and (4). By assuming $\eta = 0.04$, we obtained $\gamma = 1.25$, $\nu = 0.633$, $\xi_0 = 3.13 \times 10^{-8}$ cm, and $T_c = 44.895$°C.

The neutron scattering experiments were performed at the research reactor FRJ-2 in Jülich, Federal Republic of Germany.[42]

The scattering patterns were normalized to a constant neutron flux, corrected for counter sensitivity and general background (fast neutrons, electronic noises, etc.), converted into absolute units by using a secondary standard, and desmeared according to Lake[43] including corrections for neutron wavelength distribution and instrument geometry. The scattering cross section is composed of the differential scattering cross section of critical scattering, incoherent scattering mainly from the hydrogen atoms in the isobutyric acid, and contributions by the liquid structure factor. The latter two contributions can be considered to be independent of temperature and q. At large temperature distances ($T - T_c \sim 5$–10°C),

$$\frac{d\Sigma(q)}{d\Omega} = \frac{C_0 \varepsilon^{-\gamma}}{1 + (q\xi)^2} + \frac{d\Sigma_0}{d\Omega} \tag{13}$$

where C_0, $\xi(\varepsilon)$ and $d\Sigma_0/d\Omega$ are the three adjustable parameters. Figure 1 shows a log–log plot of $g(x)$ versus x in which $g(x)$ is composed of eight scattering patterns. As Tracy and McCoy[31] have investigated the validity of the various approximations for $g(x)$ using the two-dimensional Ising model for which an exact solution for $g(x)$ exists, we note that equation (2) is better than 1% for $x < 4$ and worse than 10% for $x > 21$, while for equation (3) the Fisher–Langer approximation $g_{FL}(x)$, has an accuracy of 1% for $x > 5.4$ and an accuracy of 5% for $x > 2.5$. By taking $\alpha = 0.125$, $\nu = 0.633$, $\xi_0 = 3.13 \times 10^{-8}$ cm, $C_2 = 1.733$, and $C_3 = -2.745$ in equation (3), only two adjustable parameters C_1 and η are needed to fit all the scattering data covering a range of x values from 2 to 270 within the experimental error limits. The results are $C_1 = 0.0972 \pm 0.0051$ cm^{-1} and $\eta = 0.0858 \pm 0.0263$. In Figure 1, we have also shown $g_{FL}(x)$ with C_1 and η values representing the upper and lower bounds, respectively. All experimental data points lie within these two solid lines. By scaling, we have extended the range of x to more than two orders of magnitude. In equation (3) the higher-order terms become less important as x becomes larger. In the limit $x \to \infty$

$$g_\infty(x) = C_1 x^{-(2-\eta)} \tag{14}$$

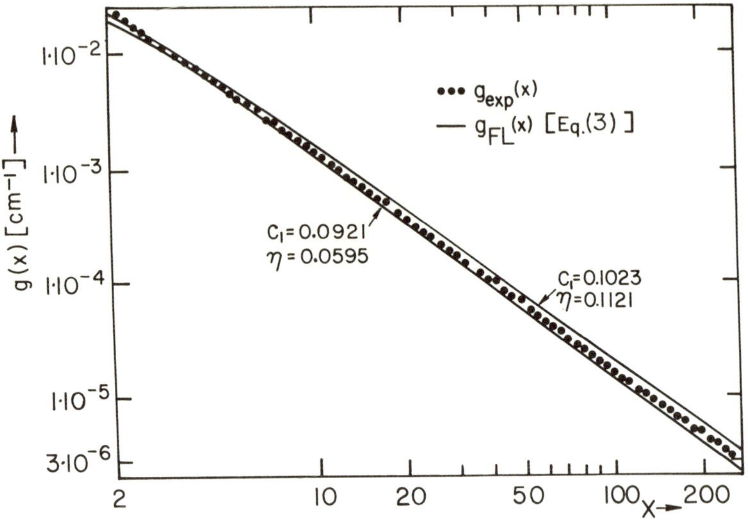

Figure 1. Log–log plot of $g(x)$ versus $x[=q\xi]$. The experimental scaling function $g_{\text{exp}}(x)$ consists of 392 data points (●) and is composed of eight critical scattering patterns measured at four different temperatures ($T - T_c = 1.001$, 0.098, 0.011, and 0.009 K). Only every fifth data point is plotted. The two solid lines represent the Fisher–Langer approximant [equation (3)] with upper and lower bounds for C_1 and η (Reference 41).

In practice, we may use equation (14) to determine η provided that the conditions $q\xi \gg 1$ *and* $qa \ll 1$ are satisfied. As $g_{\infty}(x)$ deviates less than 1% from $g_{\text{FL}}(x)$ for $x > 20$ where the statistical error of $g_{\text{exp}}(x)$ is a few percent, we can use equation (14) to fit the scattering data using only those $g_{\text{exp}}(x)$ values with x between 20 and 270. The results are $C_1 = 0.0965 \pm 0.0111$ cm^{-1} and $\eta = 0.0894 \pm 0.0325$, as shown in Figure 2. Many factors such as the scattering form factor, multiple scattering, inelastic neutron scattering due to the critical slow down, and background correction, have been considered. None could influence the value of η appreciably.

Measurements of light scattering, SAXS, and SANS have shown unequivocally a nonzero and positive value for η. The magnitude of η is still open for dispute. In the 3-methylpentane/nitroethane system,[21b] the scattering data from one fixed scattering angle after corrections were compared with $g_{\text{FL}}(x)$ using Bray's procedure with five adjustable parameters without corrections to scaling and with six adjustable parameters with corrections to scaling. Chang *et al.*[21b] found corrections to scaling to be unimportant over the temperature range of their studies $10^{-6} < \varepsilon < 2.7 \times 10^{-3}$. However, Gürmen *et al.*[39] found the value of η changes from 0.07 ± 0.01

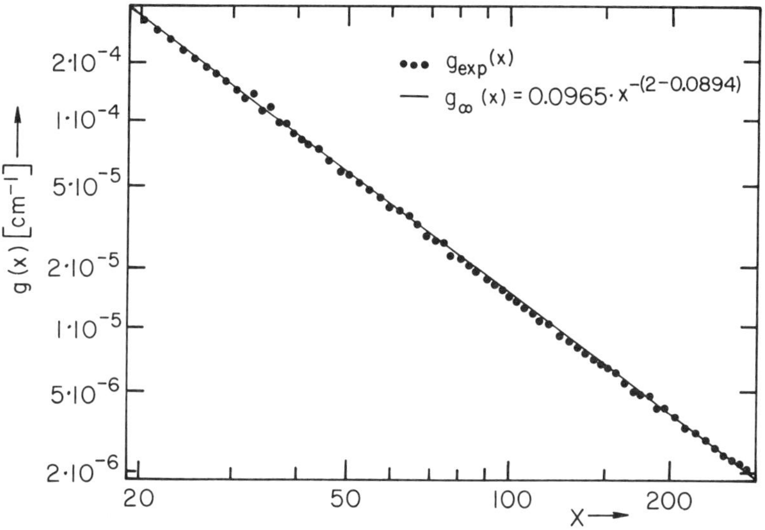

Figure 2. Log–log plot of $g(x)$ versus x. The solid line represents $g_\infty(x) = 0.0965x^{-(2-0.0894)}$ (Reference 41).

without corrections to scaling to 0.03 ± 0.02 with corrections to scaling for argon near its critical point over the temperature range $10^{-3} < \varepsilon < 6 \times 10^{-2}$. It is not unreasonable to speculate that at larger temperature distances and especially for one-component fluid systems we need corrections to scaling for $g(x)$. Nevertheless, there is no justification for concluding that there is a need for correction to the limiting scaling behavior, and that the value of η should be 0.03 based on SAXS of argon because it agrees better with the present-day accepted theory. This is not to say that η should not be 0.03 nor that corrections to scaling are not needed. In the SANS experiment, no correction to scaling was applied over the temperature range $3 \times 10^{-5} < \varepsilon < 3 \times 10^{-3}$, which was similar to that of the Chang *et al.* experiment ($10^{-6} < \varepsilon < 2.7 \times 10^{-3}$). Yet the value of η was much too large. Furthermore, by fitting $g_\infty(x)$ to the scattering data at large values of x, the η value remains greater than 0.03. The important conclusion to be made by results of recent experiments is that the critical exponent η does exist and that its value is relatively small. SANS reports a value of $2 - \eta \approx 1.914$, while the renormalization group theory (and light scattering data) indicates a value of 1.969. The agreement is, in fact, remarkable without even taking into account the high-temperature expansion limit of 1.936. It can be speculated that η from SANS should be lower.

7.2.2. Dynamic Critical Behavior

Based on the mode–mode coupling theory[44, 45] and the renormalization-group mode dynamics,[46] the critical part of the linewidth Γ_c has the form

$$\Gamma_c = R \frac{k_B T q^3}{6\pi\bar{\eta}} \left[\frac{K_0(q\xi)}{(q\xi)^3} R(q, \xi) \right] \tag{15}$$

where $\bar{\eta}$ is the macroscopic shear viscosity, $R(q\xi)$ is a weakly nonuniversal function described by Oxtoby and Gelbart,[45] and $K_0(x = q\xi) = \frac{3}{4}[1 + x^2 + (x^3 - x^{-1})\arctan x]$. Comparison of experiments with theories has been reviewed by Swinney and Henry[28] and others.[47]

7.2.2.1. Value of Numerical Constant *R*. Recent attention has been centered on the evaluation of R. The numerical constant R has the value of one in the mode–mode coupling theory and 1.2 in the renormalization-group calculations. While Chen *et al.*[48] and Sorensen *et al.*[49] reported good agreement with the theoretical value $R = 1.20$ for the critical binary fluids η-hexane/nitrobenzene and 3-methylpentane/nitroethane, respectively, Burstyn *et al.*[50] and Güttinger and Cannell,[51] using, respectively, 3-methylpentane/nitroethane and xenon, obtained $R \simeq 1.02$ and 1.01 *after background corrections.*[45] As equation (15) is valid only for the critical part of the linewidth, background correction is essential unless measurements are made in the very immediate neighborhood of the critical point where background linewidth contributions become negligible. In particular, Guttinger and Cannell[51] covered a $q\xi$ range of ~ 0.01 to ~ 10 involving a background linewidth varying from $\sim 40\%$ of the measured linewidth to a few percent of the total. They[50, 51] concluded that the experiments are in good agreement with the mode–mode coupling theory. On the other hand, Beysens *et al.*[26] have determined experimentally the universal constant R_ξ^+ concerning the amplitude of the diverging part of the specific heat at constant pressure p and critical concentration X_c, C^0_{p, X_c} in three binary fluids:

$$R_\xi^+ = (\alpha \rho_c C^0_{p, X_c}/k_B)^{1/3} \xi_0^+ \tag{16}$$

Beysens *et al.* assumed $R = 1.2$ in their computation for ξ_0^+. Furthermore, they took the relaxation time τ of the concentration fluctuations to be entirely due to critical scattering even in the temperature range $0.3 < T - T_c < 10$. Thus, the conclusion that the results are consistent with the renormalization-group calculations is open for discussion.†

† The controversy is by no means resolved, as both sides claim to have performed the correct experiments, although recent results[52] seem to confirm the smaller R (~ 1) value.

7.2.2.2. Nonexponential Decay of Critical Fluctuations. The existence of nonexponential decay of critical concentration fluctuations is another small effect which has attracted the attention of several experimental experts in the field of critical opalescence. Volochine, Bergé, and Lagues[53] raised this question back in 1970. However, the instruments at the time did not permit examination of small departures from exponential decay. Nevertheless, the prediction of a non-Lorentzian distortion to the Rayleigh line has been establish for some time.[54–56] Large departures from exponential decay predicted by the dynamic droplet model were not observed.[49, 57, 58]

A careful experimental study of the time dependence of local concentration fluctuations near the critical point of 3-methylpentane/nitroethane[21b] has revealed deviations from exponential decay of these fluctuations at temperatures within 20 mdeg from the critical temperature.[59] The experimental effects can be interpreted in terms of the frequency dependence of the critical viscosity proposed by Bhattacharjee and Ferrell.[60] The experiments took into account the effects of after-pulsing, base line, incident laser power, multiple scattering,[61, 62] and homodyning between the scattered and reflected beams. By using the cumulants method,[63] Burstyn *et al.*[59] obtained a variance ($\mu_2/\bar{\Gamma}^2$) of 0.03 at $T - T_c = 1.8$ mdeg. The measured departure from exponential decay is small indeed as any experimental error tends to increase the variance. To put the magnitude of departure from exponential decay into proper perspective, the value of 0.03 is smaller than the expected variance from an aqueous suspension of Dow "monodisperse" polystyrene latex spheres. It is even more difficult to establish $\mu_2/\bar{\Gamma}^2 = 0.016 \pm 0.007$ at $T - T_c = 11.7$ mdeg. Additional experiments by other independent workers should be helpful.

7.2.2.3. Dynamical Critical Exponent. In equation (15), we have assumed $Z = 3$, where Z is a universal dynamical scaling exponent. According to the hypothesis of dynamical scaling, the decay rate Γ_c of the order-parameter fluctuations has the effective form

$$\Gamma_c = Aq^{Z_{\text{eff}}} \tag{17}$$

for fluids in the very immediate neighborhood of the critical point. The effective q dependence of Γ_c should be greater than 3 because of the weak viscosity anomaly:

$$Z_{\text{eff}} = 3 + x_\eta \tag{18}$$

where x_η is an effective viscosity exponent.[63] Earlier results by Sorensen *et al.*[49] and by Chang *et al.*[53] for the 3-methylpentane/nitroethane system

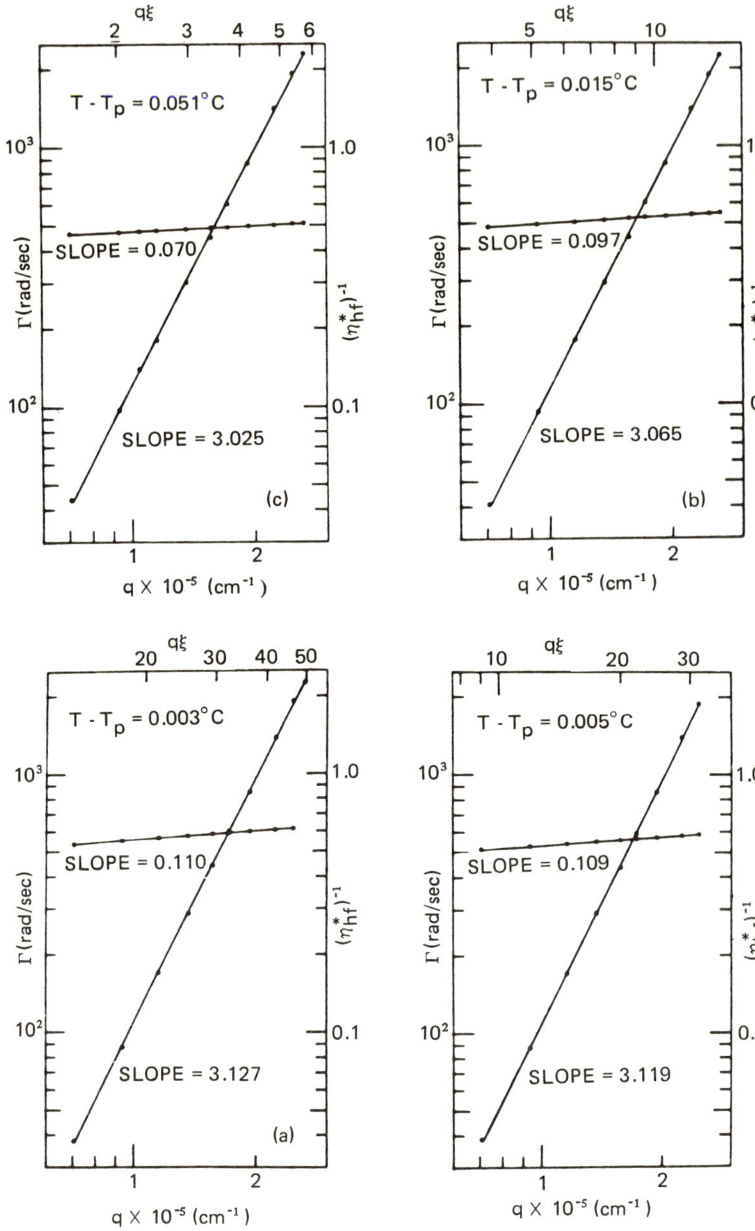

Figure 3. Log–log plots of Γ versus q (open circles) and $(\eta^*_{\mathrm{hf}})^{-1}$ vs. $q\xi$ (solid circles) for the ethanol–water–chloroform system η^*_{hf} in units of centipoise is computed from modified Kawasaki theory (Reference 65).

showed $Z_{\text{eff}} = 2.99$, while according to theory,[46, 64] $Z_{\text{eff}} = 3.065$. The $q\xi$ dependence of the high-frequency shear viscosity η^*_{hf} has been considered for the ternary liquid mixture, ethanol–water–chloroform.[65] If we took into account both the nonlocality[44b] and frequency[56] dependence of the shear viscosity, we obtained $Z = 3$ for $q\xi \gg 1$ and $Z < 3$ with $q\xi$ varying from 2 to 10, as shown in Figure 3. According to the modified Kawasaki theory, x_η has varied from 0.07 to 0.11 which is somewhat greater than $x_\eta = 0.0635 \pm 0.0004$ of the 3-methylpentane/nitroethane mixture[50a] or the latest prediction of 0.065. Recently, Burstyn and Sengers[66] obtained $Z_{\text{eff}} = 3.06 \pm 0.02$, in excellent agreement with the latest theoretical prediction of $x_\eta = 0.065$ and their own viscosity results. The agreement is again remarkable. However, there is still some controversy about the theoretical value of the dynamic exponent Z.[50b]

It is challenging and difficult to determine the magnitude of η and x_η or the existence of nonexponential decay of local critical fluctuations. The requirements for new information border on the limits of acceptable experimental errors. Definitive answers are elusive because the measurements have to be carried out using state-of-the-art instrumentation, the data analysis requires consideration of all possible sources of errors and corrections, and the experimentalists must be aware of the boundary conditions of theoretical predictions which are often expressed in physical parameters not easily accessible by experiments. It is perhaps a sign of maturity for the field indicating that exciting developments are to appear in the application of this approach to more complex systems, such as polymer solutions in the semidilute region and gels; the details appear in other chapters of this book.

7.3. DEPOLARIZED RAYLEIGH SCATTERING

Depolarized light scattering from a critical binary liquid mixture of nitrobenzene and *n*-hexane was first studied by the Fabelinskii school.[67–69] They reported dramatic variations, in contrast to the observations of Beysens *et al.*,[70, 71] who found no significant change in the linewidth of the depolarized spectrum when approaching the critical point From a spectral analysis of the depolarized light Beysens *et al.*[70, 71] were able to separate multiple scattering from the light scattered by orientation fluctuations. They attributed the discrepancy mainly to multiple scattering and took their results to be in good agreement with the Oxtoby–Gelbart theory.[72] However, the dispute has continued with Fabelinskii *et al.*,[73–75] who report a strong critical narrowing of the linewidth for several binary critical fluids. Beysens and Zalczer[76, 77] and others[78–80] show that, at most, a weak critical anomaly is related to the behavior of the shear viscosity in the region above T_c. Objectively, the comments by Beysens and Zalczer[76a] and their subsequent use of a high-performance double-pass Fabry–Perot interferometer[76b] have settled the dispute. For the nitrobenzene–*n*-hexane system, they observed two lines due to nitrobenzene. The first line is a sharp and intense Lorentzian with a linewidth behaving as the inverse shear viscosity. The second line is nearly 15 times broader with an intensity seven times weaker at $T - T_c = 1°\text{C}$. However, in their recent study of the nitroethane/isooctane mixture, Zalczer and Beysens[77] did not notice any critical anomaly in the linewidth. The couplings between the orientation and order parameter fluctuations are not adequately known to provide a reasonable explanation of the differences. The use of multipass Fabry–Perot interferometers is routine only in some laboratories and the information has limited scope. Consequently, such studies are not likely to become popular topics.

7.4. ENTROPY FLUCTUATIONS

7.4.1. Entropy Rayleigh Factor

A 213-MHz free spectral range, double-passed Fabry–Perot interferometer has been used to make relative spectral measurements of the Rayleigh line in several liquids, taking into account mainly the entropy fluctuations in the medium.[81] It should be emphasized that experimental measurements of the entropy Rayleigh factor by this method permit a check on the assumptions leading to the Rayleigh factor formulation but the details of the total isotropic scattered light due to local density fluctuations are usually not important for polymer solution studies, where the main interest is on local concentration fluctuations. As measurements of total polarized scattered intensity contain anisotropic scattering and other phenomena, such as thermal and structural relaxations, in addition to the pressure and entropy fluctuations, a precise determination of the *isotropic* Rayleigh factor R (Rayleigh plus Brillouin lines) requires measurements of the low-frequency entropy Rayleigh spectrum, even though only relative measurements have been made.

The formulations of the entropy Rayleigh factor are are as follows:

$$R_E = \frac{\pi^2}{\lambda^4}(n^2 - 1)^2 \left(\frac{1}{\rho}\frac{\partial \rho}{\partial T}\right)_p^2 \frac{k_B T^2}{\rho C_p} \qquad \text{(Einstein)} \qquad (19)$$

$$R_R = \left(\frac{n^2 + 2}{3}\right)^2 R_E \qquad \text{(Recard)} \qquad (20)$$

$$R_V = \frac{3n^2}{2n^2 + 1} R_E \qquad \text{(Vuks)} \tag{21}$$

$$R_Y = R_V\left[1 + 2\rho \frac{\partial F}{\partial \rho}\right] \qquad \text{(Yvon)} \tag{22}$$

where the correction term

$$2\rho \frac{\partial F}{\partial \rho} = 2\left(\frac{dn^2}{\partial T}\right)_p^{\text{expt}} \bigg/ \left[(n^2 - 1)\frac{n^2 + 2}{3}\left(\frac{1}{\rho}\frac{\partial \rho}{\partial T}\right)_p\right] - 1 \tag{23}$$

never exceeds 20% in pure fluids. The results on ten liquids (methanol, diethyl ether, acetone, cyclohexane, chloroform, carbon tetrachloride, benzene, nitrobenzene, bromoform, and carbon disulfide) whose refractive indices lie between 1.3 and 1.6 are in good agreement with the theories by Rocard, Vuks, and Yvon and show consistent discrepancies with the Einstein formula [equation (19)].

7.4.2. Local Entropy Fluctuations

In a critical binary mixture, the first-order signal correlation function of the scattered electric field $g^{(1)}(\tau)$ has at least two terms due to local entropy and concentration fluctuations:

$$g^{(1)}(\tau) = Ae^{-\sigma|\tau|} + Be^{-\gamma|\tau|} \tag{24}$$

Near the critical mixing point, critical opalescence occurs due to enormous increases in the amplitude of local concentration fluctuations while the amplitude of local entropy fluctuations remains approximately constant. Furthermore, as the critical temperature is approached, the linewidth γ decreases strongly while the linewidth σ decreases only very weakly. Measurements of $\sigma(\tau)$ in the presence of strong local concentration fluctuations have been accomplished by a clever cross-correlation method.[82]

By choosing a delay time range such that $\tau_M \approx 10\sigma^{-1}$, equation (24) can be approximated by

$$g^{(1)}(\tau) \approx Ae^{-\sigma|\tau|} + B[1 - \gamma|\tau| + \tfrac{1}{2}\gamma^2\tau^2 - \cdots] \tag{25}$$

Equation (25) implies that $Ae^{-\sigma|\tau|}$ is superimposed on an almost straight and slightly inclined line. Consequently, the linewidth σ can be determined correctly if $A \gtrsim B/10$. Near the critical point, $B \gg A$ such that it becomes difficult to determine σ even with the use of a high-pass filter whose cutoff frequency α is adjusted so that $\gamma \ll \alpha \ll \sigma$.

In the derivatives cross-correlation method,[82] the amplified output voltage of the photodetector $V(\tau)$ is (a) fed directly to input A of the correlator and (b) passed through a single-pole high-pass RC filter and then fed to input B of the correlator. The cross correlation function for $\tau \geq 0$ then has the form

$$g_C^{(1)}(\tau) = A \frac{\sigma}{\sigma + \beta} e^{-\sigma\tau} + \beta \frac{\gamma}{\gamma + \beta} e^{-\gamma\tau} \tag{26}$$

where $\beta = (RC)^{-1}$ and the gain factor corresponds to $(\gamma/\sigma)(\sigma + \beta)/(\gamma + \beta)$. By setting $\beta \approx \sigma$ and by retaining $\tau_M \approx 10/\sigma$, equation (26) has the approximate form

$$g_C^{(1)}(\tau) \approx \frac{1}{2} Ae^{-\sigma\tau} + \beta \frac{\gamma}{\sigma}\left[1 - \gamma\tau + \frac{1}{2}\gamma^2\tau^2\right] \tag{27}$$

equation (27) is extremely useful in retrieving σ in the presence of γ when $B \gg A$. Results of local entropy fluctuations in the neighborhood of the critical mixing point show no dramatic critical effects; but the method of detection is worth remembering.

7.5. MULTICOMPONENT FLUIDS

7.5.1. Ternary Liquid Mixtures

For a system with "hidden variables," the critical exponents become renormalized according to the formulas[83–86]

$$\alpha_x = -\alpha/(1 - \alpha) \tag{28}$$

$$\gamma_x = \gamma/(1 - \alpha) \tag{29}$$

$$\nu_x = \nu/(1 - \alpha) \tag{30}$$

Several detailed experimental works on the critical phenomena of ternary liquid mixtures (ethanol–water–chloroform[65, 87, 88] and bromobenzene–water–acetone[89]) have been reported. Earlier measurements on the renormalized values of γ and ν appear a bit too high. By using a digital photometer, as shown in Figure 4, and a flat sample cell having a 2-mm light path, we obtained $\gamma_x = 1.40 \pm 0.02$ and $\nu_x = 0.73 \pm 0.04$, which are in

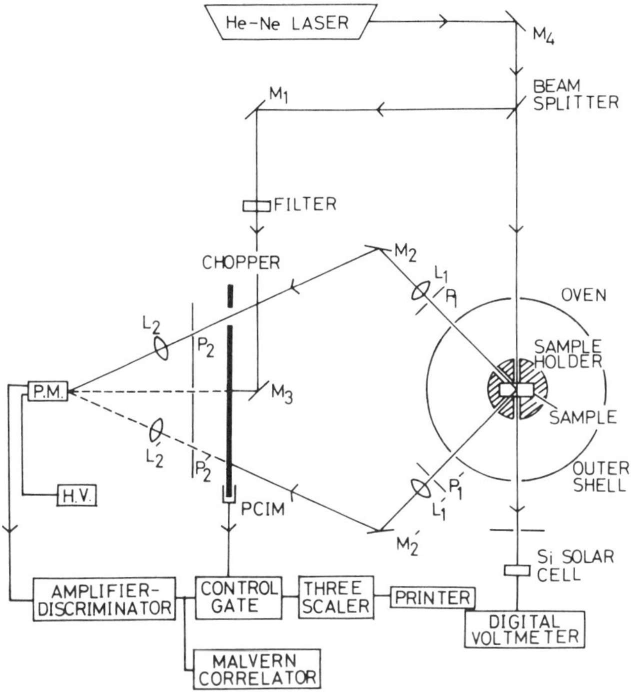

Figure 4. Experimental arrangement of a digital light-scattering photometer. The essential advantages of this system are that we can observe the scatterings from two different scattering angles and observe the laser monitoring light alternatively over short periods of time and that the only movable component is the chopper. M_1–M_4: mirrors; L_1–L_2: lenses; P_1–P_2': pinholes to define the coherence area; PCIM: photon-coupled interupter module; P.M.: ITT FW130 photomultiplier tube; H.V.: high-voltage power supply (Reference 88).

good agreement with the values of $\gamma_x = \gamma/(1 - \alpha) = 1.24/0.89 = 1.39$ and $\nu_x = \nu/(1 - \alpha) = 0.63/0.89 = 0.71$ obtained by renormalization of critical exponents according to the values listed in Table 1.

7.5.2. Binary Fluid in the Presence of Isotope Exchange

Experiments on phase diagrams and critical opalescence of a fluid mixture, isobutyric acid in D_2O, indicate that the presence of isotope exchange reactions can change the critical behavior of such a system from that of a simple binary fluid mixture.[40] Thermodynamically, we start with a system made up of two components, namely, pure isobutyric acid (ICO_2H) and pure D_2O. It should be noted that even isobutyric acid has more than two species although the amount of protons (H^+) and anions

(ICO_2^-) from dissociation of the weak acid is very small. In an isobutyric acid/D_2O fluid mixture, there are surely more than two species and the actual equilibrium concentrations of both the isobutyric acid and D_2O are not the same as the initial concentrations because of the isotope exchange reactions. Thus, the critical behavior should be considered in terms of a particular thermodynamic path with which the critical mixing point is approached.

In the one-fluid homogeneous phase region of the ICO_2H/D_2O system, we have taken the thermodynamic components to be ICO_2H and D_2O. The concentrations of the *equilibrium* species of ICO_2H and D_2O can change as a function of temperature. Therefore, the path represents the *initial* concentration used at each temperature if we were to prepare such a solution. Furthermore, the composition of the initial concentration is not unique. In principle, if we knew the isotope exchange reactions, we could prepare the same solution by mixing appropriate amounts of dissociated and exchanged species.

By using an equal-volume phase separation method, Y_c (in mass fraction of isobutyric acid) = 0.358 and T_c = 44.90°C differ from the concentration $Y_{T_{p,\max}} = 0.310$ at maximum phase separation temperature $T_{p,\max}$ of 45.11°C. In the one-phase region, $\gamma = 1.25$, $\nu = 0.633$, and $\xi_0 = 3.13$ Å, in excellent agreement with the theoretical predictions listed in Table 1. However, in the coexisting two-phase region, the critical exponents appear to be renormalized with $\gamma_x = 1.39$, $\nu_x \approx 0.76$, and $\xi_0 \approx 0.6$ Å, in agreement with the renormalized critical exponents $\gamma_x = 1.39$ and $\nu_x = 0.71$ or the results of γ_x and ν_x near the plait point of a ternary liquid mixture: ethanol–water–chloroform.

Before the introduction of $R(q, \xi)$ and R in equation (15), the simplified expression has the form

$$\Gamma_c = \frac{k_B T}{6\pi\eta_{\rm hf}^* \xi^3} K_0(x) \qquad (15a)$$

With corrections for the high-frequency viscosity ($\eta_{\rm hf}^*$), correlation function and vertex, equation (15a) has shown to be in good agreement with experiments.[47]

For the isobutyric acid/D_2O system, the universal curve $K(x)[= K_0(x)\bar{\eta}/\eta^*]$ remains applicable for the scaled linewidth $\Gamma_c \bar{\eta}/q^3$ in the one-phase region as well as along the coexistence curve. Figure 5 shows a plot of $\Gamma\bar{\eta}/q^3$ versus $q\xi$. The agreement suggests that the background linewidth is not important for this system. It should be noted that in the two-phase region along the coexistence curve, values of the critical exponents γ and ν have changed to the renormalized values of γ_x and ν_x, even though we do not understand fully why such a changeover should take place.

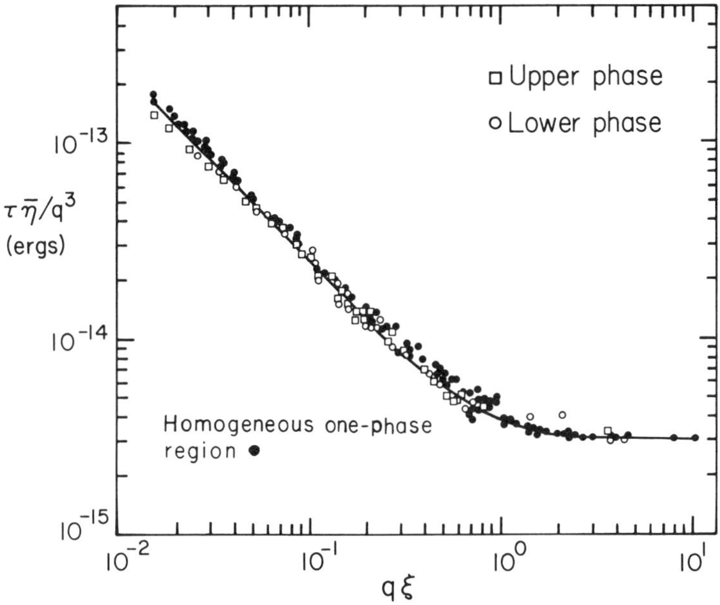

Figure 5. Plot of $\tau\bar{\eta}/q^3$ versus $q\xi$ for the isobutyric acid/D_2O mixture. (●), homogenous phase; solid line: Kawasaki universal function [B. Kalbskof, D. Woermann, B. Chu, and Erdogan Gulari, *J. Chem. Phys.* **74**(10), 5842 (1981)].

7.5.3. Tricritical Point Behavior

The simplest multicritical point, known as the tricritical point, has been observed in ordinary fluid mixtures[90–92] where symmetry breaking does not occur. A tricritical point[93] is a point at which three coexisting phases simultaneously become identical whereas an ordinary critical point is one at which two coexisting phases become identical. A classification scheme[94] and a search[95] for multicritical points in liquid mixtures have been reported. Das and Griffiths used a *five*-component mixture of ethylene glycol, water, lauryl alcohol, nitromethane, and nitroethane and observed a region of four coexisting liquid phases between 37 and 42°C. Nevertheless, there have been only very few fluid mixtures[91, 96–98] whose tricritical behavior has been investigated. The tricritical point of two such fluid systems are listed in Table 3.

In the mean field approximation, $\gamma_t = 1$, $\eta_t = 0$, and $\nu_t = 0.5$ if the path of approach to the tricritical point is at constant field; and $\gamma = 2$ and $\nu = 1$ if the path is along the constant composition. Rayleigh linewidth and intensity measurements on system I have been performed by Gollub *et al.*[99] and

Table 3. Tricritical Point of Two Four-Component Systems[a]

System I: C_6H_6–EtOH–H_2O–$(NH_4)_2SO_4$, $T_c = 48.9°C$	
$y_{salt} = 0.0175$;	$y_{H_2O} = 0.351$
$y_{EtOH} = 0.450$;	$y_{C_6H_6} = 0.181$
System II: C_6H_6–n-C_6H_{14}–H_2O–acetonitrile, $T_c = 89.68 \pm 0.03°C$	
$y_{acetonitrile} = 0.542$,	$y_{H_2O} = 0.154$
$y_{C_6H_{14}} = 0.138$,	$y_{C_6H_6} = 0.166$

[a] Minute variations in composition significantly affect the location of the phase boundaries. Therefore, measurements made in the immediate neighborhood of the tricritical point are less precise than those made near a critical point.

Kim *et al.*[100] According to equation (15), Γ_c in the hydrodynamic regime ($q\xi \ll 1$) has the form

$$\Gamma_c = \frac{k_B T}{6\pi\bar{\eta}\xi} q^2 \tag{15b}$$

where we have neglected $R(q, \xi)$ and set $R = 1$. In the measurements by Gollub *et al.*,[99] an expression

$$\frac{\Gamma}{q^2} = A\left(\frac{T - T^*}{T^*}\right)^{\nu} \tag{31}$$

was used where T^* ($\neq T_t$) was an adjustable parameter in order to account for the fact that the system was not at the tricritical composition. They obtained $\nu = 0.96 \pm 0.05$, indicating a very strong temperature dependence for ξ. From turbidity measurements, they estimated $\gamma \approx 2$, again in agreement with the mean field prediction. Based on the mean-field theory, Griffith[101] predicted two sum rules

$$\chi_\alpha^{1/2} + \chi_\gamma^{1/2} = \chi_\beta^{1/2} \tag{32a}$$

or

$$\xi_\alpha + \xi_\gamma = \xi_\beta \tag{32b}$$

and

$$\chi_\alpha^{-1/2} + \chi_\gamma^{-1/2} - \chi_\beta^{-1/2} \propto (T - T_t) \tag{33a}$$

or

$$\xi_\alpha^{-1} + \xi_\gamma^{-1} - \xi_\beta^{-1} \propto (T - T_t) \tag{33b}$$

Kim *et al.*[100] tested the sum rules using system I. After correction for double scattering the scattered intensities at a given temperature in a particular phase were analyzed according to equations (1) and (2) with

$$I(0) = C_0 \varepsilon^{-\gamma} = \text{const} \times [k_B T(\partial\varepsilon/\partial\psi)^2 \chi] \tag{34}$$

where ε, ψ, and χ are the dielectric constant, the order parameter, and the susceptibility, respectively. Near the tricritical point, $(\partial\varepsilon/\partial\psi)^2$ is replaced by the tricritical value and is assumed to be a constant. Thus, the first sum rule [equations (32a) and (32b) predicts that the ratios

$$R_I \equiv \frac{I_\alpha^{1/2} + I_\gamma^{1/2}}{I_\beta^{1/2}} \tag{35a}$$

$$R_\xi \equiv \frac{\xi_\alpha + \xi_\gamma}{\xi_\beta} \tag{35b}$$

should be constant everywhere in the three-phase region near the tricritical point and equal to 1 if we identify $(\partial\varepsilon/\partial\psi)$ to be the tricritical value. Kim *et al.* obtained $R_I = 1.2$ and $R_\xi = 1.3$, although the discrepancy can be explained partially by using actual $(\partial\varepsilon/\partial\psi)$ values in the three phases. However, the large values of R_ξ by both light scattering intensity and linewidth measurements suggest a violation of Griffiths' first sum rule beyond the estimated error limits, even though they[100] have not been able to rule out the possibility that the measurements were done sufficiently far from the asymptotic region where the rule applies. The second sum rule [equations (33a) and (33b)] was verified in coexisting phases at temperatures in a range of 1.5–1°C below the tricritical temperature.

7.6. SPINODAL DECOMPOSITION AND CRITICAL BEHAVIOR INDUCED BY SHEAR FLOW

Spinodal decomposition and critical behavior induced by external forces such as shear flow are two aspects of the field which have more promise for interesting experiments than the other topics I have discussed. I shall only refer the reader to the lectures[102, 103] presented at the NATO ASI on Scattering Techniques in August, 1980 at Wellesley College, Massachusetts. The proceedings of the conference have been published. Therefore, I have not attempted to review these two very important topics as it would have been a review of the reviews already made available by the main contributors.[102, 103]

7.6.1. Spinodal Decomposition

Recent experimental work on spinodal decomposition has been contributed mainly by Goldburg[102] and his co-workers[104–110] and by C. M. Knobler and his co-workers.[111, 112] Goldburg[118] has made a review of Cahn's linear theory of spinodal decomposition,[113] the nonlinear theory of Langer *et al.*,[114] the cluster dynamics of Binder[115] and of Binder and Stauffer,[116] and the hydrodynamic mode-coupling calculation of Kawasaki and Ohta.[117] Recent experiments on anomalous supercooling[110] near the critical point have led to needs for further refinements in detecting droplets in their initial stages of growth and in determining the droplet size distributions as a function of time.

7.6.2. Critical Behavior Induced by Shear Flow

Light scattering studies of critical fluctuations induced by shear flow have a relatively short history[103] with experimental measurements[118–121] and theoretical developments[122–124] progressing at about the same time.

In the region $S\tau > 1$, where S and τ are the shear rate and the lifetime of concentration fluctuations, respectively, the critical temperature $T_c(s)$ is lowered by the shear: $T_c(S) = T_c(0) - T_0 S^{\sigma_0}$ with $\sigma_0 = 1/3\nu$; the fluctuations become strongly anisotropic with respect to the flow direction, and the susceptibility in the flow direction is proportional to $[T - T_c(S))]^{-1}$. The qualitative agreement between experiments and theory is very good. Since the effects are observed only when $S\tau > 1$, it would be easier and interesting to study polymer mixtures first near its critical mixing point and then we may proceed to make measurements away from the critical behavior as anisotropy and reorientation of polymer molecules in solution induced by shear flow are important in other applications.

ACKNOWLEDGMENTS

I would like to thank many of my colleagues: D. Beysens, D. S. Cannell, W. I. Goldburg, K. Kawasaki, J. V. Sengers, and B. Widom, for sending me copies of their reprints and preprints, and in particular, J. V. Sengers for reading the manuscript and providing me with many valuable comments. Support of this work by the National Science Foundation, the U.S. Army Research Office, and the Petroleum Research Fund, administered by the American Chemical Society, is gratefully acknowledged.

REFERENCES

1. B. Chu, *Laser Light Scattering*, Academic Press, New York (1974).
2. H. E. Stanley, *Introduction to Phase Transitions and Critical Phenomena*, Oxford University Press, New York (1971); second edition in preparation.
3. C. Domb and M. S. Green, eds., *Phase Transitions and Critical Phenomena*, Vol. 1–6, Academic Press, New York (1972–1977).
4. A. Levelt Sengers, R. Hocken, and J. V. Sengers, *Phys. Today* **30**, 42 (1977).
5. S. C. Greer, *Acc. Chem. Res.* **11**, 427 (1978); S. C. Greer and M. R. Moldover, *Ann. Rev. Phys. Chem.* **32**, 233 (1981).
6. R. L. Scott, *Special Periodical Reports, Chemical Thermodynamics*, Vol. 2, Chap. 8, The Chemical Society, London (1978), pp. 238–274.
7. P. C. Hohenberg, in *Microscopic Structure and Dynamics of Liquids*, J. Dupuy and A. J. Dianoux, eds., Plenum Press, New York (1978), pp. 333–366.
8. K. G. Wilson and J. Kogut, *Phys. Rep.* **12C**, 75 (1974).
9. P. C. Hohenberg and B. I. Halperin, *Rev. Mod. Phys.* **49**, 435 (1977).
10. M. R. Moldover, J. V. Sengers, R. W. Gammon, and R. J. Hocken, *Rev. Mod. Phys.* **51**, 79 (1979).
11. J. V. Sengers and L. M. H. Levelt Sengers, in *Progress in Liquid Physics*, C. A. Croxton, ed., John Wiley & Sons, New York (1978), pp. 103–174. There are also several other review articles on critical phenomena in this volume.
12. K. Kawasaki and J. D. Gunton, in *Progress in Liquid Physics*, C. A. Croxton, ed., John Wiley & Sons, New York (1978), pp. 175–211.
13. F. J. Wegner, *Phys. Rev. B* **5**, 4529 (1972).
14. M. Ley-Koo and M. S. Green, *Phys. Rev. A* **16**, 2483 (1977).
15. F. W. Balfour, J. V. Sengers, M. R. Moldover, and J. M. H. Levelt Sengers, in *Proceedings of the 7th Symposium on Thermophysical Properties*, A. Cezairliyan, ed., ASME, New York (1977), p. 786; *Phys. Lett.* **65A**, 223 (1978).
16. J. M. H. Levelt Sengers and J. V. Sengers, in *Perspectives in Statistical Physics*, M. S. Green Memorial Issue of Studies in Statistical Mechanics, H. J. Raveche, ed., North-Holland, Amsterdam (1980).
17. G. A. Baker, B. G. Nickel, M. S. Green, and D. I. Meiron, *Phys. Rev. Lett.* **36**, 1351 (1976); G. A. Baker, B. G. Nickel, and D. I. Meiron, *Phys. Rev. B* **17**, 1365 (1978).
18. J. C. LeGuillou and J. Zinn-Justin, *Phys. Rev. Lett.* **39**, 95 (1977); *Phys. Rev. B* **21**, 3976 (1980).
19. M. Nakata, T. Dobashi, N. Kuwahara, M. Kaneko, and B. Chu, *Phys. Rev. A* **18**, 2683 (1978).
20. D. Beysens, *J. Chem. Phys.* **71**, 2557 (1979); D. Beysens and A. Bourgou, *Phys. Rev. A* **19**, 2407 (1979).
21. (a) R. F. Chang, H. Burstyn, J. V. Sengers, and A. J. Bray, *Phys. Rev. Lett.* **37**, 1481 (1976); (b) R. F. Chang, H. Burstyn, and J. V. Sengers, *Phys. Rev. A* **19**, 866 (1979).
22. H. Guttinger and D. S. Cannell, *Phys. Rev. A* **24**, 3188 (1981).
23. J. R. Hastings, J. M. H. Levelt Sengers, and F. W. Balfour, *J. Chem. Thermodyn.* **12**, 1009 (1980).
24. J. M. H. Levelt Sengers, B. Kamgar-Parsi, F. W. Balfour, and J. V. Sengers, *J. Phys. Chem. Ref. Data* **12**, 1 (1983).
25. P. Calmettes, *J. Phys. (Paris)* **40**, L-535 (1979).
26. D. Beysens, R. Tufeu, and Y. Garrabos, *J. Phys. (Paris)* **40**, L-623 (1979).
27. N. Kuwahara, D. V. Fenby, M. Tamsky, and B. Chu, *J. Chem. Phys.* **55**, 1140 (1971).

28. *Phase Transitions: Cargese 1980*, M. Levy, J. C. Le Guillou, and J. Zinn-Justin, eds., Plenum Press, New York (1982). For examples, see the articles by M. R. Moldover, p. 63, and J. V. Sengers, p. 95.
29. H. L. Swinney and D. L. Henry, *Phys. Rev. A* **8**, 2586 (1973).
30. M. E. Fisher and J. S. Langer, *Phys. Rev. Lett.* **20**, 665 (1968).
31. M. E. Fisher and R. J. Burford, *Phys. Rev. B* **10**, 2818 (1974).
32. C. A. Tracy and B. M. McCoy, *Phys. Rev.* **12**, 368 (1975).
33. R. A. Ferrell and D. J. Scalapino, *Phys. Rev. Lett.* **34**, 200 (1975).
34. A. J. Bray, *Phys. Lett.* **55A**, 453 (1976); *Phys. Rev. Lett.* **36**, 285 (1976); *Phys. Rev. B* **14**, 1248 (1976).
35. M. E. Fisher and A. Aharony, *Phys. Rev. B* **10**, 2818 (1974).
36. W. J. Camp, D. M. Saul, J. P. Van Dyke, and M. Wortis, *Phys. Rev. B* **14**, 3990 (1976).
37. B. Chu and W. P. Kao, *J. Chem. Phys.* **42**, 2608 (1965); S. P. Lee and B. Chu, *J. Chem. Phys.* **60**, 2940 (1974); see also B. Chu, *Ber. Bunsenges. phys. Chem.* **76**, 202 (1972).
38. H. D. Bale, J. S. Lin, D. A. Dolejsi, J. L. Casteel, O. A. Pringle, and P. W. Schmidt, *Phys. Rev. A* **15**, 2513 (1977).
39. E. Gürmen, M. Chandrasekhar, P. E. Chambley, H. D. Bale, D. A. Dolejsi, J. S. Lin, and P. W. Schmidt, *Phys. Rev. A* **22**, 170 (1980).
40. E. Gulari, B. Chu, and D. Woermann, *J. Chem. Phys.* **73**, 2480 (1980).
41. R. Schneider, L. Belkoura, J. Schelten, D. Woermann, and B. Chu, *Phys. Rev. A* **22**, 5507 (1980).
42. J. Schelten, *Kerntechnik* **14**, 86 (1972).
43. J. A. Lake, *Acta Crystallogr.* **23**, 191 (1967).
44. (a) K. Kawasaki, *Ann. Phys. (N.Y.)* **61**, 1 (1970); (b) K. Kawasaki and S. M. Lo, *Phys. Rev. Lett.* **29**, 48 (1972).
45. D. W. Oxtoby and W. M. Gelbart, *J. Chem. Phys.* **61**, 2957 (1974).
46. E. D. Siggia, B. I. Halperin, and P. C. Hohenberg, *Phys. Rev. B* **13**, 2110 (1976).
47. B. Chu, S. P. Lee, and W. Tscharnuter, *Phys. Rev. A* **7**, 353 (1973).
48. S. H. Chen, C. C. Lai, and J. Rouch, *J. Chem. Phys.* **68**, 1994 (1978).
49. C. M. Sorensen, R. C. Mockler, and W. J. O'Sullivan, *Phys. Rev. Lett.* **40**, 777 (1978).
50. (a) H. C. Burstyn, J. V. Sengers, and P. Esfandiari, *Phys. Rev. A* **22**, 282 (1980); (b) a closed-form approximant for the dynamic scaling function has been proposed by H. C. Burstyn, J. V. Sengers, J. K. Bhattacharjee, and R. A. Ferrell, *Phys. Rev. A* **28**, 1567 (1983).
51. H. Guttinger and D. S. Cannell, *Phys. Rev. A* **22**, 285 (1980).
52. K. Hamano, T. Namura, T. Kawazura, and N. Kuwahara, *Phys. Rev. A* **26**, 1153 (1982); S. H. Chen, C. C. Lai, J. Rouch, and P. Tartaglia, *Phys. Rev. A* **27**, 1086 (1983).
53. B. Volochine, P. Bergé, and I. Lagues, *Phys. Rev. Lett.* **25**, 1414 (1970).
54. R. F. Chang, P. H. Keyes, J. V. Sengers, and C. O. Alley, *Phys. Rev. Lett.* **27**, 1706 (1971).
55. R. Perl and R. A. Ferrell, *Phys. Rev. Lett.* **29**, 51 (1972); *Phys. Rev. A* **6**, 2358 (1972).
56. S. M. Lo and K. Kawasaki, *Phys. Rev. A* **8**, 2176 (1973); T. Ohta and K. Kawasaki, *Prog. Theor. Phys.* **55**, 1384 (1976).
57. D. Wonica, H. L. Swinney, and H. Z. Cummins, *Phys. Rev. Lett.* **37**, 66 (1976).
58. C. M. Sorensen, R. C. Mockler, and W. J. O'Sullivan, *Phys. Rev. A* **16**, 365 (1977).
59. H. C. Burstyn, R. F. Chang, and J. V. Sengers, *Phys. Rev. Lett.* **44**, 410 (1980); H. C. Burstyn and J. V. Sengers, *Phys. Rev. A* **27**, 1071 (1983).
60. J. K. Bhattacharjee and R. A. Ferrell, *Phys. Rev. A* **23**, 1511 (1981).
61. D. Beysens and G. Zalczer, *Phys. Rev. A* **15**, 765 (1977).
62. R. A. Ferrell and J. K. Bhattacharjee, *Phys. Rev. A* **19**, 348 (1979).
63. T. Ohta, *J. Phys. C* **10**, 791 (1977).

64. J. D. Gunton, in *Dynamical Critical Phenomena and Related Topics*, C. P. Enz, ed., Springer-Verlag, New York (1979), p. 1.
65. B. Chu and F. L. Lin, *J. Chem. Phys.* **61**, 739, 5132 (1974).
66. H. C. Burstyn and J. V. Sengers, *Phys. Rev. Lett.* **45**, 259 (1980).
67. A. K. Atakhodzhaev, L. M. Kashaeva, L. M. Sabirov, V. S. Starunov, T. M. Utarova, and I. L. Fabelinskii, *Zh. Eksp. Teor. Fiz. Pisma Red.* **17**, 95 (1973).
68. I. L. Fabelinskii, V. S. Starunov, A. K. Atakhodzhaev, L. M. Sabirov, and T. M. Utarova, *Opt. Commun.* **15**, 432 (1975).
69. G. I. Kolesnikov, V. S. Starunov, and I. L. Fabelinskii, *Zh. Eksp. Teor. Fiz. Pisma Red.* **24**, 73 (1976).
70. D. Beysens, A. Bourgou, and G. Zalczer, *Opt. Commun.* **15**, 436 (1975); *J. Phys. (Paris) Colloq. C* **1**, 225 (1976).
71. D. Beysens, A. Bourgou, and H. Charlin, *Phys. Lett.* **53A**, 236 (1975).
72. D. W. Oxtoby and W. M. Gelbart, *J. Chem. Phys.* **60**, 3359 (1974).
73. I. L. Fabelinskii, G. I. Kolesnikov, V. J. Schreiner, and V. S. Starunov, *Phys. Lett.* **59A**, 408 (1976).
74. I. L. Fabelinskii, G. I. Kolesnikov, and V. S. Starunov, *Opt. Commun.* **20**, 130 (1977).
75. I. L. Fabelinskii, V. S. Starunov, A. K. Atakhodzaev, L. M. Sabirov, and J. M. Utarova, *Opt. Commun.* **20**, 135 (1977).
76. (a) D. Beysens and G. Zalczer, *Opt. Commun.* **22**, 236 (1977); (b) *Phys. Rev. A* **18**, 2280 (1978).
77. G. Zalczer and D. Beysens, *J. Chem. Phys.* **72**, 348 (1980).
78. J. R. Petrula, H. L. Strauss, K. Q. H. Lao, and R. Pecora, *J. Chem. Phys.* **68**, 623 (1978).
79. G. D. Phillies, P. J. Chappell, and D. Kivelson, *J. Chem. Phys.* **68**, 4031 (1978).
80. I. M. Arefev, *Opt. Commun.* **10**, 277 (1974).
81. D. Beysens, *J. Chem. Phys.* **64**, 2579 (1976).
82. P. Calmettes and C. Laj, *Phys. Rev. Lett.* **36**, 1372 (1976); *Opt. Commun.* **19**, 257 (1976).
83. M. E. Fisher, *Phys. Rev.* **176**, 257 (1968).
84. M. E. Fisher and P. E. Scesney, *Phys. Rev. A* **2**, 825 (1970).
85. B. Widom, *J. Chem. Phys.* **46**, 3324 (1967).
86. J. C. Wheeler and B. Widom, *J. Am. Chem. Soc.* **90**, 3024 (1968).
87. F. L. Lin, D. Thiel, and B. Chu, *Phys. Lett.* **47A**, 479 (1974).
88. K. Ohbayashi and B. Chu, *J. Chem. Phys.* **68**, 5066 (1978).
89. C. S. Bak, W. I. Goldburg, and P. N. Pusey, *Phys. Rev. Lett.* **25**, 1420 (1970).
90. See references given in R. B. Griffiths and B. Widom, *Phys. Rev. A* **8**, 2173 (1973); B. Widom, *J. Phys. Chem.* **77**, 2196 (1973).
91. J. C. Lang, Jr. and B. Widom, *Physica (Utrecht)* **81A**, 190 (1975).
92. J. L. Creek, C. M. Knobler, and R. L. Scott, *J. Chem. Phys.* **67**, 366 (1977).
93. R. B. Griffiths, *J. Chem. Phys.* **60**, 195 (1974).
94. R. B. Griffiths, *Phys. Rev. B* **12**, 345 (1975).
95. B. K. Das and R. B. Griffiths, *J. Chem. Phys.* **70**, 5555 (1979).
96. P. Bocko, *Fluid Phase Equilibria* **4**, 137 (1979).
97. S. I. Sinegubova and N. I. Nikurashina, *Russ. J. Phys. Chem.* **51**, 1230 (1977); A. I. Gei, S. I. Sinegubova, and N. I. Nikurashina, *Russ. J. Phys. Chem.* **51**, 1231 (1977).
98. N. I. Nikurashina, S. I. Sinegubova, and N. B. Trifonona, *Russ. J. Phys. Chem.* **52**, 1231 (1978).
99. J. P. Gollub, A. A. Koenig, and J. S. Huang, *J. Chem. Phys.* **65**, 639 (1976).
100. M. W. Kim, W. I. Goldburg, P. Esfandiari, and J. M. H. Levelt Sengers, *J. Chem. Phys.* **71**, 4888 (1979); *Phys. Rev. Lett.* **44**, 80 (1980).
101. M. Kaufman, K. K. Bardhan, and R. B. Griffiths, *Phys. Rev. Lett.* **44**, 77 (1980).

102. W. I. Goldburg, *The Dynamics of Phase Separation near the Critical Point: Spinodal Decomposition and Nucleation*, NATO Advanced Study Institute on Scattering Techniques Applied to Supramolecular and Nonequilibrium Systems, S. H. Chen, B. Chu, and R. Nossal, eds., Plenum Press, New York (1981).
103. D. Beysens, *New Critical Behavior Induced by Shear Flow in Binary Fluids*, NATO Advanced Study Institute on Scattering Techniques Applied to Supramolecular and Nonequilibrium Systems, S. H. Chen, B. Chu, and R. Nossal, eds., Plenum Press, New York (1981).
104. J. S. Huang, W. I. Goldburg, and A. W. Bjerkaas, *Phys. Rev. Lett.* **32**, 921 (1974).
105. W. I. Goldburg and J. S. Huang, in *Fluctuations, Instabilities and Phase Transitions*, T. Riste, ed., Plenum Press, New York (1975), p. 87.
106. M. W. Kim, A. J. Schwartz, and W. I. Goldburg, *Phys Rev. Lett.* **41**, 657 (1978).
107. W. I. Goldburg, C.-H. Shaw, J. S. Huang, and M. S. Pilant, *J. Chem. Phys.* **68**, 484 (1978).
108. W. I. Goldburg, A. J. Schwartz, and M. W. Kim, *Prog. Theor. Phys. Suppl.* **64**, 477 (1978).
109. Y. C. Chou and W. I. Goldburg, *Phys. Rev. A* **20**, 2105 (1979).
110. A. J. Schwartz, S. Krishnamurthy, and W. I. Goldburg, *Phys. Rev. A* **21**, 1331 (1980).
111. N.-C. Wong and C. M. Knobler, *J. Chem. Phys.* **66**, 4707 (1977); **69**, 725 (1978); *Phys. Rev. Lett.* **43**, 1733 (1979).
112. R. G. Howland, N.-C. Wong, and C. M. Knobler, *J. Chem. Phys.* **73**, 522 (1980).
113. J. W. Cahn, *Trans. Metall. Soc. AIME* **242**, 166 (1968); J. W. Cahn and J. E. Hilliard, *J. Chem. Phys.* **28**, 258 (1958).
114. J. S. Langer, M. Baron, and H. D. Miller, *Phys. Rev. A* **11**, 1417 (1975).
115. K. Binder, *Phys. Rev. B* **15**, 4425 (1977).
116. K. Binder and D. Stauffer, *Phys. Rev. Lett.* **33**, 1006 (1974); *Adv. Phys.* **25**, **343** (1976).
117. K. Kawasaki and T. Ohta, *Prog. Theor. Phys.* **59**, 362 (1978).
118. W. I. Goldburg, in *Light Scattering near Phase Transitions*, H. Z. Cummins and A. P. Levanyuk, eds., North-Holland, Amsterdam (1983).
119. D. Beysens, M. Gbadamassi, and L. Boyer, *Phys. Rev. Lett.* **43**, 1253 (1979); D. Beysens and M. Gbadamassi, *J. Phys. (Paris) Lett.* **40**, L565 (1979).
120. D. Beysens, in *Ordering in Strongly Fluctuating Condensed Matter Systems*, T. Riste, ed., NATO ASI 1979, Plenum Press, New York (1979).
121. D. Beysens and M. Gbadamassi, *Phys. Lett.*, **77A,** 171 (1980).
122. A. Onuki and K. Kawasaki, *Prog. Theor. Phys. Suppl.* **64**, 436 (1978); *Phys. Lett.* **A72**, 233 (1979); *Ann. Phys. (N.Y.)* **121**, 456 (1979).
123. K. Kawasaki and A. Onuki, in *Lecture Notes in Physics 104: Dynamical Critical Phenomena and Related Topics*, C. Enz, ed., (Springer, Heidelberg (1979).
124. A. Onuki and K. Kawasaki, *Prog. Theor. Phys.* **53**, 122 (1980).

8

Laser Light Scattering in Micellar Systems

Norman A. Mazer†

Department of Physics
Massachusetts Institute of Technology
Cambridge, Massachusetts 02139

Department of Medicine
Brigham and Women's Hospital
Boston, Massachusetts 02115

8.1. INTRODUCTION

Amphiphiles (i.e., molecules possessing both hydrocarbon and polar moieties) display a variety of aggregative properties when dissolved in aqueous and nonaqueous solvents. Such behavior in general reflects the tendency for amphiphiles to associate into macromolecular structures in which solvent contact is minimized with those moieties that interact poorly with the solvent and maximized with those moieties that interact favorably. In the case of aqueous systems, such amphiphile aggregates were first termed micelles by McBain[1] in 1913. Early views on micelle structure and thermodynamics were developed by Hartley,[2] who investigated the properties of aqueous solutions of paraffin chain salts nearly 50 years ago. Using a variety of physical chemical techniques, including diffusion methods, Hartley proposed that micelles in dilute solution had a spherical structure in which the hydrocarbon chains formed a liquidlike core inside the micelle with the polar head groups of the amphiphiles located at the micelle surface (see Figure 1). Hartley attributed the driving force for micellization to the unfavorable interaction between hydrocarbon and water, the so-called "hydrophobic effect"[3], and further recognized that this force would be opposed, in the case of ionic amphiphiles, by the electrostatic interactions in the micellar surface. These views have been largely confirmed by many

† Present address: Sandoz Ltd., Pharmaceutical Development, Clinical Biophysics Research, CH-4002 Basel, Switzerland.

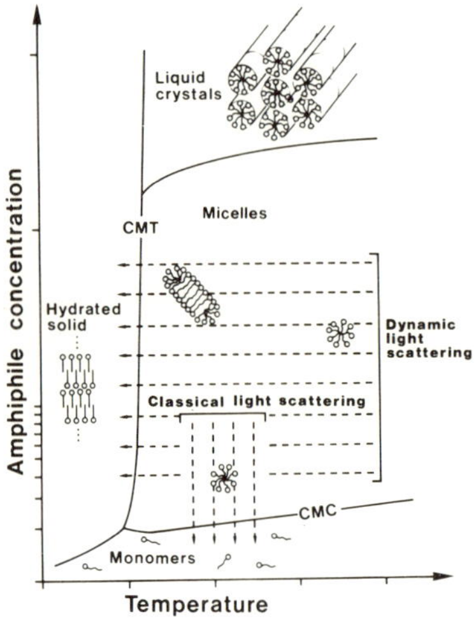

Figure 1. Schematic phase diagram for an ionic alkyl chain amphiphile. Micellar phase exists at temperatures above the critical micellar temperature (CMT), and at concentrations greater than the CMC. Spherical micelles, as envisioned by Hartley,[2] are formed near the CMC and at high temperature. Rodlike micelles are formed at low temperature and at high concentration. Dynamic light scattering methods can probe the micellar phase at high concentrations, whereas classical light scattering is useful near the CMC. Also shown are the regions where the hydrated solid and hexagonal liquid crystalline phases appear.

subsequent experimental studies and provide the starting point of modern thermodynamic treatments of micelle formation. An excellent review of the current literature in both aqueous and nonaqueous amphiphile systems can be found in the monographs of Lindman and Wennerström[4, 5] and of Eicke.[6]

The use of light scattering methods for investigating micellar systems was pioneered by Debye in the 1940s. In a review paper entitled "Light scattering in soap solutions"[7], Debye illustrated the kind of information obtainable from classical light scattering methods (i.e., measurements of the mean intensity of the scattered light, $\bar{I}$). From the dependence of $\bar{I}$ on amphiphile (or detergent) concentration, C, he showed that the critical concentration at which micelles begin to form, denoted CMC, could be readily determined. For $C < \text{CMC}$, the system contains only small detergent monomers (M.W. $\lesssim 500$) which scatter the light weakly, whereas for $C > \text{CMC}$ the monomers cooperatively associate to form large micellar aggregates (M.W. $\gtrsim 10{,}000$), which scatter the light strongly. A discontinuity in the slope of $\bar{I}$ vs. C thus defines experimentally the location of the CMC. For the concentration region above the CMC, Debye assumed a parallelism between the scattering of light from micellar solutions and from polymer (or protein) solutions.

Through this analogy the concentration dependence of $\bar{I}(C)$ could be used to deduce the micellar weight M and second osmotic virial coefficient of the systems, A_2, according to the relation[7]

$$\frac{C - \text{CMC}}{\bar{I}(C) - \bar{I}(\text{CMC})} = \alpha\left[\frac{1}{M} + 2A_2(C - \text{CMC})\right] \tag{1}$$

where α is an experimental constant that depends on the refractive increment of the micelles dn/dc, incident wavelength λ, and scattering volume. Implicitly assumed in Equation (1) are that the micellar parameters M and A_2 are constants, independent of concentration above the CMC, and that the concentration of detergent monomers in equilibrium with the micelles remains equal to the CMC. Debye and co-workers[8] later showed that additional information on micellar size (and shape) could be obtained if the micelles were sufficiently large to produce a measurable angular variation in $\bar{I}$.

Using Debye's general approach, numerous investigators, notably Mysels,[9] Huisman,[10] Kratohvil,[11] and others,[12,13] have deduced CMC, M and A_2 for a variety of micellar systems. Most studied among these is the anionic detergent sodium dodecyl sulfate (SDS), in which the dependence of these parameters on added salt concentration has been investigated[9,13] in order to gain insight into the electrostatic effects which govern both micelle formation and intermicellar interactions (as inferred by A_2). While such studies have provided important knowledge about micellar systems and have also led to advances in the theory and practice of classical light scattering, it must be acknowledged that the assumptions inherent in the Debye equation have severely limited the information obtainable by classical light scattering techniques. In particular the assumption that the properties of the micelle (i.e., M and A_2) remain constant above the CMC has been shown by Mukerjee,[14] Israelachvilli,[15] and others[16] to be untrue for most systems. The deductions of M and A_2 can thus only provide estimates of micellar weight and interactions that have been extrapolated to the CMC. However, as the interaction effects are often sizeable and can obscure the variation of M with C, the possibilities for systematic errors in the extrapolations do exist, as has been pointed out by Mukerjee.[14] For these reasons the classical light scattering methods have been unable to provide reliable information on how micellar systems behave at concentrations much greater than the CMC, where transitions in micellar size, shape, and structure may be occurring.

While all techniques for studying macromolecules (or micelles) are to varying degrees influenced by macromolecular interactions, it has become appreciated[17–19] that the diffusion coefficients obtained from dynamic light scattering experiments[20] are much less affected than many other parameters (i.e., the scattered intensity $\bar{I}$, osmotic pressure, or sedimentation coefficient). Using the Stokes–Einstein relation,[21] one can therefore calculate from the diffusion coefficient the apparent hydrodynamic radius of the micelles which, under many conditions, will provide a valid estimate of micellar size even at concentrations substantially greater than the CMC.[22,23] In addition dynamic light scattering can also provide measures of the polydispersity of micellar systems,[22] information which has been difficult to obtain by conventional techniques. Finally, by combining both dynamic light scattering and classical light scattering experiments on the same micellar systems, it becomes possible to make deductions concerning micellar shape,[22] aggregation number,[22] and micellar interactions[24] in regions of the micellar phase diagram (see Figure 1) that were previously inaccessible by classical light scattering alone. By these new approaches, modern laser light scattering is becoming a powerful tool for studying micelles and is being applied to a wide variety of synthetic[25–29] and biological[30–34] micellar systems where the concentration regions of interest are often much greater than the CMC. The information which has been derived from these applications, moreover, is providing new insights into the structure, thermodynamics, and interactions of micellar systems and is also yielding valuable clues to the physiological actions of various biological amphiphiles.

In this chapter, a brief summary of the experimental and theoretical basis by which laser light scattering methods provide information on micellar size, polydispersity, and shape will first be given. Following that, I shall present a review of the major applications of these methods to the areas of

- Aqueous synthetic detergent systems;
- Biological micelles;
- Microemulsion and inverted micellar systems.

In addition to presenting the major experimental results obtained in these studies, a brief discussion of the new theoretical insights which have developed from them will also be given. In so doing it is hoped that the reader will see the significant impact that laser light scattering techniques have recently made in elucidating some of the fascinating aspects of molecular aggregation displayed by amphiphilic molecules.

8.2. THEORETICAL ASPECTS OF DEDUCING MICELLAR SIZE, POLYDISPERSITY, AND SHAPE

The application of dynamic light scattering methods[20] for measuring translational diffusion coefficients, D, in micellar systems is relatively straightforward.[22, 35] As these systems are generally polydisperse, the autocorrelation function of the scattered light intensity will contain a sum of exponentials[36]:

$$R(\tau) = \left(\sum G_i e^{-D_i q^2 \tau} \right)^2 \qquad (2)$$

where D_i is the diffusion coefficient and G_i is the relative intensity scattered by the ith micellar species. The magnitude of the scattering vector, q, will be given by

$$q = \frac{4\pi \sin(\theta/2)}{\lambda/n} \tag{3}$$

where θ is the scattering angle, λ the incident wavelength, and n the index of refraction of the medium. If the polarizability per unit mass of all micelles is constant, then G_i will equal[22]

$$G_i = \frac{C_i M_i P_i(q)}{\sum C_i M_i P_i(q)} \tag{4}$$

where C_i is the concentration (weight/volume), M_i the micellar weight, and $P_i(q)$ the scattering form factor of the ith species.

Together, equations (2)–(4) relate the autocorrelation function to the micellar size distribution C_i, micellar weights M_i, diffusion coefficients D_i, and form factors $P_i(q)$. Alternatively, one could express the spectrum of the scattered light in terms of these parameters and in that case would obtain a sum of Lorentzians. In either case, one wishes to extract information on the set of D_i's and G_i's characterizing the sample. For autocorrelation function measurements, this is most generally done by performing a cumulants analysis of the data[36, 37] which provides information on the moments of D_i:

$$\overline{D^j} = \sum G_i D_i^j \tag{5}$$

Typically one is able to measure the first 2 or 3 moments in a micellar system[22, 25, 30] and thus deduce the mean diffusion coefficient, $\bar{D}$, and characterize the micellar polydispersity in terms of the variance, V, and skewness, S, which are defined as

$$V = 100 \frac{(\overline{D^2} - \bar{D}^2)^{0.5}}{\bar{D}} \% \tag{6a}$$

$$S = 100 \frac{\overline{D^3} - 3\overline{D^2}\bar{D} + 2\bar{D}^3}{(\overline{D^2} - \bar{D}^2)^{1.5}} \% \tag{6b}$$

From equations (4) and (5) it can be seen that $\bar{D}$ will approximate the so-called Z-average diffusion coefficient under conditions where the $P_i(q)$ values remain close to unity (this is guaranteed if $\bar{D}$ is extrapolated to

$\theta = 0$). In any case the $\bar{D}$ value can be used to provide information on micellar size, by using the Stokes–Einstein relation to deduce the apparent mean hydrodynamic radius $\bar{R}_h$[22]

$$\bar{D} = \frac{kT}{6\pi\eta\bar{R}_h} \tag{7}$$

where kT is the thermal energy, and η is the viscosity of the *solvent*. If it can be assumed that the diffusion coefficient of each micellar species, D_i, is related to the hydrodynamic radius $(R_h)_i$, of that species by the Stokes–Einstein relation, then the mean value, $\bar{R}_h$, defined by equation (7) will be equal to (6):

$$\bar{R}_h = \frac{\sum C_i M_i P_i(q)}{\sum C_i M_i P_i(q)(R_h)_i^{-1}} \tag{8}$$

The validity of equation (8) rests on the assumption that one can neglect the effects of intermicellar interactions on the diffusion coefficients D_i and on the mean value $\bar{D}$. A large body of experimental[17–19] and theoretical analysis [17, 38, 39] supports this belief under conditions where the macromolecules (or in the present case, micelles) interact as hard spheres. A brief development of the theory is now given. For a monodisperse system of nonaggregating macromolecules, the dependence of the mutual translational diffusion coefficient measured by dynamic light scattering on particle concentration C (in units of weight/volume) will be given by a generalized Stokes–Einstein relation[18, 38]

$$D(C) = \frac{\partial\pi/\partial C}{f(C)}(1 - vC) \tag{9}$$

where $\partial\pi/\partial C$ is the osmotic compressibility, $f(C)$ the hydrodynamic friction factor, and $(1 - vC)$ is a reference frame correction factor (v is the partial specific volume of the macromolecule). The concentration dependences of $\partial\pi/\partial C$ and $f(C)$ both arise from macromolecular interactions and can be expressed as virial expansions in the concentration, with coefficients that need not be the same. To first order these expansions are given by[17, 23]

$$\frac{\partial\pi}{\partial C} = kT(1 + 2A_2 MC) \tag{10a}$$

$$f(C) = f_0(1 + B_f C) = 6\pi\eta R_h(1 + B_f C) \tag{10b}$$

where A_2 is the second osmotic virial coefficient [also appearing in the "Debye equation" (1)], M the macromolecular weight, f_0 the frictional factor at infinite dilution (equal to $6\pi\eta R_h$), and B_f the virial coefficient of the frictional interactions.

Substituting (10a, b) into (9) one obtains

$$D(C) = \frac{kT}{6\pi\eta R_h}\left[\frac{1 + 2A_2 MC}{1 + B_f C}(1 - vC)\right] \tag{11}$$

The quantity in brackets thus accounts (to first order) for the effect of macromolecular interaction on the concentration dependence of D. In the case of hard-sphere interaction $2A_2 M$ is rigorously calculated[17] to be $8v$, while B_f has been shown in various theories[39, 40] to range from $6v$ to $7v$. As a result of the near identity of these quantities, the bracketed term remains close to unity, even for C values as large as 10 g/dl. Such behavior is in fact observed experimentally for a variety of macromolecules that interact primarily as hard spheres.[18, 41] When repulsive electrostatic interactions are considered it is typically found[17] that $2A_2 M$ can substantially exceed $8v$, with only a slight increase in B_f. As a result $D(C)$ will increase with concentration. However, such electrostatic effects can be virtually eliminated[17] for charged macromolecules by adding neutral electrolyte to the systems in a quantity sufficient to screen the electrostatic field ($\sim$0.4 M). Under these conditions the bracketed term will again approximate unity and the variation of $D(C)$ [or $\bar{D}(C)$ in polydisperse micellar systems] should reflect the concentration dependence of $\bar{R}_h$. It should be noted that aggregation theory[14–16] predicts that in systems containing a single amphiphilic compound, $\bar{R}_h$ can only increase or remain constant as a function of C. Thus an observed increase in $\bar{D}$ can never reflect a decrease in $\bar{R}_h$, but must result from repulsive interactions. Conversely a decrease in $\bar{D}(C)$ is generally an indication of micellar growth with concentration,[25, 30] although in some particular cases it could result from micellar interactions that are attractive in nature.[29] While in general it must be appreciated that micellar growth and interactions can both contribute to $\bar{D}(C)$, it can still be seen from equation (11) that the contribution of the latter will always be reduced (relative to the $\bar{I}$ measurements) due to the tendency of the thermodynamic and frictional interactions to cancel each other.

Deductions of micellar shape[22, 30, 31, 42] can often be made by combining the values of $\bar{R}_h$ with information obtained from measurements of the scattered intensity, $\bar{I}$, and its angular dependence. At concentrations large compared to the CMC (where the contributions of monomers can be neglected) equation (1) implies that $\bar{I}$ will be given by

$$\bar{I} = \alpha^{-1}\mathrm{CMP}(q)\left(\frac{1}{1 + 2A_2 MC}\right) \tag{12}$$

where the scattering form factor $P(q)$ has been introduced to give the angular dependence of the intensity.[8, 20] Although Equation (12) neglects polydispersity and the so-called preferential interactions[11, 43] between micelles and electrolyte, it nevertheless provides a useful starting point in analyzing $\bar{I}$ data. For example in systems where the variables of temperature or added salt exert strong effects on the micellar size distribution, such effects will primarily influence $\bar{I}$ through the product $M \cdot P$ of equation (12). The bracketed quantity, which accounts for micellar interactions, is only weakly influenced at constant C and therefore can be regarded as a constant factor. Thus for such systems the variation of $\bar{I}$ at constant C can be compared with theoretical predictions of the product $M \cdot P$ that are derived from the variation of $\bar{R}_h$ using different models of micellar shape. As will be seen later, this approach has been successfully employed to demonstrate a sphere-to-rod transition in the SDS system, at concentrations much greater than the CMC.[22] A similar means for deducing micellar shape is useful for micelles that are sufficiently large to produce an angular dependence in $\bar{I}$. This dependence can be measured either as the form factor $P(q)$ or as the dyssymmetry ratio $d(\theta)$ (defined as the ratio of P at supplementary angles: θ, $\pi - \theta$). From these measurements, one can deduce the mean radius of gyration of the micelles, $\bar{R}_g$, which will be related to the size, shape, and polydispersity of the micelles in a different manner than $\bar{R}_h$. In the case where $P(q)$ is measured, $\bar{R}_g$ can be derived by the limiting behavior at small q according to the relation[20]

$$P^{-1}(q) \simeq 1 + \tfrac{1}{3}q^2\bar{R}_g^2 \tag{13}$$

This relation is valid for any micellar shape in the limit $q\bar{R}_g \lesssim 1$. In practice the range of q values that are experimentally accessible ($\sim 5 \times 10^{-4}$ to 3×10^{-3} Å^{-1}) determine a minimum detectable $\bar{R}_g$ of about 100 Å. If $\bar{R}_g$ can be detected, then one can again compare the variation of $\bar{R}_g$ with $\bar{R}_h$ as means for deducing micellar shape.[42] Having deduced the micellar shape one can then use the $\bar{R}_h$ values to estimate mean aggregation numbers $\bar{n}$ for the micelles. Such estimates require assumptions about the density and hydration of the micelles, and are admittedly indirect. Nevertheless even if the absolute uncertainties in $\bar{n}$ are as large as $\pm 30\%$, they can still provide useful information on micellar structure in cases where $\bar{n}$ varies by 1 or 2 orders of magnitude under the influence of added NaCl, temperature, or detergent concentration.[22]

In sum, we have seen that a considerable amount of information on micellar size, shape, polydispersity, and structure is potentially obtainable from the combination of dynamic and classical light scattering measurements. Moreover this combined technique can give useful information even at concentrations large compared to the CMC, provided that the micellar interactions can be regarded as hard-sphere effects. In cases where the interaction effects are much stronger, quantitative information about the interactions can also be deduced.

8.3. APPLICATIONS OF LASER LIGHT SCATTERING TO MICELLAR SYSTEMS

8.3.1. Aqueous Synthetic Detergent Systems

8.3.1.1. Size, Shape, and Polydispersity in SDS Solutions. Following McQueen and Hermans,[44] who were the first to determine the diffusion coefficient of SDS micelles with dynamic light scattering, my colleagues and I began a systematic investigation of the size, shape, and polydispersity of SDS micelles over a wide region of the micellar phase. In our first study[22] deductions of $\bar{R}_h$ were made as a function of temperature (10–85°C), and added NaCl concentration (0.15–0.6 M) for SDS concentrations (1.7–6.9 × 10^{-2} M) that were large compared to the CMC. The results, shown in Figures 2A and 2B, illustrate that with increasing temperature, $\bar{R}_h$ approaches a minimum value of ~25 Å, independent of the NaCl or SDS concentration. The closeness of this value to the length of an extended SDS anion (~23 Å) is suggestive of a spherical structure compatible with Hartley's model for small micelles (Figure 1). Conversely with decreasing temperature, $\bar{R}_h$ is seen to increase in a manner which depends strongly on

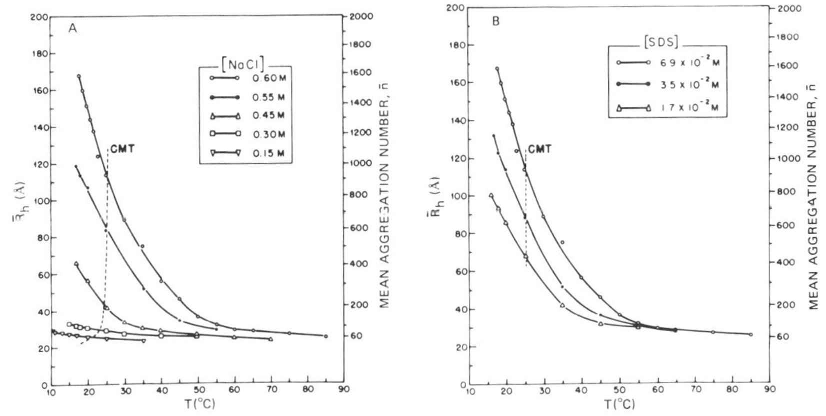

Figure 2. Mean hydrodynamic radius $\bar{R}_h$ of SDS micelles as functions of temperature, NaCl, and SDS concentration from Reference 22. Detergent concentration is constant at 6.9 × 10^{-2} M in (A). NaCl concentration is constant at 0.6 M in (B). Right-hand axes indicate a scale of mean aggregation numbers $\bar{n}$. Reprinted with permission of *J. Phys. Chem.* Copyright 1976, American Chemical Society.

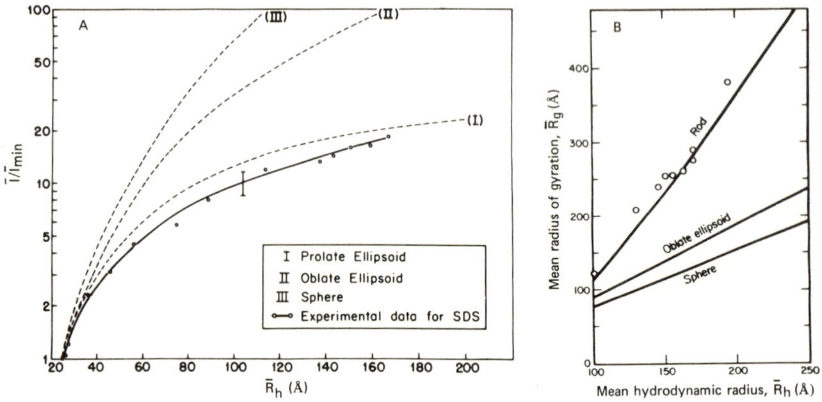

Figure 3. Deductions of micellar shape in an SDS system. In (A) measurements of the scattered intensity versus $\bar{R}_h$ are compared with theoretical calculations (see Reference 22), whereas in (B), measurements of the mean radius of gyration $\bar{R}_g$ versus $\bar{R}_h$ are shown (see Reference 42). Both studies suggest that the large micelles have a rodlike structure. Reprinted with permission of *J. Phys. Chem.* (References 22, 42). Copyright 1976, 1978 American Chemical Society.

NaCl and SDS concentrations. At high NaCl concentrations (>0.45 *M*) $\bar{R}_h$ increases markedly and can exceed 150 Å, whereas at low NaCl very little micellar growth is observed (Figure 2A). Similarly at high NaCl concentration (0.6 *M*) the degree of micellar growth increases with SDS concentration (Figure 2B). It is notable that the temperature dependence of $\bar{R}_h$ shows no discontinuities as one passes below what is known as the critical micellar temperature (CMT). In this region of the phase diagram (Figure 1) the detergent monomers will precipitate into a hydrated solid phase at equilibrium. However, before such precipitation occurs (i.e., in metastable systems), micelle formation will proceed in exactly the same manner as above the CMT.(22) This view, which was initially based on the ideas of Murray and Hartley,(45) has now received direct experimental support from the NMR and conductivity studies of Franses *et al.*(46)

A deduction of the shape of these large micelles was obtained from measurements of the temperature dependence of $\bar{I}$ obtained in 0.6 *M* NaCl, SDS = 6.9×10^{-2} *M*. These data(22) were analyzed by plotting the ratio $\bar{I}(T)$ to $\bar{I}_{min}$ (the intensity measured at 85°C, corresponding to the small presumably spherical micelles) versus the corresponding $\bar{R}_h$ value. From equation (12) this intensity ratio will correspond approximately to the ratio $(M \cdot P)/M_{min}$, where $M \cdot P$ is the product of micellar weight and form factor at the temperature T. On this basis one can calculate theoretical predictions for the variation of intensity ratio versus $\bar{R}_h$ based on different models of micellar shape.(22) Figure 3A shows the experimental results(22) along with theoretical curves assuming I, rodlike; II, disklike; and III, spherical shapes. (In curves I and II the semiminor axes remain equal to 25 Å.) The data are found to agree closely with the rodlike shape, thus suggesting a sphere-to-rod transition of the micellar structure in the SDS system at high NaCl concentrations. This finding is similar to results found by various techniques (NMR,[47, 48] viscosity,[49] classical light scattering[8]) in other ionic detergent systems where rodlike structures have been deduced, and is consistent with the properties of the phase diagram for SDS/water systems[50] where a hexagonal phase (composed of rodlike structures, see Figure 1) is observed to form at very high SDS concentrations (>1 M). Subsequent studies[51–53] using other techniques confirmed our observation that an abrupt growth of the SDS micelles occurs in the presence of high NaCl concentration (>0.4 *M*) at detergent concentrations much greater than the CMC. Nevertheless, in view of continuing controversies concerning the shape of large, nonspherical micelles,[54, 55] we felt that additional confirmation for the rodlike shape would be useful. This was provided in the studies of Young *et al.* 1978,[42] where measurements of the angular dyssymmetry of $\bar{I}$ were made as a function of temperature in the high salt regime. The dyssymmetries were found to be quite large (> 1.20 at $\theta = 45°$) under conditions where large rods were anticipated, and permitted deductions of the mean radius of gyration, $\bar{R}_g$, as a function of temperature. The variation of $\bar{R}_g$ plotted as a function of $\bar{R}_h$ was then compared with theoretical predictions for different micellar shapes (Figure 3B) as had been done previously for the intensity ratios. Again the data were found to be in excellent agreement with the rodlike structure. Although it was subsequently discovered[42] that the SDS samples employed in these two early studies[22, 42] contained appreciable quantities of longer chain length alkyl sulfates (29% C_{14} and 5% C_{16}) and thus gave $\bar{R}_h$ values that were greater than those for "pure" SDS micelles (24, 25), the evidence for a sphere-to-rod transition in the alkyl sulfate system was strongly implied. Recently this evidence has been questioned by Briggs *et al.*,[56] who have analyzed the effect of polydispersity on the theoretical dependence of $\bar{R}_g$ vs. $\bar{R}_h$, a contribution that was not included in the theoretical curves of Young *et al.*[42] Their calculations show that the experimental data fall below the curve expected for polydisperse rigid rods, using a model for the micellar size distributions[23, 25] that is consistent with the experimental data. An explanation for this discrepancy can now be provided (Mazer *et al.*, 1981[57]) by allowing the rodlike micelles to possess some degree of flexibility, characterized by a persistence length. The persistence length (L_p) originally introduced by Peterlin[58] to describe wormlike polymer coils has recently been used by Porte *et al.*[59] to evaluate the micellar flexibility of cetyl pyridinum bromide (CPBr) rods. In the present case, new experimental data on "pure"

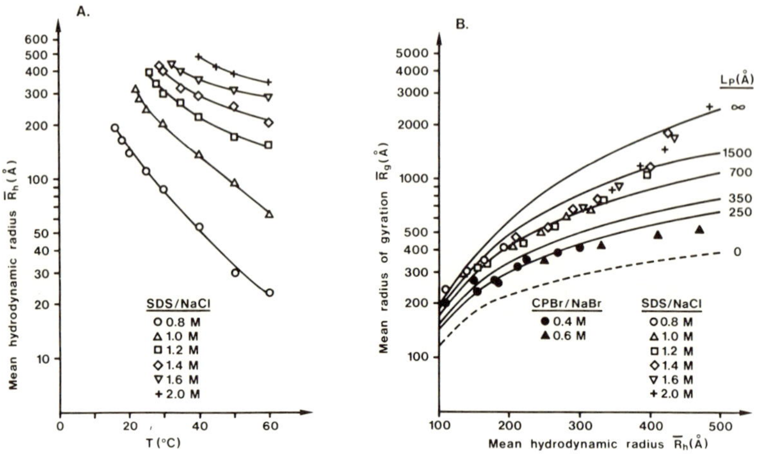

Figure 4. $\bar{R}_h$ and $\bar{R}_g$ measurements of SDS micelles at very high NaCl concentrations (0.8–2.0 *M*) from Reference 57. In A, $\bar{R}_h$ is plotted versus temperature for solutions containing 1.7×10^{-2} *M* SDS and various NaCl concentrations. In B, the $\bar{R}_g$ values are plotted versus the corresponding $\bar{R}_h$ values and compared with theoretical curves that incorporate the effect of micellar flexibility characterized by the persistence length L_p. Also shown are data for CPBr micelles from Reference 60. See text for details.

SDS solutions (1.7×10^{-2} *M*) has been obtained at very high NaCl concentrations (0.8–2.0 *M*) as a function of temperature (25–60°C), in order to characterize large micelles with $\bar{R}_h$ values spanning the range from 25 to 500 Å (Figure 4A). The corresponding variation of $\bar{R}_g$ [deduced from the scattering form factor, equation (13)] versus $\bar{R}_h$ is plotted in Figure 4B and compared with a family of theoretical curves that incorporate polydispersity but assume different values for L_p ranging from 0 (the random coil) to ∞ (the rigid rod). With increasing flexibility (i.e., decreasing L_p) the dependence of $\bar{R}_g$ vs. $\bar{R}_h$ is shifted downward from the curve corresponding to the rigid rods. Excellent agreement is found between the data and the theoretical curve calculated for a persistence length of 700 Å over the range of $\bar{R}_h$ values from 100 to 350 Å. For the very large micelles ($\bar{R}_h > 350$ Å), the $\bar{R}_g$ values exceed the predicted dependence on $\bar{R}_h$. Strong intermicellar interactions and/or the formation of new micellar structures (e.g., torus, disk, or liposomes) may explain this deviation. Nevertheless it is evident that for a wide range of NaCl concentrations and temperatures the SDS rods can be characterized by a constant persistent length of ~700 Å. It is interesting that this value is substantially larger than the persistence length deduced by Porte *et al.*[59] for the CPBr rods of 400 Å (or smaller) using magnetic anisotropy and laser light scattering. Appell and Porte[60] have also recently

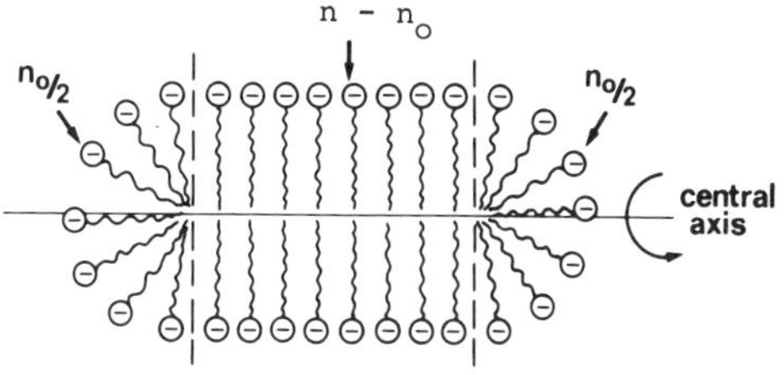

Figure 5. Prolate sphero-cylindrical model for the rodlike micelle from References 23, 25. Reprinted with permission of *J. Phys. Chem.* (Reference 25). Copyright 1980, American Chemical Society.

measured $\bar{R}_g$ vs. $\bar{R}_h$ for CPBr micelles, and their data are seen in Figure 4B to be in good agreement with the theoretical curve corresponding to L_p = 250 Å. The theoretical basis for the greater rigidity (higher L_p value) of SDS micelles relative to CPBr is presently being investigated and appears to reflect subtle differences in the electrostatic interactions between detergents and counterions at the micellar surface.[57]

In addition to providing a detailed characterization of micellar size, shape, and structure in the SDS system, the previous studies have also been useful in the development of a quantitative thermodynamic model[23, 25] of the sphere-to-rod transition. A brief discussion of this model and its application to the SDS system will now be given. The model, which is analogous to theories proposed by Mukerjee,[14] Israelachvilli,[15] and Tausk,[61] assumes that detergent molecules can form a distribution of micelles whose aggregation numbers, n, vary from n_0, $n_0 + 1$, $n_0 + 2$, ... The smallest aggregate, containing n_0 molecules, is assumed to have a spherical structure, whereas the larger aggregates are modeled as spherocylinders, a structure first proposed by Debye and Anacker.[8] As seen in Figure 5, a rodlike micelle of aggregation number n will have $(n - n_0)$ molecules in a cylindrical portion, and $n_0/2$ molecules in each of the hemispherical endcaps. In this way the rodlike micelle provides two environments for detergent molecules that are characterized by their standard chemical potentials, μ_c^0 and μ_s^0, for the cylinder and sphere, respectively. By assuming that the standard chemical potential of the entire micelle μ_n^0 depends additively on the number of molecules in each environment [i.e., $\mu_n^0 = n_0\mu_s^0 + (n - n_0)\mu_c^0$], it can be shown[25] that the mole fraction of n-mers, X_n, is to a good approximation given by

$$X_n \approx \frac{1}{K}\exp\left\{\frac{-n}{[K(X - X_B)]^{1/2}}\right\} \tag{14}$$

where X is the total mole fraction of detergent molecules in the system, X_B is a thermodynamic parameter corresponding closely to the CMC, and K is a second thermodynamic parameter defined by[23, 25]

$$K \equiv \exp[n_0(\mu_s^0 - \mu_c^0)/RT] \tag{15}$$

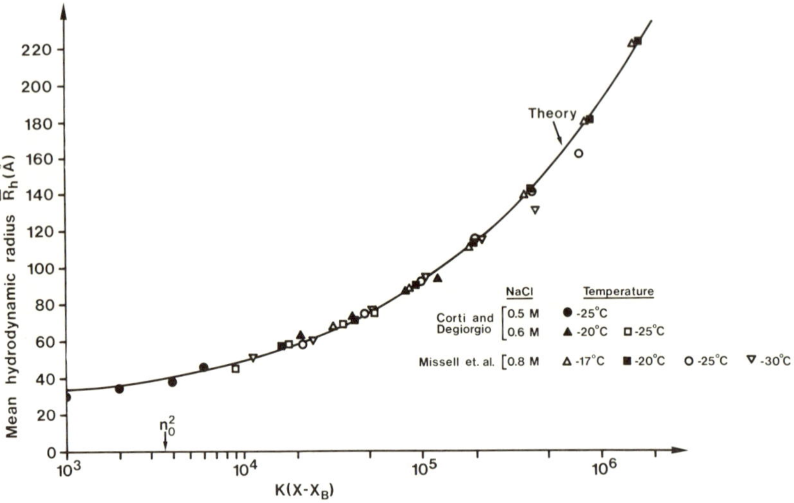

Figure 6. Theoretical dependence of $\bar{R}_h$ versus $K(X - X_B)$ for sphere-to-rod transition from Reference 25. Also shown are experimental data from References 24, 25, and 62, which are seen to be in good agreement with the theory. Reprinted with permission of *J. Phys. Chem.* (Reference 25). Copyright 1980, American Chemical Society.

From equation (14) it is seen that the micelle size distribution, X_n, will decrease exponentially with n at a rate determined by the square root of the product $K(X - X_B)$. Using this result, the number averaged aggregation number, $\bar{n}_N$, is calculated to be[25]

$$\bar{n}_N \approx n_0 + [K(X - X_B)]^{1/2} \tag{16}$$

which shows that $\bar{n}_N$ will substantially exceed n_0 only when the parameter K or concentration X is sufficiently large to make $K(X - X_B) \gg n_0^2$. To quantitatively test and apply this model to SDS, we have used equation (14) in conjunction with estimates of the hydrodynamic radius of each micellar species to calculate the dependence of $\bar{R}_h$ on the product $K(X - X_B)$ according to equation (8). This theoretical dependence (Figure 6) shows that for $K(X - X_B) \ll n_0^2$ ($\sim$3600 for SDS) $\bar{R}_h$ remains close to 25 Å, the size of the spherical micelles, but for $K(X - X_B) > n_0^2$, $\bar{R}_h$ increases appreciably. In this latter regime, temperature and NaCl concentration influence micellar growth by affecting K, whereas the SDS concentration determines the value of X. At constant temperature and NaCl concentration, the theory thus provides an explicit prediction for the dependence of $\bar{R}_h$ on X. Figure 6 shows a test of this prediction using the experimental data obtained by Missel *et al.*[25] in 0.8 M NaCl (T = 17, 20, 25, 30°C) and by

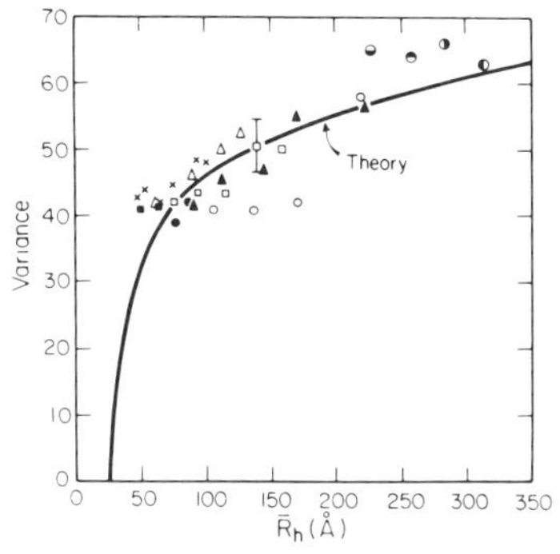

Figure 7. Theoretical and experimental dependence of the polydispersity index, variance, versus $\bar{R}_h$ for the sphere-to-rod transition from Reference 25. Reprinted with permission of *J. Phys. Chem.* (Reference 25). Copyright 1980, American Chemical Society.

Corti and DeGiorgio[24, 62] in 0.5 and 0.6 *M* NaCl ($T = 25°C$). To compare theory with experiment, a one-parameter fit was first made to the concentration dependence of $\bar{R}_h$ measured at constant temperature in order to obtain the parameter K. The experimental $\bar{R}_h$ values were then plotted as functions of $K(X - X_B)$ using that value of K. As seen from Figure 6, the experimental concentration dependence is in excellent agreement with the theory for values of $K(X - X_B)$ which span three orders of magnitude. A second test of the theory concerns its ability to accurately predict the polydispersity of micellar sizes associated with the sphere-to-rod transition. Figure 7 shows experimental values of the polydispersity index [variance V, see equation (6b)], plotted versus the corresponding $\bar{R}_h$ values, and compares these with the theoretical dependence predicted from the model. This comparison involves no free parameters, and the agreement is found to be quite good. The present analysis thus provides strong support for the view that under the conditions of high salt, the concentration dependence of $\bar{R}_h$ and V are quantitatively consistent with the influence of detergent concentration on the micelle size distribution expected for a sphere-to-rod transition. Moreover it is seen that in this regime the concentration dependence of $\bar{R}_h$ is governed by a single thermodynamic parameter, K, which can be readily deduced from the dynamic light scattering data.[85] In recent studies[63] new insights into the thermodynamics of the sphere-to-rod transition are being obtained from analyzing the variation of K with such variables as temperature, alkyl chain length, counterion size, and solvent additives such as urea. From this work a fuller appreciation of the contributions of electrostatic and hydrophobic interactions in the formation and growth of micelles is evolving.

8.3.1.2. Micellar Interaction in Ionic Detergent Systems. As seen from equations (11) and (12), the concentration dependence of $\bar{I}$ and D both contain contributions from micellar interactions. While such effects are discussed elsewhere in this book, it is the present aim to indicate here the

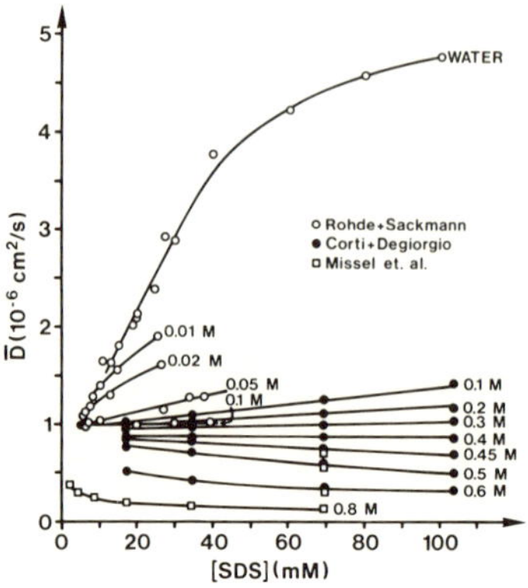

Figure 8. Concentration dependence of the diffusion coefficient of SDS micelles at various NaCl contents measured at 25°C. Data are compiled from the studies of Rohde and Sackmann,[64] Corti and Degiorgio,[24, 62] and Missel *et al.*[25, 63]

magnitude and direction of this contribution on the diffusion coefficient and to identify the experimental conditions for which the interaction effects are significant. Corti and Degiorgio[24, 62] and Rohde and Sackmann[64] have investigated micellar interactions in aqueous SDS solutions containing little or no added NaCl (0–0.2 *M*), conditions where the electrostatic repulsion between micelles should be appreciable. Their measurements of $\bar{D}$ as a function of SDS concentration at 25°C are shown in Figure 8 along with data obtained at high NaCl concentration (up to 0.8 *M*) by Missel *et al.*[25, 63] Although there are minor discrepancies between the three studies (i.e., the data obtained in 0.1 *M* NaCl), the data taken as a whole illustrate that the concentration dependence of $\bar{D}$ varies dramatically as a function of added NaCl. One can identify three regimes of behavior which roughly correspond to the ranges of NaCl concentration investigated in the three experimental studies. At the very low NaCl concentrations (0–0.1 *M*) studied by Rohde and Sackmann,[64] $\bar{D}$ increases strongly with SDS concentration as expected for a system in which there are strong electrostatic interactions. They interpret this behavior according to the theory of Stephen,[65] which attributes the increase in $\bar{D}$ to the Coulomb interaction between the charged micelles and the rapidly diffusing counterions and coions. Although the Stephen theory can be fitted to the data, the deductions of micellar aggregation number and effective charge as functions of NaCl concentration do not agree well with other studies.[9, 10, 13] An alternative explanation for the strong increase in $\bar{D}$ follows from equation (11) where the effect of electrostatic repulsion between the micelles will be to increase the value of $2A_2 M$ (relative to B_f) and thus lead to a positive slope for $\bar{D}$ vs. C. This view is consistent with the theoretical analysis of Corti and Degiorgio,[62] and would explain the reduction in slope with added NaCl concentration in terms of the progressive screening of the electrostatic repulsion. As the electrostatic effects clearly dominate the concentration dependence of $\bar{D}$ in this low salt regime, it is virtually impossible to deduce information on the size of micelles at concentrations large compared to the CMC. It is even difficult to reliably extrapolate $\bar{D}$ to the CMC, as there appears to be considerable curvature in the concentration dependence, and the diffusion of detergent monomers may no longer be neglected. Nevertheless it seems reasonable to conclude that the micelle size close to the CMC is consistent with a hydrodynamic radius of ~24–25 Å ($\bar{D} \sim 1 \times 10^{-6}$ cm^2/sec) and is probably independent of NaCl concentration in this low salt range. Conversely in the region of high NaCl concentration (>0.45 *M*) studied extensively by Missel and co-workers, $\bar{D}$ decreases appreciably with SDS concentration. Such behavior was discussed at length in the previous section, and shown to be quantitatively consistent with the concentration dependence expected for a sphere-to-rod transition (see Figure 6). In this regime, micellar interactions appear to have little effect on the diffusion coefficient. Finally in the intermediate NaCl regime (0.2–0.45 *M*) studied by Corti and Degiorgio, the concentration dependence of $\bar{D}$ is fairly weak and changes its slope from positive to negative. In this regime, both micellar repulsion and micellar growth appear to influence the concentration dependence, the latter effect becoming more important with increasing NaCl concentration. Corti and Degiorgio have analyzed[62] their data from 0.1 to 0.5 *M* NaCl using a model that explicitly treats the hard-sphere and electrostatic repulsion between micelles, but does not allow for the possibility of a concentration-dependent micellar growth. They further assume that an attractive interaction between micelles (presumably through dispersion forces) exists and may contribute to the concentration dependence of $\bar{D}$. Their theory successfully accounts for the concentration dependence of $\bar{D}$ for NaCl concentrations between 0.1 and 0.4 *M* where the repulsive interactions dominate, but fails to predict the behavior of $\bar{D}$ at higher NaCl concentrations. Presumably this failure results from neglecting the growth of the micelles, which becomes greater with increasing NaCl concentration. In view of this omission, it seems premature to conclude, as Corti and Degiorgio have done,[62] that attractive interaction between globular micelles plays an important role in ionic micellar systems.

In summary, the present compilation of dynamic light scattering studies of SDS micelles shows that micellar interactions exert a significant influence on the concentration dependence of D only at very low NaCl concentrations ≤0.2 *M*, where electrostatic repulsion between micelles is dominant. At higher NaCl concentrations the repulsion becomes effectively screened, and the excluded volume interactions remain. For globular micelles, these interactions have little effect on $\bar{D}$, whose concentration dependence is dominated by the growth of the micelles from spheres to rods. It is

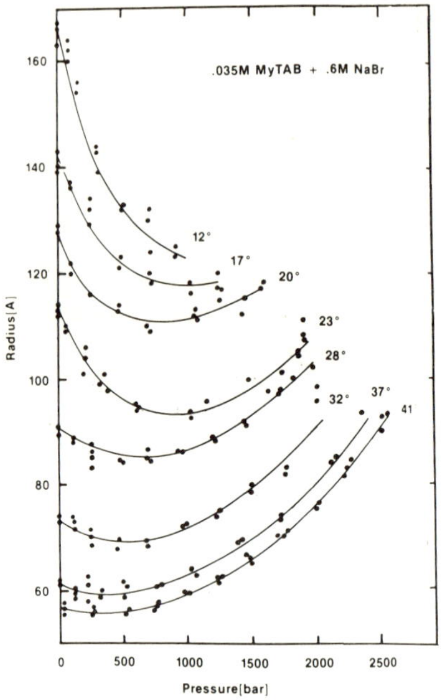

Figure 9. Influence of pressure on the $\bar{R}_h$ of MYTAB micelles at various temperatures from Reference 26. Reprinted with permission.

conceivable, however, that at sufficiently high detergent concentrations where micelles may become quite long (> 1000 Å), the excluded volume effects may no longer be negligible. Such behavior must certainly be expected as one approaches the conditions at which the liquid crystalline middle phase (a hexagonal packing of long rods) begins to form (see Figure 1).

8.3.1.3. Effect of Pressure on Ionic Micelles. Although the effect of pressure on the CMC of micellar systems has been studied for a number of years,[66,67] only recently has direct information on the pressure dependence of micellar size been obtained. Using dynamic light scattering techniques, Nicoli *et al.*[26] measured the dependence of $\bar{R}_h$ on temperature (12–41°C) and pressure (1–2500 bars) for myristyl trimethyl ammonium bromide (MYTAB) micelles, a typical cationic detergent believed to exhibit a sphere-to-rod transition.[68] Working at moderate MYTAB concentrations (0.035 *M*) in the presence of added salt (0.6 *M* NaBr), conditions where micellar growth has occurred at atmospheric pressure, Nicoli *et al.* observed an interesting nonmonotonic variation of $\bar{R}_h$ as a function of pressure. As seen in Figure 9, $\bar{R}_h$ first decreases with pressure and then increases. The pressure, P_{min}, at which $\bar{R}_h$ attains a minimum value decreases as a function of temperature from ~1000 bar at 17° to ~300 bar at 41°C. Likewise the $\bar{R}_h$ values corresponding to the minima also decrease with temperature and parallel the temperature dependence at atmospheric

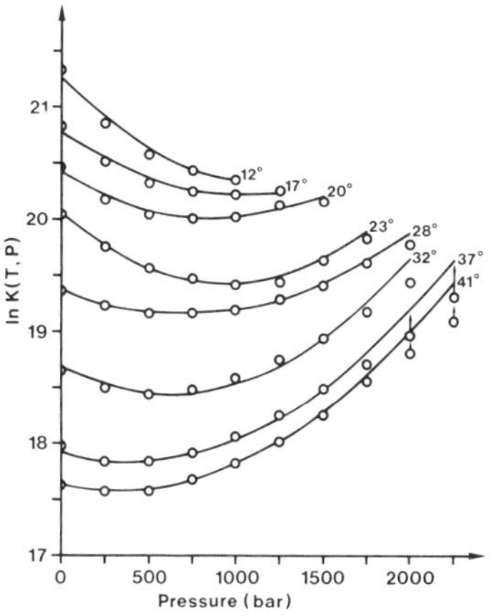

Figure 10. The thermodynamic parameter K characterizing the sphere-to-rod transition as a function of pressure and temperature for MYTAB micelles as deduced from the data in Figure 9 (Reference 69).

pressure. The latter is quite similar to the behavior of SDS micelles in the presence of high NaCl, and strongly suggests a temperature-dependent sphere-to-rod transition. In addition Nicoli *et al.* noted in their study that above a critical pressure, P_c, the detergent would precipitate from solution. In contrast to P_{min}, the P_c values increased with temperature from ~750 bar at 12°C to ~2500 bar at 41°C. These precipitation phenomena have been shown by Tanaka *et al.*[66] in other systems (such as SDS) to be completely analogous to the Krafft point (or CMT phenomenon) seen at atmospheric pressure, and represent an insolubility of the detergent monomer, rather than the micelle.[22] While the study of Nicoli *et al.* illustrates a dramatic influence of pressure on the micelle size distribution, a quantitative thermodynamic analysis of their results has not yet been presented. In principle the pressure effects provide information on the partial molar volumes of the various micellar structures (i.e., sphere and rod), as well as their respective compressibilities (Mazer *et al.*, 1981[69]). Using the theoretical model of the sphere-to-rod transition, employed previously for the SDS system, one can obtain such information from the MYTAB data. The results of such an analysis[69] are quite illuminating and will be briefly presented here. From Figure 6 it was seen that $\bar{R}_h$ depends on the product of a thermodynamic parameter K and the micelle concentration. As the latter is essentially constant in the present experiments, it must be concluded that the pressure dependence of $\bar{R}_h$ results from the pressure dependence of K. Using the functional relationship between $\bar{R}_h$ and $K(X - X_B)$ appropriate for a C_{14} detergent (i.e., $n_0 \approx 82$), one can deduce from the data of Figure 9 the dependence of K on temperature and pressure. This is done in Figure 10, which shows that at each temperature the variation of ln K with pressure can be well described by a quadratic function:

$$\ln K(T, P) = A(T)[P - P_{min}(T)]^2 + \ln K_{min}(T) \tag{17}$$

where $P_{\min}(T)$ denotes the pressure at which ln K (and hence $\bar{R}_h$) attains a minimum denoted ln $K_{\min}(T)$ and $A(T)$ describes the curvature. The physical significance of equation (17) follows from the fact that pressure dependence of ln K is related to the pressure dependence of the chemical potential difference $\mu_s^0 - \mu_c^0$ [see equation (15)] and that the latter can be related to the difference in partial molar volume $v_s - v_c$ for a monomer located in a spherical or cylindrical micellar structure (Figure 5). Quantitatively, these relationships are given by

$$v_s(T, P) - v_c(T, P) = \frac{\partial[\mu_s(T, P) - \mu_c(T, P)]}{\partial P} \tag{18a}$$

$$= \frac{RT}{n_0} \frac{\partial \ln K(T, P)}{\partial P} \tag{18b}$$

$$= \frac{2RTA(T)}{n_0} [P - P_{\min}(T)] \tag{18c}$$

where equation (18a) follows from the thermodynamic definitions of v_s and v_c, and (18b) and (18c) follow from equations (15) and (17). Equation (18c) illustrates that for $P < P_{\min}$, v_s is smaller than v_c. This is intuitively consistent with the view that the application of pressure perturbs any chemical equilibria to the state of smaller volume (to reduce the $P\Delta V$ work). Thus the decrease in $\bar{R}_h$ for $P < P_{\min}$ corresponds to the fact that for this pressure range, molecules in the spherical micellar structures have smaller volumes than those in the cylindrical structure whereas above $P_{\min}$ the reverse is true. That $v_s - v_c$ varies both with pressure and temperature and changes sign at $P_{\min}(T)$, is a reflection of the different compressibilities $(-\partial v/\partial P)_{s,c}$ and thermal expansibilities $(\partial v/\partial T)_{s,c}$ in the spherical and cylindrical states. In Figure 11 this variation in $v_s - v_c$, as deduced from equation (18c), and the temperature dependence of $A(T)$, $P_{\min}(T)$, is plotted. It is seen that the $v_s - v_c$ values range from ~ -0.4 cm^3/mole ($T = 12°$C, $P = 1$ bar) to ~ 0.6 cm^3/mole ($T = 41°$C, $P = 2000$ bar). These values are approximately an order of magnitude smaller than the differences in partial molar volume between monomer and micellar state $v_1 - v_m$ (as deduced from the variation of CMC with pressure[66] and by direct measurements[67]) but show the same trends with pressure and temperature. The finding that v_s is smaller than v_c at atmospheric pressure is consistent with the volumetric studies on sodium octanoate systems by Ekwall *et al.*,[70] which show that the partial molar volume per monomer in a spherical micelle is ~ 0.2 cm^3/mole smaller than in the hexagonal phase, where the monomers are assembled into cylindrical structures (Figure 1). Similar volumetric studies on the MYTAB system could be conducted to directly estimate the values of v_s and v_c, and

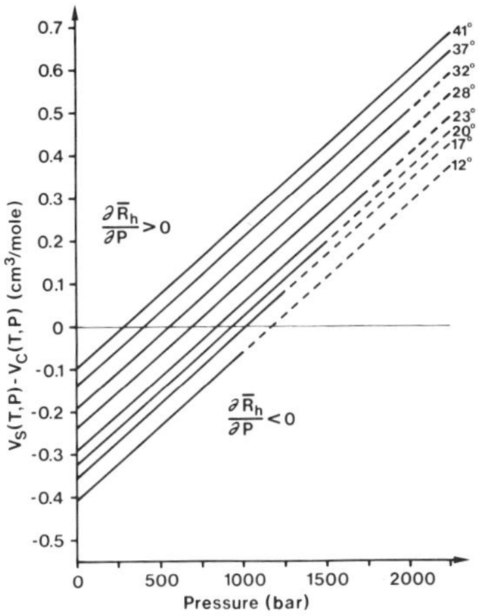

Figure 11. The difference in partial molar volume for a molecule located in a spherical, V_s, or cylindrical, V_c, micellar structure as a function of pressure and temperature, as deduced from the data in Figure 10 (Reference 69). Dashed lines are extrapolated into the pressure regime where the system has precipitated into the hydrated solid phase.

their respective compressibilities for comparison with the present deductions. A quantitative understanding of the role of hydrophobic and electrostatic interactions in determining the sign and magnitudes of $v_1 - v_m$ and $v_s - v_c$ remains to be elucidated.[66, 67]

8.3.1.4. Nonionic Micellar Systems. In contrast to ionic detergents, nonionic amphiphiles possess polar moieties that contain no charged groups. Most typical among these is the family of polyoxyethylene–alkyl ether compounds ($C_i E_j$) whose polar moiety consists of a short polymer of j ethylene oxide units, CH_2–CH_2–O–, (denoted E_j) that is attached by an ether linkage to an alkyl chain containing i carbons (denoted C_i). The concentration/temperature phase diagrams for these compounds have been investigated for a range of i and j values.[71, 73] A typical phase diagram is shown (Figure 12A,B) for the $C_{12}E_6$ detergent,[72] which has recently been studied by Corti and Degiorgio,[27] using a combination of dynamic and classical light scattering methods. As is characteristic of most nonionic detergents, the micellar phase is bounded at high temperatures by an asymmetric curve, above which the micellar solution separates into two isotropic phases. This phase boundary, which is not seen in ionic detergent systems, is commonly referred to as the "cloud point" curve,[73] since solutions become opalescent as the boundary is approached. To date there has been considerable controversy concerning the size, shape, and structure of nonionic micelles, as well as the origin of the cloud point phenomenon itself. While some investigators have interpreted classical light scattering data to indicate a temperature dependent growth[72] of the micelles as the cloud point is approached, others have postulated that such growth, if it occurs, results only from the secondary association of nonionic micelles.[74] Still

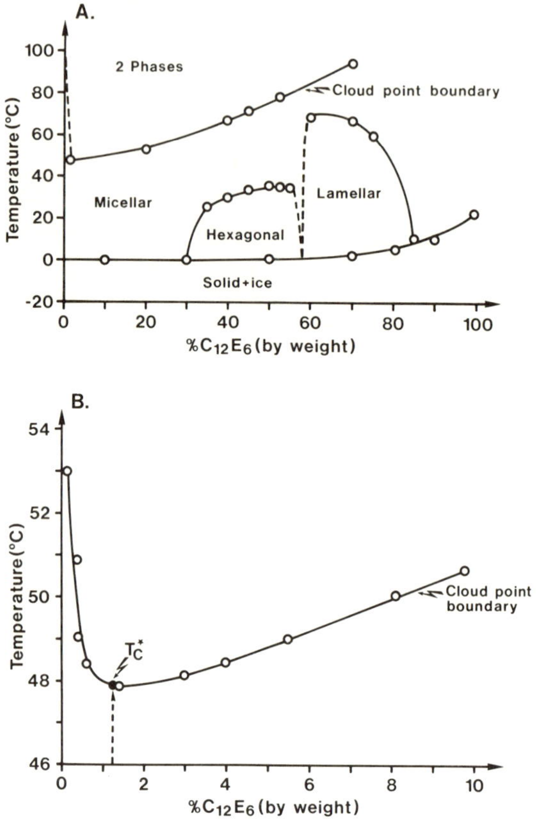

Figure 12. Concentration/temperature phase diagrams from the $C_{12}E_6$ system from Reference 72. In (A), a large-scale view of all phases between −20–100°C and 0%–100% detergent is shown. In (B) an enlarged view of the cloud point boundary is given. Dashed arrow indicates the path studied by Corti and Degiorgio[27] where critical behavior is seen. T_c^* is the critical temperature. Reprinted with permission from Reference 72.

others have suggested that the cloud-point phenomenon is a manifestation of critical opalescence[27, 75] associated with the ensuing phase separation behavior and need not be attributed to micellar growth at all. This latter view has recently received strong support from the investigations of Corti and Degiorgio, who have measured the diffusion coefficient, D, scattered intensity $\bar{I}$, and its angular variation for the $C_{12}E_6$ system as a function of temperature and concentration. Their studies show that in the vicinity of the concentration corresponding to the minimum of the cloud curve (i.e., ~1.25 g/dl) the temperature dependence of $D(T)$ and $\bar{I}(T)$ can be fitted by power laws as one approaches the "minimum" cloud point temperature, T_c^*:

$$D(T) = D_0(1 - T/T_c^*)^{\nu} \tag{19a}$$

$$\bar{I}(T) = I_0(1 - T/T_c^*)^{-\gamma} \tag{19b}$$

where ν and γ are the so-called critical exponents and D_0 and I are constants. Equations (19a) and (19b) are well known[76] to describe the behavior of binary liquid mixtures in the vicinity of a critical point, and their appearance in the present context offers strong evidence that the cloud-point phenomenon is itself an example of binary coexistence between water molecules and micelles. According to this view, the occurrence of phase separation above the cloud-point boundary is a reflection of the thermodynamic advantages for water and micelles to segregate from each other at higher temperature, and results from the overall balance between micelle–micelle, micelle–water, and water–water interactions. Below the cloud point, these interactions likewise give rise to long-range fluctuations in the local micellar concentration whose spatial extent can be characterized by a correlation range,[77] $\xi(T)$, which diverges ($\xi \rightarrow \infty$) as the temperature approaches T_c^*. Corti and Degiorgio have deduced $\xi(T)$ from the angular dependence of $\bar{I}$ and have shown that their measurements of $D(T)$ are approximately equal to $kT/6\pi\eta\xi(T)$ where η is the macroscopic shear viscosity. This relationship, which appears similar to the Stokes–Einstein relation (equation), is another hallmark of critical behavior.[76, 77] Moreover, it illustrates that in these systems the D value is not reflecting changes in micellar size, but rather shows the growth of the correlation range. In fact, Corti and Degiorgio[27] suggest that over the temperature range 30–50°C there is little indication from their data that any micellar growth is occurring. This view is consistent with the findings of Staples and Tiddy,[51] who used NMR methods to study the $C_{12}E_6$ detergent as a function of temperature. Their analysis of the proton NMR spectra gave no evidence of the signal broadening that would be expected if the micelles were growing into large asymmetric structures with a single hydrocarbon core, and at best suggested a secondary association of globular micelles. In contrast, the same NMR methods[51] did reveal a true growth of SDS micelles (i.e., spectral broadening) as a function of added NaCl and lowering temperature, consistent with the sphere-to-rod transition inferred by dynamic light scattering.[22, 25] It thus appears that there are significant differences in the solution properties of ionic and nonionic micelles and that the critical phenomenon discussed here is a unique aspect of the latter. In this regard it is noteworthy that similar behavior is seen in aqueous solutions of high molecular weight polyoxyethylene polymers,[78, 79] i.e., macromolecules

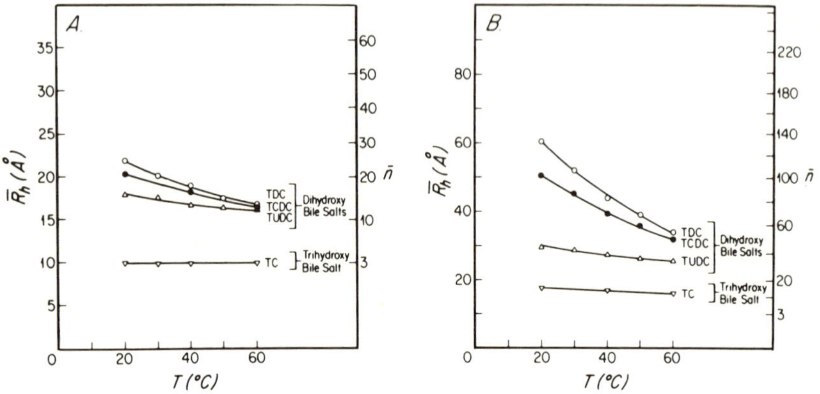

Figure 13. $\bar{R}_h$ values of bile salt micelles measured on four different bile salt species as a function of temperature (Reference 30). The NaCl concentration is 0.15 *M* in A and 0.6 *M* in (B). The bile salt concentration is 10 g/dl for all data points. Reprinted with permission of *Biochemistry* (Reference 30). Copyright 1979, American Chemical Society.

constituted of the same polar moieties as the nonionic micelles. The polymer systems have an asymmetric coexistence curve as the micelles,[78] and likewise display a temperature-dependent opalescence as the phase boundary is approached. It is tempting to attribute the critical behavior seen in nonionic micellar systems to the same molecular interactions that exist between the nonionic polymers and water. Dynamic light scattering studies of the latter systems would be valuable in exploring this hypothesis.

8.3.2. Biological Micelles

8.3.2.1. Aqueous Biliary Lipid Systems. Bile salts, lecithin, and cholesterol—the three major lipid components of bile—exhibit a rich variety of aggregative phenomena when dissolved either individually or in combination with each other in aqueous solution.[80, 81] An understanding of these phenomena is vital for elucidating the mechanisms by which these lipids are secreted and transported in bile, how they function in fat absorption, and for understanding the process of cholesterol gallstone formation. For these reasons my colleagues and I have actively investigated aqueous biliary systems using the methods of dynamic and conventional light scattering[30–32] discussed earlier. A brief summary of our experimental and theoretical results is now given.

Simple Micelle Formation in Bile Salt Solutions. Bile salts are a family of ionic steroid amphiphiles which are capable of self-associating in

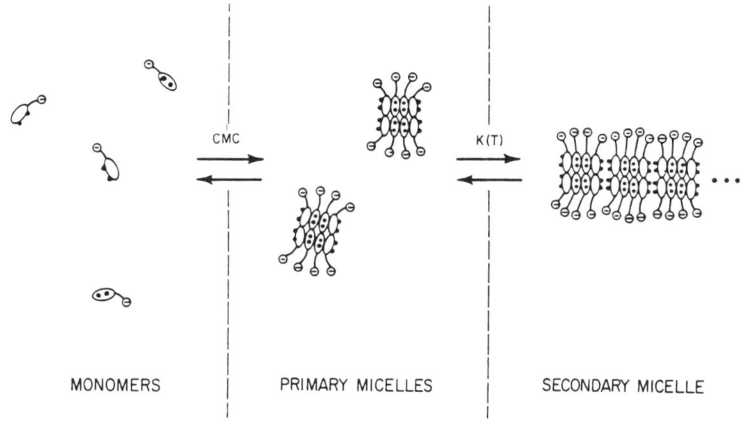

Figure 14. Schematic for the primary–secondary micelle model used to describe bile salt aggregation (Reference 30). See text for explanation. Reprinted with permission of *Biochemistry* (Reference 30). Copyright 1979, American Chemical Society.

aqueous solution. Figure 13 shows the $\bar{R}_h$ values of bile salt micelles[30] as functions of temperature and NaCl concentration of the four common bile salt species of man: taurocholate (TC), taurodeoxycholate (TDC), taurochenodeoxycholate (TCDC), and tauroursodeoxycholate (TUDC). These species differ in the number and location of their hydroxyl groups on the steroid nucleus, and their micelles are seen to vary in size in the following order:

$$\text{TDC} > \text{TCDC} > \text{TUDC} > \text{TC}$$

Although the sizes and aggregation numbers of bile salt micelles are smaller than SDS micelles, they vary with NaCl concentration, temperature, and bile salt concentration (data not shown) in much the same way.[30] It has also been shown[30] that the shape of the larger bile salt micelles, as deduced from $\bar{I}$ vs. $\bar{R}_h$ data, is consistent with a rodlike structure. These data have been interpreted in terms of a quantitative model[30] based on the primary–secondary micelle hypothesis proposed by Small.[82] The model (see Figure 14) assumes that bile salt monomers associate at the CMC to form globular primary micelles ($\bar{R}_h \sim$ 10–15 Å, $\bar{n} \sim$ 5–10). At higher concentration the primary micelles polymerize in a linear fashion to form rodlike secondary micelles. The size distribution (and hence $\bar{R}_h$) of the secondary aggregates can be related theoretically to the bile salt concentration C_{BS}, polymerization constant K, and primary micelle aggregation number n_0 in

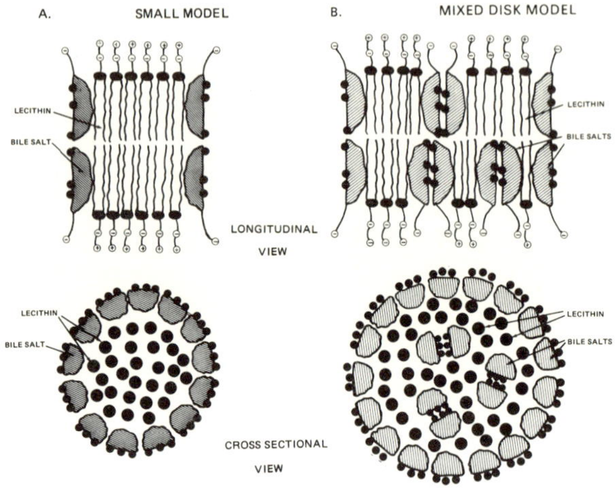

Figure 15. Schematic representations of the structure of bile salt–lecithin mixed micelles. (A) The simple bilayer disk model proposed by Small[84]. (B) The "mixed disk" model proposed by Mazer *et al.*[31, 85] In the latter model, bile salts not only coat the perimeter of the disk but are also present in the interior of the bilayer, possibly as reversed dimers. Reprinted with permission of *Biochemistry* (Reference 31). Copyright 1980, American Chemical Society.

terms of the product KC_{BS}/n_0. The concentration dependence of $\bar{R}_h$[30] and $\bar{I}$[83] are well predicted by this model, and thus permit deductions of the polymerization constant K analogous to the deduction of the thermodynamic parameter in the SDS system.[25] In the present case the unusual amphiphilic structure of this bile salt molecule makes the primary–secondary micelle picture easier to visualize than the continuous growth picture used in the SDS analysis. From the mathematical standpoint, however, both models are quite similar.

Mixed Micelle Formation in Bile Salt–Lecithin Solutions. Small postulated[84] that bile salts solubilize lecithin bilayers by dispersing them into disk-shaped aggregates whose structure is shown in Figure 15. According to Small's model, the radius of the mixed micellar disk (r) should vary with the concentrations of bile salt, C_{BS}, and lecithin, C_L, as follows[31]:

$$r = \left(\frac{2\rho}{\sigma}\right) \frac{C_L}{C_{BS} - \mathrm{IMC}} \tag{20}$$

where ρ is the number of bile salts per unit length around the perimeter of the micelle, σ is the number of lecithin molecules per unit area of bilayer, and IMC is the small concentration of bile salt monomers in equilibrium with the mixed micelles. In the limit when $C_{BS} \gg \mathrm{IMC}$ equation (20) predicts that r should vary linearly with the L/BS molar ratio (C_L/C_{BS}). This

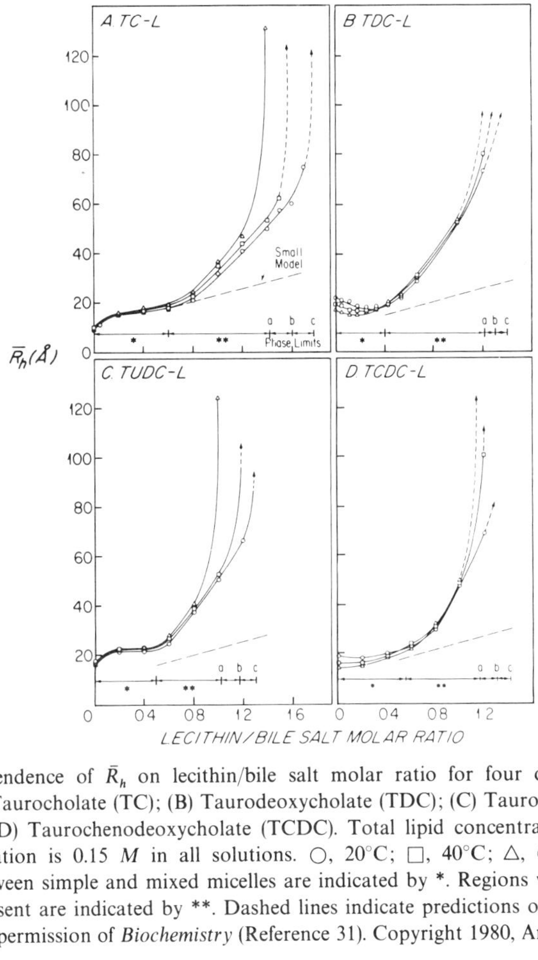

Figure 16. Dependence of $\bar{R}_h$ on lecithin/bile salt molar ratio for four different bile salt species.[31] (A) Taurocholate (TC); (B) Taurodeoxycholate (TDC); (C) Tauroursodeoxycholate (TUDC); and (D) Taurochenodeoxycholate (TCDC). Total lipid concentration 10 g/dl and NaCl concentration is 0.15 M in all solutions. ○, 20°C; □, 40°C; △, 60°C. Regions of coexistence between simple and mixed micelles are indicated by *. Regions where only mixed micelles are present are indicated by **. Dashed lines indicate predictions of Small model.[84] Reprinted with permission of *Biochemistry* (Reference 31). Copyright 1980, American Chemical Society.

prediction has been tested[31] using the four bile salt species studied previously, and the results are shown in Figure 16. At low L/BS ratios (<0.5), the $\bar{R}_h$ values show different behavior depending on bile salt species. A detailed analysis[31] of this regime suggests that both simple bile salt micelles and mixed micellar aggregates coexist in varying proportions at these low L/BS ratios. However, at high L/BS ratios (>0.6) only mixed micelles are present[31] and their sizes deviate markedly from the prediction of the Small model as the L/BS ratio increases. In fact, the $\bar{R}_h$ values increase nonlinearly and appear to diverge ($\bar{R}_h \to \infty$) as the L/BS ratios approach the phase limits for lecithin solubilization appropriate to each system.[31]

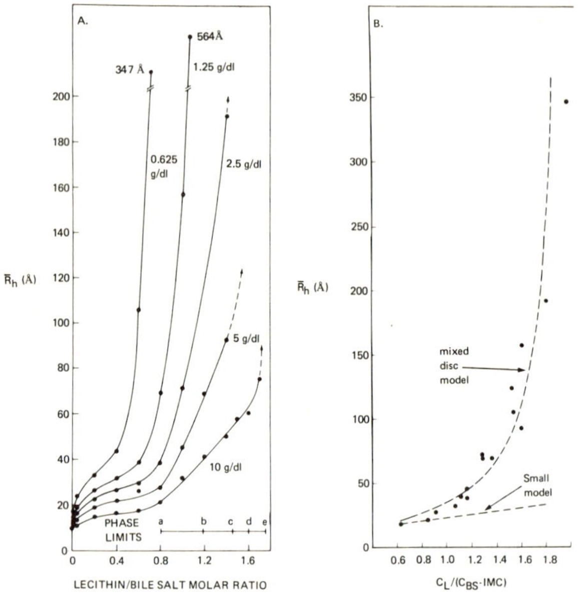

Figure 17. $\bar{R}_h$ values for TC-L mixed micelles for various total lipid concentrations in 0.15 *M* NaCl at 20°C (Reference 31). In (A) data are plotted versus the L/BS molar ratio. In (B) data are plotted versus $C_L/(C_{BS} - \text{IMC})$ and are compared with theoretical predictions based on the Small model and mixed-disk model. Reprinted with permission from *Biochemistry* (Reference 31). Copyright 1980, American Chemical Society.

This "divergence phenomenon" is further illustrated in Figure 17A, where the influence of total lipid concentration as well as L/BS ratio on $\bar{R}_h$ is shown. With decreases in total lipid concentration, the L/BS ratio corresponding to the micellar phase limit decreases and the $\bar{R}_h$ values are thus seen to "diverge" earlier.[31] Dilution thus leads to a large increase in mixed micellar size, the opposite of what was seen in pure bile salt systems.[30] Although the simple disk model proposed by Small is clearly incompatible with these data, a deduction of mixed micellar shape using scattered intensity measurements suggests that the large mixed micelles are nevertheless disklike.[31] It is thus necessary to explain why the radii of the mixed micellar disks exceed the predictions of the Small model and furthermore to understand the apparent divergence of micellar size at the phase limit.

By a simple extension of Small's model we have shown that both questions can be answered if one postulates that bile salts not only coat the perimeter of the mixed micelle but are also incorporated within the interior of the lecithin bilayer in a fixed stoichiometry (see Figure 15B—"mixed disk" model[31, 85]). If α represents the ratio of lecithin to bile salt molecules within the bilayer of the mixed micelle, the disk radius r will depend on C_L and C_{BS} according to the mixed disk model as[31]

$$r = \frac{(2\rho/\sigma')[C_L/(C_{BS} - \text{IMC})]}{1 - \alpha^{-1}[C_L/(C_{BS} - \text{IMC})]} \tag{21}$$

where σ' is the number of lecithins per unit area of a mixed bilayer (containing bile salts). Equation (21) predicts a divergence in r when C_L and C_{BS} satisfy the relationship

$$\frac{C_L}{C_{BS} - \text{IMC}} = \alpha \tag{22}$$

Physically this corresponds to the situation where all of the bile salt molecules (excluding the IMC) are included within the bilayer, leaving no additional bile salts to form the disk perimeters that are needed to solubilize the bilayer. In this way, the mixed disk model explains both the apparent divergence of the mixed micelle size and the origin of the phase limit. It should be appreciated that this explanation is based on structural features of the mixed micelles, rather than the critical phenomena discussed earlier in the nonionic system. At L/BS ratios greater than the phase limit, our experimental results[31, 32] suggest that the bilayer structures can exist as metastable liposomes ($\bar{R}_h$ 200–700 Å), i.e., a disk-to-liposome transition can take place rather than a true phase separation. Details of this interesting transition are being investigated.[83] Dynamic light-scattering studies of the size and stability of liposomal systems produced by other methods have previously been reported by Chen *et al.*[86] and Ostrowsky and Sornette.[87]

Quantitative support for the mixed disk model comes from the success of fitting the experimental data of Figures 16 and 17 with theoretical predictions based on equation (21). This is illustrated in Figure 17B, where the data of Figure 17A, obtained at various total lipid concentrations, are replotted as a function of $C_L/(C_{BS} - \text{IMC})$. The data are now seen to fall on a single functional relationship which can be fitted quite well by the mixed disk model. Moreover, the resulting deductions[31] of the parameters: $(2\rho/\sigma)$ (17.5 Å), α (2.0 at 20°C), and IMC (3.2 mm for large mixed micelles at 20°C) are found to be in excellent agreement with independent estimates based on molecular models and other experimental data.[31, 88]

Cholesterol Solubilization and Precipitation in Model Bile Systems. Because of its relevance to the problem of cholesterol gallstone formation, we have systematically investigated the process of cholesterol solubilization in bile salt and bile salt–lecithin solutions and in addition have studied the

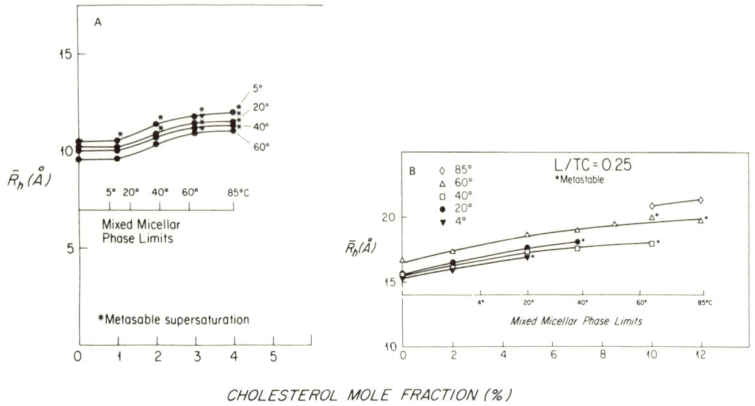

Figure 18. Influence of cholesterol incorporation on $\bar{R}_h$ in taurocholate–lecithin–cholesterol systems.(32, 35) In (A) only simple TC micelles are present. In (B) the L/TC ratio is 0.25 and simple, and mixed TC-L micelles coexist. Measurements are made at various temperatures. Systems that are metastably supersaturated with cholesterol are indicated by *. All samples contain 10 g/dl total lipid concentration in 0.15 *M* NaCl. Reprinted from Reference 35.

evolution of cholesterol precipitation from supersaturated systems.(32, 35) In Figures 18A and 18B, the influence of solubilized cholesterol on $\bar{R}_h$ is shown for simple (TC) and mixed (TC–lecithin) solutions. The small TC micelles increase only slightly 10 → 11.5 Å as the cholesterol mole fraction reaches 4% and show no abrupt change as the limit of solubilization is reached and exceeded. Similarly, in solutions where simple and mixed micelles coexist (L/TC = 0.25), only a modest increase in size (2–3 Å) occurs as up to 12% cholesterol is solubilized. These data suggest that the incorporation of cholesterol into simple and mixed micelles causes only small alterations of micellar size, and does not require any significant reorganization of micellar structure. On this basis we have analyzed(32, 89) cholesterol solubilization in terms of binding equilibria, in which the simple TC micelles possess a single binding site and the mixed TC–L micelles contain multiple sites whose number depends on the mixed micellar size.(32) By coupling the binding equilibria to the equilibrium that exists between cholesterol monomers and precipitates, one obtains the model of cholesterol solubilization(32) shown in Figure 19. This model implies that the limit of cholesterol solubilization is reached when the concentration of cholesterol monomers becomes equal to the aqueous solubility of the cholesterol monomer.

Using this concept, we can also understand the temperature dependence of cholesterol solubilization and the use of temperature variations to study cholesterol precipitation.(32, 35) At high temperature (>60°C) one is

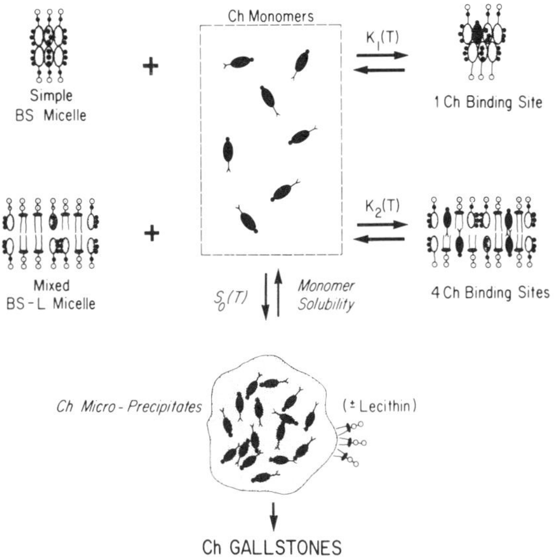

Figure 19. Schematic diagram of cholesterol solubilization in bile salt–lecithin solutions.[35] See text for explanation.

able to solubilize the maximum amount of cholesterol by virtue of the increase in monomeric solubility with temperature and the increased affinity of the binding sites. When the temperature of such a system is rapidly lowered, the concentration of cholesterol monomers will instantaneously exceed its maximum solubility. If this degree of monomeric supersaturation is greater than the critical degree needed for nucleation, a rapid homogeneous precipitation of cholesterol from the monomeric phase will occur.[32] Solubilized cholesterol will then be released from the micelles in order to repopulate the monomer concentration, leading to further precipitation. In effect then, a flow of cholesterol from micelles → monomers → precipitate occurs. Experimental studies[32] of such precipitation phenomena are illustrated in Figure 20 where the scattered intensity $I(t)$ and the $\bar{R}_h$ values of the microprecipitates have been monitored as a function of time after making temperature "drops" ΔT of varying depth. The data show an induction time, t_{ind}, associated with the initiation of precipitation which becomes larger as the degree of supersaturation becomes smaller (i.e., varies inversely with ΔT). In addition we see that the final size of the microprecipitates also varies inversely with ΔT from 500 to 1900 Å. Such findings[32] are consistent with a homogeneous nucleation process.[90] By employing the concepts of nucleation theory, we have been able to deduce from these and other data[32] the interfacial surface energy, E_i, between

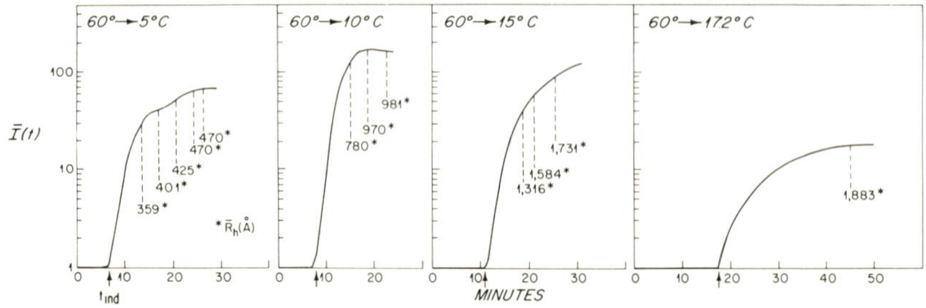

Figure 20. Temperature "drop" studies of cholesterol precipitation kinetics in model bile systems.[32, 35] $\bar{I}(t)$ represents scattered intensity from the solutions. $\bar{R}_h$ values correspond to the size of microprecipitates. Reprinted from Reference 35.

precipitate and solution. In the absence of lecithin, E_i is estimated to be greater than 12.8 erg/cm^2, whereas in the presence of lecithin E_i is ~ 5 erg/cm^2.[32] These values can account for the relative metastability of supersaturated solutions containing no lecithin and suggest that when present lecithin may act to lower the interfacial energy by coating the surface of the microprecipitates. By this coating mechanism the cholesterol precipitates can remain dispersed in solution, analogous to the solubilization of oil in microemulsion systems.[91]

Native Bile Systems. In very recent studies[92] dynamic light scattering has been used to characterize the macromolecular components present in native bile samples obtained from the dog under different physiological conditions. The nonexponentiality of the measured autocorrelation functions indicate that native bile contains roughly two macromolecular populations that differ widely in size. The minor population (whose concentration is estimated to be less than 0.1% of the total solutes) consists of large particles $\bar{R}_h \sim$ 400–500 Å, and may represent a heavy molecular weight protein, membrane vesticle, or pigment aggregate. The major population consists of small macromolecules whose $\bar{R}_h$ values are 20–35 Å, consistent with sizes expected for micellar aggregates. Evidence that these small particles are, in fact, mixed micelles, has been obtained by studying the effect of diluting native bile with physiological buffer solutions. Such studies[92] reveal a dramatic growth in the size of the small particles with dilution ($\bar{R}_h$ exceeds 200 Å), which is followed at higher dilutions by a transition to liposomes. This behavior is striking in its similarity to the effect of dilution on model bile systems (Figure 18A) and results from the equilibrium between the intermicellar concentration of bile salt monomers and the bile salts present in the mixed disk micelles. These studies thus provide compelling evidence that native bile does contain mixed micelles whose size and equilibria closely resemble the model system.

8.3.2.2. Sphingomyelin and Gangliosides. In addition to the biliary lipids there exist many other amphiphilic molecules of biological importance. In the first application of dynamic light scattering methods to biological micellar systems, Cooper *et al.*[33] studied mixed micelles formed by Triton X-100 (T), a synthetic nonionic detergent, and sphingomyelin (S), an amphiphile found in cell membranes similar in structure to lecithin. At low S/T ratios, they obtained evidence for the coexistence of pure Triton micelles and a mixed micelle of comparable size ($\bar{R}_h \sim 50$ Å) similar to the behavior of bile salt–lecithin systems at low L/BS ratios.[31] With increasing S/T ratio (above the coexistence range), $\bar{R}_h$ increases and exceeds 80 Å, which is also reminiscent of the growth of the bile salt–lecithin mixed micelles. Unfortunately, their data are not close enough to the Triton–sphingomyelin phase limit (S/T = 2.1) to indicate whether $\bar{R}_h$ is actually diverging. Nevertheless their diffusion data coupled with deductions of micellar weight and viscosity[93] suggest that the mixed micelles are disk-shaped and that the disk radius is increasing substantially with S/T ratio. In view of the similarity between the Triton–sphingomyelin and bile salt–lecithin mixed micellar systems, it would appear that the mixed disk model, developed for the latter case, may be rather general for describing the solubilization of diacyl membrane lipids by detergents.

More recently Corti *et al.*[34] have used conventional and dynamic light scattering to study the aggregative behavior of G_{M1} and G_{D1a} ganglioside molecules. These are also diacyl membrane components, but differ from sphingomyelin and lecithin in possessing hydrophilic head groups composed of sialic acid residues. In contrast to the behavior of the previous membrane components, the gangliosides are able to form micellar aggregates by themselves without the need of a solubilizing detergent. Corti *et al.*[34] find similar $\bar{R}_h$ values for the G_{M1} and G_{D1a} micelles of 63.9 and 59.5 Å, respectively, but rather different aggregation numbers of 352 and 229. In the case of G_{M1}, they suggest that the data are consistent with an oblate or disklike micellar structure (of 2 : 1 axial ratio). However, a similar analysis of the G_{D1a} data (not done by the authors) reveals that a prolate shape of axial ratio 4.5 : 1 is as good or better at fitting the data as the oblate shape for this ganglioside. The prolate shape would be more consistent with the properties of the liquid crystalline phases formed in mixed ganglioside–water systems,[94] which indicate the strong preference of hexagonal (i.e., rodlike) structures over lamellar phases. Conceivably both prolate and oblate shapes may coexist in ganglioside mixtures and their relative amounts could have important biological consequences.

8.3.3. Microemulsion and Inverted Micellar Systems

Following the applications of light scattering methods for studying single-component amphiphile–water systems and multicomponent bio-

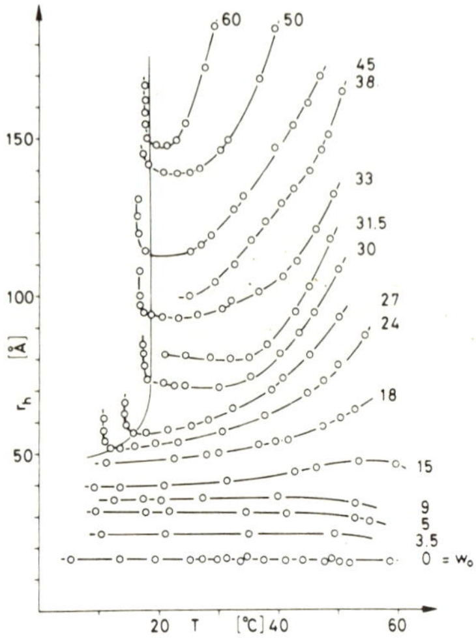

Figure 21. $\bar{R}_h$ values for AOT–H_2O–isooctane microemulsion droplets as a function of temperature and the H_2O/AOT molar ratio from Reference 28. Reprinted with permission of *J. Phys. Chem.* Copyright 1979, American Chemical Society.

logical micellar systems, a number of groups[28, 29, 95–97] have begun studying the technologically important microemulsion systems.[91] These systems are optically transparent mixtures containing water, hydrocarbon oil, surfactant, and in some cases a medium chain alcohol which is termed a cosurfactant. The composition–temperature phase diagrams of these systems have been studied for many years, and typically display three isotropic regions corresponding to[91] (1) water-rich systems (L_1 phase), (2) oil-rich systems (L_2 phase), and (3) a transition region where oil, water, and surfactant are present in comparable amounts (surfactant phase).

In the L_1 and L_2 regions it is generally believed that the major component (i.e., water or oil, respectively) exists as a continuous phase and that the minor component is solubilized in small microdroplets whose surfaces are coated by the surfactant (and cosurfactant if present).[91] Compelling evidence for this long-held view has recently been provided for L_2 microemulsions in studies of the aerosol–OT/H_2O/isooctane system by Zulauf, Eicke, and co-workers.[28, 98] Using laser light scattering, they have deduced the hydrodynamic radius of the small water droplets and inverted AOT micelles present in these systems, at various H_2O/AOT molar ratios and temperatures (see Figure 21). These data illustrate a sizable growth of the microemulsion droplets with increasing H_2O/AOT ratio, and a further increase in particle size as a function of temperature for systems with large H_2O/AOT ratios. Zulauf and Eicke[28] have shown that the minimum $\bar{R}_h$ values measured at each molar ratio are quantitatively consistent with a so-called "equipartition model" in which the droplets contain a homogeneous water core covered by a monolayer of AOT molecules (having a constant surface density, independent of radius). Such a model is conceptually similar to Small's model of the mixed micelle in the bile salt/lecithin/water system[84] and predicts that the droplet radii should increase linearly with the H_2O/AOT molar ratio [similar to equation (20)]. At a constant composition the further growth of the microemulsion droplets with temperature and the eventual phase separation that occurs are also reminiscent of the divergence behavior[31] seen in the biliary lipid system and could be explained by a generalization of the "mixed disk" model. If one allows AOT molecules to partition between the surface and interior of the water droplets in a manner that is temperature dependent, then the resulting decrease in AOT molecules available to coat the surface will necessitate an increase in droplet size relative to the "equipartition model." This is exactly equivalent to the increase in disk radius predicted in the mixed disk model.[31] In the present case, the generalized "mixed droplet" model would also be consistent with the experimental fact that the aqueous solubility of the AOT molecule increases appreciably with rising temperature.[99] Although discussed qualitatively by Zulauf and Eicke,[28] the predictions of a "mixed droplet" model have not yet been evaluated quantitatively.

In further studies of microemulsion systems, Vrij and co-workers[95] and Cazabet *et al.*[29] have studied quarternary systems in regions of the phase diagram where oil is believed to be the continuous phase (i.e., the L_2 region). These studies have employed both intensity and diffusion coefficient measurements and have focused on the effects of droplet concentration as a means for gaining insight into the interactions between droplets and the properties of the system as it approaches the transition region of the phase diagram. The dependence of $\bar{I}$ and D value on the volume fraction of droplets is shown in Figure 22 for a microemulsion system studied by Cazabat *et al.*[29] The data are remarkable in showing a nonmonotonic behavior in both quantities, and have been analyzed theoretically by the authors in terms of a simple model for the interactions between droplets which assumes a close-range hard-sphere repulsion and a longer-range attractive potential. The model provides an excellent fit to the $\bar{I}$ data (see Figure 22) and qualitatively explains the behavior of D. The analysis suggests that the maximum and minimum in $\bar{I}$ and D, respectively, are both related to the behavior of the osmotic compressibility of the system $\partial\Pi/\partial\phi$, which varies strongly with ϕ. This basic relationship was previously mentioned in our discussion of equations (11) and (12). In the present case, $\partial\pi/\partial\phi$ reflects both the attractive and repulsive interactions at low ϕ values ($\phi < 0.1$), whereas at higher values ($\phi \approx 0.2$–0.4) it becomes dominated by the repulsion effects. At still higher ϕ values a maximum appears in the dependence of D on ϕ, which the authors attribute to a structural change

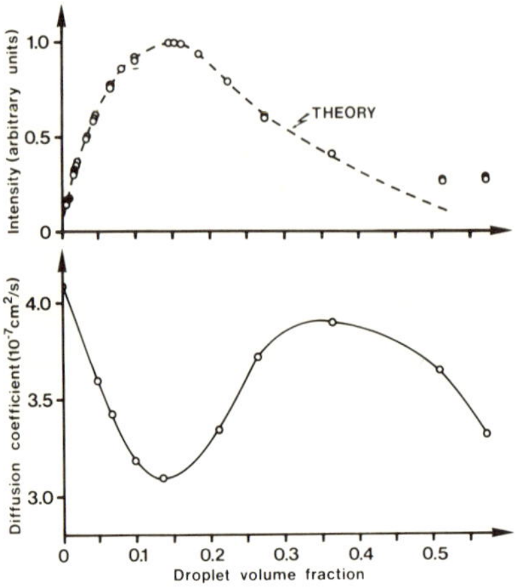

Figure 22. Dependence of scattered intensity and diffusion coefficient on droplet volume fraction for a quaternary microemulsion system studied in Reference 29. Dashed curve is a theoretical prediction based on a model of droplet interactions. Reprinted with permission from Reference 29.

associated with the transition region of the system. As yet a microscopic understanding of the transition region, where the system transforms from oil droplets to water droplets (or vice versa), remains to be elucidated. This fascinating inversion behavior, which some speculate to evolve through a bicontinuous oil/water system,[100, 101] remains an active area of investigation where dynamic light scattering studies may provide new insights.

8.4. SUMMARY

This chapter has attempted to illustrate how modern laser light scattering methods can provide detailed information on the size, shape, structure, interactions, and thermodynamic properties of the macromolecular aggregates that form in micellar systems. A discussion of the theoretical issues underlying such deductions has been given, and a review of the application of laser light scattering to the study of synthetic micelles, biological micelles, and microemulsion systems has been presented. While it would be impossible to cite every relevant study in these rapidly growing areas, it is hoped that those studies which have been chosen illustrate the diversity of phenomena seen in micellar systems, as well as the remarkable utility of laser light scattering methods for probing these systems. It is also hoped that the present review has indicated areas where further experimental and theoretical work concerning both the light scattering technique and its application to micellar systems would be fruitful. Such studies could have great impact in the areas of physical chemistry, medicine, and technology where a better understanding of micellar phenomena is vital.

ACKNOWLEDGMENTS

It is a pleasure to thank Professors George Benedek of MIT and Martin Carey of Harvard Medical School, who introduced me to the fields of laser light scattering and micelles, respectively, and in whose laboratories much of the work presented here was performed. I also wish to specially thank Professor Björn Lindman of the University of Lund, Sweden, for making possible a six-month stay in his group, during which time this manuscript was written and prepared, and for innumerable discussions with him and his colleagues in Scandinavia that have enlarged my view of micellar phenomena. I am likewise grateful to Professor Werner Känzig, of the ETH, Zürich, for enabling me to work in his research group, where the recent studies on SDS micelles and Native Bile systems were performed. Many thanks to my dear colleagues and friends at MIT, PBBH, ETH, and LU for making this work productive and enjoyable.

REFERENCES

1. J. W. McBain, Colloids and their viscosity (discussion), *Trans. Faraday Soc.* **9**, 99 (1913).
2. G. S. Hartley, *Aqueous Solutions of Paraffin Chain Salts*, Hermann, Paris (1936).
3. C. Tanford, *The Hydrophobic Effect*, Wiley, New York (1972).
4. H. Wennerström and B. Lindman, Micelles. Physical chemistry of surfactant association, *Phys. Rep.* **52**, 1–86 (1979).
5. B. Lindman and H. Wennerström, Amphiphile aggregation in aqueous solution, *Top. Curr. Chem.* **87**, 1–83 (1980).
6. H.-F. Eicke, Surfactants in nonpolar solvents, *Top. Curr. Chem.* **87**, 84 (1980).
7. P. Debye, Light scattering in soap solutions, *Ann. N.Y. Acad. Sci.* **51**, 575 (1949).
8. P. Debye and E. W. Anacker, Micelle shape from dyssymetry measurements, *J. Phys. Colloid Chem.* **55**, 644 (1959).
9. K. J. Mysels and L. H. Princen, Light scattering by some lauryl sulfate solutions, *J. Phys. Chem.* **63**, 1696 (1959).
10. H. F. Huisman, Light scattering of solutions of ionic detergents, *Proc. Kon. Ned. Akad. Wet., Ser. B* **67**, 367, 376, 388, 407 (1964).
11. J. P. Kratohvil and H. T. Dellicolli, Measurement of the size of micelles: the case of sodium taurodeoxycholate, *Fed. Proc.* **29** (4), 1335 (1970).
12. W. Prins and J. J. Hermans, Light scattering by solutions of some sodium alkyl-*l*-sulfates, *Proc. Kon. Ned. Akad. Wet., Ser. B* **59**, 298 (1956).
13. M. F. Emerson and A. Holtzer, On the ionic strength dependence of micelle number III, *J. Phys. Chem.* **71**, 1898 (1967).
14. P. Mukerjee, The size distribution of small and large micelles: A multiple equilibrium analysis, *J. Phys. Chem.* **76**, 565 (1972).
15. J. N. Israelachvilli, D. J. Mitchell, and B. W. Ninham, Theory of self-assembly of hydrocarbon ampliphiles into micelles and bilayers, *J. Chem. Soc. Faraday Trans. 2* **72**, 1525 (1976).
16. E. Ruckenstein and R. Nagarajan, On critical concentrations in micellar solutions, *J. Colloid Interface Sci.* **57**, 388 (1976).

17. P. N. Pusey, Macromolecular diffusion, in *Photon Correlation and Light Beating Spectroscopy*, H. Z. Cummins and E. R. Pike, eds., Plenum, New York (1974).
18. G. D. J. Phillies, G. B. Benedek, and N. A. Mazer, Diffusion in protein solutions at high concentration: A study by quasielastic light scattering spectroscopy, *J. Chem. Phys.* **65**, 1883 (1976).
19. R. J. Cohen and G. B. Benedek, The functional relationship between polymerization and catalytic activity of beef liver glutamate dehydrogenase. I. Theory, *J. Mol. Biol.* **108**, 151 (1976).
20. B. J. Berne and R. Pecora, *Dynamic Light Scattering with Applications to Chemistry, Biology, and Physics*, Wiley, New York, (1976).
21. A. Einstein, *Investigations on the Theory of the Brownian Movement*, Dover, New York (1956).
22. N. A. Mazer, G. B. Benedek, and M. C. Carey, An investigation of the micellar phase of sodium dodecyl sulfate in aqueous sodium chloride solutions using quasielastic light scattering spectroscopy, *J. Phys. Chem.* **80**, 1075 (1976).
23. N. A. Mazer, M. C. Carey, and G. B. Benedek, Size, shape, and thermodynamics of sodium dodecyl sulfate micelles using quasielastic light scattering spectroscopy, in *Micellization, Solubilization, and Microemulsions, Vol. 1*, K. L. Mittal, ed., pp. 359–382, Plenum, New York (1977).
24. M. Corti and V. Degiorgio, Investigation of aggregation phenomena in aqueous sodium dodecyl sulfate solutions at high NaCl concentration by quasielastic light scattering, in *Solution Chemistry of Surfactants*, K. L. Mittal, ed., p. 377, Plenum, New York (1979).
25. P. J. Missel, N. A. Mazer, G. B. Benedek, C. Y. Young, and M. C. Carey, Thermodynamic analysis of the growth of soldium dodecyl sulfate micelles, *J. Phys. Chem.* 1044 (1980).
26. D. F. Nicoli, D. R. Dawson, and H. W. Offen, Pressure dependence of micelle size by photon correlation spectroscopy, *Chem. Phys. Lett.* **66**, 291 (1979).
27. M. Corti and V. Degiorgio, Critical behavior of a micellar solution, *Phys. Rev. Lett.* **45**, 1045 (1980).
28. M. Zulauf and H.-F. Eicke, Inverted micelles and microemulsions in the ternary system H_2O/aerosol–OT/isooctane as studied by photon correlation spectroscopy, *J. Phys. Chem.* **83**, 480 (1979).
29. A. M. Cazabat, D. Langevin, and A. Pouchelson, Light scattering study of water–oil microemulsions, *J. Colloid Interface Sci.* **73**, 1 (1980).
30. N. A. Mazer, M. C. Carey, R. F. Kwasnick, and G. Benedek, Quasielastic light scattering studies of aqueous biliary lipid systems: Size, shape, and thermodynamics of bile salt micelles, *Biochemistry* **18**, 3064 (1979).
31. N. A. Mazer, G. B. Benedek, and M. C. Carey, Quasielastic light scattering studies of aqueous biliary lipid systems: Mixed micelle formation in bile salt–lecithin solutions, *Biochemistry* **19**, 601 (1980).
32. N. A. Mazer, Quasielastic light scattering studies of micelle formation, solubilization, and precipitation in aqueous biliary lipid systems, Ph.D. thesis, Massachusetts Institute of Technology, Cambridge, MA (1978).
33. V. G. Cooper, S. Yedgar, and Y. Barenholz, Diffusion coefficients of mixed micelles of Triton X-100 and sphingomyelin and of sonicated sphingomyelin liposomes, measured by autocorrelation spectroscopy of Rayleigh scattering light, *Biochim. Biophys. Acta* **363**, 86 (1974).
34. M. Corti, V. Degiorgio, R. Ghidoni, S. Sonnino, and J. Tettamanti, Laser-light scattering investigations of the micellar properties of gangliosides, *Chem. Phys. Lipids* **26**, 225 (1980).
35. N. A. Mazer, M. C. Carey, and G. B. Benedek, Quasielastic light scattering studies of

model bile systems, in *Lasers in Biology and Medicine*, F. Hillenkamp, R. Pratesi, and C. A. Sacchi, eds., Plenum, New York (1980), pp. 127–150.
36. D. E. Koppel, Analysis of macromolecular polydispersity in intensity correlation spectroscopy: The method of cumulants, *J. Chem. Phys.* **57**, 4814 (1972).
37. N. A. Mazer, Data analysis for light scattering studies of macromolecular polydispersity: An extension of the method of cumulants, S.B. thesis, Massachusetts Institute of Technology (1973).
38. G. D. J. Phillies, Effects of intermacromolecular interactions on diffusion. 1. Two-component solutions, *J. Chem. Phys.* **60**, 976 (1974).
39. G. K. Batchelor, Brownian diffusion of particles with hydrodynamic interaction, *J. Fluid Mech.* **74**, 1 (1976).
40. C. W. Pyun and M. Fixman, Frictional coefficient of polymer molecules in solution, *J. Chem. Phys.* **41**, 937 (1964).
41. B. D. Fair and A. M. Jamieson, Effect of electrodynamic interactions on the translational diffusion of bovine serum albumin at finite concentrations, *J. Colloid Interface Sci.* **73**, 130 (1980).
42. C. Y. Young, P. J. Missel, N. A. Mazer, G. B. Benedek, and M. C. Carey, Deduction of micellar shape from angular dissymmetry measurements of light scattered from aqueous sodium dodecyl sulfate solutions at high sodium chloride concentrations, *J. Phys. Chem.* **82**, 1375 (1978).
43. J. P. Kratohvil, The concentration dependence of micelle aggregation and the shape of micelles of sodium dodecyl sulfate and hexadecyltrimethylammonium bromide, *Chem. Phys. Lett.* **60**, 238 (1979).
44. D. McQueen and J. Hermans, Determination of the translational diffusivity of detergent micelles from the spectrum of scattered light, *J. Colloid Interface Sci.* **39**, 358 (1975).
45. R. C. Murray and G. S. Hartley, Equilibrium between micelles and simple ions, with particular reference to the solubility of long-chain salts, *Trans. Faraday Soc.* **31**, 183 (1935).
46. E. I. Franses, H. T. Davis, W. G. Miller, and L. E. Scriven, Nature of large aggregates in supercooled aqueous solutions of sodium dodecyl sulfate, *J. Phys. Chem.* **84**, 2413 (1980).
47. U. Hendriksson, L. Odberg, J. C. Eriksson, and L. Westman, Nitrogen-14 nuclear magnetic relaxation in aqueous micellar solutions of n-hexadecyltrimethylammonium bromide and chloride, *J. Phys. Chem.* **81**, 76 (1977).
48. J. Ulmium and H. Wennerström, Proton NMR bandshape for large aggregates; micellar solutions of hexadecyltrimethylammonium bromide, *J. Mag. Res.* **28**, 309 (1977).
49. D. Stigter, Intrinsic viscosity and flexibility of rodlike detergent micelles, *J. Phys. Chem.* **70**, 1323 (1966).
50. V. Luzzati, H. Mustacchi, A. Skoulios, and F. Husson, La structure des colloides d'association. I. Les phases liquide–crystallines des systemes amphiphile-eau, *Acta Crystallogr.* **13**, 660 (1960).
51. E. J. Staples and G. J. T. Tiddy, Nuclear magnetic resonance technique to distinguish between micelle size changes and secondary aggregation in anionic and nonionic surfactant solutions, *J. Chem. Soc. Faraday I* **74**, 2530 (1978).
52. S. Hayashi and S. Ikeda, Micelle size and shape of sodium dodecyl sulfate in concentrated NaCl solutions, *J. Phys. Chem.* **84**, 744 (1980).
53. P. Llanos and R. Zana, Use of pyrene examer formation to study the effect of NaCl on the structure of sodium dodecyl sulfate micelles, *J. Phys. Chem.* **84**, 3339 (1980).
54. K. L. Mittal, ed., *Micellization, Solubilization and Microemulsions*, Vol. 1, Plenum, New York (1977); see discussion on pp. 419–428.

55. J. E. Liebner and J. Jacobus, Charged micelle shape and size, *J. Phys. Chem.* **81**, 130 (1977).
56. J. Briggs, D. F. Nicoli, and R. Ciccolello, Light scattering from polydisperse SDS micellar solutions, *Chem. Phys. Lett.* **73**, 149 (1980).
57. N. Mazer, P. Missel, H. Roder, and W. Kanzig, Flexibility of rodlike SDS micelles, unpublished results.
58. A. Peterlin, Light scattering by non-Gaussian macromolecular coils, in *Electromagnetic Scattering*, M. Kerker, ed., Pergamon, Elmsford, New York (1963), pp. 357–375.
59. G. Porte, J. Appell, and Y. Poggi, Experimental investigations on the flexibility of elongated atylpyridinium bromide micelles, *J. Phys. Chem.* **84**, 3105 (1980).
60. J. Appell and G. Porte, An investigation on the micellar shape using angular dissymmetry of light scattered by cetylpyridinium bromide, *J. Colloid Interface Sci.* **81**, 85 (1981).
61. R. J. M. Tausk and J. Th. G. Overbeek, Physical chemical studies of short-chain lecithin homologues. IV. A simple model for the influence of salt and the alkyl chain length on the micellar size, *Biophys. Chem.* **2**, 175 (1974).
62. M. Corti and V. Degiorgio, Quasi-elastic light scattering study of intermicellar interactions in aqueous sodium dodecyl sulfate solutions, *J. Phys. Chem.* **85**, 711 (1981).
63. P. J. Missel, N. A. Mazer, M. C. Carey, and G. B. Benedek, Thermodynamics of the sphere-to-rod transition in alkyl sulfate micelles, in *Solution Behavior of Surfactants—Theoretical and Applied Aspects*, K. L. Mittal and E. J. Fendler, eds., Plenum, New York (1982).
64. A. Rohde and E. Sackmann, Quasielastic light scattering studies of micellar sodium dodecyl sulfate solutions at the low concentration limit, *J. Colloid Interface Sci.* **70**, 494 (1979).
65. M. J. Stephen, Spectrum of light scattered from charged macromolecules in solution, *J. Chem. Phys.* **55**, 3878 (1971).
66. M. Tanaka, S. Kaneshina, K. Shin-no, T. Okajama, and T. Tomida, Partial molal volumes of surfactant and its homologous salts under high pressure, *J. Colloid Interface Sci.* **46**, 132 (1974).
67. E. Vikingstad, A. Skauge, and H. Hoiland, Effect of pressure and temperature on the partial molal volume and compressibility of sodium decanoate micelles, *J. Colloid Interface Sci.* **72**, 59 (1979).
68. G. Lindblom, B. Lindman, and L. Mandell, Effect of micellar shape and solubilization on counter-ion binding studied by 81 Br NMR, *J. Colloid Interface Sci.* **42**, 400 (1973).
69. N. Mazer, J. Peetermans, and G. Benedek, Effect of pressure on the sphere-to-rod micelle transition, unpublished results.
70. P. Ekwall, H. Eikrem, and L. Mandell, The properties and structures of aqueous sodium caprylate solutions. I. The densities and partial specific volumes, *Acta Chem. Scand.* **17**, 111 (1963).
71. J. C. Lang and R. D. Morgan, Nonionic surfactant mixtures. I. Phase equilibria in $C_{10}E_4$–H_2O and closed-loop coexistence, *J. Chem. Phys.* **13**, 5849 (1980).
72. R. R. Balmbra, J. S. Clunie, J. M. Corkil, and J. F. Goodman, Effect of temperature on the micelle size of a homogeneous non-ionic detergent, *Trans. Faraday Soc.* **58**, 1661 (1962).
73. K. Shinoda and H. Arai, The correlation between phase inversion temperature in emulsion and cloud point in solution of nonionic emulsifier, *J. Phys. Chem.* **68**, 3485 (1964).
74. P. H. Elworthy and C. B. Macfarlane, Chemistry of nonionic detergents. Part V. Micellar structures of a series of synthetic nonionic detergents, *J. Chem. Soc.*, 907 (1963).
75. K. W. Kerrmann, J. G. Brushmiller, and W. I. Courchene, Micellar properties and critical opalescence of dimethylalkylphosphine oxide solutions, *J. Phys. Chem.* **70**, 2909 (1966).

76. G. Benedek, Spectra of light scattering by critical fluctuations, in *Light Scattering Spectra of Solids*, G. Wright, ed., Springer-Verlag, Berlin, (1968).
77. K. Kawasaki, Mode coupling and critical dynamics, in *Phase Transitions and Critical Phenomena*, Vol. 5a, C. Domb and M. S. Green, eds., Academic, London (1976).
78. R. Kjellander and E. Florin, Water structure and changes in thermal stability of the system poly(ethylene oxide)–water, *J. Chem. Soc. Faraday Trans. 1* **77**, 2053 (1981).
79. S. Saeki, N. Kuwahara, M. Nakata, and M. Kaneko, Upper and lower critical solution temperatures in poly(ethylene glycol) solutions, *Polymer* **17**, 685 (1976).
80. M. C. Carey and D. M. Small, The characteristics of mixed micellar solutions with particular reference to bile, *Am. J. Med.* **49**, 590 (1970).
81. M. C. Carey and D. M. Small, Physical chemistry of cholesterol solubility in bile—relationship to gallstone formation and dissolution in man, *J. Clin. Invest.* **61**, 998 (1978).
82. D. M. Small, Size and structure of bile salt micelles, influence of structure, concentration, counterion concentration, pH, and temperature, *Adv. Chem. Ser. No.* **84**, 31 (1968).
83. P. Schurtenberger, N. Mazer, and W. Kanzig, Micelle-to-vesicle transition in bile salt–lecithin solutions, *J. Phys. Chem.* (in press).
84. D. M. Small, Physical-chemical studies of cholesterol gallstone formation, *Gastroenterology* **52**, 607 (1967).
85. N. A. Mazer, M. C. Carey, R. F. Kwasnick, and G. B. Benedek, Quasielastic light scattering spectroscopic studies of aqueous bile salt, bile salt–lecithin and bile salt–lecithin–cholesterol solutions, in *Micellization, Solubilization and Microemulsions*, Vol. 1, K. L. Mittal, ed., Plenum, New York (1977), p. 383.
86. F. A. Chen, A. Chrzenszcyk, and B. Chu, Quasielastic laser light scattering of phosphatidylcholine vesicles, *J. Chem. Phys.* **64**, 3403 (1976).
87. N. Ostrowsky and D. Sornette, Stability and fusion of visicles, in *Light Scattering in Liquids and Macromolecular Solutions*, V. Degiorgio, M. Corti, and M. Giglio, ed., Plenum, New York (1980), p. 125.
88. C. Gahwiller, C. von Planta, D. Schmidt, and H. Steffen, Untersuchungen über die grosse, struktur und dynamik von gallensaure/lecithin–mischmicellar, *Z. Naturforsch.* **C32**, 748 (1977).
89. N. A. Mazer, G. B. Benedek, and M. C. Carey, What determines the solubilization of cholesterol in bile? *Gastroenterology Abst.*, Nov. (1978).
90. A. S. Walton, *The Formation and Properties of Precipitates*, Interscience, New York (1967).
91. K. Shinoda and S. Friberg, Microemulsions: colloidal aspects, *Adv. Colloid Interface Sci.* **4**, 281 (1975).
92. N. Mazer, P. Schurtenberger, W. Kanzig, M. Carey, and R. Preisig, Quasielastic light scattering (QLS) studies of native dog bile—comparison with model systems, *Gastroenterology Abst.* May (1981).
93. S. Yedgar, Y. Barenholz, and V. G. Cooper, Molecular weight, shape and structure of mixed micelles of Triton X-100 and sphingomyelin, *Biochim. Biophys. Acta* **363**, 98 (1974).
94. W. Curatolo, D. M. Small, and G. G. Shipley, Phase behavior and structural characteristics of hydrated bovine brain gangliosides, *Biochim. Biophys. Acta* **468**, 11 (1977).
95. A. A. Calje, W. G. M. Agterob, and A. Vrij, Light scattering of a concentrated W–O micro emulsion; Application of modern fluid theories, in *Micellization, Solubilization and Microemulsions*, Vol. 2, K. L. Mittal, ed., Plenum, New York (1977), p. 779.
96. R. Finsy, A. Devriese, and H. Lekkerkerker, Light scattering study of the diffusion of interacting particles, *J. Chem. Soc., Faraday Trans. 2* **76**, 967 (1980).
97. E. Gulari, B. Bedwell, and S. Alkhafaji, Quasi-elastic light scattering investigation of microemulsions, *J. Colloid Interface Sci.* **77**, 202 (1980).

98. H.-F. Eicke and R. Kubik, The optical matching phenomenon in water–oil microemulsions, *Ber. Bunseges Phys. Chem.* **84**, 36 (1980).
99. J. Rogers and P. A. Winsor, Change in the optic sign of the lamellar phase (G) in the aerosol OT–water system with composition or temperature, *J. Colloid Interface Sci.* **30**, 247 (1969).
100. L. E. Scriven, Equilibrium bicontinuous structures, in *Micellization, Solubilization, and Microemulsions*, Vol. 2, K. L. Mittal, ed., Plenum, New York (1977), p. 877
101. B. Lindman, N. Kamenka, T.-M. Kathopoulis, B. Brun, and P.-G. Nilsson, Translational diffusion and solution structure of microemulsions, *J. Phys. Chem.* **84**, 2485 (1980).

9

Light Scattering from Polymer Gels

Toyoichi Tanaka

Department of Physics and Center for Materials Science and Engineering
Massachusetts Institute of Technology
Cambridge, Massachusetts 02139

9.1. INTRODUCTION

Polymer molecules in a solution undergo random thermal motions. The motions give rise to space and time fluctuations of the polymer concentration. In the previous chapters we have seen that such concentration fluctuations (and thus the random thermal motions) can be observed using the technique of laser light scattering spectroscopy.

If the polymer solution is so dilute that the interaction among individual molecules is negligible, the random motions of the polymer can be described as a three-dimensional random walk. The random walk is an ideal diffusion process characterized by a diffusion coefficient, D, which means a sinusoidal concentration fluctuation with a wave vector $\mathbf{q}$ decays exponentially with a decay time $\tau = (Dq^2)^{-1}$. As the interaction among the individual polymers becomes significant the random motion of the polymers deviates from a simple diffusion process, and a sinusoidal concentration fluctuation does not necessarily follow the above decay law. In turn, from a detailed analysis of the nonideal concentration fluctuations, it is possible to study the interaction among polymers.

In this chapter we consider polymer gels in which the interaction among polymers is, in a sense, infinitely large. Namely, all the polymers are cross-linked by covalent bonds to form a three-dimensional network. It is not easy to describe the random motions of cross-linked polymers if we start from the ideal diffusive motions of individual polymers and gradually modify them by introducing interactions. It is by far easier to take a roundabout point of view and start by considering the cross-linked polymer network itself. It is interesting to observe that the collective random motions in a cross-linked polymer network can be described by a pure diffusion process, just as for the opposite case of an infinitely dilute solution of polymers.

In this chapter, using a continuum model we first describe the random motions in polymer gels, and derive the correlation function of light scattered from the gels. A comparison is made between the gel mode and diffusion process of an ideal polymer solution. Second, the viscoelastic parameters introduced in the continuum model will be connected to the properties at the molecular level. Finally the phase transition of polymer gels will be briefly sketched, and the random motion of polymer networks near the phase transition will be discussed.

9.2. COLLECTIVE MODES IN GELS

9.2.1. Collective Diffusion in a Gel

Consider gels to consist of a cross-linked polymer network that gives elasticity to the gel, and a liquid that occupies the space between the network chains.[1] We are concerned with the structural fluctuations of the polymer network rather than those of the gel liquid. The gel liquid is treated as an incompressible fluid. Since the wavelength of light is usually much larger than the average distance between neighboring cross-links, we treat the polymer network as a continuous medium. The network structure and its fluctuations are then uniquely described by a displacement vector $\mathbf{u}(\mathbf{r}, t)$ which represents the displacement of a point $\mathbf{r}$ on the polymer network at time t from its average location.

We consider the small deformation of a unit cube of polymer network of average density ρ. The displacement vector $\mathbf{u}(\mathbf{r}, t)$ then obeys the following linear equation:

$$\rho(\partial^2/\partial t^2)\mathbf{u} = \nabla \cdot \tilde{\sigma} - f(\partial/\partial t)\mathbf{u} \tag{1}$$

where $\tilde{\sigma}$ is the stress tensor whose component σ_{ik} gives the force along the k axis on a unit plane perpendicular to the i axis. Equation (1) is nothing but a representation of Newton's second law. The term on the left represents the mass times the acceleration of the unit cube. The terms on the right-hand side represent the forces exerted on the cube. The first term is the net force of the internal stresses and is expressed as the difference of the internal stresses on each two opposing walls of the cube. Thus, the net force is given by the divergence of the stress tensor. The second term is the drag friction on the network by the gel liquid. The friction is assumed to be proportional to the relative velocity $\partial\mathbf{u}/\partial t$ of the network and the liquid. The friction coefficient f is assumed to be proportional to the viscosity η of the gel liquid.

The stress tensor $\tilde{\sigma}$ is related to the displacement vector $\mathbf{u}$ by the following:

$$\sigma_{ik} = K\nabla \cdot \mathbf{u}\delta_{ik} + 2\mu(u_{ik} - \tfrac{1}{3}\nabla \cdot \mathbf{u}\delta_{ik}) \tag{2}$$

where

$$u_{ik} \equiv \tfrac{1}{2}[\partial u_k/\partial x_i + \partial u_i/\partial x_k] \tag{3}$$

The first term in equation (2) represents the stress produced by a volume change, while the second term represents the stress caused by a shear deformation. The coefficients K and μ are the bulk and shear moduli of the polymer network. Substituting equations (2) and (3) into equation (1), an equation of motion for the displacement vector is obtained

$$f\frac{\partial\mathbf{u}}{\partial t} = \left(K + \frac{\mu}{3}\right)\nabla(\nabla \cdot \mathbf{u}) + \mu\Delta\mathbf{u} \tag{4}$$

In most cases the acceleration term in equation (1) is much smaller than the other terms; it is neglected here. The equation of motion has three solutions corresponding to one longitudinal and two transverse modes. Each solution can be expressed by a diffusion equation

$$\frac{\partial u_l}{\partial t} = \frac{K + \frac{4}{3}\mu}{f}\frac{\partial^2 u_l}{\partial x_l^2} \qquad \text{(longitudinal)} \tag{5}$$

$$\frac{\partial u_t}{\partial t} = \frac{\mu}{f}\frac{\partial^2 u_t}{\partial x_l^2} \qquad \text{(transverse)} \tag{6}$$

where x_l represents the coordinate along the wave vector of the mode, and u_l and u_t are components of the displacement vector along and perpendicular to the wave vector. The normal modes in a gel are thus described by a diffusion equation with a diffusion coefficient D that is the ratio of the appropriate elastic modulus E, and a friction coefficient f

$$D = E/f \tag{7}$$

where

$$E = \begin{cases} K + \frac{4}{3}\mu & \text{(longitudinal)} \quad (8) \\ \mu & \text{(transverse)} \quad (9) \end{cases}$$

9.2.2. Comparison between Diffusion of Polymers and Gels

It is interesting to compare the diffusion processes in the cases of individual polymers in solution and the polymer network of a gel. Equation (7) for the diffusion coefficient of a polymer network can be used to describe polymers in solution if we consider that the system consists of an elastic "gas" of polymers immersed in a viscous liquid. The elastic modulus of such a "gas" is given by $E = NkT$, where N is the number of polymers per unit volume. The friction coefficient is provided by the Stokes formula, $f = N6\pi\eta a$, where η is the liquid viscosity and a is the hydrodynamic radius of the polymer. Substituting these expressions into the formula $D = E/f$, we produce the Stokes–Einstein relation for the diffusion coefficient of polymers in solution:

$$D = kT/6\pi\eta a \tag{10}$$

In general, diffusion process is observed when an elastic medium is immersed in a viscous fluid. The phonons that propagate in an elastic medium are overdamped in a viscous fluid, becoming diffusive modes.

There is a fundamental difference, however, between the diffusion process in a solution of polymers and in gels. The difference results from the cross-linking between polymers in gels. To illustrate this difference consider the interesting experiment carried out by Munch and his colleagues.[2] They prepared a polymer solution consisting of two types of polymers having different degrees of polymerization. They also prepared three types of gels, two consisting of only one type of polymer, and one made by cross-linking the two types of polymers together. They measured and compared the correlation functions of light scattered from the solution and the gels. For the solution, the correlation function obtained was a sum of two exponential functions corresponding to the separate and independent diffusion of the two types of polymer. For the gel consisting of both types of polymer, the correlation function was a single exponential with a diffusion coefficient somewhere in between the diffusion coefficients observed in gels each consisting of only one of the polymers. The gel consisting of two types of polymers resembles a crystal lattice consisting of two different kinds of masses arranged alternately. In such a crystal there exist two types of normal modes corresponding to acoustic and optical phonons. The acoustic phonon depends on the wave vector and represents an overall density wave in which the two types of masses move together. The optical phonon is independent of the wave vector and represents a relative motion of one kind of mass with respect to the other. In the case of gel consisting of two types of polymers only the mode corresponding to the acoustic phonon can be seen by laser light scattering spectroscopy, because the mode corresponding to the optical phonon is of a much higher frequency.

In summary, the normal modes in a solution consisting of two types of polymers are two diffusive modes, each mode corresponding to the diffusion of each polymer type. In the gel consisting of two types of polymers the normal modes are an ideal diffusive mode and an optical mode. This explains why in the gel the correlation function is single-exponential, while in the polymer solution, the correlation function is double-exponential.

9.2.3. Light Scattering from Collective Diffusion Modes in a Gel

The longitudinal and transverse diffusive modes in the gel give rise to polarized and depolarized light scattering, respectively.[1] The correlation function of the scattered electric field is proportional to the spatial Fourier transform of the correlation function of the displacement vector. The correlation function of the electric field of the scattered light is given by

$$\langle E_{\text{pol}}(\mathbf{q},t)E_{\text{pol}}(\mathbf{q},0)\rangle = \frac{I_0}{c}\left(\frac{\omega_0}{c}\right)^4 \frac{\sin^2\phi}{4\pi R^2}\left(\frac{\partial\varepsilon}{\partial\rho}\right)_T^2 \rho^2 \frac{LkT}{K+\frac{4}{3}\mu} \times \exp\left[-\frac{(K+\frac{4}{3}\mu)q^2t}{f}\right] \tag{11}$$

for polarized scattering and by

$$\langle E_{\text{dep}}(\mathbf{q},t)E_{\text{dep}}(\mathbf{q},0)\rangle = \frac{I_0}{c}\left(\frac{\omega_0}{c}\right)^4 \frac{\sin^2\phi}{4\pi R^2}\left(\frac{\partial\varepsilon_D}{\partial u_{xy}}\right)_T^2 \frac{LkT}{\mu}\exp\left(\frac{-\mu q^2 t}{f}\right) \tag{12}$$

for depolarized scattering. Here I_0 is the incident intensity of the light, L is the illuminated length, and ϕ is the angle between the polarization of the incident light and the scattering direction. ε and ε_D are the diagonal and off-diagonal elements of the dielectric tensor. The quantity $(\partial\varepsilon/\partial\rho)_T$ can be obtained by measuring the index of refraction n of the gel while changing the gel concentration ρ, since $\varepsilon = n^2$. The quantity $(\partial\varepsilon_D/\partial u_{xy})_T$ can be obtained by applying a shear deformation u_{xy} to the gel and measuring the depolarization of a light passing through the gel along the z axis. The ratio of the depolarized vs. polarized light intensity measured in the forward direction is given by $I_D/I_P = (\varepsilon_D/\varepsilon)^2$, and therefore we can obtain ε_D as a function of u_{xy}.

In self-beating spectroscopy we obtain a time correlation function not of the electric field but of the intensity of the scattered light. However, because of the Gaussian properties of the electric field, the correlation

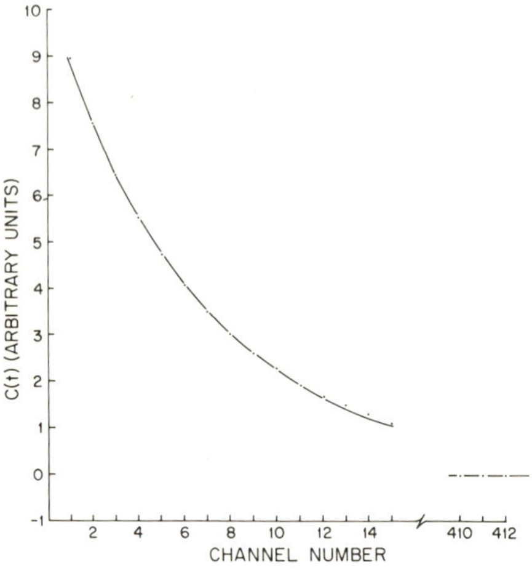

Figure 1. A measured correlation function of light scattered at a 90° angle from a 5% polyacrylamide gel at 25°C. The correlation function decays exponentially.

function of the intensity is given by the square of the correlation functions in equation (11) or (12). Thus the decay constants $\Gamma = 1/\tau$ become $2[(K + \frac{4}{3}\mu)/f]q^2$ and $(2\mu/f)q^2$ for polarized and depolarized scattering, respectively.

Thus from the time correlations of the scattered light, we can obtain the ratios $(K + \frac{4}{3}\mu)/f$ and μ/f. From the intensity measurements of the scattered light, which can be obtained by putting $t = 0$ in equations (11) and (12), we can obtain the quantities $K + \frac{4}{3}\mu$ and μ. Consequently, by using only polarized and depolarized scattering techniques, we can in principle determine all the viscoelastic parameters of a gel, K, μ, and f.

The exponential form of the correlation function of scattered light and the q^2 dependence of the decay rate are the characteristic feature of a diffusion process. These were experimentally verified by Tanaka, Hocker, and Benedek on polyacrylamide gels (Figures 1 and 2). The collective diffusion modes were also confirmed by Munch and his colleagues,(2, 3) and Hecht and Geissler.(4, 5)

9.2.4. Comparison between Light Scattering and Swelling of Gels

The collective diffusion process not only governs the microscopic concentration fluctuations in the gel, but also the kinetics of macroscopic swelling or shrinking of gels.(6) Measurements of the change in the radius of spherical acrylamide gels show that the characteristic time of swelling is proportional to the square of the radius. This corresponds to the q^2 dependence of the decay rate in the concentration fluctuations of the gel. It is remarkable that the diffusion coefficient obtained from the ratio between

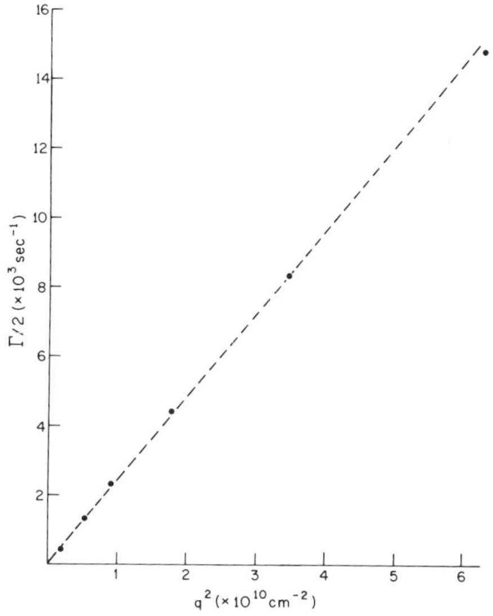

Figure 2. The relaxation rate Γ of the correlation function of light scattered by a 5% polyacrylamide gel is found to be linear in q^2 where $\mathbf{q}$ is the scattering vector. The exponential decay of the correlation function and the q^2 dependence of the decay rate show that the concentration fluctuations of the gel are mathematically described by a diffusion equation.

the characteristic time of swelling and the square of the radius coincides with the value obtained by light scattering measurements.

Comparison of the longitudinal modulus determined by deswelling experiments and light scattering were performed by Hecht and Geissler.[4, 5] They observed a good agreement between macroscopic and microscopic determinations. It is important to notice that the ordinary "unidirectional" stretching experiments do not give the longitudinal modulus, because such stretching does not involve a volume change and thus gives only a shear modulus, (Figures 3, 4 and 5). It is necessary to swell or deswell the gel as Hecht and Geissler did to obtain the longitudinal modulus. With these experiments the existence and characteristics of the longitudinal collective diffusive mode in gels have been confirmed.

So far no experiments have been reported on depolarized light scattering from gels. Presumably the depolarizability of the gels typically studied is very small. It would not be very difficult, however, to make gels which would give rise to depolarized light scattering. One could look at gels made of polymers whose solutions give depolarized light scattering.

Recently Nossal and his colleagues have devised a technique of incoherent light scattering with which to determine the shear modulus of a gel.[7, 8] They gave mechanical disturbance to a gel using a sound speaker, which generated standing shear waves having nodes at the cell surfaces. Light is incoherently scattered by inhomogeneous structures embedded in the gel which move together with the standing shear waves. The spectros-

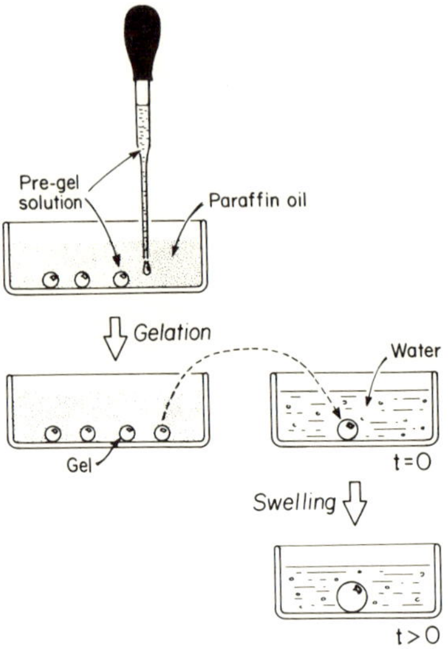

Figure 3. Preparation of spherical polyacrylamide gels in paraffin oil in a plastic dish. The spherical gel was transferred into water in which the gel swelled until it reached equilibrium size.

copy of the scattered light provides the characteristic frequencies of the standing waves and thereby one can determine the shear modulus of the gel.

9.3. KIRKWOOD–RISEMANN-TYPE EXPRESSION OF DIFFUSION COEFFICIENT

In the previous section, the collective diffusive mode in gels was described in terms of the elasticity of the polymer network and the friction

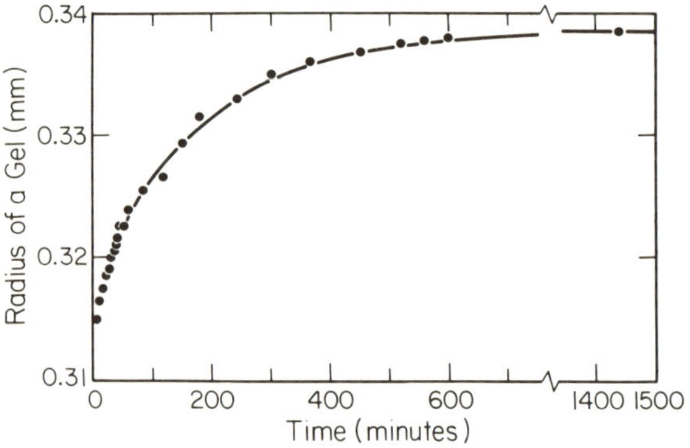

Figure 4. The radius of the spherical gel increases with time. Except at the very beginning the radius increases exponentially: $r = r_0 + \Delta r\ [1 - \exp(-t/\tau)]$, where τ represents the characteristic time of swelling.

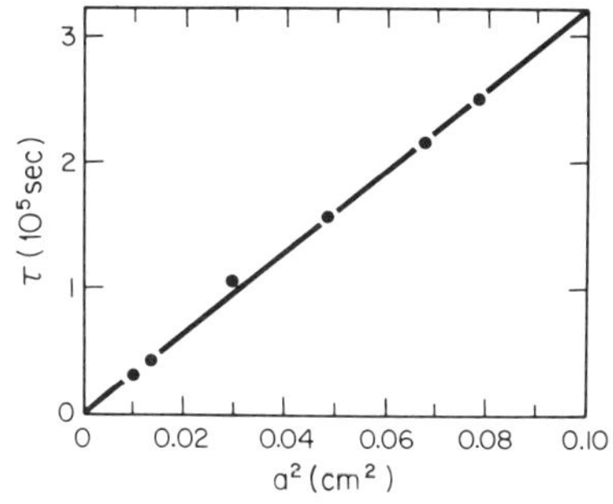

Figure 5. The characteristic time τ of the swelling of spherical polyacrylamide gels is a linear function of the square of the equilibrium radius a.

coefficient between the network and the gel fluid. We now examine how these quantities are expressed microscopically in terms of the chemical structure of the gel network. We first express the viscoelastic coefficients and the collective diffusion coefficient in terms of the spatial correlation function of the polymer network concentration fluctuations.[9] The relation will be a manifestation of the fluctuation–dissipation theorem established by Kubo.[10] Kirkwood and Risemann were the first to express the friction and diffusion coefficients in terms of the spatial correlation function.[9] They considered the case of polymer solutions in good solvents. Kawasaki[11] and Ferrel derived a similar formula describing the dynamics of binary mixtures near the critical point. The theories are known as mode-coupling and mode-decoupling, respectively. Such formulas were used by de Gennes to explain the dynamic scaling phenomena in polymer solutions in good solvents.[9] In the next section similar formulas will be derived and used to describe the dynamic behavior of a gel near the phase transition. The generality of the expressions is a result of the general assumption for the model, namely, the system is a mixture of solute and incompressible fluid. The solute can be a fluid, individual polymers, or a polymer network.

Let us now derive the diffusion coefficient starting from the linear response expressions for the elastic modulus K and the friction coefficient f

$$K^{-1} = \frac{1}{\phi kT} \int \langle \Delta\phi(\mathbf{r})\Delta\phi(\mathbf{o})\rangle \, d\mathbf{r} \equiv \frac{1}{kT} \int g(\mathbf{r}) \, d\mathbf{r} \tag{13}$$

and

$$\begin{aligned} f^{-1} &= \frac{1}{\phi} \int \langle \Delta\phi(\mathbf{r}) \frac{1}{6\pi\eta_s r} \Delta\phi(\mathbf{o})\rangle \, d\mathbf{r} \\ &= \int \frac{g(\mathbf{r})}{6\pi\eta_s r} \, d\mathbf{r} \end{aligned} \tag{14}$$

where ϕ is the volume fraction of the polymer network and η_s is the shear viscosity of the solvent.[9, 12] The correlation function $g(\mathbf{r})$ represents the conditional probability of finding a network monomer at position $\mathbf{r}$ when another monomer is at the origin $\mathbf{o}$. Equation (13) for the elastic modulus corresponds to the zero-wave vector response function of the network volume fraction when it is disturbed by an impulse of excess osmotic pressure. Equation (14) represents the net velocity response of a monomer in the network when an equal force is applied to each monomer of the network. The factor $1/6\pi\eta_s r$ corresponds to the diagonal element of the Oseen tensor averaged in the direction of the separation vector $\mathbf{r}$. The Oseen tensor represents the velocity response at a position r when a force is applied at the origin. Use of the solvent viscosity for the Oseen tensor is valid when the polymer network has a low concentration. From equations (13) and (14) we can construct the collective diffusion coefficient:

$$D = \frac{\int (kT/6\pi\eta_s r) g(\mathbf{r})\, d\mathbf{r}}{\int g(\mathbf{r})\, d\mathbf{r}} \tag{15}$$

This equation indicates that the collective diffusion coefficient is the average of $kT/6\pi\eta_s r$ with the spatial correlation of the volume fraction $[g(\mathbf{r})]$ as a weight factor. The viscoelastic parameters needed to characterize the dynamics of the collective diffusion of the gel can be calculated with knowledge of the static spatial correlation function $g(\mathbf{r})$.

9.3.1. Gels in Good Solvent

The spatial correlation function $g(\mathbf{r})$ for semidilute polymer solutions was given by Edwards,[13] taking into account the excluded volume effect, and revised by de Gennes by including the so-called correlation effects.[9] De Gennes also showed that the resulting correlation function can be applied to a cross-linked gel which is swollen in a good solvent. The result is expressed in the Yukawa function

$$g(\mathbf{r}) = c\,\frac{\xi}{r}\exp\left(-\frac{r}{\xi}\right) \tag{16}$$

Here the parameter ξ represents the average distance between neighboring cross-links in the polymer network. It may be called the mesh size of the network. For such a spatial correlation function, the collective diffusion coefficient is in exact correspondence with the Stokes–Einstein formula:

$$D = \frac{kT}{6\pi\eta_s \xi} \tag{17}$$

The elastic coefficient and the friction coefficient are

$$K = \frac{kT}{4\pi c \xi^3} \tag{18}$$

and

$$f = \frac{3\eta_s}{2c\xi^2} \tag{19}$$

Using the scaling arguments these parameters can be expressed in terms of the network concentration ϕ or the number, N, of polymer units between cross-links. The readers are recommended to the book by de Gennes, which clearly explains such procedures.[9] The results are

$$K \sim \phi^{1/4} \sim N^{1/5} \tag{20}$$

$$f \sim \phi^{-1/2} \sim N^{2/5} \tag{21}$$

and

$$D \sim \phi^{3/4} \sim N^{-3/5} \tag{22}$$

9.3.2. Light Scattering from Gels in Good Solvents

Munch and his colleagues carried out systematic studies of the diffusion coefficient in polystyrene gels consisting of polymers having a monodisperse molecular weight. They found a scaling between D and ϕ with an exponent slightly lower than the prediction of 0.75 in equation (22). Geissler and Hecht determined the exponents for the longitudinal modulus and diffusion coefficients using light scattering and deswelling experiments. They obtained

$$E \sim \phi^{2.25}$$

and

$$D \sim \phi^{0.77}$$

Thus the scaling theories predict very well the dependence of the viscoelastic parameters on the concentration of polymer network.

9.4. PHASE TRANSITION IN GELS

It has recently been found that drastic changes in the state of the gel can be brought about by small changes in the external conditions.[14–16]

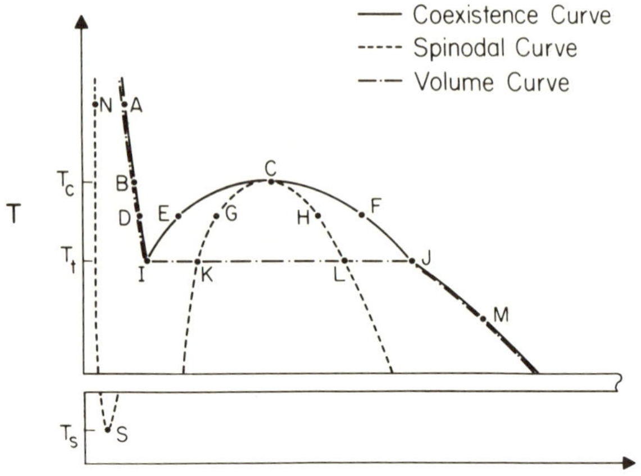

Figure 6. A sketch of the phase diagram of polyacrylamide gels. The vertical axis represents either the temperature or the quality of the solvent (gel fluid). The horizontal axis represents increasing network concentration. The volume curve represents the equilibrium network concentration in a large volume of solvent.

For example, when the temperature is lowered, the polymer network loses its elasticity and therefore becomes increasingly compressible. When a certain critical temperature is reached, the elasticity falls to zero and the compressibility becomes infinite.[17] At the same time the effective size of the pores in the polymer network increases, and at the critical temperature in the pores reach macroscopic size, (Figure 6). The gel can also swell or shrink by a factor of as much as several hundred when the temperature is varied. Under some conditions the swelling or shrinking is discontinuous, so that an infinitesimal change in temperature can cause a large change in volume. Moreover, temperature is not the only factor that can give rise to such transformations; they can also be brought about by altering the composition, the pH, or the ionic strength of the solvent in which the gel is immersed or by imposing an electric field across the gel.

The varied responses of the gel to changes in external conditions can now be understood in the context of phase transitions and critical phenomena.[18–21] Just as many substances can exist as a liquid or as a vapor under different circumstances, so can a gel sometimes have two phases, which are distinguished by different configurations of the polymer network. The discontinuous change in the volume and other properties of the gel constitute an abrupt transition between the phases, analogous to the boiling of a liquid. At higher temperatures (and under various other conditions) the two phases of the gel can no longer be distinguished; in a similar way the distinction between liquid and vapor disappears at high temperature and pressure. The set of conditions where the phases of the gel first separate is

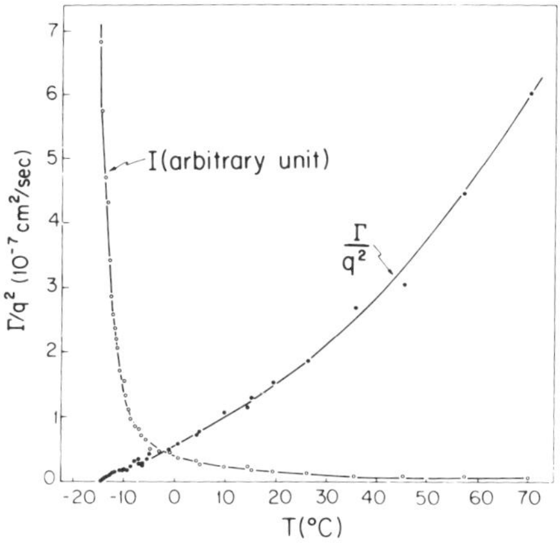

Figure 7. The intensity, I, and the relaxation rate Γ of laser light scattered by a 2.5% polyacrylamide gel. Here Γ is divided by the square of the scattering vector $|\mathbf{q}|$.

the critical point of the gel, and there the polymer network exhibits large-scale fluctuations in density, pore size, and other properties.

Critical Phenomena. When we fix the volume fraction of the polymer network at its critical value and approach the critical point by lowering the temperature we observe critical divergence and critical slowing down of the concentration fluctuations of the polymer network.[18] Both the elastic modulus and the friction coefficient diminish at the spinodal line, (Figures 7 and 8). At this point, the compressibility of the network (the inverse of the elastic modulus) diverges, allowing thermal agitations to cause huge concentration fluctuations in space and time.[19] The spatial concentration correlation function is given by the Ornstein–Zernike formula

$$g(\mathbf{r}) = C\,\frac{\xi}{r}\exp\left(-\frac{r}{\xi_s}\right) \tag{23}$$

equation (23) is similar to the spatial correlation function for gels in good solvents except for the difference in interpretation of the characteristic length, ξ_s, in the exponential of equation (23). Here, ξ_s represents the distance over which the network motions are correlated and involves many cross-linking units and polymers. Again, using the Kirkwood–Risemann or Kawasaki–Ferell formula, equations (13), (14), and (15), we have

$$D = \frac{kT}{6\pi\eta_s\,\xi_s} \tag{24}$$

$$K = \frac{kT}{4\pi C\xi\xi_s^2} \tag{25}$$

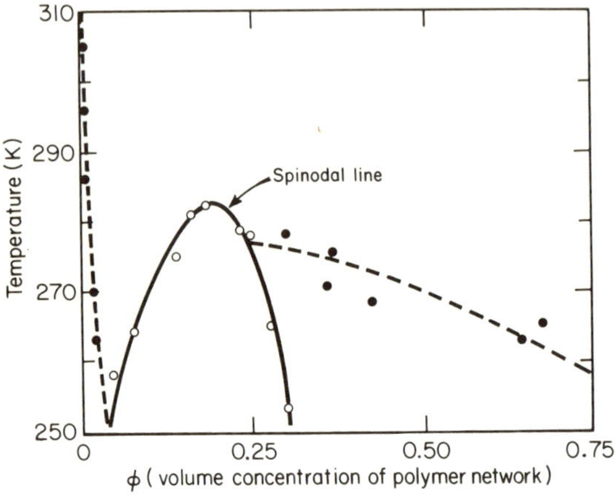

Figure 8. The spinodal line of a polyacrylamide gel where the fluctuations of light scattered by the gel diverge and slow down infinitely. The gel was originally dried and then mixed with definite amounts of a 44% acetone–water mixture. In such a way its concentration was adjusted. The solid circles denote the equilibrium concentrations of the gel when immersed in a large volume of a 44% acetone–water mixture.

and

$$f = \frac{3\eta_s}{2C\xi\xi_s} \tag{26}$$

As we approach the spinodal point the correlation length, ξ_s, increases and diverges. The viscoelastic parameters and the diffusion coefficient diminish with the divergence of ξ_s as shown in equations (24), (25), and (26). The formula were verified on acrylamide gels in water using laser light scattering spectroscopy.[20] From the amplitude of the correlation function of scattered light the elastic modulus, K, and its temperature dependence was determined. From these measurements, the correlation length was found to have the form

$$\xi_s = \xi_0 \left(\frac{T}{T_s - 1} \right)^{-1/2} \tag{27}$$

where T_s is the spinodal temperature and ξ_0 is a constant. Using equations (24) and (27) it is possible to predict the diffusion coefficient and its temperature dependence

$$D(T) = \frac{kT}{6\pi\eta_s RT\xi_0} \left(\frac{T}{T_s - 1} \right)^{-1/2} \tag{28}$$

This formula satisfactorily describes the measured diffusion coefficient as a function of temperature. However, more precise determination of the spinodal temperature in gels with well-characterized structure is needed to confirm the mean field exponent shown in equation (27).

9.5. CONCLUSION

We have seen how light scattering spectroscopy can provide important information on the physics of gels. There are interesting and important studies yet to be done in this field. For example, a comparison between polymer solutions and gel is an interesting subject. Patterson recently reported the similarity in their light scattering properties at high concentrations.[21] It is more desirable, however, to carry out such comparisons not at high concentrations in a good solvent, but rather in a very dilute swollen gel state or near the critical point where the diffusive modes in polymer solutions and gels may be quite different. The time change in the diffusion mode as the polymer network is formed during the gelation process is also an interesting question. The precise determination of the various critical exponents is yet to be done. Light scattering from polyelectrolyte gels and the effect of pH, ions, and electric field are also exciting problems.

REFERENCES

1. T. Tanaka, L. O. Hocker, and G. B. Benedek, *J. Chem. Phys.* **59**, 5151 (1973); J. A. Marquesce and J. M. Deutch, *J. Chem. Phys.* **75**, 5239 (1981).
2. J. P. Munch, S. Candau, R. Duplessix, C. Picot, J. Herz, and H. Benoit, *J. Polym. Sci.* **14**, 1097 (1976).
3. P. M. Munch, S. Candau, and G. Hild, *J. Polym. Sci.* **15**, 11 (1977).
4. E. Geissler and A. M. Hecht, *Macromolecules* **13**, 1276 (1980); **14**, 185 (1981); *J. Phys. (Paris)* **39**, 955 (1978); *J. Phys. (Paris) Lett.* **40**, L-173 (1979).
5. A. M. Hecht and E. Geissler, *J. Chem. Phys.* **73**, 4077 (1980).
6. T. Tanaka and D. J. Fillmore, *J. Chem. Phys.* **70**, 1214 (1979).
7. R. A. Gelman and R. Nossal, *Macromolecules* **12**, 311 (1979).
8. R. Nossal, *J. Appl. Phys.* **50**, 3105 (1979).
9. P. G. de Gennes, *Scaling Concepts in Polymer Physics*, Cornell University Press, Ithaca (1979).
10. R. Kubo, *J. Phys. Soc. Jpn* **12**, 570 (1957).
11. K. Kawasaki, *Ann. Phys. (N.Y.)* **61**, 1 (1970).
12. H. Yamakawa, *Modern Theory of Polymer Solutions*, Harper and Row, New York (1972); P. J. Flory, *Principles of Polymer Chemistry*, Cornell University Press, Ithaca (1953).
13. S. F. Edwards, *Proc. Phys. Soc.* **88**, 265 (1966).
14. T. Tanaka, *Sci. Am.* **244**, 124 (1981).
15. T. Tanaka, *Phys. Rev. Lett.* **40**, 820 (1978).

16. T. Tanaka, D. J. Fillmore, S-T. Sun, I. Nishio, G. Swislow, and A. Shah, *Phys. Rev. Lett.* **45**, 1636 (1980).
17. T. Tanaka, S. Ishiwata, and C. Ishimoto, *Phys. Rev. Lett.* **38**, 771 (1977).
18. S. J. Candau, J. P. Munch, and G. Hild, *J. Phys.* **41**, 1031 (1980).
19. A. Hochberg, T. Tanaka, and D. Nicoli, *Phys. Rev. Lett.* **43**, 217 (1979).
20. T. Tanaka, *Phys. Rev. A* **17**, 763 (1978).
21. K. Dusek and D. Patterson, *J. Polym. Sci. A*-2 **6**, 1209 (1968).
22. G. Patterson, Abstract No. 27, *Symposium International sur les Macromolecules, Strasbourg* (1981).

10

Biological Applications†

Victor A. Bloomfield

Department of Biochemistry
University of Minnesota,
St. Paul, Minnesota 55108

10.1. INTRODUCTION

In 1964, Pecora[1] showed theoretically the utility of quasielastic light scattering (QLS) for characterizing the translational, rotational, and internal dynamics of polymer chains. In 1967, the theory was brought to experimental realization in two quite different types of biological systems. Benedek and co-workers[2] used QLS to measure translational diffusion coefficients of proteins and nucleic acids, and Berge *et al.*[3] measured the motility of spermatozoa. These two types of biological applications—macromolecular characterization and physiological analysis—have both developed enormously since the late 1960s, but the major themes emerged early.

This chapter will therefore survey the many ways in which QLS has been used to ascertain the size, shape, and dynamics of biological macromolecules and supramolecular assemblies, and to characterize flow and other properties in physiological and biomedical situations. The macromolecules to be considered include proteins, nucleic acids, viruses, and other nucleoprotein complexes, polysaccharides, and phospholipid vesicles. Macromolecular characterization comprises both structural analysis of individual biopolymers and determination of their conformational transitions and intermolecular interactions. The physiological and biomedical applications include characterization of cell surfaces, muscle, protoplasmic

† Portions of this chapter are reprinted with permission from the Annual Review of Biophysics and Bioengineering, Vol. 10. © 1981 by Annual Reviews Inc.

streaming, ciliary beating, and cell motility, and such analytical applications as immunoassay and blood flow velocimetry.

The first part of this review will emphasize general principles, strategy, and tactics in the QLS investigations of biological systems, while the second part will cover specific applications, particularly from the mid-1970s up to mid-1981. There have been many recent useful books and review articles,[4–21] to which the reader is directed for additional views of this literature. Several of my coauthors in this volume have relieved me of the responsibility of discussing in detail topics which might otherwise have been covered in this chapter. These include micelles (Mazer, Chapter 8), rotational diffusion measured by depolarized scattering (Zero and Pecora, Chapter 3), and charge and concentration effects (Pusey and Tough, Chapter 4, and Schaefer and Han, Chapter 5).

10.2. PHYSICAL PRINCIPLES OF QUASIELASTIC LIGHT SCATTERING

Light is scattered from refractive index fluctuations. In a polymer solution, these fluctuations are associated with the polarizability difference between solute and solvent. If the scattering volume contains N macromolecules (this generic term also included aggregates, cells, etc.), and is irradiated with light of angular frequency ω_0, the scattered electric field at time t is

$$E_s(q, t) = \sum_{j=1}^{N} A_j(t)\exp\{i[\omega_0 t - \mathbf{q} \cdot \mathbf{r}_j(t)]\} \tag{1}$$

Here $\mathbf{q}$ is the scattering vector (often denoted $\mathbf{K}$ or $\mathbf{k}$ by other authors), equal to the difference between the scattered and incident wave vectors. Its magnitude is

$$|\mathbf{q}| = q = (4\pi n/\lambda_0)\sin(\theta/2) \tag{2}$$

where n is the refractive index of the solution, λ_0 the wavelength of the light *in vacuo*, and θ the scattering angle. The jth particle is located at $\mathbf{r}_j(t)$ and has scattering amplitude $A_j(t)$. The scattered field will vary in time owing to translational motion (changes in $\mathbf{r}_j$), rotational or internal motions (changes in A_j), and changes in the occupation number N.

10.2.1. Autocorrelation Function

The average intensity of the scattered light, neglecting constants of proportionality, is $\langle I \rangle = \langle |E_s| \rangle^2$, where the angular brackets denote a time average. The dynamics of the system are reflected in the normalized field autocorrelation function

$$g^{(1)}(\tau) = \langle E_s^*(t)E_s(t + \tau)\rangle / \langle I \rangle \tag{3}$$

This may be measured directly in a heterodyne scattering experiment, in which the scattered beam is mixed on the surface of the photodetector with a stronger reference beam (generally an unscattered portion of the incident laser light). However, it is much more common experimentally, save in electrophoretic light scattering or laser Doppler velocimetry experiments (see below), to perform a homodyne experiment, in which the scattered light mixes with itself, and thus to measure the intensity autocorrelation function, written in normalized form as

$$g^{(2)}(\tau) = \langle E_s^*(\tau)E_s(t)E_s^*(t + \tau)E_s(t + \tau)\rangle/\langle I\rangle^2 \tag{4}$$

When the scattered field is Gaussian with zero mean, as will normally be the case for a solution of diffusing macromolecular scatterers, then these autocorrelation functions are connected by the Siegert relation

$$g^{(2)}(\tau) = 1 + |g^{(1)}(\tau)|^2 \tag{5}$$

The scattered light is detected by a photomultiplier tube, whose average photocurrent $\langle i\rangle$ is proportional to $\langle I\rangle$. In early QLS experiments, an analog measurement of the fluctuation photocurrent was made: $g^{(2)}(\tau) = \langle i(t)i(t + \tau)\rangle/\langle i\rangle^2$. Thus QLS is sometimes called "intensity fluctuation spectroscopy." It is now more common, however, to count scattered photons directly (hence, another synonym for QLS: photon correlation spectroscopy). If n_i photons are detected in the ith measurement interval, then

$$g^{(2)}(\tau) = \left(\sum_{i=1}^{N} n_i n_{i+p}\right)/\langle n\rangle^2 \tag{6}$$

where $\tau = p\,\Delta T$; ΔT is the length of the measurement interval, or channel time; N_c is the number of channels; and $\langle n\rangle$ is the average number of photons counted in time ΔT.

10.2.2. Power Spectrum

QLS measurements can also be made in the frequency domain. The power spectrum $S^{(i)}(q, \omega)$ and autocorrelation function $g^{(i)}(q, \tau)(i = 1$ or $2)$ are related by the Wiener–Khinchine theorem:

$$S^{(i)}(q, \omega) = \int_{-\infty}^{\infty} g^{(i)}(q, \tau)\exp(j\omega\tau)\, d\tau \tag{7}$$

Since frequency shifts of the scattered light away from the incident frequency ω_0 are related to the motions of the scatterers, the power spectral version of QLS is often called "Doppler shift spectroscopy," though this name is not really appropriate save for scatterers in uniform motion. While power spectral measurements are useful in velocimetry, they are less common than in the early days of the field because of the greater efficiency of photon correlation instrumentation.

In the most common experimental situation, it is assumed that number fluctuations are negligible, translation and internal modes are independent, and the stochastic processes undergone by the system are stationary, hence independent of the time. If in addition there is no polydispersity, so that all N particles are identical, then equations (1) and (3) yield

$$\begin{aligned} g^{(1)}(\tau) &= \exp(i\omega_0 \tau)\langle A^*(0)A(\tau)\rangle\langle \exp - i\mathbf{q} \cdot [\mathbf{r}(\tau) - \mathbf{r}(0)]\rangle/\langle |A(0)|^2\rangle \\ &= \exp(i\omega_0 \tau)C_A(\tau)C_\phi(\tau) \end{aligned} \tag{8}$$

where $A(\tau)$ is the scattering amplitude per particle. If the scatterers are spherical, or small compared to q^{-1}, the amplitude autocorrelation function

$$C_A(\tau) = \langle A^*(0)A(\tau)\rangle/\langle |A(0)|^2\rangle = 1$$

Then $g^{(1)}(\tau)$ gives information on the translational motion of the particles, through the phase autocorrelation function $C_\phi(\tau)$. This is related to the intermediate structure factor $G_s(R, \tau)$, which gives the probability of finding a particle at position R at time τ, if it was at the origin at $\tau = 0$:

$$C_\phi(\tau) = \langle \exp - i\mathbf{q} \cdot [\mathbf{r}(\tau) - \mathbf{r}(0)]\rangle = \int G_s(\mathbf{R}, \tau)\exp(-i\mathbf{q} \cdot \mathbf{R})\, d^3\mathbf{R} \tag{9}$$

10.2.3. Translational Diffusion

$G_s(\mathbf{R}, t)$ is the solution of Fick's second law of diffusion, $\partial G/\partial\tau = D\nabla^2 G$, subject to the boundary condition $G_s(\mathbf{R}, 0) = \delta(\mathbf{R})$. This solution is well known to be

$$G_s(R, t) = (4\pi D\tau)^{-3/2} \exp(-R^2/4D\tau) \tag{10}$$

When inserted into equation (9), it yields

$$C_\phi(\tau) = \exp(-q^2 D\tau) \tag{11}$$

Thus in the normal homodyne scattering experiment, equations (5), (8), and (11) yield

$$g^{(2)}(\tau) = 1 + |C_A(\tau)|^2 \exp(-2q^2 D\tau) \tag{12}$$

This equation asserts the major use of QLS: the rapid determination of translational diffusion coefficients. The measurement is rapid compared to traditional synthetic boundary techniques because a typical particle need diffuse only a distance comparable to q^{-1} or λ_0, rather than a macroscopic distance, to produce a measurable phase shift in the scattered light. According to the Einstein relation $\langle x^2 \rangle = 2Dt$, the characteristic time t_D to diffuse an rms distance $\langle x^2 \rangle^{1/2} = q^{-1}$ is $t_D \approx (q^2 D)^{-1}$, in accord with equation (11).

10.2.4. Uniform Translational Motion

If a particle moves with constant velocity **v** then

$$G_s(\mathbf{R}, t) = (\mathbf{R} - \mathbf{v}\tau) \tag{13}$$

yielding

$$C_\phi(\tau) = \exp(-i\mathbf{q} \cdot \mathbf{v}\tau) \tag{14}$$

Since complex phase information is lost in homodyne detection, according to equation (5), heterodyne methods must be employed in velocimetry studies and electrophoretic light scattering (see below).

In the power spectral model, uniform translational motion leads to a spectral shift $\Delta\omega = \omega - \omega_0 = \mathbf{q} \cdot \mathbf{v}$. If in addition diffusion is superimposed upon the drift velocity, then the power spectrum is a Lorenzian whose maximum is shifted by $\mathbf{q} \cdot \mathbf{v}$ with half-width at half-height $q^2 D$:

$$S^{(1)}(q, \omega) = \frac{1}{\pi} \frac{q^2 D}{(\omega \pm \mathbf{q} \cdot \mathbf{v})^2 + (q^2 D)^2} \tag{15}$$

10.2.5. Rotational and Internal Motions

The amplitude autocorrelation function $C_A(\tau)$ will differ significantly from unity if the scale of internal motions is not very much smaller than the wavelength of light, since only then can light scattered from different parts of the same molecule fluctuate significantly out of phase. More precisely, rotational diffusion, and internal motion of flexible macromolecules, will influence the autocorrelation function when qL is not $\ll 1$, where L is comparable to molecular dimensions. Since the scattering angle can range from near 0° to near 180° (less than 30° or more than 150° is difficult without special optics and sample purity), and since laser wavelengths currently in use range from the near uv (363.8 nm) to the far red (702 nm), L can in principle be as small as 20 nm. In practice, few measurements have been made by polarized QLS of rotational diffusion coefficients D_R for molecules shorter than 100 nm. However, with depolarized QLS and high-frequency interferometric techniques, the rotational motion of small proteins and even small organic molecules may be studied.[22–24] A potentially important but unexploited technique is to use resonance-enhanced depolarized scattering from diphenylpolyenes.[25] This may increase the normally very weak scattering intensity by $>10^3$, enabling diphenylpolyenes to be used as scattering probes of micelles, membranes, and cell surfaces.

The form of the field autocorrelation function for combined translational and rotational diffusion is[4, 26–29]

$$g^{(1)}(\tau) = \exp(i\omega_0 \tau) \sum_{l=0}^{\infty} B_l \exp\{-[q^2 D + l(l+1)D_R]\tau\} \quad (16)$$

A usefully simplified form of the coefficients in this equation, for cylindrically symmetric anisotropic particles in the Rayleigh–Debye approximation, has been given by Aragon.[30] This equation indicates why it is often difficult to obtain reliable measurements of D_R from QLS alone. The autocorrelation function is a sum of exponentials, with weights B_l depending on the optical and structural anisotropy of the scatterers, whose relaxation times depend on both D and D_R, while the first term depends only on D. Resolution of such a sum of exponentials into its components is notoriously difficult. Consequently, D_R is often preferably obtained from transient electric birefringence decay or similar techniques.

If the macromolecule is flexible, its internal dynamics will be described by one or more relaxation times τ_R, which will contribute to the decay of the autocorrelation function in single or multiple-exponential fashion. This will be discussed in detail in the section on nucleic acids.

Equation (16) assumes that translational and rotational modes are independent and separable. This assumption, generally embracing internal modes as well, is convenient and usually sufficiently accurate; but it is not rigorously true for nonspherical or flexible particles. More complete theories, including coupling between modes, have been developed to various degrees of approximation.[31] When tested against data on tobacco mosaic virus, these theories indicate a smaller degree of anisotropic translational diffusion than would be expected from hydrodynamic calculations.

10.2.6. Number Fluctuations

Number fluctuations can be observed only in very dilute solutions of strong scatterers, such as cells, in very small scattering volumes. This follows from the well-known fact that for Gaussian fluctuations, $\delta N/$

$N \sim N^{-1/2}$. Thus if relative fluctuations at the 1% level are to be observed, $N < 10^4$. The time scale for number fluctuations is generally much slower than for phase fluctuations, since the scatterers must traverse the length of the scattering volume. If that length is l_v, then the Einstein relation indicates that diffusional number fluctuations will have a characteristic time $t_D \sim l_v^2/2D$. If the scatterers have a systematic velocity v, e.g., in bacterial motility, then the characteristic fluctuation time is $t_v \sim l_v/v$.

10.2.7. Transport Coefficients and Molecular Structure[32]

D and D_R are most commonly measured to obtain molecular dimensions. Where numerical values of these and other transport parameters are reported, they will generally be extrapolated to infinite dilution in water at 20°C. Thus the standard subscript and superscript qualifiers (e.g., $D^\circ_{20,\,w}$) will be omitted. For spheres, the familiar Stokes–Einstein equation is

$$D = k_B T/f = k_B T/6\pi\eta_0 R_h \tag{17}$$

where k_B is Boltzmann's constant, T the absolute temperature, f the translational friction coefficient, η_0 the solvent viscosity, and R_h the hydrodynamic radius. The corresponding rotational relation is

$$D_R = k_B T/f_R = k_B T/8\pi\eta_0 R_h^3 \tag{18}$$

Extensions to ellipsoids of revolution,[33] cylindrical rods,[34] and subunit arrays[32] are well known.

D can also be combined with the sedimentation coefficient S in the Svedberg equation

$$M = \mathrm{SRT}/D(1 - \bar{v}_2\rho) \tag{19}$$

to obtain the molecular weight M.

The electrophoretic mobility, μ, which may be measured by Doppler shift spectroscopy, is related to the charge Ze of the scatterers by

$$\mu = Ze\phi(\kappa R)/f \tag{20}$$

where $\phi(\kappa R)$ is a function of ionic strength, through the Debye–Hückel reciprocal length κ, and molecular size R. For large particles, such as cells, μ is proportional to surface charge density.

For random coils, the longest (and dominant) relaxation time is

$$\tau_R = A\eta_0[\eta]M/k_B T \tag{21}$$

where M is the molecular weight, $[\eta]$ the intrinsic viscosity, and A a numerical coefficient.

In summary, we see that QLS measurements of D, D_R, μ, and τ_R can provide a great amount of information about macromolecules. Application to particular biological macromolecules will be described below. In addition, we shall describe many applications to biopolymer interactions and collective motions.

10.3. INSTRUMENTATION AND DATA ANALYSIS

10.3.1. Instrumentation

A standard QLS apparatus consists of a cw laser light source, focusing optics, scattering cell mounted on a goniometer, collection optics, photomultiplier tube, discriminator and amplifier, correlator, oscilloscope and/or chart recorder for graphical output, and minicomputer for statistical data analysis and experimental control. The optical components of the system are fixed to an optical bench and generally mounted on a vibration isolation table. Several authors have reviewed general considerations of system design and criteria for choice of specific components.[8, 5, 35, 36] Just a few points of special interest will be mentioned here.

Most early QLS studies employed the 632.8-nm line of a He–Ne laser. While this is still common, Ar^+ lasers with a choice of intense lines, especially at 488 and 514 nm, are now often preferred. To obtain even larger values of q^2, desirable in studying internal motions of DNA, this laser has been operated in the near uv at 363.8 nm.[37] Although tunable dye lasers have linewidths much larger than the spectral broadenings of interest in QLS, these result from rapid phase fluctuations, which are unimportant in homodyne experiments. Hence a jet stream dye laser has been used over the 580- and 610-nm range in QLS studies of hemoglobin.[38] In fact, QLS experiments may even be carried out with a conventional mercury arc lamp light source,[39] though the lower temporal and spatial coherence lead to a much poorer signal/noise ratio.

It is difficult to measure scattering much below 30° or above 150° without special attention to optical design factors such as stray light and reflections from cell surfaces. Lastovka[40] has overcome these limitations with a vengeance, devising an optical apparatus which covers the very-small-angle regime from 0.003° to 0.15°. QLS may also be carried out on very small volumes under the optical microscope.[41]

Correlator design has advanced greatly.[35, 13] By either full multiplication techniques for slower correlation times, or clipping or scaling techniques for more rapid processes, real time correlation functions may be obtained with essentially full efficiency of photocurrent signal utilization, with channel times as short as 100 nsec. Even faster times may be attained by a photoelectron time-of-arrival method,[42] bridging the gap between digital correlation and Fabry–Perot interferometric techniques. A limiting factor below 1 μsec is photomultiplier afterpulsing. Several authors have thoroughly analyzed the statistical aspects of optical homodyne detection.[43–45]

Instrumentation for electrophoretic light scattering differs from QLS in several respects: heterodyne rather than homodyne detection; use of a spectrum analyzer rather than a correlator to measure Doppler shifts directly; need to work at low scattering angles to enhance resolution between components; and scattering cells designed to provide intense, homogeneous fields and avoid bubble formation and electroosmosis. These matters have been addressed in several reviews and papers.[46–53] With a laser cross-beam intensity correlation technique, it is possible to measure electrophoretic mobilities by homodyne QLS.[54] An extensive discussion of the theory and practice of photon correlation velocimetry has been given by Pike.[55]

Application of oscillating electric or gravitational fields may provide additional information on macromolecular dynamics. Schmitz has considered QLS by flexible polymers in the presence of a sinusoidal electric field. This will affect both center-of-mass[56] and internal[57] motions. Vibrational oscillation of the scattering cell gives rise to a damped oscillating autocorrelation function which may be analyzed to obtain the sedimentation coefficient of the scattering particle.[58]

10.3.2. Polydispersity

When the scattering solution contains macromolecules with a range of diffusion coefficients D_i, the autocorrelation function will be a sum of exponentials with relaxation times $(q^2 D_i)^{-1}$, each weighted by the relative scattering power of the ith component. Similar sums of exponentials will appear even in monodisperse solutions, if rotation or internal flexibility is important. When the relaxation times are not widely separated, accurately determining their distribution and weights is a notoriously difficult problem.[59] Several stratagems have been employed.

Least satisfactory is force-fitting to a single exponential, since this yields an ill-defined average relaxation time.

Perhaps most commonly employed is the cumulant analysis procedure developed by Koppel[28] in which the logarithm of the autocorrelation function $g^{(1)}(\tau)$ is fitted to a power series in $(-\tau)$. The first cumulant, or coefficient of the linear term in this expansion, is $K_1 = q^2 \bar{D}_z$. The z-average diffusion coefficient is related to the molecular weight M_i, weight concentration c_i, and scattering form factor $P_i(q)$ of the ith species:

$$D_z = \sum D_i M_i c_i P_i(q) / \sum M_i c_i P_i(q) \qquad (22)$$

Combination of $\bar{D}_z$ with the weight-average sedimentation coefficient[60] yields the weight-average molecular weight according to the Svedberg equation. The second cumulant K_2, the coefficient of $(-\tau)^2$, gives the variance of D. It is generally dominated by polydispersity rather than by statistical noise in $g(\tau)$. The normalized second cumulant, K_2/K_1^2, is often termed the "quality factor." Values of 0.02 or less are generally obtained for clean, monodisperse preparations. It is usually difficult to obtain statistically significant values for higher cumulants. The validity of cumulant analysis has been demonstrated on several occasions by scattering from known mixtures of polystyrene latex spheres.[61–63] For particles sufficiently large that form factor corrections are significant, the angular dependence of K_1 and K_2 can give additional information about the moments of the size distribution.[64, 65] McCally and Bargeron[66] have obtained expressions for K_1 and K_2, valid to $O(q^2)$, for Pearson III and V and Schulz distributions of polymer size. For relatively small vesicles, the number-average radius can also be obtained from $\bar{D}_z$ and its variance.[67] This method has been extended to obtain $\bar{M}_n$ and $\bar{M}_w$ for polymers for which D is known as a function of M.[68]

If there is reason to believe that the polymer size distribution obeys a particular functional form, such as Schulz,[69–73] 2-exponential,[69] or log normal,[69] the autocorrelation function may be analyzed directly to obtain the parameters of the distribution. More commonly, the form of the distribution is unknown. Then a variety of techniques for numerically inverting the autocorrelation function have been proposed to obtain the shape of the distribution, not just its moments. If $G(\Gamma)$ is the distribution of relaxation rates Γ (for translation, $\Gamma = q^2 D$), then the observed autocorrelation function is just the Laplace transform of the distribution:

$$g^{(1)}(\tau) = \langle e^{-\Gamma\tau} \rangle = \int_0^\infty G(\Gamma) \exp(-\Gamma\tau)\, d\Gamma \qquad (23)$$

Thus, in principle one may obtain $G(\Gamma)$ by inverse Laplace transformation.[74] Specific computational schemes to accomplish the inversion include Fourier[75] and eigenfunction expansion[76] methods, a nonlinear least-squares procedure which constrains the distribution to be the smoothest non-negative one[77, 78] and a least-squares method using a histogram approximation.[79, 80] These schemes all require high-quality data and considerable computation, but perform well when tested against known unimodal and bimodal distributions.

It often happens that the autocorrelation function from the molecule of interest rides upon a broad, slowly decaying background from a small amount of dust, aggregated material, or other very large impurity. When the product of diffusion coefficient and intensity of scattering of the impurity is much less than the product for the species of interest, $\bar{D}_z$ decreases in the same proportion as $\bar{M}_w$ increases.[81] The rather unpredictable effects of dust on the autocorrelation function are reviewed by Cummins and Pusey.[82] A derivative cross-correlation method, devised to examine entropy fluctuations in the presence of much stronger but slower concentration fluctuations in a critical binary mixture,[83] may be useful for examining diffusion of small molecules in the presence of larger ones.

Diffusion coefficients of macromolecules may be measured rapidly and accurately by thermal diffusion[84] as well as by QLS. Since the two methods measure different averages of D in heterogeneous solutions, they may be combined to characterize polydispersity.[85]

Combination of QLS with band transport in a stabilizing density gradient is often a more accurate, though slower, way than QLS on equilibrium mixtures to characterize particle distributions. Either band sedimentation[86, 87] or flotation,[88] or band electrophoresis[89] may be used; the latter gives size and charge distributions. Diffusional broadening may distort these distributions in suspensions of small particles.[90] This approach has been used to characterize continuous distributions of radiocolloids,[91, 92] but may be of even greater use for solutions containing several discrete components.

10.3.3. Concentration Effects

There have been roughly 30 papers published in the last five years on concentration effects on the determination of D by QLS. Despite this large volume of theoretical and experimental work, many controversial points remain. It is beyond my scope to review these in detail. Instead, I shall give my impression of the current state of the topic, and list a few papers where more detail and additional references can be obtained.[1, 3, 93, 95, 96]

Two diffusion coefficients must be distinguished: the mutual diffusion coefficient D_m and the tracer diffusion D_t. D_m measures the flow due to a concentration gradient, even a microscopic one, and is the quantity measured by QLS. D_t measures the diffusion of labeled molecules through a uniform solution. $D_m = D_t$ at infinite dilution, but they may diverge strongly at finite concentration.[97, 98] D_t is given by the Einstein relation $k_B T/f$, and D_m by

$$D_m = (\partial\pi/\partial c_2)_{P,T}(1 - \phi)/f$$

The first term is the isothermal osmotic compressibility, which, expanded in a viral series in the polymer concentration c_2, is $(\partial\pi/\partial c_2) = k_B T(1 + 2BM_2 c_2 + \cdots)$. The second term converts from a solvent-fixed to a cell-fixed reference frame; $\phi = v_2 c_2$ is the polymer volume fraction (v_2 is the partial specific volume). The friction factor may also be expanded in a viral series: $f = f^0(1 + k_f c_2 + \cdots)$. Collecting terms to first order in c_2 gives

$$D_m = (kT/f^0)[1 + (2BM_2 - k_f - \bar{v}_2)c_2 + \cdots]$$

For hard spheres, $2BM_2 = 8$, $k_f = 6.55$,[99] and v_2 is around 0.75 cm^3/g for most proteins. Hence the coefficient of c_2 is quite small, around 0.7, and D_m varies relatively little with concentration. By contrast, the concentration dependence of D_t is given by k_f alone, so D_t decreases strongly with increasing c_2.

10.3.4. Charge Effects

Since most biopolymers are polyions, a common source of concentration dependence in QLS is charge effects. Understanding of these effects is difficult and incomplete, because of the multicomponent nature of biopolymer solutions (polymer, water, and added salt), the long range of the Coulombic potential, and the need to consider three characteristic lengths: the fluctuation wavelength q^{-1}, the Debye length κ^{-1}, and the average interparticle distance $\rho^{-1/3}$, where ρ is the polymer number density.

Perhaps the most common experimental situation is scattering from a solution of a relatively small ($M < 10^6$) protein, nucleic acid, or other polyion. Since the polymer is small, its scattering is weak; hence ρ must be fairly high, and $\rho^{-1/3} \ll q^{-1}$. Under these circumstances it is found for a variety of polymers, including BSA,[100–102] tRNA,[103] and poly (L-lysine HBr),[104–106] that the apparent D measured by QLS rises sharply as the added salt concentration is decreased below about 0.01 M, that the total scattered intensity decreases in a roughly proportional way, and that D extrapolated to zero polymer concentration is less than kT/f^0 in high salt.

The first of these observations is explained reasonably well by a theory of Lee and Schurr.[104] The polyions, counterions, and coions are modeled as point charges diffusing in their common electrostatic potential. The linearized equations of motion for the fluctuations about the mean ionic concentrations are solved to give three normal modes, of which only one is expected to contribute significantly to relaxation of polyion concentration fluctuations. The value of the apparent diffusion coefficient, D_{app}, lies between D_p for the polyion and D_s for the salt. In the limit of excess salt, so that $Zc_p \ll cs_s$, where Z is the polyion charge,[105, 19]

$$D_{\text{app}} = D_p[1 + (1 - D_p/D_s)Z_c^2/2c_s]$$

Comparison of this theory with a computer simulation of polyion diffusion[107] would be of value. Other theories, based on polyion interactions through slow charge fluctuations[108] or a hard-sphere repulsive potential,[109] also give reasonable agreement with experiments on BSA.

The approximate inverse proportionality between D_{app} and scattered intensity stems from the presence of the osmotic compressibility $\partial\pi/\partial c_2$ in the numerator of equation (23) and in the denominator of the intensity equation.[110, 76, 102] The decline of D_p^0 in low salt has been attributed[111] to electrolyte friction of the translating polyion, due to energy dissipated during formation and relaxation of the ion atmosphere.

At very low salt concentrations, or in the absence of added salt, when $\kappa^{-1} \approx \rho^{-1/3}$, long-range ordering of the polyions in solution may occur. This has been observed in suspensions of R17 virus[112] and polystyrene latex spheres.[113–116] A distinct transition between ordered and disordered suspensions may be observed as a function of temperature, leading to the characterization of the ordered phase as a "colloidal crystal."[116] For these large scattering particles, $\rho^{-1/3} \approx q^{-1}$, so that Bragg scattering maxima and minima are observed. The theory of scattering from these ordered colloidal systems has been developed and reviewed by several authors,[110, 117–120] and they have been modeled by Monte Carlo techniques.[121] It is likely that long-range ordering is also responsible for the transition from the ordinary to the extraordinary phase in dilute solutions of poly (L-lysine).[104–106] This transition, which we[122] have also observed in solutions of short DNA segments isolated from mononucleosomes, is characterized by a strong decrease in scattering intensity, a rapid rise in critical salt concentration with increasing polyion density, and a low D_{app} independent of salt concentration. The nature of this transition is under investigation from several directions.[123–125]

Long-range electrostatic interactions can also complicate interpretation of electrophoretic light scattering experiments in polyelectrolyte solutions with little or no added salt. In a solution with only two ionic species (polyion and associated counterions), only a single electrophoretic Doppler shift frequency is predicted[126] and observed.[127] This frequency is not determined by the electrophoretic mobility of either ion separately. Linearized fluctuation transport theory should not be applied in the regime $\rho^{-1/3} \leq \kappa^{-1}$.[128] These difficulties will disappear when the salt concentration is raised, though then new problems with Joule heating may emerge.

It has also been speculated that linear fluctuation theory might not apply when ρ is not much greater than q^3, since then the number of molecules in the fluctuation volume is not $\gg 1$. However, a test of the wave-vector and concentration dependence for BSA with ρq^{-3} ranging from 2 to 27,000 showed no variation in D over that range.[129]

10.4. MACROMOLECULAR CHARACTERIZATION AND INTERACTIONS

In this section we survey the application of QLS to the major classes of biopolymers, both as individual macromolecules and as multimolecular complexes.

10.4.1. Proteins

10.4.1.1. Molecular Weight and Size of Globular Proteins. Determinations of protein molecular weight by empirical calibration of hydrodynamic radius R vs. M for denatured coils, or by combination of D with intrinsic viscosity $[\eta]$ (Scheraga–Mandelkern equation), are standard procedures in protein chemistry. They have been adapted to QLS measurement of R and D and applied to BSA, pepsinogen, aldolase, chymotrypsinogen, and ovomucoid.[130] Reasonably linear log–log correlations were obtained for the two procedures. Disagreement between them for M of ovomucoid indicates branching in this glycoprotein.

Small amounts of large aggregates can often seriously perturb light scattering measurements of small proteins. By approximating lysine-rich histone and its aggregates as spheres of two quite different sizes, subtracting the slowly decaying exponential contribution of the aggregate to $g^{(2)}(\tau)$, and using the amplitude of the slow decay to correct a Zimm plot of total scattering intensity, good values for the unassociated histone ($D = 8.5 \times 10^{-7}$ cm^2/sec, $M = 1.1 \times 10^4$) were obtained.[131] Aggregation may have afflicted earlier measurements on α-chymotrypsinogen.[132]

An early study of human plasma lipoproteins[133] led to D and R of 0.62×10^{-7} cm/sec and 396 Å for very low density lipoproteins; 1.8×10^{-7} cm^2/sec and 132 Å for low-density lipoproteins; and 4.4×10^{-7} cm^2/sec and 55 Å for high-density lipoproteins. More recent work on VLDL,[88] using flotation to obtain the distribution of molecular weights and sizes, yielded D from 0.50 to 0.25×10^{-7} cm^2/sec, corresponding to R from 425 to 870 Å. $\bar{M}_w$, D, R, and the radius of gyration of Lp(a)-lipoprotein were determined by QLS and total intensity scattering.[134]

10.4.1.2. Shape and Flexibility of Rodlike Proteins. According to equation (16), rodlike molecules whose length L is comparable to q^{-1} should show rotational contributions to the autocorrelation function. One such molecule is rat tail collagen, which is about 3000 Å long and 136 Å in diameter. An initial QLS study[135] found spectral contributions from both single rods and aggregates. D^0 was somewhat higher than theoretically anticipated for a rod of this size, and the concentration dependence was anomalous. The expected rotational contribution to D at high angles was not observed, perhaps due to impeded rotation at the high concentrations used. A subsequent study[136] using depolarized forward scattering, gave $D_R = 1082\ \text{sec}^{-1}$ for the monomer but also detected large aggregates. This gives a length according to Broersma's equation[34] about 13% less than the electron microscopic value. Flexibility is a possible explanation of this discrepancy. See also the discussion of fd phage, below.

Another rodlike protein aggregate is pyocin F1, a bacteriocin resembling the tail of bacteriophage λ. QLS analysis gave a length of 1480 Å, in good agreement with the sum of 1055 Å for the rod and 430 Å for the fiber determined from electron microscopy.[137] The most thoroughly studied rodlike proteins are the muscle proteins, which will be discussed in a later section.

10.4.1.3. Conformational Changes and Denaturation. Since D is inversely proportional to R, QLS is a sensitive probe of size changes associated with denaturation and other conformational transitions. The technique was used to study the temperature dependence of D for the two major antifreeze glycoproteins from an Antarctic fish.[138] These glycoproteins, with M's of 10,500 and 17,500, are >500 times as active in lowering the freezing temperature of water than would be expected if their action were simply colligative. No conformational changes were detected at -0.2°C that might be related to ΔT_f. Further lowering of T close to freezing caused decreases in R for both active and inactive glycoprotein components.

Combined use of QLS and total intensity scattering can distinguish between changes in size and molecular weight. This approach was used to probe the behavior of glyceraldehyde-3-phosphate dehydrogenase.[139] The enzyme aggregates readily, dissociates maximally between 22 and 26°C, and exhibits a conformational change with T in the presence of ATP that is distinct from changes in state of association.

Cannell and Dubin[34] designed a differential laser spectometer which can measure very small (3 parts in 10^3) changes in D. They applied this to determination of the effect of succinate and carbamyl phosphate on the translational diffusion coefficient of aspartate transcarbamylase.[140] Consistent with earlier sedimentation results, D decreases by $4.1 \pm 0.6\%$ in the presence of these ligands, indicating negligible change in protein partial specific volume under these conditions.

Even more subtle changes have been measured in the size of hemoglobin as a function of oxygenation.[141] The diffusion coefficients of oxy- and methemoglobin tetramers were the same within experimental error of $\pm 0.2\%$, while D for deoxyhemoglobin was $1.5 \pm 1.0\%$ larger. This is in good agreement with crystallographically determined changes in dimensions as a function of oxygenation. To perform this work successfully, several subtle factors had to be carefully controlled, including extrapolation to zero laser power (to minimize absorption, especially by methemoglobin, with attendant localized solution heating and convection), and correction for reversible dissociation of oxyhemoglobin tetramers into dimers.

Larger size changes are observed when globular proteins are denatured.[142] R increases by 18% for lysozyme and ribonuclease, and by 26% for chymotrypsinogen. Analysis of the pH dependence of the transition temperature indicated that the lysozyme charge changed by about 2 protonic charges during unfolding. QLS measurements of D as a function of temperature for ribonuclease in 1 M guanidine HCl enabled determination of the thermodynamic parameters for the folding–unfolding transition, using a two-state model.[143] Evidence was also obtained for a pretransition swelling of the protein.

10.4.1.4. Interactions. The globular proteins, especially bovine serum albumin (BSA) and hemoglobin, have served as testing grounds for theories of ionic, hard-sphere, and hydrodynamic interaction effects on D as measured both by QLS and traditional methods. For BSA, the basic experimental observations are that D decreases with increasing protein concentration, increases with decreasing ionic strength, and decreases as the pH is lowered below 5.[100, 144, 145, 101, 102, 96] The decrease of D as pH goes from 5 to 2 represents an expansion of the molecule.[101] The other effects are due to intermolecular interactions, the basic theory of which has been summarized above. It is important to realize that, when hard-sphere contributions to the viral expansion are computed, the hydrated radius of the protein must be used.[146] Other aspects of the concentration and charge dependence, such as the proper averaging of long-range electrostatic and hydrodynamic forces,[147–149] and attractive short-range electrodynamic interactions,[146, 145] are still controversial.

Hemoglobin (Hb) behaves similarly, in that D decreases linearly with increasing protein concentraction. Equation (23) is obeyed if a hard-sphere model is assumed and if a hydrodynamic interaction term is included.[150] No significant differences were found between D^0 or the concentration dependence of D for oxy HbA and oxy HbS.[151] QLS in concentrated, strongly absorbing samples such as Hb can potentially be complicated by multiple scattering, convection, thermal lensing, or temperature increases. In unliganded Hb solutions, only the last effect is important.[152] It can be dealt with by extrapolation to zero laser power. Absorption of He–Ne laser light by the 630-nm band of methemoglobin may have been responsible for the anomalous positive intercepts in plots of QLS linewidth vs. q^2 which were attributed to the kinetics of dimer–tetramer interaction.[153] Other studies on oxyHb and COHb did not observe these anomalies.[154] Concentration-dependent interaction effects on D of normal human oxyHb were studied as a function of T between 13 and 37°C.[155] A transition was observed at approximately 22°C, independent of c, postulated to be due to a quaternary rearrangement of the Hb chains which changed the effective volume and hydration sphere of the tetramer. These observations are not consistent with earlier calorimetric work.[156] QLS and electrophoretic light scattering studies on human COHb showed that tetramers dissociated to dimers above pH 10.[157] The dissociation constant was $< 10\ \mu$M at pH 9.6, increasing to $> 500\ \mu$M at pH 10.6. The electrophoretic mobilities of Hb_4 and Hb_2 were indistinguishable, implying an increase of 2.8 and 4.4 net negative charges per dimer upon dissociation.

Some intriguing results emerged when oxyHbA and oxyHbS were studied as a function of pH and ionic strength as well as protein concentration.[158] When pH was varied from 6.3 to 7.4 at 0.15 M salt, D^0 decreased monotonically. This was attributed to an increase in the thickness of the hydration shell with increasing pH. Both $\langle D \rangle$ and the normalized variance of D were measured over a range of conditions, with a decrease $\langle D \rangle$ and concomitant increase in variance being plausibly attributed to aggregation. While no indication of aggregation was observed for either oxyHbA or oxyHbS at high salt at protein concentrations below 10%, there were indications of aggregation of oxyHbA at very low ionic strength.

10.4.1.5. Association and Polymerization. If care is taken to average properly over all species in solution, or to disentangle their contributions to the autocorrelation function in some other way, QLS is a useful method for investigating protein association reactions. It becomes particularly powerful when combined with total scattering intensity measurements of M. The behavior of β-lactoglobulin A at pH 4.58 is consistent with a simple dimer–octamer equilibrium model.[159] The octamers behave hydrodynamically like close-packed spheres. Neither dimer nor octamer changes size between 3.5 and 25°C, and the dimer is invariant between pH 5.60 and 4.58.

The membrane-active peptide antibiotic gramicidin A exists as a monomer in dimethylsulfoxide, but as a dimer in methanol and dioxane. It is presumably also associated in other low-polarity media such as the interior of membrane bilayers.[160]

Blood-clotting proteins undergo a variety of association reactions in the presence of metal ions and membrane surfaces. A simple model system is prothrombin fragment 1 (F1) whose carbohydrate portion can be removed by anhydrous HF, producing aglyco-F1.[161] The aglyco-F1 retains a slow, Ca-dependent conformational transition monitored by intrinsic protein fluorescence. Both F1 and aglyco-F1 undergo a rapid self-association in the presence of Ca^{2+}, but aglyco-F1 undergoes a secondary self-association with kinetics identical to the fluorescence change. The carbohydrate seemingly masks this secondary association site in the native molecule. In a subsequent investigation of the comparative effects of Ca, Mg, Mn, and three lanthanide ions,[162] it was found that six ions were bound per F1 monomer in each case; but the order of filling of the sites, as reflected in fluorescence and association changes, was different. The parent prothrombin self-associated little in the presence of Ca or Mn, but lanthanides induced trimer formation. The conversion of prothrombin to thrombin has been followed by QLS,[163] but the extensive dimerization reported for prothrombin at low concentrations may reflect irreversible aggregates. Prothrombin, after it is cleaved to thrombin, acts by hydrolyzing fibrinogen in fibrin, which in turn forms clots. Human fibrinogen, carefully prepared to avoid aggregation, has $D = 2.04 \pm 0.09 \times 10^{-7}$ cm^2 s^{-1}, compatible with an elongated rigid molecule.[164] Measurements over a wide range of T showed thermal denaturation at 40°C and a second thermal instability at 3°C. QLS study of fibrin intermediate polymers, stabilized against lateral aggregation, gave $D = 0.37 \pm 0.05 \times 10^{-7}$ cm^2/sec. Deviation of the autocorrelation relaxation rate vs. q^2 plot from linearity at scattering angles above 60^0 was interpreted as a contribution from rotational diffusion, leading to $D_R = 142 \pm 32$ sec.$^{-1}$ These values are consistent with a rigid rod 4500 Å long and 90 Å in diameter, though flexibility could not be ruled out.[165]

An interesting series of papers[166–168] deals with the functional relationship between polymerization and catalytic activity of beef liver glutamate dehydrogenase (GDH). The enzyme is postulated to exist in active (x) or inactive (y) conformational forms, with activity independent of whether the monomer is free or incorporated in polymer. Three models for the polymerization equilibria are considered, differing in the values of the xy binding constants in the polymer: I, $K_{xy} = K_{yx} = 0$; II, $K_{xy} = K_{yx} = (K_{xx}K_{yy})^{1/2}$; III, $K_{xy} = K_{xx}$, $K_{yx} = K_{yy}$. Polymerization affects activity by altering the balance in the total number of x and y molecules in the system. Enzyme activity was measured by the rate of NADH production, while QLS measurement of the mean and variance of D (first and second cumulants) was used to determine the polymer size distribution, taking into account scattering form factors for long rods, and using an equivalent prolate ellipsoid model for the length dependence of D. Excellent agreement with model II was obtained, enabling fitting of data with three independent parameters. Another study of the polymerization of GDH by QLS[169] found equivalent and identical association sites, but did not measure enzyme activity. A recent exchange[170, 171] debates whether independent kinetic evidence is consistent with a coupling between polymerization and activity.

In a study of tubulin at low temperature where assembly into microtubules does not occur,[172] several components were observed. One has the diffusion coefficient expected for tubulin dimer, the others are significantly larger. One of these is in equilibrium with the dimer. When the solution is warmed to 36°C and cooled back to 4°C, another large, metastable component is formed.[173] Relation of these large species to 26-member closing rings observed as intermediates in microtuble formation is unclear. QLS from the dilute solutions of microtubules with lengths $>1\ \mu$m gave an angular dependence of the autocorrelation function consistent with translational diffusion.[174] However, D did not correlate with lengths measured by electron microscopy. These solutions may be behaving as a network of entangled polymers.

QLS and total scattering intensity have been used to study the *in vitro* fibrillogenesis of type I collagen.[175] Initiation of aggregation appears to involve at least two steps. In the first, $\langle D \rangle$ decreases but there is no change in the intensity of 90° scattering. In the second, $\langle D \rangle$ decreases further, and the intensity increases. Theoretical calculations of diffusion coefficients and scattering form factors for long rods modeled as prolate ellipsoids indicate that the step I aggregates are staggered linear dimers and trimers, while the step II aggregates are both longer and thicker (laterally associated). The effects of temperature and collagen concentration on fibrillogenesis are proposed to be due mainly to their influence on the concentration of linear aggregates.

10.4.2. Nucleic Acids

10.4.2.1. tRNA. Several studies have examined the change in structure of tRNA as a function of solution conditions. When the ionic strength is reduced from 0.2 *M* to 0.1 *M* in the presence of 1.0 mM Mg^{2+}, D increases by 11% for both unfractionated *E. coli* tRNA and yeast $tRNA^{Phe}$.[103] This implies a conformational transition to a more compact form which is not observed in 10 mM Mg^{2+}. Similar behavior was observed for five other elongator tRNA species, but not for the initiator $tRNA^{fMet}$ from *E. coli*.[176] This transition is pictured in terms of a "block and hinge" model. Below 0.1 *M* salt D decreases sharply, probably due to partial unfolding. Analysis of the diffusion virial coefficients suggests that the net charge per $tRNA^{Phe}$ decreases from -10 to -8 when $[Mg^{2+}]$ increases from 1 to 10 mM.

The structure of yeast phenylalanyl tRNA also seems to depend upon its state of acetylation, in a magnesium-dependent fashion.[177] When the phenylalanine-$tRNA^{Phe}$ is allowed to deacetylate spontaneously at neutral pH, D increases by 18% at high Mg^{2+} concentrations (2 to 10 mM), but no change in D is observed at lower Mg^{2+} concentrations (0.5 to 0.0 mM). Thus the aminoacylated form is more extended than the free tRNA. The magnesium dependence of the change in D behaves in a highly cooperative fashion, consistent with a two-state transition in which several additional Mg^{2+} ions are bound to the aminoacylated form. Despite this binding of additional cations, however, the tRNA concentration dependence of D indicates that at higher $[Mg^{2+}]$, the aminoacylated form has a larger negative charge than the nonacylated form, perhaps as the result of changes induced in the hydration layer or elsewhere in the molecule. There is no charge difference in low Mg^{2+}.

Bulk yeast tRNA undergoes a complex series of changes in D and D_R during thermal unfolding.[178] Rotational diffusion for this small molecule was measured by depolarized scattering using a photoelectron time-of-arrival technique with a delay time resolution of 1 nsec. This technique compares reasonably with Fabry–Perot interferometry.[179] The two diffusion coefficients were interpreted by Perrin's equation for prolate ellipsoids to obtain equivalent semimajor and semiminor axes. Before melting, the axial ratio is 3 : 1. As T increases, a more compact intermediate is observed, followed by a highly elongated, 180 : 22 Å molecule at 70°C. The tRNA appears to become smaller and more spherical as salt is increased from 0.2 to 1.0 *M*. Similar techniques were used to determine the influence of NaCl concentration on the size and shape of initially salt-free tRNA in solution.[180] A surprising progression is reported: the salt-free molecule is relatively compact, expands as [NaCl] increases to 5 mM, then becomes more compact again. In addition, an extrafast relaxation component was observed in low-salt and salt-free conditions. This was attributed to modulation of intermolecular interactions by charge fluctuation related to the ionization kinetics of the nucleotide bases. These intriguing results should clearly be explored further with fractionated tRNA and using a more complete theoretical treatment.

It may be pertinent in this regard that Patkowski *et al.*[181] have observed ordering in salt-free and low-salt tRNA solutions. They used small-angle x-ray scattering to determine the apparent intermolecular distance within ordered tRNA clusters, while angular dissymmetry and QLS measurements gave estimates of the static correlation length or apparent cluster diameter. This is ca. 500–1000 Å in 1 mM salt. Both the solution order and cluster size decrease with decreasing tRNA concentration or increasing ionic strength. This study illustrates the complex dynamics that may occur in such polyelectrolyte solutions, and underlines the importance of taking charge and concentration effects into account when attempting to interpret light scattering data in terms of macromolecular conformation.

QLS was used to analyze the dimerization of two tRNA molecules, *E. coli* $tRNA^{Glu}$ and yeast $tRNA^{Phe}$, which interact through complementary anticodons.[182] The equilibrium constant at 22°C in 0.1 *M* salt, derived from the concentration dependence of the effective *D*, varied from 1×10^{-5} without Mg^{2+} to 1.5×10^{-6} in 10 mM Mg^{2+}. ΔH and ΔS were obtained from the *T* dependence. The analysis of dimerization was accomplished using a procedure that should be of use in numerous similar situations. It involves analytical least-squares fitting of the best single exponential decay constant, giving an apparent *D*, to a model with two species of different *D*'s (in this case, monomer and dimer). The analysis takes into account concentration dependence of the *D*'s due to charge or other virial effects, and also the fraction of unreactive species.

A similar analysis was used to detect ligand-induced conformational changes in phenylalanyl-tRNA synthetase of *E. coli* K10.[183] The ligands studied were Mg^{2+}, L-phenylalanine, MgATP, $tRNA^{Phe}$, modified $tRNA^{Phe}$, yeast $tRNA^{Phe}$, and noncognate $tRNA^{Tyr}$ and $tRNA^{Ile}$. Analysis of *D* for the synthetase alone indicates that it is a highly extended molecule, as is not uncommon for proteins that bind to nucleic acids. Binding of Mg^{2+} causes further expansion, while subsequent binding of *E. coli* $tRNA^{Phe}$ to the Mg-saturated enzyme leads to recontraction. Of the other tRNA's investigated, only $tRNA^{Tyr}$ did not cause a similar change, while the other small substrates had no effect. The changes in D_{app} with reactant concentrations allowed construction of titration curves, yielding dissociation constants generally in agreement with those obtained by other techniques.

The physiologically less relevant association between bulk tRNA and BSA in 1 mM Na cacodylate buffer exhibits some intriguing features.[184] Aggregates form: those in the presence of excess BSA are twice as large as those in the presence of excess tRNA. The correlation function was fitted to an equilibrium diffusion-reaction model for a simple unimolecular isomerization. The association rate constant was two orders of magnitude lower than theoretical for a diffusion controlled reaction; the discrepancy was attributed to electrostatic repulsions between the negatively charged macromolecules. However, since the dynamic model is admittedly oversimplified, and since a variety of thermodynamic nonidealities can occur in such polyelectrolyte solutions,[181] such detailed interpretation may be premature. In fact, theoretical considerations[185, 186] indicate that it should be very difficult to observe macromolecular association kinetics by QLS, except under unusual circumstances.

10.4.2.2. DNA. The linear dimensions of high molecular weight DNA molecules in solution are comparable to or greater than the laser wavelengths used in QLS experiments. Hence internal motions of the DNA chain may be probed as well as translational diffusion. Obtaining this additional dynamical information is one of the attractive potentials of QLS; but extracting it has not been easy, due both to theoretical complexities and to a range of experimental pitfalls, including polydispersity, relatively low scattering power of DNA, and contamination of solutions with dust and polycations. QLS and low-angle total intensity scattering on sheared, length fractionated calf thymus DNA with $\bar{M}_w = 3.75 \pm 0.15 \times 10^5$ gave a Z-average $R_g = 207 \pm 10$ nm, yielding a persistence length of 66 ± 6 nm,[187] in good agreement with other studies.

Recent developments in DNA methodology, including restriction enzymes, sequencing or high-resolution gels for length determination, and preparative gel electrophoresis or reverse phase chromatography for separations, have made it possible to prepare quantities of short, monodisperse DNA for physical studies. Mandelkern *et al.*[188] have used QLS and transient electric birefringence of restriction fragments from the plasmid pBR 322, ranging in length from 64 to 267 base pairs, to determine the dimensions of DNA in solution. Combining translational and rotational diffusion coefficients, and extrapolating the results to zero length to remove effects of helix flexibility, they found a hydrodynamic diameter of 22–26 Å (the higher value according better with the field-free birefringence decay values), and a rise per base pair of 3.34 ± 0.1 Å. These are in good agreement with the values obtained by fiber diffraction, taking into account a monolayer of water around the phosphates, and indicate that, in these respects at least, the structure of DNA in solution is not significantly different from that in the fiber. The persistence length was found to be 450 ± 50 Å, rather lower than generally found in these low ionic strength (0.2 to 10 mM for transient electric birefringence, 50 mM for QLS) solutions.

Most recent work seems to agree that a plot of the apparent diffusion coefficient of high M DNA [obtained from the decay rate $q^2 D_{\text{app}}$ of $g^{(2)}(t)$ as a function of q^2] has plateau regions at high and low q, with a smooth intervening transition region. The autocorrelation function is fitted well by single exponential decays in the two plateaus, but may exhibit multi-exponential behavior in the transition region. The low-q D_{app} is to be identified with D for the molecular center of mass, hence depends on M; while the high-q D_{app} is that for segmental diffusion, and therefore should be M independent. This interpretation was first obtained for a model in which has since been justified in terms of the more standard Rouse–Zimm model independent segments move in a long-range harmonic potential.[189, 190] It of polymer dynamics.[191–193] Lin and Schurr[191] presented a computational scheme to obtain the three parameters of the Rouse–Zimm model—the number of Gaussian subchains $N + 1$, the rms subchain extension b, and the subchain diffusion coefficient kT/f—from the two plateau diffusion coefficients and q^2 at the midpoint of the transition. Another scheme is presented in a later paper.[194]

There is much current interest in the polyelectrolyte behavior of DNA,[195] and QLS determination of chain parameters as a function of ionic strength should shed light on this behavior. Clean preparations of monodisperse viral DNA from ϕ29 or λ phages show little variation with pH from 7.85 to 10.25, with salt concentration from 0.01 M NaCl to 5.4 M NH_4Cl, or with molecular weight or GC content.[194] Introduction of single strand breaks has no effect at pH 9.46, but produces large changes at pH 10–10.25, suggesting the introduction of titratable joints at those pH's.[194] These joints appear to be associated with bound polycations, either deliberately introduced or present as contaminants.[196] The effect of ionic strength has also been analyzed[197] by fitting to a two-term exponential, $1/\tau_1 = Dq^2$, $1/\tau_2 = Dq^2 + \tau_{\text{int}}$. τ_{int} is an internal relaxation time, found to be inversely proportional to q^2 as predicted by Rouse–Zimm theory, and independent of ionic strength over the range 0.01 to 0.10 M NaCl. However, the transition region between the two extreme q^2 regimes is reported to be strongly dependent on salt, suggesting that the internal dynamics of DNA are ionic-strength dependent.[198] Further work is clearly needed to reconcile these differences in detail, but in general it appears that DNA internal dynamics are moderately affected by ionic strength. One study also reports that b and f are significantly different for DNA in 0.4 M NH_4Cl compared to other salts at comparable concentrations.[199]

Some of the early work on QLS of high molecular weight DNA reported very long relaxation time tails on the correlation functions.[200, 201] These slow decays are not observed in more recent work;[202, 203, 194] they presumably are the result of dust or polycation-induced aggregates in inadequately clarified solutions. They did, however, provoke an intriguing theoretical model of the coupling of internal modes with anisotropic translational diffusion in congested solutions,[200] which may be of value in understanding QLS from concentrated solutions. A somewhat different theoretical formulation of this problem has been used[204] to interpret the concentration dependence of QLS from polyadenylic acid solutions,[205] suggesting that the faster of two relaxation times may reflect a semiconcentrated scattering regime where $qL > 4.4$ and $\rho^{-1/3} > L$ (L is the largest characteristic dimension of the polymer coil). However, this may also simply indicate the onset of the transition between low-q and high-q limits in the Rouse–Zimm theory.[199]

At high salt concentrations, the CD spectrum of DNA undergoes a characteristic change. With clean ϕ29 viral DNA, this is not associated with a collapse in tertiary structure.[206] The contraction in hydrodynamic radius found in calf thymus DNA[207] may be due to adsorbed proteins or polycations.[206] Polycations such as the polyamines spermidine and spermine can cause abrupt condensation of viral DNA to toroidal structures whose hydrodynamic radii are comparable to those of phage heads.[208] Condensation occurs, when the data are interpreted by counterion condensation theory,[195] at ionic conditions such that 89%–90% of the DNA phosphate charge is neutralized. Less dramatic reduction in hydrodynamic volume was earlier observed with DNA in the presence of histones at 0.8 M NaCl in 2 M urea.[37] Viral ϕ29 DNA, condensed with spermidine or its homologs, exhibits more irregular, polydisperse structures at pH 10.2 than at neutral pH, suggesting a change in the mode of triamine binding under alkaline conditions.[209]

Measurements of QLS as a function of T through the helix-coil transition region for two viral DNA's showed substantial oscillations in D_{app}, which were attributed to sequential melting of regions of heterogeneous base composition along the DNA.[210, 211] Intermolecular association is another possible explanation.

Electrophoretic light scattering of calf thymus DNA found that the mobility of the denatured form was 15% lower than that of the native molecule.[212] Agreement with these results for native DNA was obtained using a sinusoidal, rather than pulsed square-wave, electric field.[56]

10.4.2.3. Chromatin. Current studies on chromatin indicate that this nucleohistone complex consists of alternating nuclease-accessible and resistant regions, visualized electron microscopically as beads on a string. The "beads" are termed core particles of ν bodies, and consist of 140 bp DNA wrapped around an octameric histone aggregate. The "string" is spacer DNA of variable length. The way in which the core-spacer repeats, or mononucleosomes, are arranged in space to form high orders of structure is of much interest in understanding chromosome structure and function. Partial nuclease digests of chicken erythrocyte chromatin may be fractionated into short oligonucleosomes. Analysis of S and D as a function of number of nucleosomes, using subunit hydrodynamic theory, suggests that the multimer structure is either a helix or a flexible coil with attractive interactions between nucleosome units.[213] Further analysis, using different size spherical subunits for the core and spacer, is consistent with a helical conformation up to 20 repeat units.[214] If the constraint is lifted that the linker DNA lies along the line of centers of the core particles then the data can be represented by a flexible helix.[215] There is a transition at about 8 nucleosome units in which the core particle separation goes from the fully extended 236 Å to a compact 135 Å. This flexible second-order folded structure should be responsive to environmental conditions.

Analysis of somewhat longer oligonucleosomes also gives evidence for helical superstructure, though the helix parameters vary somewhat, depending perhaps on solution conditions and preparation. Crothers and coworkers[216] used QLS and transient electric birefringence to determine D and D_R for fragments of the 30-nm chromatin fiber isolated from calf thymus nuclei, containing 9, 12, and 17 kilobase pairs, corresponding to 46, 61, and 82 nucleosomes. The samples were stabilized by cross-linking with dimethylsuberimidate in 0.1 M salt to ensure retention of a compact configuration. The diffusion coefficients, when interpreted according to suitable hydrodynamic expressions for cylinders, yielded 33 ± 3 nm for the fiber diameter, and a length of 1.5 ± 0.1 nm per nucleosome. Assuming a superhelical pitch of 11 nm, corresponding to touching nucleosomes, this leads to 7.5 ± 0.5 nucleosomes per superhelical turn. There is some indication of flexibility for the longest oligomer.

Recent work from our laboratory[127] has identified two classes of chromatin from chicken erythrocyte nuclei, distinguished by their solubility properties at physiological ionic strength. We have made QLS and total scattering intensity measurements on fragments, produced by micrococcal nuclease digestion, over a range of molecular weights from 8 to 53 nucleosome equivalents.[218] Hydrodynamic radii were compared with those computed for helical models with various assumed geometries, using subunit hydrodynamic theory;[32] radii of gyration, from Zimm plots, were computed for the same models. An excellent fit for both types of chromatin was obtained with the following helix parameters in 0.1 M salt: 2.2 nm rise/nucleosome and 25.3 nm pitch, corresponding to 11.5 nucleosomes/turn; 19 nm radius from helix center to nucleosome center, or about 48 nm outer diameter. With these dimensions, nucleosomes directly above each other by one helix turn do not touch; the model is both thicker and more expanded axially than those described above. The data rule out close packed solenoidal or superbead structures. Consistent with the presumably weak bonding interactions between subunits, the superhelix expands markedly (D decreases) as the ionic strength is lowered.

10.4.2.4. Ribosomes. There are two states of the large *E. coli* ribosomes: 45S and 50S. The diffusion coefficients of these are 1.79×10^{-7} and 1.91×10^{-7} cm^2/sec. Therefore, assuming no charge in partial specific volume, the transition takes place without change in molecular weight.[219] D_R of 70S ribosomes, determined by depolarized QLS, gave a hydrodynamic radius slightly larger than that obtained from D_T, implying nonspherical shape.[220]

10.4.3. Viruses

QLS determinations of D have been combined with measurements of S and $\bar{v}$ to obtain molecular weights and sizes for several viruses, including type 5 adenovirus,[221] MS2[222] and ϕNS11[223] bacteriophages, and infectious pancreatic necrosis virus.[224] Comparative measurements of size and polydispersity of several insect viruses have been made by QLS, resistive-pulse analysis (RP), and electron microscopy (EM).[225] While polydispersity is resolvable by RP and measurable by EM, it is lumped into a weighted average size by QLS. The electrophoretic mobilities of TMV[226] and several RNA tumor viruses[227] are generally in good agreement with free boundary electrophoresis, and have been used to calculate surface charge densities.

Several studies have been directed at vesicular somatitis virus (VSV), which has glycoprotein spikes projecting from a lipid bilayer envelope. Defective *T* particles of various types are produced by viral RNA deletions; these have been characterized by D, S, $\bar{v}$, and EM.[228] QLS was used to monitor aggregation and thermolability of some mutants in the G protein (glycoprotein) and M protein (the other viral envelope protein) of VSV.[229] Similar behavior was observed in all G-protein mutants, but not in M-protein mutants, upon changing salt concentration or adding 2% ethanol. It was concluded that thermolability and aggregation are both the result of an irreversible conformational change in the G protein, causing a change in the VSV surface. The lengths of the spikes of VSV and the Sindbis virus are estimated by comparing effective diameters measured by QLS and by the resistive-pulse technique.[230] The former measures essentially out to the tips of the spikes, but RP determines only the nonconducting diameter, to which the spikes make a negligible contribution. The spike lengths are 175 ± 25 Å for VSV and 70 ± 20 Å for Sindbis.

Depolarized QLS for TMV gives D_R of 323 ± 17 sec^{-1} at pH 10, reasonable agreement with previous determinations, but with puzzling pH variations.[231]

Day and co-workers have carried out an interesting series of studies of filamentous bacteriophages fd, Pf1, and Xf. These all appear in the electron microscope as long, thin, somewhat flexible rods, and contain single-stranded circular DNA. By measuring D by QLS and D_R by transient signal averaging electric birefringence, the length and diameter of the equivalent rigid cylinder can be calculated using Broersma's theory. For fd, the dimensions are 8950 ± 200 Å by 90 ± 10 Å.[232] Combining these values with measurements of M for the phage, protein subunits, and DNA,[233, 232] it is possible to compute the ratio of nucleotides to protein subunits. For fd this ratio is definitely nonintegral, 2.30 ± 0.11. For the two other filamentous phages the ratio seems to be nearly integral: one in Pfl and two in Xf.[234] From measurements of this sort it is also possible to calculate the number of protein subunits/repeat of the virus helix, a number difficult to obtain from x-ray scattering patterns alone.[235, 236] For Xf, there are 27 subunits/repeat, with an uncertainty of $\pm 15\%$.

Though fd phage appears somewhat flexible in electron micrographs, its EM contour length agrees well with the length determined by transient electric birefringence decay[232] and its QLS autocorrelation function obeys T/η scaling.[237] These observations suggest that the virus behaves like a rigid rod in solution. However, analysis of the high-angle QLS behavior[238] indicates the presence of a fast-relaxing component which is plausibly associated with internal flexibility.

Perhaps the most complex structurally of all viruses are the T-even bacteriophages. QLS has been used to study several steps in the assembly and conformational change of these viruses. The kinetics of joining of heads and tails of bacteriophage T4D were followed by measuring the time variation of $g^{(2)}(\tau)$.[239] Knowing D for heads alone and for head–tail complexes (the tails alone contribute negligibly to the scattering intensity), it was possible to compute the fraction of unreacted heads as a function of time. The second-order rate constant at 22°C is $1.02 \times 10^7\ M^{-1}$ sec.$^{-1}$, about 1/500 that predicted by simple von Smoluchowski theory for a diffusion controlled reaction. The discrepancy is largely due to orientation factors. The ability to measure such a high rate constant depends on the strong scattering power of the phage heads, permitting concentrations in the range $10^{-10}\ M$ and thus slowing the bimolecular reaction to a half-life of several hundred seconds.

Tail fiber attachment of T4D was studied[240] by an instrument that combines QLS with band electrophoresis.[89] A maximum of six fibers are attached to the head–tail complex; the apparatus permitted resolution of all seven peaks expected for unfibered and 1–6 fibered species. Electrophoretic mobility became less negative as fibers were added successively, due both to increased friction and reduced net charge. Analysis of population distribution as a function of fiber input demonstrated that the attachment reaction is random and noncooperative. The tail fibers will reorient from a retracted state at low pH and T to an extended state at high pH and T. The increased friction, hence lower D in the fiber-extended state, permitted thermodynamic analysis of the reaction using QLS and sedimentation velocity to determine the fraction of phage in each state.[241] For T2L phage, the reaction is highly cooperative, with $H^0 = -139$ kcal/mol phage, $S^0 =$

−247 e.u., and 8 ± 2 protons taken up during the extended ⇌ retracted reaction. This reaction is also a function of ionic strength. High salt favors the extended form, presumably due to screening of attractive Coulombic interactions between the fibers and the rest of the phage. However, the position of equilibrium is a function of T2L concentration, and the fibers can extend fully only at low concentrations, presumably due to intermolecular excluded volume effects.[242]

Although M and $\bar{v}$ of the phage do not change appreciably during the fiber transition, S/D for the fiber-extended form of T2L was 18% higher than for the retracted form, when D was measured by QLS.[242] This discrepancy was explained with the realization that eccentric rotation of the strongly scattering phage head with respect to the center of frictional resistance introduces significant rotational contributions to $g^{(2)}(\tau)$, especially in the fiber-extended form.[243–246] When the appropriate correction for rotation was made, the molecular weights from S/D come into agreement. Recent advances in hydrodynamic modeling allow accurate calculation of these off-center rotation effects for asymmetric particles.[247]

Depolarized forward scattering allows determination of D_R without the application of external fields. After minimization of double-scattering effects, either by lowering the concentration or by optimizing geometry,[248] D_R may be obtained with good accuracy. For phages T4B (fibers retracted) and T7, $D_R = 258 \pm 12\ \text{sec}^{-1}$ and $4528 \pm 100\ \text{sec}^{-1}$, respectively.[249, 250]

10.4.4. Polysaccharides and Proteoglycans

Biopolymers containing complex carbohydrate chains are plentiful and important, but have been difficult to study physically because of their structural complexity and polydispersity. The work described in this section therefore generally represents preliminary forays into largely unexplored territory.

Chu and co-workers have presented a series of studies on meningococcal polysaccharides. These are immunogenic, and can immunize against meningococcal meningitis infection. The potencies of the A and C groups have been correlated with molecular size, lending potential practical value to QLS measurements of size and polydispersity. Initial measurements of two fractions of the group C polysaccharide showed very low $\bar{D}_z$ and very high normalized second cumulant, indicating extensive aggregation and polydispersity.[251] Aggregates are broken up to some extent by the addition of EDTA, or at higher T with long equilibration times, but some very large material remains. Stable large aggregates are formed at high ionic strength.[252] Combination of QLS with measurements of total scattering intensity, refractive index increments, viscosity, and sedimentation was used to characterize in some detail the structure and thermodynamic interactions of the C-1 fraction.[253] It has $\bar{M}_w = 5.15 \times 10^5$, $R_G = 345$ Å, hydrodynamic radius 240 Å, and behaves like a random-coiled polymer with average spherical shape. $\bar{M}_w$, and therefore aggregation, is a minimum at 0.2 *M* KCl. This study used a cumulant analysis of QLS data. More detail about the size distribution is obtained from a histogram analysis.[79, 80] When this approach was applied to group B meningococcal polysaccharide, two size fractions were detected with hydrodynamic radii 298 and 50 Å and relative integrated intensities 82% and 18% in 0.4 *M* buffer. In pure water, the relative scattering powers remained nearly the same, but the sizes changed to 402 and 29 Å.[254]

Dextran appears to behave as a typical random coil, as judged by the *M* dependence of *D* and intrinsic viscosity.[255] Cellulose trinitrate of high *M* and moderate polydispersity, on the other hand, exhibits some unusual behavior.[256] It has a large, positive second viral coefficient, a *D*-*M* exponent in the range -0.8 to -1.0 characteristic of a freely draining coil, and no evidence in the QLS spectrum for internal motions that would have been expected for such a large polymer.

Xanthan gum is a polysaccharide with extensive short-chain branching. It undergoes a thermal transition in low-salt aqueous solutions. QLS shows that *D* increases with *T* near the transition, with a concomitant increase in polydispersity.[257] This is interpreted as the breakdown of a multimolecular aggregate. The variation of *D* with *c* in the region where chain entanglements are expected does not agree with recent theoretical predictions (see below); this may be due to chain stiffness in the xanthan molecule.[257]

Proteoglycans have a linear protein core from which glycosaminoglycan chains are pendant. They are abundant in mammalian connective tissue, where numerous proteoglycan subunits are attached to a central linear hyaluronate molecule. The proteoglycan subunit fraction A1–D1–D1 isolated from bovine nasal septum has been studied by QLS, sedimentation, and intrinsic viscosity.[258] Size and shape deduced from these measurements are consistent with the accepted model of a prolate ellipsoid with semiaxes 1900 by 400 Å only if considerable segmental flexibility is allowed. Considerable self-association occurs in the absence of the denaturant guanidinium chloride. At low salt, solution behavior becomes strongly nonideal, perhaps due both to long range ordering and macromolecular congestion. Titration with hyaluronate yields a sigmoidal increase of the Stokes radius, indicative of noncovalent complex formation.

One type of glycosaminoglycan emanating from the protein core of proteoglycans is chondroitin-6-sulfate. *D* of this molecule does not vary over a wide range of salt types or concentrations, indicating a relatively stiff wormlike chain.[259] In 3 M NaCl, two stages of self-association occur as T is raised. This temperature dependence suggests aggregation through hydrophobic interactions.

Chondroitin sulfate has also been used[260] to check some predictions of counterion condensation theory.[195] Electrophoretic light scattering measurement of polyion mobility, combined with tracer determination of counterion self-diffusion, confirms that the counterion–polyion interaction is largely electrostatic, hence that counterion substitution depends mainly on the charge difference. However, second-order effects are noticeable, particularly when the polyion has a small residual charge.

Although hyaluronic acid is so large and polydisperse that QLS studies are hard to interpret, its effect on the microviscosity of aqueous solutions can be monitored by measuring its effect on D or polystyrene spheres as solution probes.[261] The microviscosity drops rapidly as the salt concentration increases, leveling out above 0.05 M. 930-Å spheres yield a smaller microviscosity than 1830-Å ones, indicating a sieve effect. The 930-Å spheres also reflect an apparent expansion with dilution of the hyaluronic acid chains. See the section on gels and entangled solution for some related studies.

10.4.5. Vesicles and Protein–Membrane Complexes

QLS has been used extensively to characterize the size and size distribution of phospholipid vesicles, and to monitor the efficiency of new preparation methods.[262] The size and molecular weight of synthetic surfactant vesicles, prepared by sonication, decreased exponentially with irradiation time to limiting values.[263] Among the procedures used to determine the size distribution are fitting to a 2-exponential function,[264, 265] which is effective in accounting for a small fraction of large contaminants, and approximation of the distribution as a first order spline function.[266] D_Z and its relative dispersion have been computed for vesicles modeled as spherical shells[267] taking into account the scattering form factor for hollow shells[268]; a method has been presented to convert these parameters into number-averaged size and size dispersion.[67]

Vesicles change size or fuse in response to a variety of treatments. Pure phosphatidylserine (PS)[269, 270] and mixed PS–phosphatidylcholine[271] vesicles undergo fusion as a function of T with a sharp maximum depending on $[Ca^{2+}]$. There is a plausible correlation between the fusion peak and the phase transition T for PS in the presence of Ca^{2+} prior to fusion.

The intact photoreceptor membrane from bovine rod outer segment behaves in the bleached state as a very homogeneous spherical vesicle, with $R = 0.49 \pm 0.07$ μm, in dilute aqueous sucrose.[272] This is derived from the native disk shape by osmotic swelling. The response of these membranes to increasing osmolality (sucrose concentration *m*) depends on whether they are bleached or not.[273] When they are modeled as oblate ellipsoidal shells, the axial ratio *p* increases rapidly with *m* in the bleached state, less rapidly in the unbleached state. At fixed *m*, *p* increases upon bleaching. In all cases the semimajor axis remains unchanged, implying a significant decrease in volume and surface area upon photobleaching.

Isolated zymogen granules of the mouse pancreas also respond in size to solution osmolality, though to a smaller extent than predicted by the van't Hoff relation.[274] Aggregation of the granules is produced by increasing [Ca^{2+}] or low pH. These studies were performed by QLS under a microscope, permitting use of only about 50 μl of sample per experiment.

Preliminary measurements on DPPC vesicles as a function of *T* near the lipid phase transition suggest a reversible conformational change of the vesicle walls and a limited onset of fusion.[275] However, the variance of *D* was high even at low *T*, and did not increase in the transition region, suggesting extensive polydispersity, which may obscure interpretation of these experiments. In a test of mechanisms of tissue damage by laser irradiation, DPPC vesicles were subjected to high power picosecond pulses from a Nd: glass laser.[276] The vesicles were unaffected by single pulses with energy densities up to 9 mJ/cm^2, but showed irreversible response to pulse trains of about 100 pulses with total density >600 mJ/cm^2 per train. The high local electric field strengths accompanying the laser light may convert single spherical vesicles to larger, more stable multilamellar aggregates.

QLS has been used to prove the structure of complexes of the vitamin K-dependent blood-clotting proteins prothrombin and factor X with membranes.[277] When these proteins bind to very small monodisperse phospholipid vesicles in the presence of Ca^{2+}, the change in Stokes radius suggests that they must bind at one end of the elongated protein and extend radially outward into solution. When proteins can be bound either directly to phospholipids or to receptors incorporated in vesicles, this seems to be a generally useful approach to the structure of extrinsic protein–membrane complexes. For example TMV, covalently conjugated with several hundred molecules of different hormones, binds to hormone receptors in adrenal cortex membrane vesicles. QLS and transient electric birefringence enabled elucidation of specific binding effects.[278]

One of the most complex and biologically relevant protein–membrane systems yet studied by dynamic light scattering is vesicles isolated from guinea pig tubules, containing (Na^+, K^+)-ATPase.[279] Laser Doppler electrophoresis was used to determine the changes in vesicle size and surface charge density attending the binding of various ions and substrates. There is one net negative charge per protein unit that contributes to the electrophoretic mobility; its pK of 3.9 suggests an aspartic acid residue. Although Na^+ and K^+ have different functional roles in the ATPase system, there is no difference in their effect on the surface potential. Deviations from the simple Debye relation between surface potential and ionic strength probably reflect the complex surface structure of these biological membranes. The binding constant for Mg^{2+} is higher than found by other methods, and no difference was observed between Mg^{2+} and Ca^{2+}. ATP and ADP behave similarly: neither binds to the ATPase in the absence of Mg^{2+}, though the binding becomes weaker with higher $[Mg^{2+}]$. The vesicle size, originally 204 nm diameter, shrinks in area by 35% upon binding of ATP or ADP. Since a corresponding increase in surface charge density is not observed, it is postulated that nucleotide binding may lead to protein patching.

The decrease in D and increase in R have been used to deduce the thickness, density, and adsorption isotherms of proteins adsorbed to polystyrene latex spheres.[280, 281] The proteins include human and bovine serum albumin, α-globulin, and glycoproteins from human erythocyte membranes.

10.4.6. Micelles

While QLS is a useful way to study micelle structure in solution, it is beset, as are other physical techniques, with a difficult complication. Most biologically interesting detergent or bile salt micelles are highly charged, leading to substantial intermolecular interaction effects at moderate concentrations. However, extrapolation to infinite dilution is difficult both because many micelles, being rather small, scatter weakly; and because dynamic association–disassociation equilibria may change with dilution.

Sodium dodecyl sulfate (SDS) micelles have been studied by QLS and total intensity scattering[282–284] at concentrations substantially above the cmc, over a range of T and ionic strength. The micelles grow, as judged both by average Stokes radius R and by radius of gyration R_G determined by angular dissymetry, as T is lowered and salt is raised. The polydispersity, measured from the variance of D, is about 70% of the mean. The minimum micellar radius is about 25 Å, corresponding to the stretched-out length of a hydrated SDS chain. Under conditions of more extensive growth, micelle shapes were deduced by comparing R and R_G with theoretical models for spheres, oblate, and prolate ellipsoids. A prolate or rodlike shape was consistent with the scattering data. These studies were performed without extrapolation of scattering to infinite dilution, so they may be open to objections that concentration dependence of D was not taken into account.

However, at the relatively high ionic strengths employed, 0.15 to 0.6 *M* NaCl, long-range electrostatic interactions should be well screened.

A study of SDS micelles at the low concentration limit, near the cmc, was performed as a function of SDS and NaCl concentrations.(285) The salt dependence was analyzed using Stephen's theory. It was concluded that the aggregation number increases from 27 to 95 as [NaCl] rises from 0 to 0.5 *M*, and then plateaus. A rather indirect argument concluded that the micelles were oblate spheroids. Addition of very small amounts of dodecanol to SDS near the cmc causes a great enhancement of turbidity, due to formation of emulsion droplets with *R* about 500 Å.(286)

Lecithin has two cmc's, the first at very low concentrations. A second transition at higher concentration appears to be associated with the elongation of spherical micelles into cylindrical or ellipsoidal aggregates.(287) The failure of the QLS relaxation rate to extrapolate to zero with q^2 was taken as evidence of rotation of rodlike micelles, but might be due instead to restricted motion in these crowded (20%) solutions.

QLS measurements on gangliosides G_{M1} and G_{D1a} enable determination of an upper limit to the cmc, $<1 \times 10^{-6}$ *M* for both.(288) *M* and *R* are 532,000 and 64 Å for G_{M1}, 417,000 and 60 Å for G_{D1a}; a 3 : 1 mixture gave intermediate values. Ca^{2+} did not affect these values. The size of myristyltrimethylammonium micelles varies markedly with *T* and salt at high pressures in the range 1–3000 bar.(289)

Bile salt micelles, and their mixed micelles formed with lecithin and cholesterol, show very complex phase behavior as a function of composition, salt, and *T*. The order of size of pure bile salt micelles is taurodeoxycholate (tauroDC) > taurochenoDC > tauroursoDC > taurocholate.(290) Taurocholate shows little dependence of *R* on bile salt concentration dependence of *R*, attributed to strong intermicellar charge effects, at 0.15 *M* NaCl.(291) Another group finds little *c* dependence at 0.15 *M* NaCl, and a positive dependence of *R* on *c* at 0.6 *M* NaCl.(290) Pure bile salt micelles grow as *T* is lowered, but growth is inhibited by urea. The large micelles are rodlike and polydisperse, interpreted as a two-stage process in which small, globular primary micelles polymerize into rodlike secondary micelles.(290) If lecithin (L) is mixed with bile salt (BS) at low L/BS ratio, QLS and total intensity measurements(292, 293) suggest that pure BS micelles and mixed micelles coexist in a highly disperse solution. At high L/BS, only mixed micelles are observed, whose size increases with L/BS ratio and appears to diverge as the phase boundary is approached. These mixed micelles are proposed to have the shape of a mixed bilayer disk surrounded on the perimeter by pure BS. There are conflicting reports(291, 293, 294) whether or not addition of cholesterol changes micellar size.

The reader is referred to Chapter 8 in this volume by Mazer for a more thorough discussion of this complicated area.

10.5. PHYSIOLOGICAL AND BIOMEDICAL APPLICATIONS

In this section we survey a wide variety of applications of QLS, which attest to the versatility of this technique.

10.5.1. Cataracts

Cataracts, it has been proposed,[(295)] are the result of light scattering from high molecular weight aggregates (HM) of lens protein. The concentration of HM in human lenses increases with age, and these aggregates are found mainly in the nuclear region of nuclear sclerotic cataractous lenses.[(296)] Using QLS to measure D in intact human and bovine lenses, it was found that at 37°C in intact lenses, D is only moderately less than D of α-crystallin protein of calf lens in solution.[(297)] However, as T is decreased, $\bar{D}$ decreases sharply. It appears that the phenomenon of cold cataract in the calf lens is due to long-range correlations and high turbidity associated with a first-order phase separation of the protein–water mixture in the lens. Cold cataract appears at nearly the same T at which $\bar{D} = 0$. In cataractous human lenses D is 5.5 times smaller than in the normal lens, indicative of substantial protein aggregation. Assuming spherical aggregates, with $M \sim R^3 \sim D^{-3}$, this implies $\bar{M} \approx 5 \times 10^8$, close to the value determined biochemically. In a more detailed study,[(298)] it was assumed that two size ranges of proteins occurred in the normal human lens. Low molecular weight aggregates had $D = 2.15 \times 10^{-7}$ cm^2/sec, with a variance of 0.88; HM had $D = 0.4 \times 10^{-7}$ cm^2/sec, with a very large variance of 1.40. With these values it was deduced that the concentration of HM in normal lenses increases monotonically with age, from 0 in the infant to 3% at 60 years. The percent concentration determined by QLS is in excellent agreement with biochemical determinations.

10.5.2. Immunoassay

Detection by QLS of the initial stages of antigen–antibody agglutination can serve as a very sensitive analytical probe. In the initial study,[(299)] polystyrene latex spheres were coated with bovine serum albumin, and rabbit antibody to BSA was added to cause agglutination, which was detected by the decrease in D. The strong scattering by the aggregated spheres greatly amplifies the protein agglutination. The procedure was also applied to measurement of mouse immunoglobulin IgA,[(300)] where good agreement with radioimmunoassay was found. A protein concentration of only 5 to 10 ng/ml was required, suggesting that picomolar amounts of protein can be detected. This paper also presents an analysis of the roles of electrostatic repulsion and van der Waals attraction of the latex particles in specific and nonspecific agglutination. QLS immunoassay has also been run in the agglutination-inhibition mode for human chorionic gonadotropin and human luteinizing hormone.[301] This has advantages over the direct mode in eliminating the requirements of large antigen size (to prevent nonspecific aggregation) and more than one haptenic group per antigen (to allow multivalent agglutination). Recently, it has been shown that anisotropic total intensity scattering measurements are an even more sensitive immunoassay than QLS.[302]

Another approach is to measure the decrease in electrophoretic mobility of antigen-coated polystyrene beads upon their binding of antibody.[303]

10.5.3. Cell Surfaces

Electrophoretic light scattering (ELS) has become a convenient and useful way to determine electrophoretic mobilities of cells. This gives the distribution of surface charge density σ in a cell population, and the change of σ with chemical treatment, disease state, or other variables. Compared with ELS of small proteins, ELS of cells is technically straightforward, because negligible diffusional broadening allows adequate resolution at conveniently high scattering angles, and because cell scattering power is high.[304, 305] Because diffusion is generally negligible, the breadth of the ELS spectrum reflects directly the breadth of the mobility distribution.

The greatest effort has been directed at human lymphocytes from normal subjects and leukemic patients. For a comprehensive review see Uzgiris.[53] An early study showed that T lymphocytes have a higher mobility than B lymphocytes, and that the T-cells are subdivided into two populations, denoted fast and slow.[306] Similar electrophoretic and sedimentation heterogeneity occurs in mouse thymus lymphocytes.[307] Leukemic cells have a lower mode mobility than normal cells in low salt[308] but not in physiological saline,[309] and their distribution is more symmetrical.[308] When normal T and B cells are treated with neuraminidase, an enzyme which cleaves *N*-acetylneuraminic acid (a major source of surface charge), their relative mobilities are reversed and their electrophoretic distinguishability enhanced. Lymphoblasts of patients with acute lymphocytic leukemia differ from normal cells in their responses to neuraminidase treatment and ionic strength changes, implying a different surface structure.[309] In chronic lymphocytic leukema (CLL) patients, and in postoperative cancer patients, there is a decrease in fast T-cells and an increase in slow T-cells compared with normal donors.[51] There appear to be correlations in normal subjects between the numbers of fast and slow

T-cells and the number of cells with high and low affinities for rosette formation with sheep red blood cells; but these correlations may not hold in some cancer patients.[310, 51, 311] In another study,[312] the lymphocytes of CLL patients had a single mobility peak corresponding most closely to fast T-cells; but the surface properties were more like B-cells.

When pokeweed, a mitogen, binds to rat thymus lymphocytes, the electrophoretic mobility, hence σ, decreases; and the distribution broadens, implying a variability in binding capacity among the cells.[313] Another mitogen, the plant lectin concanavalin A (con A) also decreases μ of lymphocytes.[314] Lymphocytes that have been sensitized by tuberculin can be detected by their altered μ distributions after incubation with the antigen PPD.[315]

Lymphokines are soluble factors, produced by antigen- or lectin-stimulated lymphocytes, which mediate the expression of cellular immunity. Although the lymphokine leukocyte inhibitory factor (LIF) and macrophage migration inhibitory factor (MIF) inhibit migration of polymorphonuclear lymphocytes and macrophages, respectively, neither LIF nor MIF causes a change in μ of these cells.[316, 317] Such changes are produced, however, by other stimulated lymphocyte supernatant components.

Substantial changes in electrokinetic surface properties occur when guinea pig peritoneal macrophages and eosinophils are reacted with IgG immune complexes,[318] and when rat serosal mast cells are immunologically activated.[319] Cross-linking of cell surface receptors of peritoneal macrophages with con-A causes a great increase in the width of the μ-distribution, implying a heterogeneity of cell response to cross-linking.[320]

ELS studies comparing Ca^{2+} and Mg^{2+} indicate that the specific role of Ca^{2+} in exocytosis is not due to its ability to decrease electrostatic repulsion between negatively charged membranes, either in chromaffin granules[321] or in synaptic vesicles and synaptosomal membranes.[322] It has also been found that red cell surface charge is not a function of cell age.[323]

QLS study of the role of dextran in inducing aggregation of the oral bacterium *Streptococcus mutans* indicates that aggregation is inhibited by the negatively charged cell surfaces, while agglutination occurs by polymer bridging.[324]

10.5.4. Monolayers, Films, and Membranes

QLS presents some attractive features for the analysis of the dynamics of interfaces. It can probe surface tension, surface elasticity of films, and surface viscosity. It is nonperturbing, which is particularly important for fragile surfaces or interfaces near a critical point, and it can probe a wide frequency range. However, it suffers from substantial difficulties: poor signal/noise, sensitivity to external vibrations, and complex theoretical interpretation. These problems have inhibited its wide application. However, recent experimental refinements[325–327] demonstrate the potentially high accuracy of the technique.

For propyl stearate monolayers, QLS measurement of film elasticity agrees within a factor of 4 with that obtained from $\pi - A$ isotherms. Film viscosity values, on the order 10^{-7} kg/sec, increase with film density.[328] Similar measurements have been made on monolayers of myristic, pentadecanoic, and arachidic acids.[329] The variation of film elasticity with monolayer density suggests a phase transition in the monolayer.

Total intensity scattering from soap films gives the second derivative, with respect to film thickness, of the interaction free energy between the two film surfaces. This in turn can be interpreted in terms of double-layer repulsive and attractive forces.[330] The QLS beat frequency spectrum gives the shear viscosity and surface tension.[331] Heterodyne QLS of the stretching mode of 50 nm free liquid films shows maxima in the MHz region, indicating the propagating character of the mode. The peak frequencies are related to the surface tension of the film.[332]

10.5.5. Gels and Entangled Solutions

QLS presents one major advantage over traditional mechanical techniques for elucidating the viscoelastic properties of gels: it is nonperturbing. Particularly with the loose gels often encountered in biological systems, the very act of applying a probe may distort the gel structure, while QLS measures spontaneous thermal fluctuations or the response to low-amplitude environmental vibrations. Among the questions to be answered about a particular gel are: Is it really a permanently cross-linked network, or just an extensively entangled, highly concentrated solution? What are the density, functionality, and chemical nature of its cross-links? What are the elastic and frictional properties of the chains connecting network junctions, as compared with equivalent free chains, and what are their interactions with solvent? The information content of QLS alone is insufficient to answer these questions; but QLS does provide information on elastic and frictional properties of gel networks which, with the help of suitable theoretical models, may provide molecular insight.

There are currently three major models for QLS of gels: continuum, harmonically bound particle, and Rouse–Zimm chain. The continuum model, originated by Tanaka *et al.*,[336] is most appropriate when $q^{-1} \gg$ typical cross-link spacing in the network. It predicts an exponential decay rate $G_l q^2/f$ or $G_t q^2/f$ for longitudinal or transverse waves, measured by polarized or depolarized scattering, respectively. f is the frictional coefficient per unit value of the fiber network for motion relative to gel liquid, G_l is the longitudinal compressional modulus of the fiber network, and G_t is its shear modulus. In principle, G alone can be measured by total intensity scattering; but in practice S/N is too poor, so that either G or f must be measured by a macroscopic technique so that the other can be determined by QLS. Rubber elasticity theory has been used to express the moduli in terms of network functionality and the length and swelling of the elastic chain elements.[337] The scattered intensity decreases with equilibrium gel concentration according to scaling theory, while the fluctuations from swollen networks with controlled amounts of pendant chains increase with decreasing cross-linking density.[338] The moduli depend on the solvents in which the polymer is prepared and swollen.[339]

The theory of a continuous elastic solid has been used to interpret QLS of standing displacement waves set up in cuvettes subjected to mechanical vibrations.[340–342] A recent extension of the theory takes into account viscous and internal energy dissipation.[343] The resonant peaks observed depend on the cuvette dimensions and on the longitudinal sound speed, from which the shear elastic modulus of the bulk gel can be determined. They also depend on whether the material is a soft gel such as agarose,[340, 342] which does not stick to the wall, or a hard gel such as polyacrylamide,[341] which does. The modulus varies according to the third power of the polymer concentration, in good agreement with theoretical predictions. The mechanical excitation method has been used to monitor the time evolution of the modulus of fibrin gels cross-linked with bloodclotting factor XIII.[344] The modulus varies linearly with the number of cross-links until maximum rigidity is achieved.

The harmonically bound particle model for QLS by gels has been refined to include a proper statistical average over initial positions, and to consider additional static scattering due to spatial structuring of the scattering particles.[345] In contrast to the continuum model, this theory predicts a reduction in the initial amplitude $g^{(2)}(0)$ of the gel relative to sol, correlated with the constraints on motion in the gel state. These theoretical differences are consistent with different experimental observations on what appear to be the same systems, 2.5% and 5.0% polyacrylamide gels.[336, 345] Clearly, further work is required to reconcile these discrepancies.

Ideally, $g^{(2)}(\tau)$ would be derived from a detailed molecular theory of the elastic network and its frictional interaction with solvent. Though such a detailed theory is still out of reach, general consideration of Rouse–Zimm theory shows that when q becomes sufficiently large, the decay rate always takes the simple form q^2kT/f, where f is the frictional coefficient of a single bead in the chain.[193] When this result is incorporated in the viscoelastic continuum model,[336] the apparent long-wavelength diffusion coefficient of the gel is $D_G = 2kT/f(1 - 2/\phi)$, where ϕ is the network functionality.[193]

QLS of 0.5% calcium alginate gels showed a small degree of spectral broadening, with long range (~0.1 mm), long term (~10 min) fluctuations superimposed on the autocorrelation function.[346, 347] These macroscopic fluctuations are attributed to very small bulk oscillations of the gel which modulate the speckle pattern in the limit of small spectral broadening or long relaxation times.[348] The diffusion coefficients of dextran fractions and globular proteins within these gels were determined from the heterodyne beat spectrum. Molecules with $2R < 10$ nm were virtually unimpeded, while those with $2R > 30$ nm had D about 1/3 the free solution value. The implies large interstitial spaces within the gel, with junction zones consisting of an association of hundreds of chains.[346, 347]

Comparison of D and collective diffusion coefficients in aqueous polyacrylamide gels indicates that solvent viscosity does not depend on polymer concentration.[349] Interpretation of such experiments may be aided by a theory of QLS of a spherical Brownian particle immersed in a general viscoelastic medium.[350]

Entangled solutions represent an intermediate regime between dilute polymer solutions and gels. The theory of dynamical behavior and QLS of such systems is based on the reptation concept of de Gennes,[351, 352, 353] in which each polymer chain is imagined to crawl in an evanescent tube of neighboring polymers. A variety of behavior can result, depending on the relative values of the characteristic length of a polymer chain, the concentration of chains (which determines the dimensions of the tube), and q^{-1}. It would lead us too far afield to review this rapidly developing subject in detail, which has thus far been directed mainly to synthetic macromolecules; but it is likely to be important in future understanding of concentrated biopolymer solutions, so the interested reader is directed to some recent theoretical and experimental articles.[354–359] The molecular arrangement of cervical mucus is thought to be an entangled solution rather than a cross-linked network.[360]

10.5.6. Muscle

QLS studies of essentially monodisperse, native muscle thin filaments as a function of T and ionic strength show clear evidence of flexibility of the filaments[233] associated with the removal of tropomyosin. This is deduced from the lack of T/η scaling at high scattering angles; such scaling would have been expected for rigid rods. This qualitative conclusion is buttressed by comparison with a quantitative theory of scattering from very long, semiflexible filaments.[238] The lack of scaling is explicitly associated with internal flexing modes of motion, whose relative contribution at high angles will change if the flexibility changes with T or other solution conditions. However, it is claimed that environmentally dependent changes in the longest, overall rotational relaxation time of the filament cannot be completely ruled out with the data available. Previous investigations of F-actin flexibility have been obscured by polydispersity of the preparation.[362, 363] Comparison of the scattering and viscosity behavior of F-actin filaments in the presence of heavy meromyosin and myosin subfragment S-1, as a function of ATP, indicates that at maximal actin activation the physical properties are much closer to dissociated than to fully complexed systems. Thus most of the HMM and S-1 are in a refractory state unable to bind actin.[362]

QLS from intact single frog skeletal muscle fibers[364] showed a decay rate of about 70 msec at $\theta = 30°$ that varied linearly with the projection of the scattering vector on the fiber axis. This result, together with the amplitude $g^{(2)}(0)$, is consistent with a model of isometrically contracting fibers in which the scattering material has relative axial velocities of 1–2 m/sec with relative displacements > 0.1 μm. These decay rates are considerably slower than those observed in whole muscle or small fiber bundles.[365] If there exist structural fluctuations on the scale of the myofibrillar sarcomere, that arise from force imbalances due to independent crossbridge cycling, they must be relatively slow and of small amplitude. Carlson[366] has discussed various models of structural fluctuations in the steady state of muscular contraction, and Fujime *et al.*[367, 368] have presented a simple model for QLS, diffraction intensity changes, and electro-optic effects in striated muscle fibers.

In contrast to skeletal muscle, which shows no fluctuations at rest, resting cardiac muscle shows QLS with a decay rate varying directly with a Ca-dependent force.[369]

10.5.7. Biological Velocimetry

Laser Doppler spectroscopy[55, 370] has been used in two biological applications of velocimetry: blood flow and protoplasmic streaming. We note some clever instrumental developments: a laser Doppler microscope,[371] a velocity-tracking circuit using real-time spectrum analysis,[372] and a cross-correlation technique for transforming photon counting measurements on fluid flow to the frequency domain.[373]

The basic Doppler equation relating frequency shift to velocity is simply $\Delta\omega = \mathbf{q} \cdot \mathbf{v}$. Detection is carried out in the heterodyne mode, with the local oscillator signal often coming from scattering from vessel or cell wall or surrounding stationary tissue.

10.5.7.1. Blood Flow. Laser Doppler velocimetry (LDV) of blood flow in retinal vessels, first accomplished in rabbits, was carried out on human after the apparatus was adapted to short measuring times (<1 sec) and low power density (0.05 W/cm^2).[374] Velocities of about 2 cm/sec were measured. Much of the signal comes from multiple scattering, which broadens the spectrum,[375] so an empirical correlation of $\Delta\omega$ vs. v is required. A high-frequency cutoff indicates a maximum red blood cell speed; this can be obtained using short analysis times or from angular analysis of the scattered light.[376] An analog photocurrent processing method[377] was adapted to give continuous monitoring of velocity, enabling analysis of pulsatile flow.

Flow in vessels not accessible to external observation, such as the femoral vein, can be monitored by LDV using a narrow fiber optic catheter.[378] Even this minor intrusion can be avoided by simply measuring the autocorrelation function of light transmitted and reflected through the skin of the tissue of interest, either directly[379] or through a fiber optic guide which gives improved S/N.[380] The laser light is completely depolarized and randomized in direction by multiple scattering from the tissue. A theory of the correlation function for scattering of such light by red blood cells or similar particles moving with a specified velocity distribution has been constructed[380]; but data analysis must still be approached empirically. The ability to monitor blood flow changes nonintrusively and in real time should be useful in assessing response to drugs, burn damage, skin graft viability, and muscle perfusion.

10.5.7.2. Protoplasmic Streaming. QLS has numerous advantages for the characterization of protoplasmic streaming[380]: it observes objectively all motion in the scattering volume, the distribution of Doppler shifts is directly related to the distribution of velocities, it is rapid, different velocity components can be observed by changing the cell orientation relative to $\mathbf{q}$, and submicroscopic particles can be observed. In work on the common freshwater alga *Nitella flexilis*, it has been found that diffusion is negligible, particles of different sizes have the same velocity, the streaming velocity increases linearly with T to an optimum at 34°C and the v distribution becomes narrower, and laser light causes a very local inhibition of protoplasmic streaming.[381] In later work,[382] it was shown that the intense low-frequency part of the spectrum is due to amplitude modulation of the scattered light by the array of chloroplasts in the cell, and that photobleaching of these allows determination of a corrected frequency and velocity distribution. The tonoplast membrane seems to be flowing along with the endoplasm and vacuolar sap which it separates. At certain concentrations, ATP and cytochalasin B affect the velocity histogram, while colchicine does not.

In the slime mold *Physarum polycephalum*, rhythmic wall contractions can be monitored along with streaming velocities, and can be observed even in the absence of streaming.[383, 384] Major conclusions are that a small fraction of protoplasmic particles move with velocities much higher than apparent from optical microscopy; that the v distribution at the maximum of the oscillation cycle is more sharply peaked than parabolic, suggesting a viscosity gradient across the plasmodium tubes; that transverse motion is nearly as fast as longitudinal; and that streaming is inhibited by high-intensity laser light, which causes the irradiated protoplasm to gel.[384]

Preliminary work has also been reported on cytoplasmic streaming in the higher plant *Elodes canadensis*,[385] and on particle motion during microtubule assembly.[386]

10.5.8. Motility

QLS has been used for some time to measure motility of bacteria and spermatozoa. Early work is well reviewed by Cummins.[387] It was originally thought that the characteristic decay of the autocorrelation function was due to translational swimming effects, and motile microorganisms were modeled as point particles which maintained a constant velocity v for times long compared to the correlation time $(\mathbf{q} \cdot \mathbf{v})^{-1}$. The method of splines was adapted to determination of swimming speed distributions,[388] and corrections were made by various workers for a percentage of dead, diffusing particles. At low scattering angles, the mean and variance of the speed distribution can be reliably obtained.[389]

It soon was recognized, however, that bacteria or sperm heads are too large to be treated as point particles, and that their rotational motion must be considered in analyzing $g^{(2)}(\tau)$ at higher scattering angles. A series of papers by Chen *et al.* models micron-size bacteria as coated ellipsoids to compute scattering angle dependent form factor contributions in the Rayleigh–Gans–Debye (RGD) approximation, and taking successively into account translation,[390, 391] rotation,[392] and alternating straight line and twiddling motion in chemotaxis.[393] While not rigorous for such large particles, the RGD approximation is adequate to within about 10%.[394]

Rotation of large, hydrodynamically anisotropic particles can be studied directly by QLS from a solution of such particles undergoing hydrodynamic shear.[395] The ideas and equations are similar to those for translational velocimetry. Since the particle scattering form factor varies with its periodically oscillating orientation in the flow field, the power spectrum contains a set of lines whose frequency maxima are proportional to the shear rate and to the anisotropy. The breadth of the lines is approximately equal to the rotational diffusion coefficient, so resolution is best with large particles. Agreement with the known dimensions of *E. coli* is satisfactory.

Bull sperm heads are even larger and more asymmetric than *E. coli*, hence rotational effects are even more severe. Early work[396–398] must be reexamined in that light, though the fraction of motile sperm may be estimated with confidence.[399] It appears from model calculations that the autocorrelation function is a direct measure of mean head rotation frequency,[400, 401] and that scattering from the midpiece must also be included.[400] Forward depolarized QLS from sea chestnut spermatozoa gave an average rotational velocity of 230 rad/sec, attributed mainly to scattering from the flagella.[402, 403]

However, since human sperm heads are less anisotropic than bull sperm heads, it is possible by working at low angles to obtain correlation functions that are mainly determined by translational swimming velocity except at the highest speeds.[404] Data were collected at 11.82° on sperm that had been diluted with clear seminal fluid. This procedure maintains sperm viability, while eliminating multiple scattering complications. The data, when analyzed by the method of splines,[388] agreed well with videomicroscopic results save above apparent speeds of 80 μm/sec, where contribution from head rotation evidently distorts the distribution.

Progesterone, at concentrations 1000 × physiological, reduces the apparent swimming speed of sperm, suggesting a contraceptive mechanism in addition to its action on secretory cells of the cervix.[405]

Number fluctuation spectroscopy of motile microorganisms should give mainly translational motion,[406] though even with this technique orientation fluctuations have been detected.[407] Autocorrelation analysis of films of the number of cells occupying a microscopic field gives translational mobility alone.[408]

Heterodyne QLS can be used to determine the frequency of ciliary beating, and the effects of agents such as Ca^{2+} on this frequency, with results in good agreement with conventional but less convenient methods such as high-speed cinematography.[397]

10.6. CONCLUSION

QLS has been applied to an enormous variety of biochemical and biological systems. Though most early work was quite properly devoted to simpler model systems, there have been many recent applications to increasingly complex and realistic situations. This is bound to continue, particularly as QLS apparatus, complete with microprocessor data analysis, becomes commercially available. While this expansionist trend is to be encouraged, some words of caution are in order.

Concentration and charge effects, and behavior of fibers, gels, and entangled solutions, are still not well understood. Extraction of intramolecular motions from QLS autocorrelation function tends to be based on oversimplified theoretical models. In general, interpretation of QLS is extremely model dependent, and models which properly reflect the complexities of biological behavior are hard to devise and to solve. QLS often cannot distinguish between alternative models. As is usually the case in biophysics, many techniques must be focused on a problem to get a complete, unambiguous solution.

Careful attention must be paid to sample quality, aggregation, and biological activity. Light scatterers are often better instrumentalists than they are biochemists; and close collaboration with those who can assess the chemical and biological homogeneity of a preparation is generally necessary. Simple transfer of sample from the biochemistry lab to the laser lab is rarely sufficient. By the same token, QLS often detects aggregation hidden to other techniques.

Technical developments to be looked for include better S/N in depolarized QLS, extension to submicrosecond relaxation times, more convenient ELS apparatus, broader implementation of procedures to characterize the distribution of relaxation times, and a variety of specialized analytical applications.

REFERENCES

1. R. Pecora, *J. Chem. Phys.* **40**, 1604–1614 (1964).
2. S. B. Dubin, J. H. Lunacek, and G. Benedek, *Proc. Nat. Acad. Sci. USA* **57**, 1164–1171 (1967).
3. P. Berge, B. Volochine, R. Billard, and A. Hamelin, *C.R. Acad. Sci. Ser. D* **265**, 889 (1967).
4. B. J. Berne and R. Pecora, *Dynamic Light Scattering: With Applications to Biology, Chemistry, and Physics*, Wiley, New York (1976).
5. Chu, B. *Laser Light Scattering*, Academic, New York (1974), 317.
6. H. Z. Cummins and E. R. Pike, eds., *Photon Correlation Spectroscopy and Velocimetry*, Plenum, New York (1976), p. 590.
7. V. A. Bloomfield, *Ann. Rev. Phys. Chem.* **28**, 233–259 (1977).
8. V. A. Bloomfield and T. K. Lim, *Meth. Enz.* **48**, Part F, 415–494 (1978).
9. V. A. Bloomfield, *Ann. Rev. Biophys. Bioeng.* **10**, 421–450 (1981).
10. F. D. Carlson, *Ann. Rev. Biophys. Bioeng.* **4**, 243–264 (1975).
11. S. H. Chen and M. Holz, *Med. Res. Eng.* **12**, 19–25 (1977).
12. S. H. Chen and A. V. Nurmikko, in *Spectroscopy in Biology and Chemistry—Neutrons, Lasers, and X-Rays*, S. H. Chen and S. Yip, eds., Academic, London (1975).
13. B. Chu, *Phys. Scr.* **19**, 458–470.
14. L. De Maeyer, K. Gnadig, J. Hendrix, and B. Saleh, *Q. Rev. Biophys.* **9**, 83–107 (1976).
15. W. H. Flygare, *Chem. Soc. Rev.* **6**, 108–138 (1977).

16. A. M. Jamieson and M. E. McDonnell, *Adv. Chem. Ser.* **174**, 163–205 (1979).
17. C. S. Johnson, Jr., and D. A. Gabriel, Laser Light Scattering, in *Spectroscopy in Biochemistry*, J. E. Bell, ed., CRC, Boca Raton (1980), Chap. 10.
18. R. Pecora, 1979, *Macromol. Chem., Suppl. 2, Curr. Top. Synth. Structu. Polym.* 73–80 (1979).
19. J. M. Schurr, *CRC Crit. Rev. Biochem.* **4**, 371–431 (1977).
20. I. W. Shepherd, *Rep. Prog. Phys.* **38**, 565–620 (1975).
21. H. J. V. Tyrrell and P. J. Watkiss, *Ann. Rept. Chem. Soc. A*, **73**, 35–52 (1976).
22. D. R. Bauer, J. I. Brauman, and R. Pecora, *Ann Rev. Phys. Chem.* **27**, 443–463 (1976).
23. D. R. Bauer, S. J. Opella, D. J. Nelson, and R. Pecora, *J. Am. Chem Soc.* **97**, 2580–2582 (1975).
24. R. Pecora, *J. Polym Sci. Polym. Symp.* **65**, 79–84 (1978).
25. D. R. Bauer, B. Hudson, and R. Pecora, *J. Chem. Phys.* **63**, 588–589 (1975).
26. T. J. Herbert, *J. Colloid Interface Sci.* **69**, 122–127 (1979).
27. T. J. Herbert, *J. Colloid Interface Sci.* **72**, 157–158 (1979).
28. I. S. Jones, *Chem. Phys. Lett.* **46**, 485–487 (1977).
29. R. Pecora and S. R. Aragon, *J. Chem. Phys.* **66**, 2506–2516 (1977).
30. S. R. Aragon, *J. Chem. Phys.* **73**, 1576–1580 (1981).
31. J. M. Rallison and L. G. Leal, *J. Chem. Phys.* **74**, 4819–4825 (1981).
32. J. Garcia de la Torre and V. A. Bloomfield, 1981, *Qt. Rev. Biophys.* **14**, 81–139 (1981).
33. F. Perrin, *J. Phys. Radium* [7] **5**, 497–511; **7**, 1–11 (1936).
34. S. Broersma, *J. Chem. Phys.* **32**, 1626–1632, 1632–1635 (1960); **74**, 6989–6990 (1981).
35. B. Chu, *Pure Appl. Chem.* **49**, 941–962 (1977).
36. V. DeGiorgio, in *Photon Correlation Spectroscopy and Velocimetry*, H. Z. Cummins and E. R. Pike, eds., Plenum, New York (1977), pp. 142–163.
37. K. L. Wun and W. Prins, *Biopolymers* **14**, 111–117 (1975).
38. C. R. Jones and C. S. Johnson, Jr., *J. Chem. Phys.* **65**, 2020–2021 (1976).
39. E. Jakeman, P. N. Pusey, and J. M. Vaughan, *Opt. Commun.* **17**, 305–308 (1976).
40. J. B. Lastovka, *Bell Syst. Tech. J.* **55**, 1225–1293 (1976).
41. T. J. Herbert and J. D. Acton, *Appl. Opt.* **18**, 588–590 (1979).
42. E. Gulari and B. Chu, *Rev. Sci. Instrum.* **48**, 1560–1567 (1977).
43. E. Jakeman, C. J. Oliver, and E. R. Pike, *Adv. Phys.* **24**, 349–405 (1975).
44. P. N. Pusey, in *Photon Correlation Spectroscopy and Velocimetry*, H. Z. Cummins and E. R. Pike, eds., Plenum, New York (1977), pp. 45–141.
45. K. Soda, I. Nisho, and A. Wada, *J. Appl. Phys.* **47**, 729–735 (1976).
46. W. H. Flygare, B. R. Ware, and S. L. Hartford, Electrophoretic light scattering, in *Molecular Electroptics*, C. T. O'Konski, ed., Academic, New York (1976), Part 1, pp. 321–366.
47. D. D. Haas, and B. R. Ware, *Anal. Biochem.* **74**, 175–188 (1976).
48. R. Mohan, R. Steiner, and R. Kaufmann, *Anal. Biochem.* **70**, 506–525 (1976).
49. B. A. Smith and B. R. Ware, in *Contemporary Topics in Analytical and Clinical Chemistry*, Vol. 2, D. M. Hercules, G. M. Hieftje, L. R. Snyder, and M. A. Evenson, eds., Plenum, New York (1978), pp. 29–54.
50. E. E. Uzgiris and D. H. Cluxton, *Rev. Sci. Instrum.* **51**, 42–46 (1980).
51. E. E. Uzgiris, J. H. Kaplan, T. J. Cunningham, S. H. Lockwood, and D. Steiner, in *Electrophoresis*, N. Catsimpoolas, ed., Elsevier/North-Holland, New York (1978), pp. 427–440.
52. B. R. Ware, 1974, *Adv. Colloid Interface Sci.* **4**, 1–44 (1974).
53. E. E. Uzgiris, *Adv. Colloid Interface Sci.* **14**, 75–171 (1981).
54. J. Josefowicz and F. R. Hallett, *Appl. Opt.* **14**, 740–742 (1975).

55. E. R. Pike, in *Photon Correlation Spectroscopy and Velocimetry*, H. Z. Cummins and E. R. Pike, eds., Plenum, New York (1977), pp. 246–343.
56. K. S. Schmitz, *Chem. Phys. Lett.* **63**, 259–264 (1979).
57. K. S. Schmitz, *Chem. Phys. Lett.* **42**, 137–140 (1976).
58. A. Wada, I. Nishio, and K. Soda, *Rev. Sci. Instrum.* **50**, 458–463 (1979).
59. R. I. Jenrich and M. L. Ralston, *Ann. Rev. Biophys. Bioeng.* **8**, 195–238 (1979).
60. R. J. Goldberg, *J. Phys. Chem.* **57**, 194–202 (1953).
61. C. B. Bargeron, *J. Chem. Phys.* **60**, 2516–2519 (1974).
62. C. B. Bargeron, *J. Chem. Phys.* **61**, 2134–2138 (1974).
63. J. C. Brown and P. N. Pusey, *J. Chem. Phys.* **62**, 1136–1144 (1975).
64. G. A. Brehm, and V. A. Bloomfield, *Macromolecules* **8**, 663–665 (1975).
65. M. Schmidt, W. Burchard, and N. C. Ford, *Macromolecules* **11**, 452–454 (1977).
66. R. L. McCally and C. B. Bergeron, *J. Chem. Phys.* **67**, 3151–3156 (1977).
67. J. C. Selser and Y. Yeh, *Biophys. J.* **16**, 847–848 (1976).
68. J. C. Selser, *Macromolecules* **12**, 909–916 (1979).
69. T. A. King and M. F. Treadaway, *J. Chem. Soc. Faraday Trans. II* **73**, 1616–1626 (1977).
70. G. Meyerhoff and J. Raczek, *Macromol. Chem.* **177**, 1199–1214 (1976).
71. V. J. Morris, G. J. Brownsey, and B. R. Jennings, *J. Chem. Soc. Faraday Trans.* **75**, 141–147 (1979).
72. R. Pecora and S. R. Aragon, *J. Chem. Phys.* **64**, 2395–2404 (1976).
73. B. E. A. Saleh and J. Hendrix, *Chem. Phys.* **12**, 25–30 (1976).
74. T. Tanaka, *Soc. Polym. Sci., Jpn* **7**, 62–71 (1975).
75. S. W. Provencher, *Biophys. J.* **16**, 27–41 (1976).
76. S. W. Provencher, *J. Chem. Phys.* **64**, 2772–2777 (1976).
77. S. W. Provencher, *Markromol. Chem.* **180**, 201–209 (1979).
78. S. W. Provencher, J. Hendrix, and L. DeMaeyer, *J. Chem. Phys.* **69**, 4273–4276 (1978).
79. B. Chu, E. Gulari, and E. Gulari, *Phys. Scr.* **19**, 476–485 (1979).
80. E. Gulari, Y. Tsunashimo, and B. Chu, *J. Chem. Phys.* **70**, 3965–3972 (1972).
81. D. B. Sellen, *Polymer* **16**, 773–776 (1975).
82. H. Z. Cummins and P. N. Pusey, in *Photon Correlation Spectroscopy and Velocimetry*, H. Z. Cummins and E. R. Pike, eds., Plenum, New York (1977), pp. 164–199.
83. P. Calmettes and C. Laj, *Opt. Commun.* **19**, 257–260 (1976).
84. M. Giglio and A. Vendramini, *Phys. Rev. Lett.* **38**, 26–30 (1977).
85. M. Corti, V. DeGiorgio, M. Giglio, and A. Vendramini, *Opt. Commun.* **23**, 282–284 (1977).
86. D. E. Koppel, *Biochemistry* **13**, 2712–2719 (1974).
87. M. A. Loewenstein and M. H. Birnboim, *Biopolymers* **14**, 419–430 (1975).
88. S. T. Kunitake, E. Loh, V. N. Schumaker, S. K. Ma, C. M. Knobler, J. P. Kane, and R. L. Hamilton, *Biochemistry* **17**, 1936–1942 (1978).
89. T. K. Lim, G. J. Baran, and V. A. Bloomfield, *Biopolymers* **16**, 1473–1488 (1977).
90. G. J. Wei and V. A. Bloomfield, *Biophys. Chem.* **9**, 97–103 (1979).
91. T. K. Lim, V. A. Bloomfield, and G. Krejcarek, *Int. J. Appl. Radiation Isotopes* **30**, 531–536 (1979).
92. G. J. Wei and V. A. Bloomfield, *Anal. Biochem.* **101**, 245–253 (1980).
93. S. A. Allison, E. L. Chang, and J. M. Schurr, *Chem. Phys.* **38**, 29–41 (1978).
94. A. R. Altenberger, *Chem. Phys.* **15**, 269–277 (1976).
95. G. D. J. Phillies, *J. Chem. Phys.* **60**, 976–982, 983–989 (1974).
96. P. R. Wills, *J. Chem. Phys.* **70**, 5865–5874 (1979).
97. R. S. Hall, and C. S. Johnson, Jr. *J. Chem. Phys.* **72**, 4251–4253 (1980).
98. G. D. J. Phillies, G. B. Benedek, and N. A. Mazer, *J. Chem. Phys.* **65**, 1883–1892 (1976).
99. G. K. Batchelor, *J. Fluid Mech.* **52**, 245–268 (1972).

100. P. Doherty and G. B. Benedek, *J. Chem. Phys.* **61**, 5246–5434 (1974).
101. T. Raj and W. H. Flygare, *Biochemistry* **13**, 3336–3340 (1974).
102. M. B. Weissman, R. C. Pan, and B. R. Ware, *J. Chem. Phys.* **70**, 2897–2903 (1979).
103. T. Olson, M. J. Fournier, D. H. Langley, and N. C. Ford, Jr., *J. Mol. Biol.*, **102**, 193–203 (1976).
104. W. I. Lee and J. M. Schurr, *J. Polym. Sci.* **13**, 873–888. (1975).
105. S.-C. Lin, W. I. Lee, and J. M. Schurr, *Biopolymers* **17**, 1041–1064 (1978).
106. J. M. Schurr and S. C. Lin, *Proc. Symp. Dynamic Prop. Polyion Systems*, September 9–11, 1978, Kyoto, Japan (Katchalsky Symposium).
107. D. L. Ermak, *J. Chem. Phys.* **62**, 4197–4203 (1975).
108. G. D. J. Phillies, *Macromolecules* **9**, 447–450 (1976).
109. R. Schor and E. N. Serrlallach, *J. Chem. Phys.* **70**, 3012–3015 (1979).
110. B. J. Ackerson, *J. Chem. Phys.* **64**, 242–246 (1976).
111. J. M. Schurr, *Chem. Phys.* **45**, 119–132 (1980).
112. D. W. Schaefer and B. J. Berne, *Phys. Rev. Lett.* **32**, 1110–1113 (1974).
113. J. C. Brown, P. N. Pusey, J. W. Goodwin, and R. H. Ottewill, *J. Phys.* **8**, 664–682 (1975).
114. P. N. Pusey, *J. Phys.* **11**, 119–135 (1978).
115. D. W. Schaefer, *J. Chem. Phys.* **66**, 3980–3984 (1977).
116. D. W. Schaefer and B. J. Ackerson, *Phys. Rev. Lett.* **35**, 1448–1451 (1975).
117. B. J. Ackerson, in *Photon Correlation Spectroscopy and Velocimetry*, H. Z. Cummins and E. R. Pike, eds., Plenum, New York (1977), pp. 440–443.
118. B. J. Berne, *NATO Adv. Study Inst. Ser. Ser. B* **B23**, 344–385 (1977).
119. P. N. Pusey, *J. Phys.* **8**, 1433–1440.
120. P. N. Pusey, *Phil. Trans. R. Soc. London* **293**, 429–439 (1979).
121. W. van Megen and I. Snook, *J. Chem. Phys.* **66**, 813–817 (1977).
122. A. W. Fulmer, J. A. Benbasat, and V. A. Bloomfield, *Biopolymers* **20**, 1147–1158 (1981).
123. N. B. Martin, J. B. Tripp, J. H. Shibata, and J. M. Schurr, *Biopolymers*, **18**, 2127–2133 (1979).
124. J. M. Schurr, *Biopolymers* **18**, 1831–1833 (1979).
125. D. Stigter, *Biopolymers* **18**, 3125–3127 (1979).
126. L. Friedhoff and B. J. Berne, *Biopolymers* **15**, 21–28 (1976).
127. T. Yoshimura, A. Kikkawa, and N. Suzuki, *Opt. Commun.* **15**, 277–280 (1975).
128. M. B. Weissman and B. R. Ware, *J. Chem. Phys.* **68**, 5069–5076 (1978).
129. Y. S. Oh and C. S. Johnson, Jr., *J. Chem. Phys.* **74**, 2717–2720 (1981).
130. M. E. McDonnell and A. M. Jamieson, *Bioplymers* **15**, 1283–1299 (1976).
131. F. C. Chen and B. Chu, *J. Chem. Phys.* **65**, 4508–4511 (1976).
132. T. Raj and W. H. Flygare, *Biopolymers* **16**, 545–549 (1977).
133. S. K. Davi, *J. Chem. Soc.* **70**, 700–708 (1974).
134. D. Eigner, J. Schurz, G. Jurgens, and A. Holasek, *FEBS Lett.* **106**, 165–168 (1979).
135. G. C. Fletcher, *Biopolymers* **15**, 2201–2217 (1976).
136. J. C. Thomas and G. C. Fletcher, *Biopolymers* **18**, 1333–1352 (1979).
137. K. Kuroda, M. Kageyama, T. Madea, and S. Fujime, *J. Biochem.* **85**, 21–28 (1979).
138. A. I. Ahmed, R. E. Feeney, D. T. Osuga, and T. Yeh, *J. Biol. Chem.* **250**, 3344–3347 (1975).
139. R. Gabler, N. C. Ford, and W. Westhead, *Biophys. J.* **15**, 747–751 (1975).
140. S. B. Dubin and D. S. Cannell, *Biochemistry* **14**, 192–195 (1975).
141. A. H. Sanders, D. L. Purich, and D. S. Cannell, *J. Mol. Biol.* **147**, 583–595 (1981).
142. D. F. Nicoli and G. B. Benedek, *Biopolymers* **15**, 2421–2437 (1976).
143. C.-C. Wang, D. H. Cook, and R. Pecora, *Biophys. Chem.* **11**, 439–442 (1980).
144. B. D. Fair, D. Y. Chao, and A. M. Jamieson, *J. Colloid Interface Sci.* **66**, 323–330 (1978).
145. B. D. Fair and A. M. Jamieson, *J. Colloid Interface Sci.* **73**, 130–135 (1979).

146. S. S. Alpert, *J. Chem. Phys.* **65**, 4333–4334 (1976).
147. J. L. Anderson and C. C. Reed, *J. Chem. Phys.* **64**, 3240–3249.
148. J. L. Anderson and C. C. Reed, *J. Chem. Phys.* **65**, 4336–4337 (1976).
149. G. D. J. Phillies, *J. Chem. Phys.* **65**, 4334–4335 (1976).
150. S. S. Alpert and G. Banks, *Biophys. Chem.* **4**, 287–296 (1976).
151. C. R. Jones, S. S. Johnson, Jr., and J. T. Penniston, *Biopolymers* **17**, 1581–1593 (1978).
152. R. S. Hall, Y. Oh and C. S. Johnson, Jr., *J. Phys. Chem.* **84**, 756–767 (1980).
153. E. E. Uzgiris and D. C. Golibersuch, *Phys. Rev. Lett.* **32**, 37–40 (1974).
154. D. D. Haas, R. V. Mustacich, B. A. Smith and B. R. Ware, *Biochem. Biophys. Res. Comm.* **59**, 174–180 (1974).
155. W. B. Veldkamp and J. R. Votano, *Biopolymers*, **19**, 111–124 (1980).
156. P. Privalov, N. N. Khechinashvili, and B. Atanasov, *Biopolymers* **10**, 1865–1890.
157. D. D. Haas and B. R. Ware, *Biochemistry* **17**, 4946–4950 (1978).
158. K. J. LaGattuta, V. S. Sharma, D. F. Nicoli, and B. K. Kothari, *Biophys. J.* **33**, 63–80 (1981).
159. B. Chu, A. Yeh, F. C. Chen, and B. Weiner, *Biopolymers* **14**, 93–109 (1975).
160. F. Rondelez and J. D. Litster, *Biophys. J.* **27**, 455–460 (1979).
161. C. H. Pletcher, R. M. Resnick, G. J. Wei, V. A. Bloomfield, and G. L. Nelsestuen, *J. Biol. Chem.* **255**, 7433–7438 (1980).
162. G. L. Nelsestuen, R. M. Resnick, G. J. Wei, C. H. Pletcher, and V. A. Bloomfield, **20**, 351–358 (1981).
163. G. P. Agarwal, J. G. Gallagher, K. C. Aune, and C. D. Armeniades, *Biochemistry* **16**, 1865–1870 (1977).
164. G. R. Palmer, O. G. Fritz, and F. R. Hallett, *Biopolymers* **18**, 1647–1658 (1979).
165. G. R. Palmer and O. G. Fritz, *Biopolymers* **18**, 1659–1672 (1979).
166. R. J. Cohen and G. B. Bendek, *J. Mol. Biol.* **108**, 151–178 (1976).
167. R. J. Cohen, J. A. Jedziniak, and G. B. Benedek, *Proc. R. Soc. London Ser. A* **345**, 73–88 (1975).
168. R. J. Cohen, J. A. Jedziniak, and G. B. Benedek, *J. Mol. Biol.* **108**, 179–199 (1976).
169. M. Jullien and D. Thusius, *J. Mol. Biol.* **101**, 397–416 (1976).
170. R. J. Cohen, and G. B. Benedek, *J. Mol. Biol.* **129**, 37–44 (1979).
171. D. Thusius, *J. Mol. Biol.* **115**, 243–247 (1977).
172. J. S. Gethner, G. W. Flynn, B. J. Berne, and F. Gaskin, *Biochemistry* **16**, 5776–5781 (1977).
173. J. S. Gethner, G. W. Flynn, B. J. Berne, and F. Guskin, *Biochemistry*, **16**, 5781–5785 (1977).
174. J. S. Gethner and F. Gaskin, *Biophys. J.* **24**, 505–515 (1978).
175. F. H. Silver, K. H. Langley, and R. L. Trelstad, *Biopolymers* **18**, 2523–2535 (1979).
176. K. W. Rhee, R. O. Potts, C.-C. Wang, M. J. Fournier, and N. C. Ford, *Nucleic Acids Res.* **9**, 2411–2420 (1981).
177. R. O. Potts, N. C. Ford, Jr., and M. J. Fournier, *Biochemistry* **20**: 1653–1659 (1981).
178. A. Patkowski, S. Jen, and B. Chu, *Biopolymers* **17**, 2643–2662 (1978).
179. M. Kochi, E. Gulari, and B. Chu, *J. Chem. Phys.* **71**, 4172–4174 (1979).
180. A. Patkowski and B. Chu, *Biopolymers* **18**, 2051–2072 (1979).
181. A. Patkowski and B. Chu, *J. Chem. Phys.* **73**, 3082–3087 (1980).
182. C. C. Wang, M. J. Fournier, and N. C. Ford, Jr., *Biopolymers*, **20**, 155–168 (1980).
183. E. Holler, C.-C. Wang, and N. C. Ford, Jr., *Biochemistry* **20**, 861–867 (1981).
184. A. Patkowski, E. Gulari, and B. Chu, *J. Chem. Phys.* **73**, 4178–4184 (1980).
185. V. A. Bloomfield and J. A. Benbasat, *Macromolecules* **4**, 609–613 (1971).
186. J. A. Benbasat and V. A. Bloomfield, *Macromolecules* **6**, 593–597 (1973).

187. D. Jolly and H. Eisenberg, *Biopolymers* **15**, 61–95 (1976).
188. M. Mandelkern, J. G. Elias, D. Eden, and D. M. Crothers, *J. Mol. Biol.* **152**, 153–161 (1981).
189. W. I. Lee and J. M. Schurr, *Chem. Phys. Lett.* **23**, 603–607 (1973).
190. J. M. Schurr, *Q. Rev. Biophys.* **9**, 109–134 (1976).
191. S.-C. Lin and M. Schurr, *Biopolymers* **17**, 425–461 (1978).
192. J. M. Schurr, *Biopolymers* **16**, 461–464 (1977).
193. J. M. Schurr, *Chem. Phys.* **30**, 243–247 (1978).
194. J. C. Thomas, R. D. Holder, S. A. Allison, and J. M. Schurr, *Biopolymers* **19**, 1451–1474 (1980).
195. G. S. Manning, *Q. Rev. Biophys.* **11**, 179–246 (1978).
196. S.-C. Lin, J. C. Thomas, S. A. Allison, and J. M. Schurr, *Biopolymers* **20**, 209–230 (1981).
197. M. Caloin, B. Wilhelm, and M. Daune, *Biopolymers* **16**, 2091–2104 (1977).
198. K. S. Schmitz, *Biopolymers* **18**, 479–484 (1979).
199. N. Parthasarathy and K. S. Schmitz, *Biopolymers* **19**, 1655–1666 (1980).
200. W. I. Lee, K. S. Schmitz, S.-C. Lin, and J. M. Schurr, *Biopolymers* **16**, 583–599 (1977).
201. R. Pecora and K. S. Schmitz, *Biopolymers* **14**, 521–542 (1975).
202. F. C. Chen and B. Chu, *J. Chem. Phys.* **66**, 2235–2237 (1977).
203. F. C. Chen, A. Yeh, and B. Chu, *J. Chem. Phys.* **66**, 1290–1305 (1977).
204. M. Adam and M. Delsanti, *Macromolecules* **10**, 1229–1237 (1977).
205. P. Mathiez, G. Weisbuch, and C. Mouttet, *Biopolymers* **18**, 1465–1478 (1979).
206. J. C. Thomas and J. M. Schurr, *Biopolymers* **19**, 215–218 (1980).
207. N. Parthasarathy, K. S. Schmitz, and M. K. Cowman, *Biopolymers* **19**, 1137–1151 (1980).
208. R. W. Wilson and V. A. Bloomfield, *Biochemistry* **18**, 2192–2196 (1979).
209. S. A. Allison, J. C. Herr, and J. M. Schurr, *Biopolymers* **20**, 469–488 (1981).
210. R. L. Schmidt, J. A. Boyle, and J. A. Mayo, *Biopolymers* **16**, 317–326 (1977).
211. R. L. Schmidt, M. A. Whitehorn, and J. A. Mayo, *Biopolymers* **16**, 327–340 (1977).
212. S. L. Hartford and W. H. Flygare, *Macromolecules* **8**, 80–83 (1975).
213. B. R. Shaw and K. S. Schmitz, *Biochem. Biophys. Res. Commun.* **73**, 224–232 (1976).
214. K. S. Schmitz, *Biopolymers* **16**, 2619–2633 (1977).
215. B. R. Shaw and K. S. Schmitz, in *Chromatin Structure and Function, Part B*, C. A. Nicolini, ed., Plenum, New York (1979), pp. 427–439.
216. K. S. Lee, M. Mandelkern, and D. M. Crothers, *Biochemistry* **20**, 1438–1445 (1981).
217. A. W. Fulmer and V. A. Bloomfield, *Proc. Nat. Acad. Sci. USA* **78**, 5968–5972 (1981).
218. A. W. Fulmer and V. A. Bloomfield, *Biochemistry*, **21**, 985–992 (1982).
219. C. C. Han, I. N. Serdyuk, and H. Yu, *J. Res. Nat. Bur. Stand.* **84**, 1–8 (1979).
220. J. Bruining and H. M. Fijnaut, *Biophys. Chem.* **9**, 345–353 (1979).
221. C. J. Oliver, K. F. Shortridge, and G. Belyavin, *Biochim. Biophys. Acta* **437**, 589–598 (1976).
222. P. Nieuwenhuysen and J. Clauwaert, *Biopolymers* **17**, 2039–2040 (1978).
223. Y. Sakaki, T. Madea, and T. Oshima, *J. Biochem.* **85**, 1205–1211 (1979).
224. P. Dobos, R. Hallett, D. T. C. Kells, O. Sorensen, and D. Rowe, *J. Virol.* **22**, 150–159 (1977).
225. R. W. DeBlois, E. E. Uzgiris, D. H. Cluxton, and H. M. Mazzone, *Anal. Biochem.* **90**, 273–288 (1978).
226. S. L. Hartford and W. H. Flygare, *Macromolecules* **8**, 80–83 (1975).
227. L. Rimai, I. Salmeen, D. Hart, L. Liebes, and M. A. Rich, *Biochemistry* **14**, 4621–4627 (1975).
228. S. L. Hartford, J. A. Lesnaw, W. H. Flygare, R. MacLeod, and M. E. Reichmann, *Proc. Nat. Acad. Sci. USA* **72**, 1202–1205 (1975).

229. P. M. Keller, E. E. Uzgiris, D. H. Cluxton, and J. Lenard, *Virology* **87**, 66–72 (1978).
230. B. I. Feuer, E. E. Uzgiris, R. W. Deblois, D. H. Cluxton, and J. Lenard, *Virology* **90**, 156–161 (1978).
231. J. C. Thomas and G. C. Fletcher, *Biopolymers* **17**, 2753–2756 (1978).
232. J. Newman, H. L. Swinney, and L. A. Day, *J. Mol. Biol.* **116**, 593–606 (1977).
233. S. A. Berkowitz and L. A. Day, *J. Mol. Biol.* **102**, 531–547 (1976).
234. R. L. Wiseman and L. A. Day, *J. Mol. Biol.* **102**, 549–561 (1977).
235. F. C. Chen, G. Koopmans, R. L. Wiseman, L. A. Day, and H. L. Swinney, *Biochemistry* **19**, 1373–1376 (1980).
236. R. L. Wiseman, S. A. Berkowitz and L. A. Day, *J. Mol. Biol.* **102**, 549–561 (1976).
237. J. Newman and D. D. Carlson, *Biophys. J.* **29**, 37–48 (1980).
238. T. Maeda and S. Fujime, *Macromolecules* **14**, 809–818 (1981).
239. J. Akisiote-Benbasat and V. A. Bloomfield, *J. Mol. Biol.* **95**, 335–357 (1975); *Biochemistry* **20**, 5018–5025 (1981).
240. G. J. Baran and V. A. Bloomfield, *Biopolymers* **17**, 2015–2028 (1978).
241. J. B. Welch, III, and V. A. Bloomfield, *Biopolymers* **17**, 1987–1999 (1978).
242. J. B. Welch, III, and V. A. Bloomfield, *Biopolymers* **17**, 2001–2014 (1978).
243. J. Greve, P. C. Hopman, and G. Koopmans, *Biopolymers* **18**, 1551–1553 (1979).
244. P. C. Hopman, G. Koopmans, A. P. M. van de Fliert, and J. Greve, *Int. J. Biol. Macromolecules* **2**, 143–148 (1980).
245. G. Koopmans, B. J. Vander Meer, P. C. Hopman, and J. Greve, *Biopolymers* **18**, 1533–1542 (1979).
246. R. W. Wilson and V. A. Bloomfield, *Biopolymers* **18**, 1543–1549 (1979).
247. R. W. Wilson and V. A. Bloomfield, *Biopolymers* **18**, 1205–1211.
248. G. Koopmans, P. C. Hopman, and J. Greve, *J. Phys. A.: Math. Gen.* **12**, 581–590 (1979). (1979).
249. P. C. Hopman, G. Koopmans, and J. Greve, in *Electro-Optics and Dielectrics of Macromolecules and Colloids*, B. R. Jennings, ed., Plenum, New York (1979), pp. 197–202.
250. P. C. Hopman, G. Koopmans, and J. Greve, *Biopolymers* **19**, 1241–1255 (1980).
251. F. C. Chen, W. Tscharnuter, D. Schmidt, B. Chu, and T. Y. Liu, *Biopolymers* **12**, 2281–2292 (1974).
252. S. Fujime, F. C. Cehn, and B. Chu, *Biopolymers* **16**, 945–963 (1977).
253. Y. Tsunashima, K. Moro, B. Chu, and T. Y. Lin, *Biopolymers* **17**, 251–265 (1978).
254. E. Gulari, B. Chu, and T. Y. Lin, *Biopolymers* **18**, 2943–2961 (1979).
255. D. B. Sellen, *Polymer* **16**, 561–564 (1975).
256. D. B. Sellen, *Polymer* **16**, 169–172 (1975).
257. J. G. Southwick, M. E. McDonnell, A. M. Jamieson, and J. Blackwell, *Macromolecules* **12**, 305–311 (1979).
258. H. Reihanian, A. M. Jamieson, L. H. Tang, and L. Rosenberg, *Biopolymers* **18**, 1727–1747 (1979).
259. M. E. McDonnell and A. M. Jamieson, *J. Colloid Interface Sci.* **63**, 218–225 (1978).
260. H. Magdelenat, P. Turq, M. Chemla, R. Menez, and M. Drifford, *Biopolymers* **18**, 187–201 (1979).
261. F. R. Hallett and Al. L. Gray, *Biochim. Biophys. Acta* **343**, 648–655 (1974).
262. Y. Barenholz, D. Gibbes, B. J. Litman, J. Goll, T. E. Thompson, and F. D. Carlson, *Biochemistry* **16**, 2806–2810 (1977).
263. U. Herrman, and J. H. Fendler, *Chem. Phys. Lett.* **64**, 270–274 (1979).
264. F. C. Chen, A. Chrzeszczyk, and B. Chu, *J. Chem. Phys.* **64**, 3404–3409 (1976).
265. F. C. Chen, A. Chrzeszczyk, and B. Chu, *J. Chem. Phys.* **66**, 2237–2238 (1977).
266. J. H. Goll and G. B. Stock, *Biophys. J.* **19**, 265–273 (1977).

2-e fitting.

267. J. C. Selser, Y. Yeh, and R. J. Baskin, *Biophys. J.* **16**, 337–356 (1976).
268. R. Pecora and S. R. Aragon, *Chem. Phys. Lipids* **13**, 1–10 (1974).
269. E. P. Day, J. T. Ho, R. K. Kunze, and S. T. Sun, *Biochim. Biophys. Acta* **470**, 503–508 (1977).
270. S. T. Sun, E. P. Day, and J. T. Ho, *Proc. Nat. Acad. Sci. USA* **75**, 4325–4328 (1978).
271. S. T. Sun, C. C. Hsang, E. P. Day, and J. T. Ho, *Biochim. Biophys. Acta* **557**, 45–52 (1979).
272. T. Norisuye, W. F. Hoffman, and H. Yu, *Biochemistry* **15**, 5678–5682 (1976).
273. T. Norisuye and H. Yu, *Biochim. Biophys. Acta* **471**, 436–452 (1977).
274. A. Warashina, *Biochim. Biophys. Acta* **672**, 158–164 (1981).
275. N. Ostrowsky and C. Hesse-Bezot, *Chem. Phys. Lett.* **52**, 141–144 (1977).
276. A. P. Bruckner, J. M. Schurr, N. B. Martin, and E. L. Chang, *Appl Opt.* **18**, 1876–1877 (1979).
277. T. K. Lim, V. A. Bloomfield, and G. L. Nelsestuen, *Biochemistry* **16**, 4177–4181 (1977).
278. R. Schwyzer and V. M. Kriwaczek, *Pure Appl. Chem.* **51**, 831–835 (1979).
279. P. Schlieper, R. Mohan, and R. Kaufmann, *Biochim. Biophys. Acta.* **644**, 13–23 (1981).
280. B. D. Fair and A. M. Jamieson, *J. Colloid Interface Sci.* **73**, 130–135 (1979).
281. E. E. Uzgiris and H. P. M. Frogageot, *Biopolymers* **15**, 257–263 (1976).
282. N. A. Mazer, G. B. Benedek, and M. C. Carey, *J. Phys. Chem.* **80**, 1075–1085 (1976).
283. N. A. Mazer, M. C. Carey, and G. B. Benedek, in *Micellization, Solubitization, and Microemulsions*, K. L. Mittal, ed., Plenum, New York (1977) Vol. 1, pp. 359–381.
284. C. Y. Young, P. J. Missel, N. A. Mazer, G. B. Benedek, and M. C. Carey, *J. Phys. Chem.* **82**, 1375–1378 (1978).
285. A. Rhode and E. Sackmann, *J. Colloid Interface Sci.* **70**, 494–505 (1979).
286. M. Corti and V. Degiorgio, *Chem. Phys. Chem.* **49**, 141–4.288 (1977).
287. R. W. Duke, and D. B. Dupre, *Chem. Phys. Lett.* **44**, 309–312 (1976).
288. M. Corti, V. DeGiorgio, R. Ghidoni, S. Sonnino, and G. Tettamanti, *Chem. Phys. Lipids* **26**, 225–238 (1980).
289. D. F. Nicoli, D. R. Dawson, and H. W. Offen, *Chem. Phys. Lett.* **66**, 291–294 (1979).
290. N. A. Mazer, M. C. Carey, R. F. Kwasnick, and G. B. Benedek, *Biochemistry* **18**, 3064–3075 (1979).
291. R. T. Holzbach, S. Y. Oh, M. E. McDonnell, and A. M. Jamieson, in *Micellization, Solubilization, and Microemulsions*, K. Mittal, ed., Plenum, New York (1977), Vol. 1, pp. 403–418.
292. N. A. Mazer, G. B. Benedek, and M. C. Carey, *Biochemistry* **19**, 601–615 (1980).
293. N. A. Mazer, R. E. Kwasnick, M. C. Carey, and G. B. Benedek, in *Micellization, Solubilization, and Microemulsions*, K. L. Mittal, ed., Plenum, New York (1977), Vol. 1, pp. 383–402.
294. M. C. Carey, J.-C. Montet, M. C. Phillips, M. J. Armstrong, and N. A. Mazer, *Biochemistry* **20**, 3637–3648 (1981).
295. G. B. Benedek, *Appl. Opt.* **10**, 459–473 (1971).
296. J. A. Jedziniak, J. H. KInoshita, E. M. Yates, and G. M. Benedek, *Exp. Eye Res.* **20**, 367–369 (1975).
297. T. Tanaka and G. B. Benedek, *Invest. Opthalmol.* **14**, 449–456 (1975).
298. J. A. Jedziniak, D. F. Nicoli, H. Baram, and G. B. Benedek, *Invest. Opthalmol. Visual Sci.* **17**, 51–57 (1978).
299. R. J. Cohen and G. B. Benedek, *Immunochemistry* **12**, 349–351 (1975).
300. G. K. von Schulthess, R. J. Cohen, N. Sakota, and G. K. Benedek, *Immunochemistry* **13**, 955–962 (1976).
301. G. K. von Schulthess, R. J. Cohen, and G. B. Benedek, *Immunochemistry* **13**, 963–966 (1976).

302. G. K. von Schulthess, M. Giglio, D. S. Cannell, and G. B. Benedek, *Mol. Immunol.* **17**, 81–92 (1980).
303. E. E. Uzgiris, *J. Immunol. Methods* **10**, 85–96 (1976).
304. E. E. Uzgiris, in *Cell Electrophoresis*, A. Preece and D. Sabolovic, eds., Elsevier/North-Holland, Amsterdam (1979).
305. B. R. Ware, *ACS Symp. Ser.* **85**, 102–117 (1978).
306. J. H. Kaplan and E. E. Uzgiris, *J. Immunol.* **117**, 115–123 (1976).
307. J. Y. Josefowicz, B. R. Ware, A. L. Griffith, and N. Catsimpoolas, *Life Sci.* **21**, 1483–1488 (1977).
308. B. A. Smith, B. R. Ware, and R. S. Weiner, *Proc. Nat. Acad. Sci. USA* **73**, 2388–2391 (1976).
309. B. A. Smith, B. R. Ware, and R. A. Yankee, *J. Immunol.* **120**, 921–926 (1978).
310. J. H. Kaplan, E. E. Uzgiris, and S. H. Lockwood, *J. Immunol. Methods* **27**, 241–255 (1979).
311. E. E. Uzgiris, J. H. Kaplan, T. J. Cunningham, S. H. Lockwood, and D. Steiner, *Eur. J. Cancer* **15**, 1275–1280 (1979).
312. J. H. Kaplan and E. E. Uzgiris, in *Prevention and Detection of Cancer II Detection*, H. E. Nieburgs, ed. Marcel Dekker, New York (1979), Vol. 2.
313. J. Josefowicz, and F. R. Hallett, *FEBS Lett.* **60**, 62–65 (1975).
314. J. H. Kaplan and E. E. Uzgiris, *J. Immunol. Methods* **7**, 337–346 (1975).
315. E. E. Uzgiris and J. H. Kaplan, *J. Immunol.* **117**, 2165–2170 (1976).
316. H. R. Petty, B. A. Smith, B. R. Ware, and R. E. Rocklin, *Cell Immunol.* **54**, 435–444 (1980).
317. H. R. Petty, B. R. Ware, H. G. Remold, and R. E. Rocklin, *J. Immunol.* **124**, 381–387 (1980).
318. H. R. Petty, R. L. Folger, and B. R. Ware, *Cell Biophys.* **1**, 29–37 (1979).
319. H. R. Petty, B. R. Ware, and S. I. Wasserman, *Biophys. J.* **30**, 41–50 (1980).
320. H. R. Petty and B. R. Ware, *Proc. Nat. Acad. Sci. USA* **76**, 2278–2282 (1979).
321. D. P. Siegel, B. R. Ware, D. J. Green, and E. W. Westhead, *Biophys. J.* **22**, 341–346 (1978).
322. D. P. Siegel and B. R. Ware, *Biophys. J.* **30**, 159–172 (1980).
323. S. J. Luner, D. Szklarek, R. J. Knox, G. V. F. Seaman, J. Y. Josefowicz, and B. R. Ware, *Nature* **269**, 719–721 (1977).
324. V. Ryan, T. R. Hart, and R. Schiller, *Biophys. J.* **31**, 113–125 (1980).
325. D. Byrne and J. C. Earnshaw, *J. Phys. D. Appl. Phys.* **10**, 1207–1211 (1977).
326. D. Byrne and J. Earnshaw, *J. Phys. D.* **12**, 1113–1144 (1979).
327. S. Hard and K. Johansson, *J. Colloid Interface Sci.* **60**, 467–472 (1977).
328. S. Hard and H. Lofgren, *J. Colloid Interface Sci.* **60**, 529–539 (1977).
329. D. Bryne and J. C. Earnshaw, *J. Phys. D.* **12**, 1145–1158 (1979).
330. W. A. B. Donners, J. B. Rijnbout, and A. Vrij, *J. Colloid Interface Sci.* **61**, 249–260 (1976).
331. H. M. Fijnant and A. Vrij, *Nature Phys. Sci.* **246**, 118–119 (1973).
332. J. G. H. Joosten and H. M. Fijnaut, *Chem. Phys. Lett.* **60**, 483–485 (1979).
333. L. E. Moore, M. Tufts, and M. Soroka, *Biochim. Biophys. Acta* 386–394 (1975).
334. P. R. Dragsten, W. W. Webb, J. A. Paton, and R. R. Capranica, *Science* **185**, 5–7 (1974).
335. A. N. Popper, N. L. Clarke, and J. A. Mann, Jr., *Biophys. J.* **15**, 307–318 (1975).
336. T. Tanaka, L. O. Hocker, and G. B. Benedek, *J. Chem. Phys.* **59**, 5151–5159 (1973).
337. J. P. Munch, S. Candau, R. Duplessix, C. Picot, J. Herz, and H. Benoit, *J. Polym. Sci.* **14**, 1097–1109 (1976).
338. S. J. Candau, C. Y. Young, T. Tanaka, P. Lemarechal, and J. Bastide, *J. Chem. Phys.* **70**, 4695–4700 (1979).

339. J. P. Munch, P. Lemarachal, and S. Candau, *J. Phys.* (*Paris*) **38**, 1499–1509 (1977).
340. S. I. Brenner, R. A. Gelman, and R. Nossal, *Macromolecules* **11**, 202–207 (1978).
341. R. A. Gelman and R. Nossal, *Macromolecules* **12**, 311–316 (1979).
342. R. Nossal and S. L. Brenner, *Macromolecules* **11**, 207–212 (1978).
343. R. Nossal, *J. Appl. Phys.* **50**, 3105–3112 (1979).
344. R. A. Gelman, J. A. Gladner, and R. Nossal, *Biopolymers* **19**, 1259–1270 (1980).
345. K. L. Wun, and F. D. Carlson, *Macromolecules* **8**, 190–194 (1975).
346. W. Mackie, D. B. Sellen, and J. Stucliffe, *J. Polymer Sci*: *Polymer Symp.* **61**, 191–197 (1977).
347. W. Mackie, D. B. Sellen, and J. Sutcliffe, *Polymer* **19**, 9–16 (1978).
348. D. B. Sellen, *Polymer* **19**, 1110 (1978).
349. E. Geissler and A. M. Hecht, *J. Phys. Lett.* (*Orsay*) **40**, 173–176 (1979).
350. K. Moro and N. Saito, *J. Phys. Soc. Jpn* **40**, 1415–1420 (1976).
351. F. Brochard and P. G. deGennes, *Macromolecules* **10**, 1157–1161 (1977).
352. P. G. deGennes, *Macromolecules* **9**, 587–593 (1976).
353. P. G. deGennes, *Macromolecules* **9**, 594–598 (1976).
354. B. Chu, T. Nose, *Macromolecules* **12**, 347–348 (1979).
355. A. M. Hecht and E. Geissler, *J. Chem. Phys.* **66**, 1416–1421 (1977).
356. J. Klein, *Macromolecules* **11**, 852–858 (1978).
357. T. Nose and B. Chu, *J. Chem. Phys.* **70**, 5332–5334 (1979).
358. T. Nose and B. Chu, *Macromolecules* **12**, 1122–1129 (1979).
359. H. Reihanian and A. M. Jamieson, *Macromolecules* **12**, 684–687 (1979).
360. W. I. Lee, R. J. Verdugo, R. J. Blandau, and P. Gaddum-Rosse, *Gynecol. Invest.* **8**, 254–266 (1977).
361. J. Newman and F. D. Carlson, *Biophys. J.* **29**, 37–48 (1980).
362. A. B. Fraser, E. Eisenberg, W. W. Kielley, and F. D. Carlson, *Biochemistry* **14**, 2207–2214 (1975).
363. T. Maeda and S. Fujime, *J. Phys. Soc. Jpn* **42**, 1983–1981 (1977).
364. R. C. Haskell and F. D. Carlson, *Biophys. J.* **33**, 39–62 (1980).
365. R. F. Bonner and F. D. Carlson, *J. Gen. Phys.* **65**, 555–581 (1975).
366. F. D. Carlson, *Biophys. J.* **15**, 633–649 (1975).
367. S. Fujime and S. Yoshino, *Biophys. Chem.* **8**, 305–315 (1978).
368. S. Yoshino, Y. Umazume, R. Natori, S. Fujime, and S. Chiba, *Biophys. Chem.* **8**, 317–326 (1978).
369. D. L. Lappe and E. G. Laketta, *Science* **207**, 1369–1371 (1980).
370. B. R. Ware, *Applications of Laser Velocimetry in Biology and Medicine*, **2**, 102–117 (1977).
371. H. Mishina and T. Asakura, *Opt. Commun.* **11**, 99–102 (1974).
372. R. V. Mustacich and B. R. Ware, *Rev. Sci. Instrum.* **47**, 108–111 (1976).
373. T. S. Durrani and C. A. Greated, *Appl. Opt.* **14**, 778–786 (1975).
374. T. Tanaka, R. Riva, and I. Ben-Sira, *Science* **186**, 830–832 (1974).
375. G. T. Feke and C. E. Riva, *J. Opt. Sci. Am.* **68**, 526–531 (1978).
376. G. T. Feke, C. E. Riva, B. Eberli, and V. Benary, *Electro-Opt. Syst. Design* **32**, 40–44 (1978).
377. M. D. Stern, D. L. Lappe, P. D. Bowen, J. E. Chimosky, G. A. Holloway, Jr., H. R. Keiser, and R. L. Bowman, *Am. J. Physiol.* **232**, H441–H448 (1977).
378. T. Tanaka and G. B. Benedek, *Appl. Opt.* **14**, 189–196 (1975).
379. M. D. Stern, *Nature* **254**, 56–58, (1975).
380. R. F. Bonner, P. Bowen, R. L. Bowman, and R. Nossal, *Proc. Tech. Program Electro-Opt./Laser 78 Conf. Expo.* (1978), pp. 539–550.
381. R. V. Mustacich and B. R. Ware, *Biophys. J.* **16**, 373–388 (1976).

382. R. V. Mustacich and B. R. Ware, *Biophys. J.* **17**, 229–241 (1977).
383. K. H. Langley, S. A. Newton, N. C. Ford, Jr., and D. B. Sattelle, in *Photon Correlation Spectroscopy and Velocimetry*, H. Z. Cummins and E. R. Pike, eds., Plenum, New York (1977), pp. 519–525.
384. R. V. Mustacich, and B. R. Ware, *Protoplasma* **91**, 351–367 (1977).
385. J. C. Earnshaw, in *Photon Correlation Spectroscopy and Velocimetry*, H. Z. Cummins and E. R. Pike, eds., Plenum, New York (1977), pp. 461–464.
386. D. B. Sattelle, G. M. Langford, K. H. Langley, *NATO Adv. Study Inst. Ser B* **B23**, 543–549 (1976).
387. H. Z. Cummins, *NATO Adv. Study Inst. Ser., Ser B* **B23**, 200–225 (1976).
388. G. B. Stock, *Biophys. J.* **16**, 535–540 (1976).
389. G. B. Stock, *Biophys. J.* **22**, 79–96 (1978).
390. S.-H. Chen, M. Holz, and P. Tartagla, *Appl. Opt.* **16**, 187–194 (1977).
391. M. Holz and S.-H. Chen, *Appl. Opt.* **17**, 1930–1937 (1978).
392. M. Holz and S.-H. Chen, *Appl. Opt.* **17**, 3197–3204 (1978).
393. M. Holz and S.-H. Chen, *Biophys. J.* **23**, 15–31 (1978).
394. M. Kotlarcyk, S.-H. Chen, and S. Asano, *Appl. Opt.* **18**, 2470–2479 (1979).
395. A. V. Lomakin and V. A. Noskin, *Biopolymers* **19**, 231–240 (1980).
396. D. F. Cooke, F. R. Hallett, and C. A. V. Barker, *J. Mechanochem. Cell Motility* **3**, 219–223 (1976).
397. F. R. Hallett, T. Craig, and J. Marsh, *Biophys. J.* **21**, 203–216 (1978).
398. H. Shimizu and G. Matsumoto, *Opt. Commun.* **16**, 197–201 (1976).
399. R. Finsy, J. Peetermans, and H. Lekkerkerker, *Biophys. J.* **27**, 187–191 (1979).
400. T. Craig, R. R. Hallett, and B. Nickel, *Biophys. J.* **28**, 457–472 (1979).
401. J. D. Harvey and M. W. Woolford, *Biophys. J.* **31**, 147–156 (1980).
402. G. Matsumoto, H. Shimizu, J. Shamada, and A. Wada, *Opt. Commun.* **22**, 369–373 (1977).
403. H. Shimizu and G. Matsumoto, *Biophys. J.* **29**, 167–176 (1980).
404. J. Frost and H. Z. Cummins, *Science* **212**, 1520–1522 (1981).
405. W. I. Lee and R. J. Blandau, *Fertil. Steril.* **32**, 320–323 (1979).
406. G. Banks, D. W. Schaefer, and S. S. Alpert, *Biophys. J.* **15**, 253–261 (1975).
407. D. W. Schaefer and B. J. Berne, *Biophys. J.* **15**, 785–794 (1975).
408. R. Nossal and T. Y. Chang, *J. Mechanochem. Cell Motility* **3**, 247–251 (1976).

Index